Air Pollution: Monitoring and Assessment

Air Pollution: Monitoring and Assessment

Editor: Bernie Goldman

Callisto Reference
www.callistoreference.com

Callisto Reference,
118-35 Queens Blvd., Suite 400,
Forest Hills, NY 11375, USA

Visit us on the World Wide Web at:
www.callistoreference.com

ISBN: 978-1-63239-944-1 (Hardback)

The publisher's policy is to use permanent paper from mills that operate a sustainable forestry policy. Furthermore, the publisher ensures that the text paper and cover boards used have met acceptable environmental accreditation standards.

Printed in China.

Cataloging-in-Publication Data

Air pollution : monitoring and assessment / edited by Bernie Goldman.
p. cm.
Includes bibliographical references and index.
ISBN 978-1-63239-944-1
1. Air--Pollution. 2. Air--Pollution--Measurement. 3. Pollution--Measurement.
4. Pollution--Risk assessment. 5. Environmental impact analysis.
6. Environmental monitoring. I. Goldman, Bernie.
TD174 .A37 2018
628.5--dc23

Table of Contents

Preface

Air pollution is the existence of harmful substances that are found in the air. It causes a number of diseases and allergies which can in some cases cause death. There are various factors that are responsible for releasing pollutants into the atmosphere. The two sources of air pollution are natural sources such as dust, smoke and anthropogenic sources such as motor vehicles, smoke from power plants, etc. Different approaches, evaluations, methodologies and advanced studies on air pollution, its monitoring and control have been included in this text. For someone with an interest and eye for detail, this book covers the most significant topics in the field of air pollution.

This book unites the global concepts and researches in an organized manner for a comprehensive understanding of the subject. It is a ripe text for all researchers, students, scientists or anyone else who is interested in acquiring a better knowledge of this dynamic field.

I extend my sincere thanks to the contributors for such eloquent research chapters. Finally, I thank my family for being a source of support and help.

Editor

1

Human Exposure Pathways of Heavy Metals in a Lead-Zinc Mining Area, Jiangsu Province, China

Chang-Sheng Qu[1,4], Zong-Wei Ma[1], Jin Yang[2], Yang Liu[3], Jun Bi[1]*, Lei Huang[1]*

1 State Key Laboratory of Pollution Control and Resource Reuse, School of the Environment, Nanjing University, Nanjing, China, **2** Department of Environmental Science & Engineering, Fudan University, Shanghai, China, **3** Department of Environmental Health, Rollins School of Public Health, Emory University, Atlanta, Georgia, United States of America, **4** Jiangsu Provincial Academy of Environmental Science, Nanjing, China

Abstract

Heavy metal pollution is becoming a serious issue in developing countries such as China, and the public is increasingly aware of its adverse health impacts in recent years. We assessed the potential health risks in a lead-zinc mining area and attempted to identify the key exposure pathways. We evaluated the spatial distributions of personal exposure using indigenous exposure factors and field monitoring results of water, soil, food, and indoor and outdoor air samples. The risks posed by 10 metals and the contribution of inhalation, ingestion and dermal contact pathways to these risks were estimated. Human hair samples were also analyzed to indicate the exposure level in the human body. Our results show that heavy metal pollution may pose high potential health risks to local residents, especially in the village closest to the mine (V1), mainly due to Pb, Cd and Hg. Correspondingly, the residents in V1 had higher Pb (8.14 mg/kg) levels in hair than those in the other two villages. Most of the estimated risks came from soil, the intake of self-produced vegetables and indoor air inhalation. This study highlights the importance of site-specific multipathway health risk assessments in studying heavy-metal exposures in China.

Editor: Jaymie Meliker, Stony Brook University, Graduate Program in Public Health, United States of America

Funding: This work was supported by the National Natural Science Foundation (41171411, 41271014 & 41201545) and the Jiangsu Natural Science Foundation (BK2012733). The funders had no role in study design, data collection and analysis, decision to publish or preparation of the manuscript.

Competing Interests: One author (JY) is employed by a commercial company, SGS-CSTC Standards Technical Services (Shanghai) Co., Ltd. However, this study was not funded by the SGS company. SGS company is a cooperative partner in this study. The authors have cooperation agreement with SGS company.

* E-mail: huanglei@nju.edu.cn (LH); jbi@nju.edu.cn (JB)

Introduction

Heavy metals are known to be persistent in the human body, with excretion half-lives that last for decades. Heavy metals can lead to a wide range of toxic effects, such as carcinogenicity, mutagenicity and teratogenicity [1,2,3,4]. Heavy metal contamination is a major environmental concern on a global scale, particularly in China, with its rapid economic development [5,6,7]. Due to large-scale production and consumption and lack of regulations, heavy metals such as lead (Pb), zinc (Zn), cadmium (Cd), mercury (Hg), and chromium (Cr) are emitted into the environment in large quantities through wastewater irrigation, solid waste disposal, sludge application, vehicular exhaust and atmospheric deposition [8]. As a result, heavy metals are present in industrial, municipal and urban runoff, and they continuously accumulate in the environment in China [9,10,11,12,13]. Since 2005, health related incidents caused by heavy metal pollution have risen sharply in China, with major accidents attracting nationwide attention [14].

Heavy metals have been found widely in various environmental media (including soil, water, air and food) around the world [5,15,16]. These species may enter the human body through inhalation of dust, direct ingestion of soil and water, dermal contact of contaminated soil and water, and consumption of vegetables grown in contaminated fields. Various studies have been conducted to evaluate population health risks due to heavy metal exposure through various exposure pathways, especially soil and food chain [17,18,19,20,21,22]. Since it is difficult to identify the key exposure route because of lack of multipathway risk analysis, media or pathway-specific approach may fail to ensure public safety [23]. Therefore, it is necessary to assess the aggregate exposure to metals concerning about different environmental media and pathways. Previous studies have also demonstrated the importance of conducting multipathway risk assessment to identify the dominant pathway of potential concern [24,25,26,27].

Human health risk assessment, as formalized in 1983 [28], has been recognized as an important tool for estimating the nature and probability of adverse health effects in humans who may be exposed to chemicals and for presenting risk information to the decision maker. The US Environmental Protection Agency (EPA) hazard quotients (HQ) are widely used to characterize non-carcinogenic health effects posed by heavy metals by comparing the exposure level with a reference dose [15,22]. However, risk assessment is a complex process that is inherently linked with uncertainty [29]. Monte Carlo simulation is commonly used to quantify uncertainty in human health risk estimates [22]. It is a probabilistic approach that works with probability distributions rather than deterministic values of each parameter to estimate a risk.

For metal toxicity monitoring and environmental risk assessment, the identification of heavy metals from biological samples such as blood, urine or hair is useful for identifying exposure.

Because the sampling is less invasive, more convenient to store and transport, and less hazardous to handle, hair has been used widely in biomonitoring environmental and occupational exposures of various pollutants [30,31,32,33,34]. Furthermore, hair sample can be a useful assessment tool in characterizing long-term exposure of the measured contaminant, whereas blood and urine often reflect most recent exposures. Hair has been used in biomonitoring of heavy metals on large cohorts in Brazil [35], determining geological source and exposure through fish consumption in Lake Victoria [36], characterizing human exposure in an abandoned mine in Portugal [32] and examining the residential exposure in an e-waste recycling area in southeastern China [31]. We focus our study on using metal content in hair as a biomonitoring indicator to identify whether differences of long-term exposure existed among residents in different villages.

The current study considered human exposure to heavy metals via drinking water, dietary intake, dermal contact and inhalation in the populations living near a lead-zinc mining area in Jiangsu Province, China. In addition to Pb and Zn, eight other major associated elements were included in this study: Ag (silver), Cd, Cr, Cu (copper), Ni (nickel), Se (selenium), Tl (thallium) and Hg (mercury). These ten metals are all priority pollutant metals (PP metals) as set by the USEPA. The aims of this study are: (1) to evaluate the potential health impacts of these metals on the general population in the mining area, (2) to provide a better understanding of each exposure pathway and (3) to help local governments to prioritize pollution control and health intervention policies to protect local population.

Materials and Methods

Ethics Statement

This study was approved by the review board of Nanjing University. All participants were informed about the objectives and methods of the study before the investigation. And written consent was obtained from all participants.

Study Areas

The studied Qixia lead-zinc mining area, with reserves of four million tons of lead and zinc, is an important mining region in Jiangsu province, which is one of the fastest-developing provinces in China. The mine has been excavated for 60 years, and the nearby area is suffering serious environmental deterioration from the wastewater and solid waste discharge. Three nearby villages at the leeward side of the mine were sampled in this study to identify the potential health risks and its sources. Village 1 (V1) is 100 m from the mine, village 2 (V2) is 600 m from the mine, and village 3 (V3) is approximately 1 km from the mine. There are 32, 38 and 30 households in V1, V2 and V3 respectively, and they have similar population structures, living conditions and lifestyle. The geographical locations of the three sampled villages are shown in Figure 1.

Sampling

The drinking water of the 3 villages is tap water, which is provided by the same water supply company. We collected 10 samples from local families at random to represent the whole area. The water samples were collected in precleaned polyethylene bottles. After acidification with HCl, the samples were placed in an ice bath, then transported to the laboratory and kept at −20°C until they were analyzed.

5 soil samples were collected at the edges of residential streets and crop fields in each village. All soil samples were taken from the upper 5 cm of ground and stored in polyethylene bags. Each sample (1 kg approximately) consisted of 5 subsamples collected in an area of 100 m^2, pooled and homogenized to form a representative sample. The samples were air-dried, and stones and coarse plant roots or residues were removed. They were then thoroughly mixed, crushed, passed through a 2-mm mesh sieve and a 0.149-mm mesh sieve, and then stored in polyethylene bottles at ambient temperature prior to chemical analysis.

From January to March 2011, air sampling was conducted simultaneously in the 3 villages. 20 samples in each village and total 60 indoor and outdoor samples were collected at different sites in and around the study area. The indoor air samples were obtained from the rooms of randomly chosen families. The air samples were collected onto filter membranes using a PM_{10} air sampler (Tianhong instruments Co. Ltd, Wuhan, China) run for 3 h at a flow rate of 100 L/min. For the detection of trace Hg, the sulfhydryl cotton enrichment method was used.

In each village, we selected 4 fields at different directions, and locally produced vegetable samples were harvested in double in a quadrant of 0.25 m^2 randomly selected from each site. In total, 24 vegetable samples were collected in the study area. Food produced out of town, including rice, pork and eggs, was collected from the markets. The samples were cut into small pieces after thoroughly washing with tap water and deionized water. After drying to constant weight at 80°C in an oven, the food samples were ground and sieved for acid digestion.

Human Exposure Survey

A questionnaire-based survey was conducted in the studied villages to determine key risk factors such as dietary behaviors, daily activities and lifestyle of local people. We invited 50 local residents to participate the survey in each village. All participants were selected randomly. In V1, the research staff obtained written informed consent from 42 participants who are permanent residents of this village, and there were 42 participants in V2 and 36 participants in V3. The environmental samples were collected from these participants' immediate surroundings.

To characterize the exposure level and metal accumulation in human body, hair samples were also taken from the above survey participants. Twenty-nine, 30 and 13 participants provided usable samples in V1, V2 and V3, respectively. The hair samples were analyzed for the 10 metals to characterize the in vivo exposure levels and potential differences among the different indigenous groups in the study area.

Analytical Methods

Except for Hg, the USEPA 200.8 method was used to determine the trace metals in the water and soil samples, which were measured with inductively coupled plasma-mass spectrometry (ICP-MS) (Agilent 7500i, Agilent Scientific Technology Ltd., USA). For the airborne particle samples, an atomic absorption spectrometer (AAS, Vario 6, Jena Co., Ltd., Germany) was used to analyze the filters loaded with heavy metals according to the standard examination method for ambient air in China. Using the USEPA 6020A method, the concentrations of metals in the food and hair samples were determined with ICP-MS. The determination of Hg in the water, soil, food and hair samples was conducted using thermal decomposition, amalgamation and atomic absorption spectrophotometry (TDA/AAS) (AMA 254, LECO Co., Ltd., USA) according to the USEPA 7473 method. Trace Hg in the air, enriched in sulfhydryl cotton, was determined using cold vapor atomic absorption spectrophotometry (CVAAS) (JKG-205, Jilin Scientific Technology Co., Ltd., China) according to the NIOSH 6009 method. The detection limit of the measurements was defined as the concentration value, which is

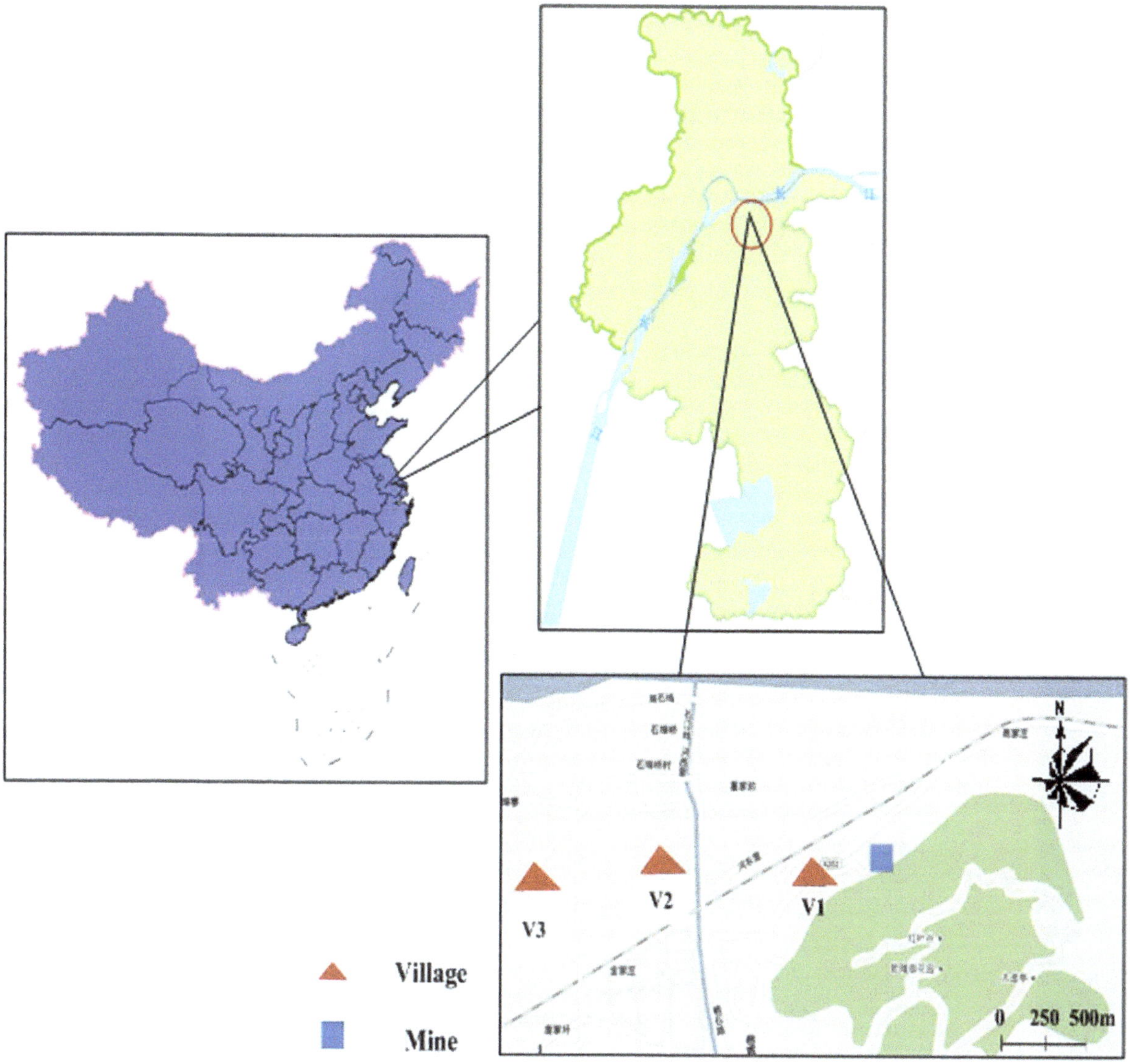

Figure 1. Locations of the study areas.

numerically equal to three times the standard deviation of 10 replicate blank measurements. Reagent blanks and standard reference materials were used in the analysis for quality assurance and quality control. The recoveries of the elements ranged from 90% to 110%.

Risk Calculation

The average daily intake (ADI) of metals by the human subjects was calculated using the following equation, which is recommended by the USEPA [37]. The equation links the time-averaged dose to the exposure medium concentration.

$$ADI = \frac{C \times IR \times EF \times ED}{BW \times AT} \quad (1)$$

where ADI is the average daily intake or dose through ingestion or inhalation (mg/kg-day); C is the chemical concentration in the exposure medium (mg/L, mg/kg, or mg/m^3); IR is the ingestion rate (L/day, kg/day, or m^3/day); EF is the exposure frequency (day/year); ED is the exposure duration (year); BW is the body weight (kg) and AT is the time period over which the dose is averaged (day).

$$ADI_D = \frac{C \times SA \times AF \times ABS \times EF \times ED}{BW \times AT} \quad (2)$$

For exposure dose through dermal contact (ADI_D) calculation, *SA* is the exposed skin surface area (cm^2), *AF* is the adherence factor (mg/cm^2/day), and *ABS* is the dermal absorption factor.

To calculate the exposure dose of each pathway, the ingestion rate of drinking water and food were obtained through the above questionnaire-based exposure survey. In addition, the ingestion rate of soil, inhalation rate of air, the adherence factor (AF), and the dermal absorption factor (ABS) were obtained from open database and literature [38,39].

The human health risks posed by heavy metal exposure are usually characterized by the hazard quotient (HQ) [37], the ratio of average daily intake to the reference dose (RfD) or the reference concentration in air (RfC) for an individual pathway and chemical.

A quotient under 1 is assumed to be safe. When the HQ value exceeds 1, there may be concerns for potential health risks associated with overexposure. The RfD and RfC values used in this study were recommended by open chemical databases and shown in Table 1. To assess the overall potential for health effects posed by more than one metal, summing HQs across metals can serve as a conservative assessment tool to estimate high-end risk rather than low-end risk to protect the public. In this study, the total HQ was used as a screening value to identify whether there is significant risk caused by metals, and any difference of total health risk existed among the 3 villages.

To accommodate the uncertainties associated within the calculation process, Monte Carlo simulation technique was used based on Crystalball software (Oracle Corporation, Vallejo, US) and considering 10,000 iterations. Before this process, distribution characteristics of each exposure parameter were tested according to the exposure survey results. A probabilistic distribution of the exposure dose and HQ values was then obtained as simulation result.

Results

Heavy Metal Levels in Different Environmental Media

The analysis results show that the tap water in the studied area is safe (unpublished data, Table S1), as all the trace heavy metal concentrations are far less than the national drinking water quality criteria.

Indoor air samples were obtained from 13, 15, and 12 households in V1, V2, and V3, respectively. The chemical analysis results (unpublished data, Table S2) show that Cu, Zn and Pb are the main pollutants. V1 was the most seriously polluted among the studied villages. The average metal levels there were 6.1 $\mu g/m^3$ for Cu, 2.5 $\mu g/m^3$ for Zn and 2.1 $\mu g/m^3$ for Pb, with a detection rate of nearly 100%. The chemical analysis results for the outdoor air show that Zn and Pb have the highest concentrations. In V1, the median metal levels were 4.2 $\mu g/m^3$ for Zn and 2.4 $\mu g/m^3$ for Pb, with a detection rate of 93.8% and 100% respectively.

The chemical analysis results of the soil samples show that 10 PP metals were detected in all samples (unpublished data, Table S3). In general, V1 is the most seriously polluted, followed by V2, and the soil quality is good in V3. In V1, the soil Pb concentration is as high as 2507 mg/kg, and the average soil zinc concentration as high as 9281 mg/kg, which is 100 times greater than that of V3.

It is well known that vegetables absorb metals from the soil. The self-produced vegetables contained higher metal concentrations in V1, especially of Zn and Pb. The average Zn concentrations in pakchoi in V1, V2 and V3 were 10.48, 5.24 and 2.34 mg/kg, respectively. The average Pb concentrations reached 0.24 and 0.08 mg/kg in V1 and V2, respectively, whereas Pb was not detected in V3. Different from the vegetables, the metals could not be detected in most of the imported egg, pork and rice samples.

Table 1. Metal reference doses (RfD).

Metal	RfD (mg/kg-d)	Source	RfC (mg/m^3)	Source
Ag	5.0E−3	IRIS[a]	–	
Cd	1.0E−3	IRIS	1.0E−5	ATSDR[b]
Cr	1.5E+0	IRIS	–	
Cu	4.0E−2	HEAST[c]	–	
Ni	2.0E−2	IRIS	9.0E−5	ATSDR
Pb	1.4E−4	Oak Ridge[d]	–	
Se	5.0E−3	IRIS	2.0E−2	Cal EPA[e]
Tl	3.0E−6	IRIS	–	
Zn	3.0E−1	IRIS	–	
Hg	1.6E−4	Cal EPA	3.0E−4	IRIS

[a]Integrated Risk Information System, U.S. EPA;
[b]The Agency for Toxic Substances and Disease Registry, U.S.;
[c]Health Effects Assessment Summary Tables, U.S. EPA;
[d]Oak Ridge National Laboratory, U.S.;
[e]California Environmental Protection Agency, U.S.

Exposure Factors

The demographic, activity and lifestyle questionnaire results are listed in Table 2. The periods that local residents have lived in this mining area range from 2 to 35 years. Local residents spend more time indoors than outdoors. The people in this rice-producing region mainly feed on rice, pakchoi, cabbage and pork. The local people almost do not eat flour, whereas the national average level was 140 g/person/day [40]. However, the consumption rate of rice for the local people was 360 g/person/day, much greater than the national average level as 238 g/person/day [40]. The consumption rate of vegetables, pakchoi dominantly, for the local people was 310 g/person/day, higher than the national average level as 276 g/person/day [40]. In this area, the vegetables are mostly self-grown, whereas other foodstuffs are mainly purchased from the market. Additionally, no difference of demographic, activity and lifestyle characteristics among these 3 villages was found at such spatial scale.

Health Risks

The health risks for the local population due to exposure to metals were evaluated using Monte Carlo simulation. The cumulative probability distribution of the total HQ values in the three villages is presented in Figure 2. For V1, the HQ ranged

Table 2. Demographic, lifestyle, and dietary characteristics of the local residents.

Characteristics	Mean	SD	Unit
Weight	62.58	10.23	kg
Activity			
Indoors	8.47	3.99	h
Outdoors	5.84	3.98	h
Sleep	9.24	2.14	h
Dietary			
Rice	360.29	71.92	g/d
Flour	6.29	17.38	g/d
Pakchoi	171.76	82.74	g/d
Cabbage	63.53	38.66	g/d
Spinach	30.44	19.36	g/d
Celery	24.26	11.49	g/d
Pork	56.85	25.01	g/d
Egg	45.44	25.97	g/d
Water ingestion	1.88	0.62	L/d
Exposure duration	35.35	18.68	a

SD: Standard deviation.

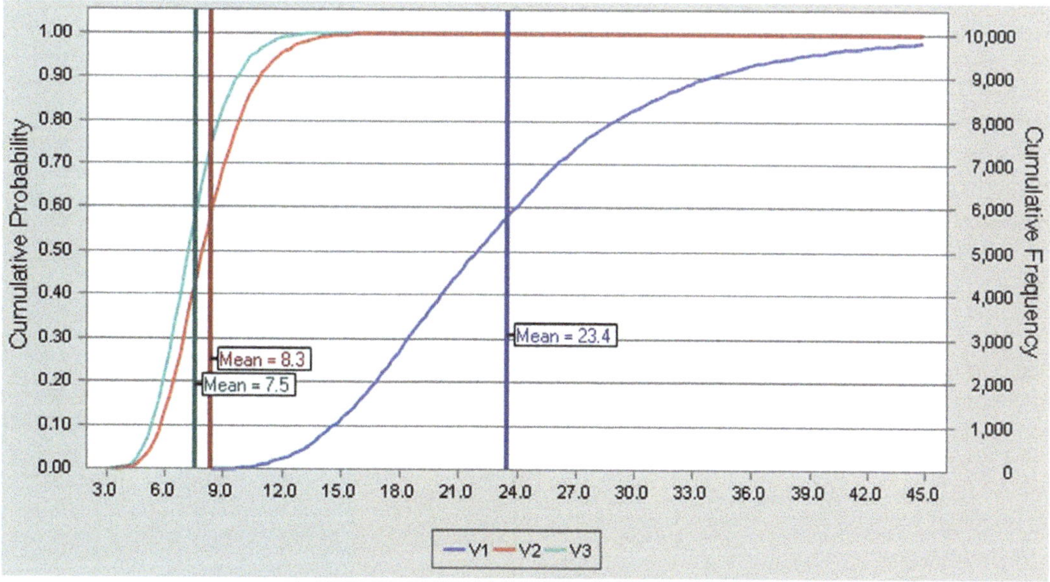

Figure 2. Cumulative probability distribution of the total HQs in the three villages. The health risks were evaluated by means of a Monte Carlo simulation based on Crystalball software for 10,000 iterations.

from 6.8 to 81.3; for V2, the HQ ranged from 2.9 to 19.3; and for V3, the HQ ranged from 3.3 to 15.8. The mean HQ values of V1, V2 and V3 were 23.4, 8.3 and 7.5, respectively. Obviously, the HQ of V1 was much higher than those of the other two.

The heavy metal HQ values from dietary sources, dust inhalation (both indoor and outdoor air), water ingestion and dermal contact, and soil ingestion and dermal contact were aggregated. The Monte Carlo simulated health risk assessment results are shown in Table 3. There is a significant difference of Pb, Cd and Hg HQs among the three villages. In V1, the average HQ value of Pb is as high as 16.2, indicating that the chronic daily intake dose exceeds the safe reference dose by 15 fold. The HQ values of Pb in V2 and V3 reach 3.6 and 3.0, respectively. The chronic daily intake dose of Cd and Hg also exceeds the corresponding limits, and the HQ in V1 was greater than those in V2 and V3.

Table 3. Hazard quotients of different metals in each village.

Metal	V1		V2		V3	
	Mean	SD	Mean	SD	Mean	SD
Ag	0.0009	0.0005	0.0003	0.0001	0.0002	0
Cd	3.32	1.91	1.49	0.843	1.81	0.812
Cr	0.001	0.0008	0.001	0.0005	0.0009	0.0005
Cu	0.189	0.046	0.164	0.039	0.18	0.046
Ni	0.42	0.15	0.41	0.15	0.35	0.15
Pb	16.2	6.96	3.63	1.33	3.01	0.92
Se	0.018	0.015	0.018	0.016	0.019	0.016
Tl	0.17	0.062	0.058	0.023	0.097	0.13
Zn	0.30	0.076	0.29	0.073	0.28	0.073
Hg	2.84	1.06	2.28	0.65	1.83	0.50

As Figure 3 shows, soil ingestion was the primary pathway to Pb exposure for V1, accounting for 40.3% of the total HQ. Food consumption contributed nearly one-third to the total HQ, and among different food sources, pakchoi were the most significant, contributing 77%. Inhalation contributed nearly one-fifth to the total HQ. Furthermore, indoor air contributed twice as much as did outdoor air. For the other two villages, the proportion of inhalation increased, whereas those of food ingestion and soil ingestion decreased sharply. This result is consistent with trend of Pb levels in the soil with increasing distance from the mine.

For Cd (Figure 4), the exposure to and uptake of this non-essential element was mainly from soil dermal contact, which contributed 47.1% to the total HQ in V1. Inhalation also contributed 44.3% to the total risk, and indoor air contributed slightly more than outdoor air. However, in V2 and V3, the total HQ decreased with increasing distance from the mine, and the contribution of soil dermal contact decreased sharply with an evident increase of inhalation. The ingestion of water and food is safe for local residents.

The local residents' average daily exposure to Hg also exceeded the safe reference dose, and the HQ value of the residents in V1 was the highest. Taking V1 as an example, the multipathway analysis (Figure 5) showed that the people of that village are exposed to Hg mainly through inhalation, with indoor air contributing the most. Food also contributed 41.6% to the total risk, mainly due to self-produced broad-leaf vegetables, such as pakchoi and cabbage, whereas the risk posed by rice from the market was not trivial. Different from Pb and Cd, soil pollution only contributes a minor fraction. V2 and V3 had similar profiles of metal levels.

Heavy Metal Levels in Hair

The metal concentrations in the hair samples are shown in Table 4. Nine metals were detected in the local residents' hair samples. The concentration of Zn was the highest of the detected metals. The geometric mean of Zn reached 240 mg/kg, which is much greater than those of the other metals. The hair samples from the residents in V1 contained higher Pb (8.14 mg/kg) and

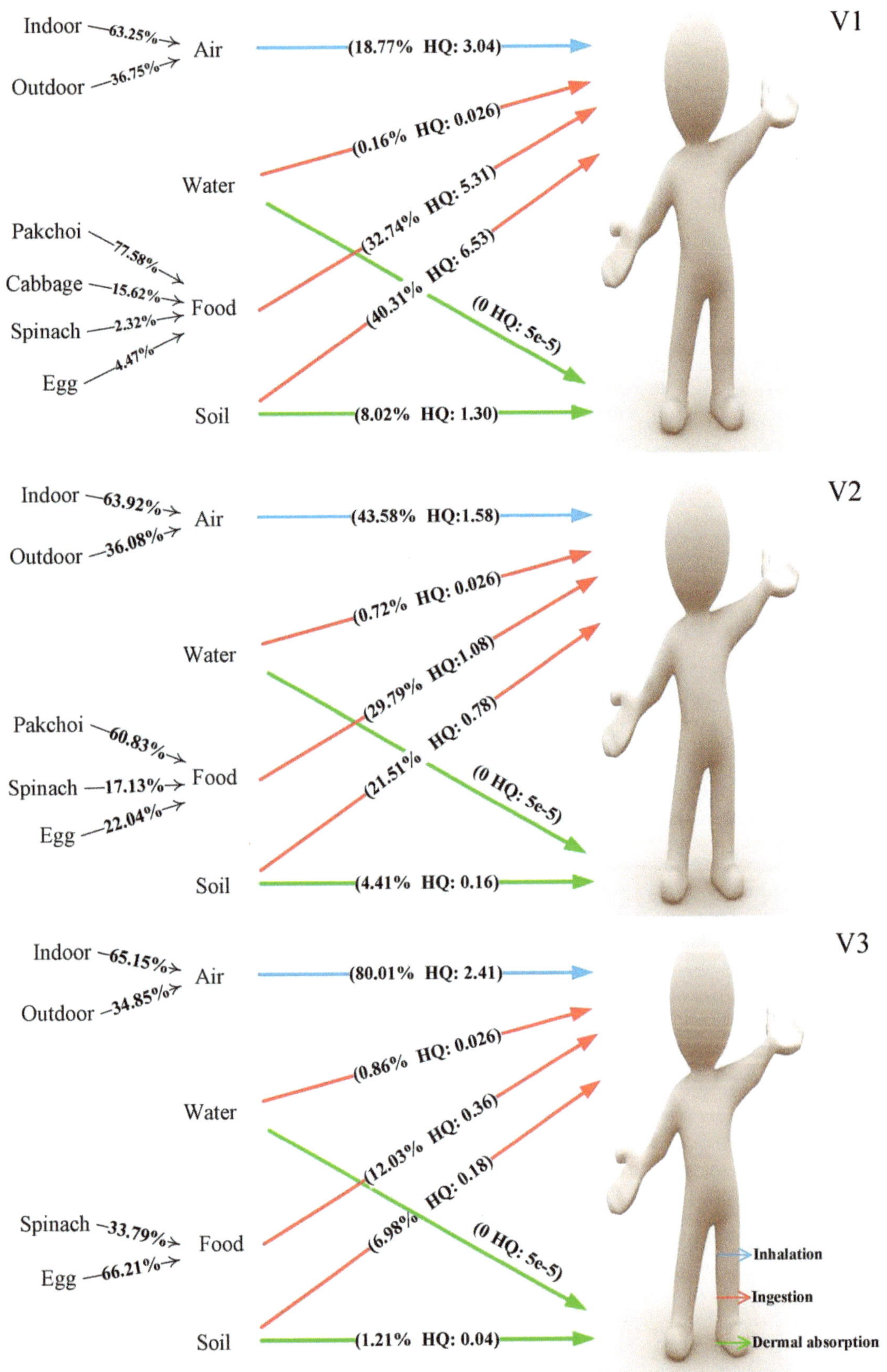

Figure 3. Multipathway analysis of HQ (Pb). Each pathway's contribution to total Pb exposure of local residents in the three studied villages was calculated based on average HQ values.

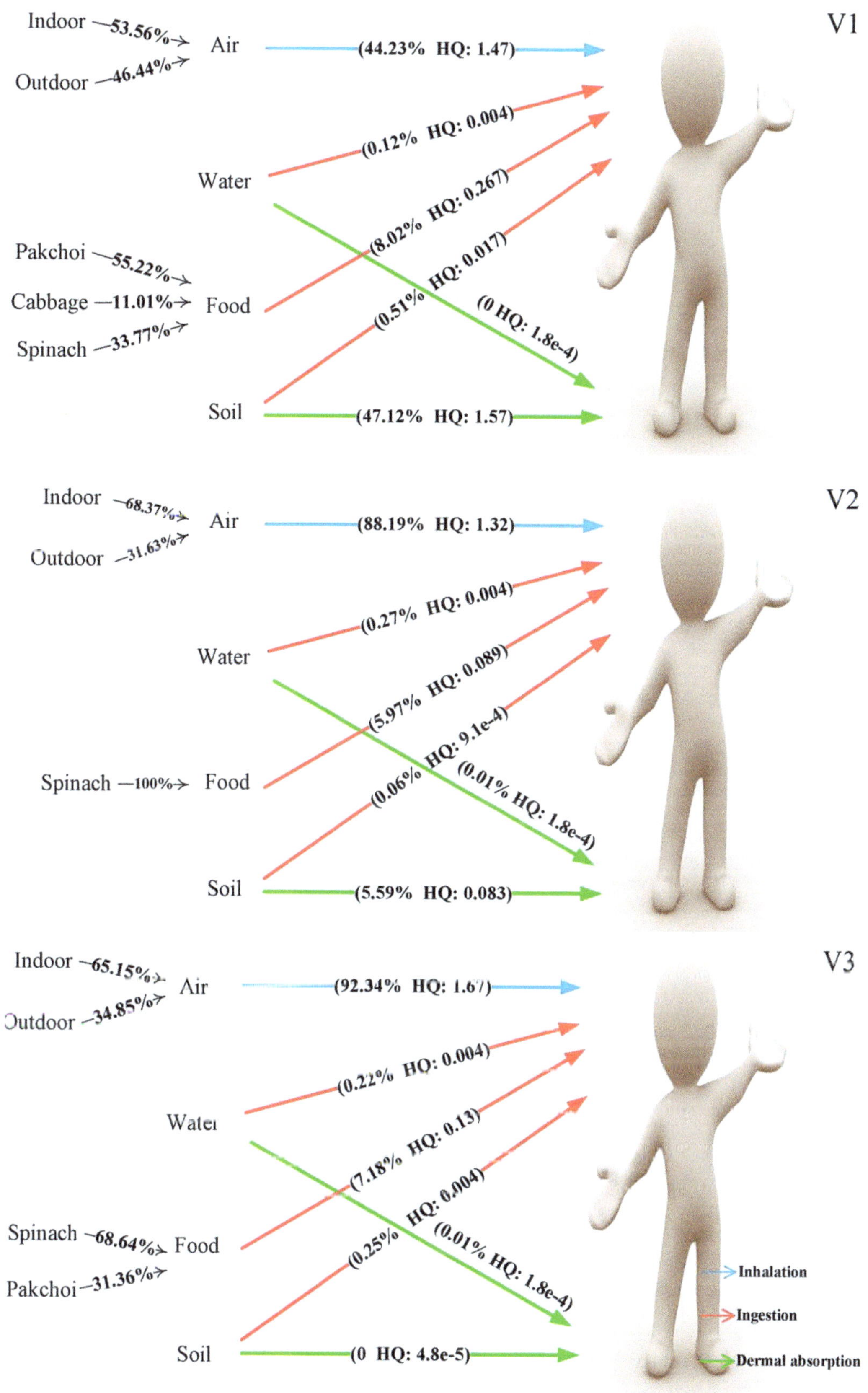

Figure 4. Multipathway analysis of HQ (Cd). Each pathway's contribution to total Cd exposure of local residents in the three studied villages was calculated based on average HQ values.

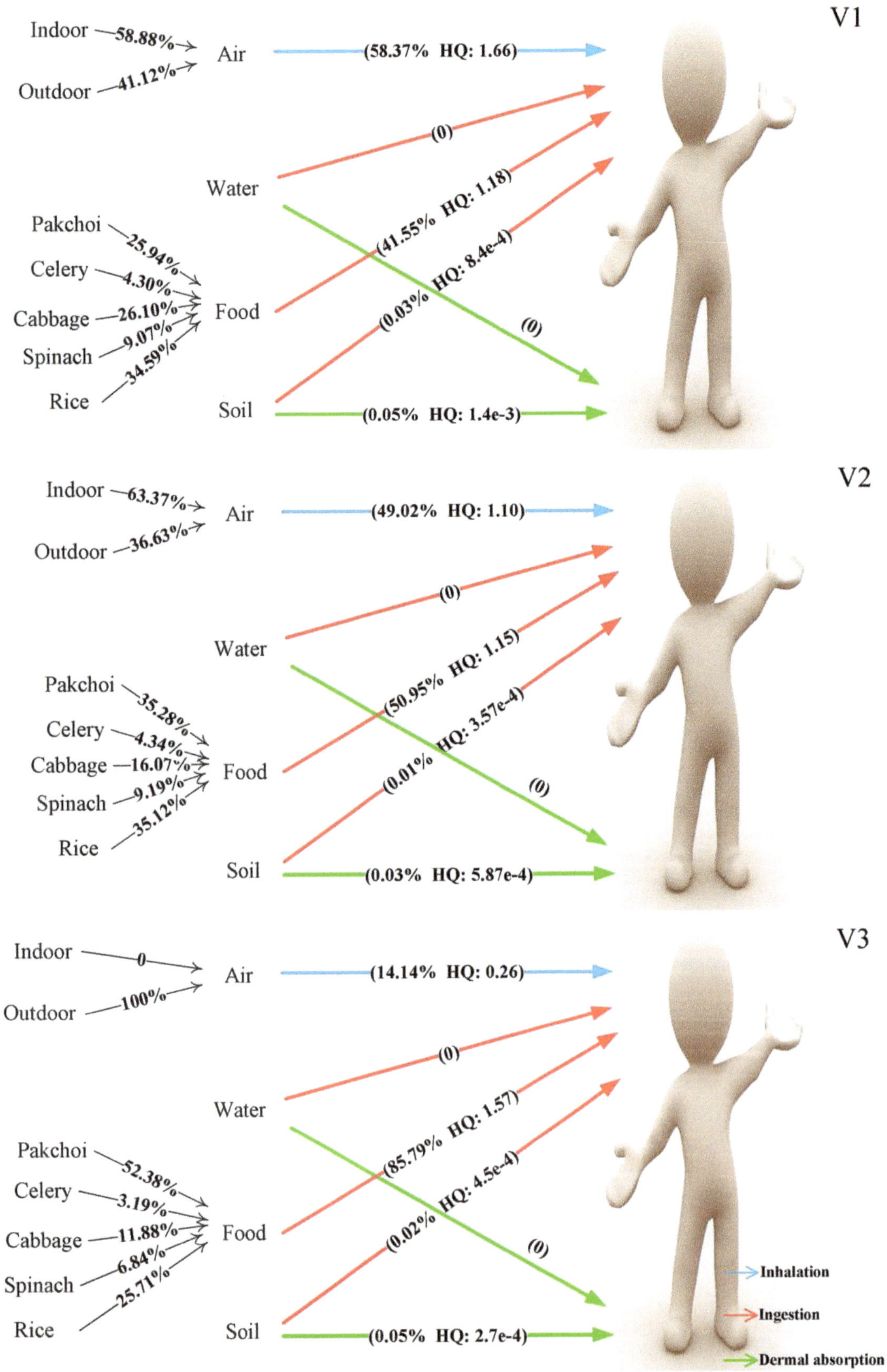

Figure 5. Multipathway analysis of HQ (Hg). Each pathway's contribution to total Hg exposure of local residents in the three studied villages was calculated based on average HQ values.

Table 4. Heavy metal levels in hair (mg/kg).

Metals	LOD	V1		V2		V3	
		Mean(SD)	Detected proportion	Mean (SD)	Detected proportion	Mean (SD)	Detected proportion
Ag	0.05	-	0	-	0	-	0
Cd	0.01	0.15(0.42)	90.0%	0.08(0.12)	63. 3%	0.14(0.17)	91.7%
Cr	0.05	1.60(1.05)	100%	2.68(2.47)	100%	2.46(5.20)	100%
Cu	0.05	9.12(7.88)	100%	11.13(13.38)	100%	8.86(3.52)	100%
Ni	0.05	0.44(0.75)	86.7%	0.73(0.89)	100%	0.71(0.90)	91.7%
Pb	0.05	8.14(12.30)	100%	2.26(4.19)	100%	3.98(4.69)	100%
Se	0.25	0.58(0.37)	46.7%	0.77(0.94)	76.7%	0.81(0.86)	75.0%
Tl	0.05	0.07(0.01)	6.7%	0.09(0.01)	6.7%	0.10(0.05)	41.7%
Zn	0.25	270.3(441.1)	100%	241(262)	100%	201.4(126.1)	100%
Hg	0.01	0.86(5.00)	100%	0.59(1.55)	100%	0.55(1.03)	75.0%

LOD: Limit of detection.

Hg (0.86 mg/kg) concentrations than those from V2 and V3 ($P<0.05$). This result is consistent with the variation of the HQ values. For the other metals, no significant differences or strong relationships with the HQ were found among the three villages.

Discussion

The environmental sample analysis results showed widespread heavy metal contamination in the air, especially of Pb, Zn and Cu. The nearest village (V1) to the mine had the heaviest pollution. Due to its respiratory, neurological, and potential carcinogenic effects, Pb is regulated by the National Ambient Air Quality Standards with seasonal mean levels not to exceed 1.0 μg/m^3 and annual mean levels not to exceed 0.5 μg/m^3 [41]. The proposed WHO guideline for Pb is 0.5 μg/m^3 [42]. The average Pb concentrations in the three villages all exceeded the limit. The highest Pb concentration occurred in the indoor air in V1, which reached 5.9 μg/m^3, and this concentration was approximately 6-fold that of the national limit.

Although there was no significant difference between indoor and outdoor air in general, the highest metal concentration was detected in indoor air. This might be attributed to various indoor heavy metal pollution emissions. Some indoor particles come from the infiltration of outdoor air or are derived from material inside the house such as household plant material, paint chips, flakes of construction materials, fuel burning and smoking [43,44,45]. A previous study found that resuspension has a significant impact on indoor particle concentrations. For example, simply walking into a room can increase the particle concentration by 100% for some supermicron particle sizes [43].

Distance from the mine was a key determinant of soil pollution, as there are significant spatial differences among these three villages. This finding was consistent with the conclusions of previous studies [46]. There was heavy Pb, Zn, Cu and Cd contamination in V1, the nearest village to the mine and the most polluted. The main body of this large mine is underground, whereas the sharp decline of heavy metal concentrations with increasing distance to the mining area suggests that it is the dominant source of metals such as Pb and Zn in the soil. This result was consistent with the previous studies, revealing the fly ash emitted from the ventilation outlet of the mine factory was the pollution source in this area [47]. Additionally, the run-off loss of solid ore deposited on open ground and dust emission from the transportation are likely important pollution sources too.

In addition, the relatively higher Pb and Zn concentrations in the self-produced vegetables were consistent with the concentrations in the top soil. In general, the highest metal concentrations were found in green leafy vegetables [18]. The deposition from dust onto the foliage of these vegetables would also add to the heavy metal content because metals firmly attach to the surfaces [48]. In this study, self-produced pakchoi and Chinese cabbage, especially in V1, had much higher Pb concentrations than did the other vegetables. In contrast, eggs, pork and rice bought from market were safer, as they had much lower metal concentrations than those of the vegetables. It was evident that outside food was safer than the self-produced food.

The calculated HQ values for all three villages exceeded the limits, indicating heavy metal pollution may pose high potential health risk to the people within the vicinity of the lead-zinc mining area. The factory itself declared to comply with strict environmental protection requirements with the goal of zero pollutants emission. The local environmental protection agency had set up hygienic buffer zone with a 200 meters safety protection distance around the mine. However, this regulation was not strictly followed, for the mine factory was only 100 meters from V1. The HQ in V1, which reached 23.4, was significantly higher than those of the other two villages. The average HQs of V2 and V3 also reached 8.3 and 7.5, respectively, indicating that the 200-m safety protection distance failed to protect the public's health. Furthermore, the estimated risks were mainly due to Pb, Cd and Hg. In V1, the average HQ value of Pb was as high as 16.2, accounting for 65% of the total HQ. Pb also contributed nearly half of the average HQ in the other two villages. The chronic daily intake dose of Cd and Hg also exceeded the safe level. Despite Zn having the highest concentration among the metals, its HQ value was below the limit, indicating that Zn does not pose a health risk to the local residents, as Zn is an essential element for the human body, and only superfluous ingestion can cause adverse health effects. It's worth noting that, different from soil exposure, which declined sharply when the distance to the mine increased, as to the pathway of inhalation, V3 had higher Pb and Cd HQ when compared to V2 in average. However, there were no significant differences ($P>0.05$). More environment examinations maybe needed in the future to assess health risk posed by Pb and Cd air pollution.

With regard to each metal, different exposure pathways made different contributions in different locations. Pb posed the highest estimated risk, which was mainly from soil ingestion, indoor air inhalation and pakchoi ingestion. However, in V1, the general population was primarily exposed to Cd via soil dermal contact and inhalation. As to Hg, food ingestion and inhalation contributed most to the HQ. In this study, the average indoor and outdoor air concentrations in V1 reached 734 and 543 ng/m^3 respectively, and were much higher than the national background level, 6 ng/m^3 [49]. Accordingly, Hg intake through inhalation is the main and most important Hg exposure source. This finding was in accord with a previous study in Guizhou Province, where atmospheric Hg concentration around power plants ranged from 556.7 to more than 1000 ng/m^3, and the local residents have the potential of developing adverse health problems because of over exposure [50]. Additionally, Hg content in rice in the study area was 0.03 mg/kg, which was lower than that of Guizhou in China (0.48 mg/kg), but much higher than that in Taiwan (0.001 mg/kg) and Shanghai (0.0047–0.0056 mg/kg) [51,52,53]. The mean concentration of vegetables measured in the study area range from 0.02 to 0.05 mg/kg, which was a bit lower than those measured at Nanning (0.024–0.077 mg/kg), but much higher than those measured at Shanghai (0.0005–0.006 mg/kg) and Zhongshan (0.0023–0.010 mg/kg) in China [54,55,56]. In general, the study area had higher Hg concentrations in rice and vegetables among the areas compared, verifying that intake of rice and vegetables is another important Hg exposure source for the residents in the study area.

Generally, the inhalation of indoor air usually contributed more than that of outdoor air because of the relatively heavier indoor air pollution and the amount of time local people spent on indoor activities. The contributions of dermal contact and ingestion of soil declined sharply with increasing distance from the mine, especially for Pb and Cd. This result was in accordance with the spatial trend of metal levels in the soil. Self-grown pakchoi was an important risk source for local residents, specifically in V1, due to its relatively high metal concentrations and large proportion in the dietary structure. These results show that dermal contact, the ingestion of soil, the inhalation of indoor air and vegetable consumption are important in reducing the overall health risk. Human exposure to soil, air and self-produced vegetables indicates the necessity of future policy responses and interventions. Local residents, especially in V1, should reduce their consumption of self-grown vegetables, maintain clean indoor air, and avoid contact with the polluted soil and dust.

Except for Ag, 9 PP metals were detected in the human hair samples. Among these detected metals, the Zn content in the hair of the sampled individuals was found to be the highest and exceeded 200 mg/kg. However, no differences were found among various locations. Consistent with the higher HQ value of the residents in V1, their hair samples exhibited higher, statistically significant Pb (8.14 mg/kg) and Hg (0.86 mg/kg) levels ($P<0.05$) than did the other two villages. The average Pb concentration in V1 was lower than that of an electronic waste recycling area in Taizhou (49.5 mg/kg), Jiangsu Province, China, but higher than that of the control sites in Ningbo (2.53 mg/kg) and Shaoxing (6.61 mg/kg) in a previous study [31]. The highest Pb concentration in hair found in V1, V2 and V3 reached 48.0, 19.7 and 18.0 mg/kg respectively, which exceeded the suggested upper limit of normal value of hair Pb (10.0 mg/kg) [57]. The HQ and corresponding Pb and Hg concentrations in the hair declined with increasing distance from the mine. It can be stated that the heavy metal pollution from the lead-zinc mine increased the exposure level of local residents, especially in V1. Human hair could be a useful biomonitoring tool to assess the extent of Pb and Hg exposure to residents in metal-polluted areas. However, for Zn, Cu, and Cd, higher levels in the hair were not detected in V1, where higher estimated exposure levels were found. This showed that hair was not biomonitoring indicator of exposure for all elements. More studies should be done to better justify use of hair as biomonitoring indicator for other metals considered in this area.

According to the above analysis, it can be concluded that mining activity may pose high potential risk to the public, and current safety protection distance failed to protect the public's health. Most of the estimated risks came from soil, self-produced vegetables and indoor air inhalation. In summary, our results suggest that the exposure of local inhabitants to heavy metals in the mining area is through multipathways. This study highlights the importance of site-specific multipathway risk assessment. However, it should also be pointed out that our risk model used conservative assumptions and we were not able to conduct a human health impacts analysis. More detailed and in-depth health investigation is necessary in the future to examine whether adverse health outcomes occur, and provide decision-making support for pollution control in this metal-polluted area accordingly.

Author Contributions

Conceived and designed the experiments: CSQ LH JB. Performed the experiments: CSQ JY ZWM. Analyzed the data: CSQ ZWM. Contributed reagents/materials/analysis tools: CSQ YL. Wrote the paper: CSQ LH.

References

1. Tong S, von Schirnding YE, Prapamontol T (2000) Environmental lead exposure: a public health problem of global dimensions. B World Health Organ 78: 1068–1077.
2. Järup L, Hellström L, Alfvén T, Carlsson MD, Grubb A, et al. (2000) Low level exposure to cadmium and early kidney damage: the OSCAR study. Occup Environ Med 57: 668.
3. Thomas LDK, Hodgson S, Nieuwenhuijsen M, Jarup L (2009) Early kidney damage in a population exposed to cadmium and other heavy metals. Environ Health Persp 117: 181.
4. Putila JJ, Guo NL (2011) Association of arsenic exposure with lung cancer incidence rates in the United States. PLoS ONE 6: e25886. doi:25810.21371/journal.pone.0025886.
5. Wang H, Stuanes AO (2003) Heavy metal pollution in air-water-soil-plant system of Zhuzhou City, Hunan Province, China. Water Air Soil Pollut 147: 79–107.
6. Nriagu JO, Pacyna JM (1988) Quantitative assessment of worldwide contamination of air, water and soils by trace metals. Nature 333: 134–139.
7. Wang QR, Dong Y, Cui Y, Liu X (2001) Instances of soil and crop heavy metal contamination in China. Soil Sediment Contam 10: 497–510.

8. Wang X, Sato T, Xing B, Tao S (2005) Health risks of heavy metals to the general public in Tianjin, China via consumption of vegetables and fish. Sci Total Environ 350: 28–37.
9. Cheng S (2003) Heavy metal pollution in China: Origin, pattern and control - a state-of-the-art report with special reference to literature published in Chinese journals. Environ Sci Pollut Res 10: 192–198.
10. Huang XF, Hu JW, Li CX, Deng JJ, Long J, et al. (2009) Heavy-metal pollution and potential ecological risk assessment of sediments from Baihua Lake, Guizhou, PR China. Int J Environ Health Res 19: 405–419.
11. Wang XQ, He MC, Xie J, Xi JH, Lu XF (2010) Heavy metal pollution of the world largest antimony mine-affected agricultural soils in Hunan province (China). J Soil Sediment 10: 827–837.
12. Zeng HA, Wu JL (2009) Sedimentary Records of Heavy Metal Pollution in Fuxian Lake, Yunnan Province, China: Intensity, History, and Sources. Pedosphere 19: 562–569.
13. Zhou JM, Dang Z, Cai MF, Liu CQ (2007) Soil heavy metal pollution around the Dabaoshan Mine, Guangdong Province, China. Pedosphere 17: 588–594.
14. Gao Y, Xia J (2011) Chromium contamination accident in China: viewing environment policy of China. Environ Sci Technol 45: 8065–8066.
15. Granero S, Domingo J (2002) Human health risks. Environ Int 28: 159–164.
16. Nadal M, Bocio A, Schuhmacher M, Domingo J (2005) Trends in the levels of metals in soils and vegetation samples collected near a hazardous waste incinerator. Arch Environ Contam Tox 49: 290–298.
17. MacIntosh DL, Spengler JD, Ozkaynak H, Tsai L, Ryan PB (1996) Dietary exposures to selected metals and pesticides. Environ Health Persp 104: 202.
18. Hough RL, Breward N, Young SD, Crout NMJ, Tye AM, et al. (2004) Assessing potential risk of heavy metal exposure from consumption of home-produced vegetables by urban populations. Environ Health Persp 112: 215.
19. Baastrup R, Sørensen M, Balstrøm T, Frederiksen K, Larsen CL, et al. (2008) Arsenic in drinking-water and risk for cancer in Denmark. Environ Health Persp 116: 231.
20. Albering HJ, van Leusen SM, Moonen EJC, Hoogewerff JA, Kleinjans JCS (1999) Human health risk assessment: A case study involving heavy metal soil contamination after the flooding of the river Meuse during the winter of 1993–1994. Environ Health Persp 107: 37–43.
21. Man YB, Sun XL, Zhao YG, Lopez BN, Chung SS, et al. (2010) Health risk assessment of abandoned agricultural soils based on heavy metal contents in Hong Kong, the world's most populated city. Environ Int 36: 570–576.
22. Mari M, Nadal M, Schuhmacher M, Domingo JL (2009) Exposure to heavy metals and PCDD/Fs by the population living in the vicinity of a hazardous waste landfill in Catalonia, Spain: Health risk assessment. Environ Int 35: 1034–1039.
23. Hu J, Dong Z (2011) Development of Lead Source-specific Exposure Standards Based on Aggregate Exposure Assessment: Bayesian Inversion from Biomonitoring Information to Multipathway Exposure. Environ Sci Technol.
24. Chadha A, McKelvey LD, Mangis JK (1998) Targeting lead in the multimedia environment in the continental United States. J Air Waste Manag Assoc 48: 3–15.
25. Vyskocil A, Fiala Z, Chénier V, Krajak L, Ettlerova E, et al. (2000) Assessment of multipathway exposure of small children to PAH. Environ Toxicol Phar 8: 111–118.
26. Lu C, Kedan G, Fisker-Andersen J, Kissel JC, Fenske RA (2004) Multipathway organophosphorus pesticide exposures of preschool children living in agricultural and nonagricultural communities. Environ Res 96: 283–289.
27. Marin C, Guvanasen V, Saleem Z (2003) The 3MRA risk assessment framework–A flexible approach for performing multimedia, multipathway, and multireceptor risk assessments under uncertainty. Hum Ecol Risk Assess 9: 1655–1677.
28. NRC (1983) Risk assessment in the federal government: Managing the process. National Research Council.
29. Li J, Liu L, Huang G, Zeng G (2006) A fuzzy-set approach for addressing uncertainties in risk assessment of hydrocarbon-contaminated site. Water Air Soil Pollut 171: 5–18.
30. Goullé JP, Mahieu L, Castermant J, Neveu N, Bonneau L, et al. (2005) Metal and metalloid multi-elementary ICP-MS validation in whole blood, plasma, urine and hair: Reference values. Forensic Sci Int 153: 39–44.
31. Wang T, Fu J, Wang Y, Liao C, Tao Y, et al. (2009) Use of scalp hair as indicator of human exposure to heavy metals in an electronic waste recycling area. Environ Pollut 157: 2445–2451.
32. Pereira R, Ribeiro R, Goncalves F (2004) Scalp hair analysis as a tool in assessing human exposure to heavy metals (S. Domingos mine, Portugal). Sc Total Environ 327: 81–92.
33. Ashe K (2012) Elevated mercury concentrations in humans of Madre de Dios, Peru. PloS ONE 7: e33305. doi:33310.31371/journal.pone.0033305.
34. Sanders AP, Flood K, Chiang S, Herring AH, Wolf L, et al. (2012) Towards prenatal biomonitoring in North Carolina: Assessing arsenic, cadmium, mercury, and lead levels in pregnant women. PloS ONE 7: e31354. doi:31310.31371/journal.pone.0031354.
35. Carneiro MF, Moresco MB, Chagas GR, Souza O, Rhoden CR, et al. (2011) Assessment of trace elements in scalp hair of a young urban population in Brazil. Biol Trace Elem Res 143: 815–824.
36. Oyoo-Okoth E, Admiraal W, Osano O, Ngure V, Kraak MH, et al. (2010) Monitoring exposure to heavy metals among children in Lake Victoria, Kenya: Environmental and fish matrix. Ecotox Environ Safe 73: 1797–1803.
37. USEPA (1989) Risk assessment guidance for Superfund volume I: human health evaluation manual (Part A). EPA/540/1-89/002. Washington, D. C.: Environmental Protection Agency, United States.
38. USEPA (1997) Exposure Factors Handbook. Washington, D. C.: Environmental Protection Agency, United States.
39. Chen SC, Liao CM (2006) Health risk assessment on human exposed to environmental polycyclic aromatic hydrocarbons pollution sources. Sci Total Environ 366: 112–123.
40. Zhai FY, He YN, Ma GS, Li YP, Wang ZH, et al. (2005) Study on the current status and trend of food consumption among Chinese population. Chin J Epidemiol 26: 485–488. (in Chinese).
41. MEP (2012) National Ambient Air Quality Standard of China. Ministry of Environmental Protection, China.
42. WHO (2000) Air Quality Guidelines for Europe (2nd Edn.). World Health Organization Regional Publications, European Series No.91.
43. Thatcher TL, Layton DW (1995) Deposition, resuspension, and penetration of particles within a residence. Atmos Environ 29: 1487–1497.
44. Rashed MN (2008) Total and extractable heavy metals in indoor, outdoor and street dust from Aswan City, Egypt. Clean–Soil Air Water 36: 850–857.
45. Rasmussen P, Subramanian K, Jessiman B (2001) A multi-element profile of house dust in relation to exterior dust and soils in the city of Ottawa, Canada. Sci Total Environ 267: 125–140.
46. Benin AL, Sargent JD, Dalton M, Roda S (1999) High concentrations of heavy metals in neighborhoods near ore smelters in northern Mexico. Environ Health Persp 107: 279.
47. Chu BB, Luo LQ (2010) Evaluation of heavy metal pollution in soils from Nanjing Qixiashang lead-zinc mines. Rock Miner Anal 29: 5–8.(in Chinese).
48. Al Jassir M, Shaker A, Khaliq M (2005) Deposition of Heavy Metals on Green Leafy Vegerables Sold on Roadsides of Riyadh City, Saudi Arabia. B Environ Contam Tox 75: 1020–1027.
49. Zhang L, Wong M (2007) Environmental mercury contamination in China: sources and impacts. Environ Int 33: 108–121.
50. Qu L, Liu P (2004) Environmental effect of buring coal for power station. J Guizhou Univ Technol 33: 92–94.
51. Chen H, Zhen C (1999) Heavy metal pollution in soils in China: status and countermeasures. AMBIO 28: 130–134.
52. Lin HS, Wong SS, Li GC (2004) Heavy metal content of rice and shellfish in Taiwan. J Food Drug Anal 12: 167–174.
53. Wang Y (1993) The status of mercury content in water, soil and crop in Shanghai. Shanghai Agric Technol 2: 26–27. (in Chinese).
54. Li J (1999) Investigation on mercury contamination in suburb soil in Nanning. Guangxi Agricul Ecol Sci 1: 80–83. (in Chinese).
55. Wang Y (1993) The status of mercury content in water, soil and crop in Shanghai. Shanghai Agric Technol 2: 26–27. (in Chinese).
56. Zhou RD, Wu HG, Huang C, Deng CG (2002) The investigation on the content of mercury, lead, cadmium, arsenic and chrome in vegetable by bargaining in Zhongshan City. Chin J Health Lab Technol 12: 582–583. (in Chinese).
57. Qin JF (2004) The upper limit of normal value of hair Pb, Cd, As, Hg in Chinese resident. Guangdong Trace Metal Sci 2004. 29–37. (In Chinese).

2

Respiratory Health before and after the Opening of a Road Traffic Tunnel: A Planned Evaluation

Christine T. Cowie[1,2]*, Nectarios Rose[1,3], Wafaa Ezz[1,2], Wei Xuan[1,2], Adriana Cortes-Waterman[1,2], Elena Belousova[1], Brett G. Toelle[1], Vicky Sheppeard[4], Guy B. Marks[1,2]

1 Woolcock Institute of Medical Research, Sydney, Australia, **2** Cooperative Research Centre for Asthma and Airways, Sydney, Australia, **3** NSW Health Department, Sydney, Australia, **4** Western Sydney and Nepean Blue Mountains Public Health Unit, Sydney, Australia

Abstract

Objective: The construction of a new road tunnel in Sydney, Australia, and concomitant reduction in traffic on a major road presented the opportunity to study the effects of this traffic intervention on respiratory health.

Methods: We made measurements in a cohort of residents in the year before the tunnel opened (2006) and in each of two years afterwards (2007–2008). Cohort members resided in one of four exposure zones, including a control zone. Each year, a respiratory questionnaire was administered (n = 2,978) and a panel sub-cohort (n = 380) performed spirometry once and recorded peak expiratory flow and symptoms twice daily for nine weeks.

Results: There was no consistent evidence of improvement in respiratory health in residents living along the bypassed main road, despite a reduction in traffic from 90,000 to 45,000 vpd. Residents living near tunnel feeder roads reported more upper respiratory symptoms in the survey but not in the panel sub-cohort. Residents living around the tunnel ventilation stack reported more upper and lower respiratory symptoms and had lower spirometric volumes after the tunnel opened. Air pollutant levels measured near the stack did not increase over the study period.

Conclusion: The finding of adverse health effects among residents living around the stack is unexpected and difficult to explain, but might be due to unmeasured pollutants or risk factors or an unrecognized pollutant source nearby. The lack of improvement in respiratory health among people living along the bypassed main road probably reflects a minimal change in exposure due to distance of residence from the road.

Editor: Stephania Ann Cormier, Louisiana State University Health Sciences Center, United States of America

Funding: The authors acknowledge funding support from the Cooperative Research Centre for Asthma & Airways and the NSW Health Department. The funders had no role in study design, data collection and analysis, decision to publish, or preparation of the manuscript, other than the contributions made by NR and VS as indicated under Competing Interests.

Competing Interests: The authors have read the journal's policy and have the following conflicts: NR was employed with the NSW Health Department on the Biostatistics Training Scheme and VS was employed in the NSW Health system, at the time the study was conducted. There are no other competing interests.

* E-mail: christine.cowie@sydney.edu.au

Introduction

Road tunnels are increasingly being constructed in major world cities to help alleviate traffic congestion. They are often located within densely populated areas, have high traffic volumes, including a substantial proportion of heavy duty vehicles, and pollutants are commonly vented to the external environment through stacks [1–3]. However, studies investigating the health impact of traffic-related air pollution (TRAP) on communities residing around road tunnels, or their ventilation stacks and portals, are scarce, probably due to difficulty in disentangling health effects associated with general traffic from effects attributable to emissions arising in the tunnels [4].

We are aware of only one study that has investigated health effects associated with exposure to road tunnel stack emissions [2]. This cross-sectional study, conducted after the tunnel was opened, found no difference in the prevalence of self-reported respiratory and irritant symptoms between three zones estimated to have varying exposure to the ventilation stack plume.

The Lane Cove Tunnel (LCT) is a 3.6 km long road tunnel that connects two motorways and is vented by two stacks located near each end of the tunnel. Modelling prior to tunnel construction predicted that areas around the tunnel feeder roads would experience increased exposure to nitrogen dioxide (NO_2) and particulate matter (PM), and areas adjacent to the bypassed main road would experience a decrease in exposure [5]. Modelling of stack emissions forecast little detectable impact on the surrounding area. The tunnel commenced operation on 25 March 2007, with an average of 43,446 vehicles per day (vpd) during May, 2007 increasing to 58,218 vpd by November, 2008, with 2.9% and 3.5% of vehicles respectively classified as heavy or articulated vehicles. Opening of the tunnel was accompanied by an approximate halving of vehicle numbers from 90,000 vpd along the bypassed main road [6].

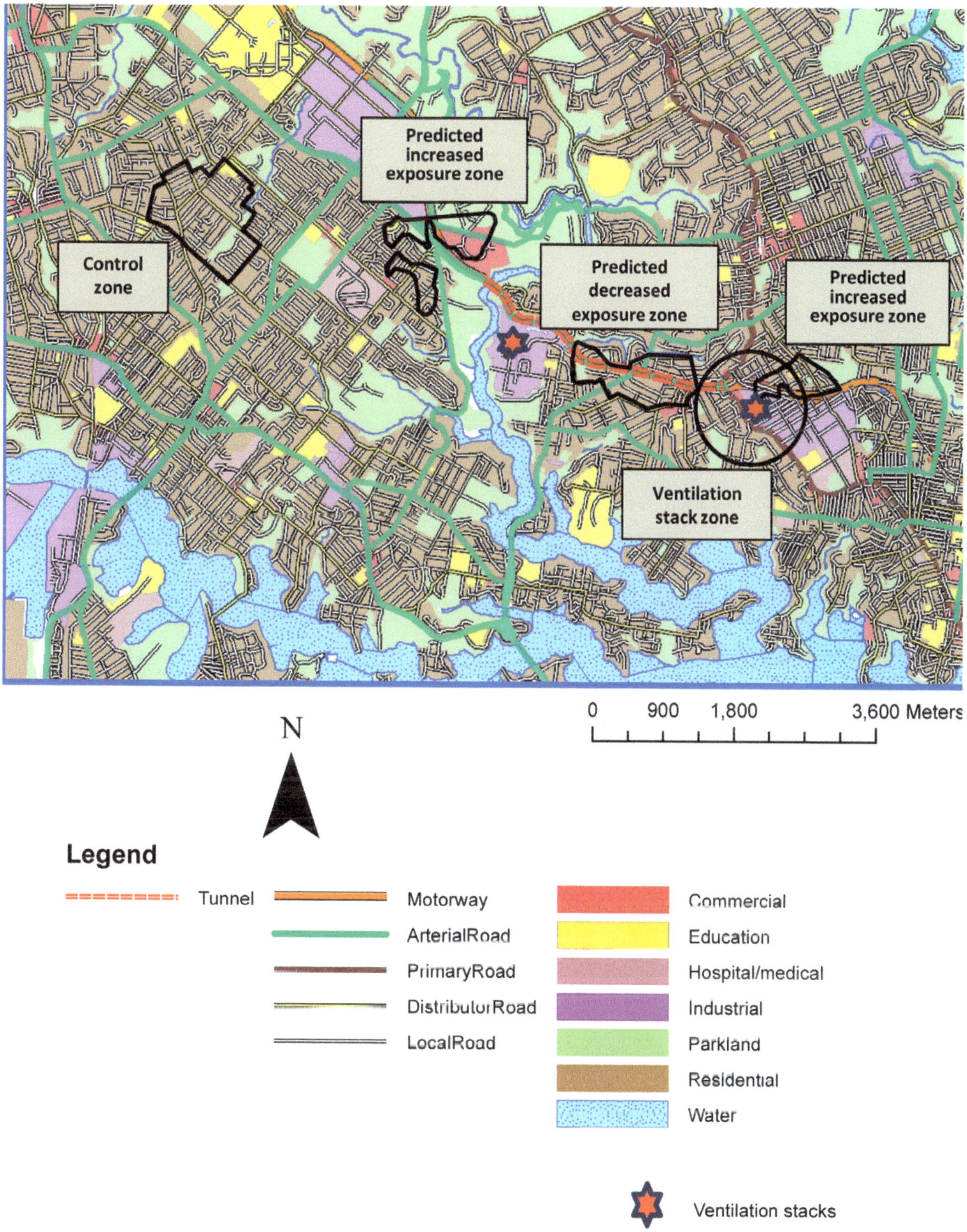

Figure 1. Map of study area showing the zones, tunnel, ventilation stacks and land use.

This study was commissioned by the NSW Health Department to inform public policy on road tunnels and their ventilation stacks and to opportunistically study respiratory health effects of a substantial traffic intervention. The principal hypothesis was that, after tunnel opening, adverse respiratory health outcomes would increase among residents near the tunnel feeder roads and

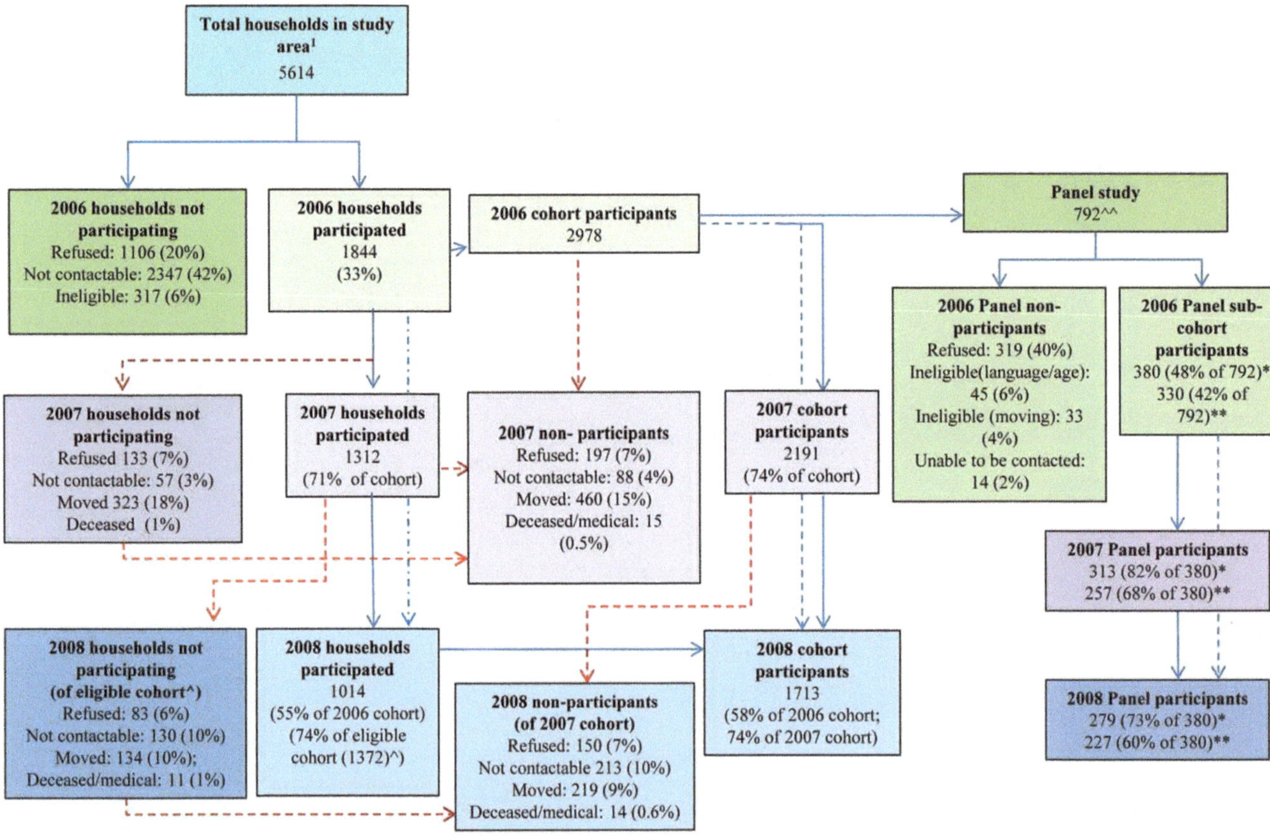

Figure 2. Participation flowchart. [1] Estimate derived from census of households within the study area. ^ Eligible cohort = 2007 cohort + additional households potentially eligible to be interviewed in 2008. ^^ Subjects approached to participate in panel study. * Subjects who took part in spirometry and eNO measurement. ** Subjects who took part in daily lung function and symptom diary. Blue arrow: Household and participant participation. Red arrow: Loss to follow-up.

decrease among residents along the bypassed main road, after adjustment for concurrent changes in a control zone. A subsidiary hypothesis was that people living around the ventilation stack would not experience a change in respiratory health status.

Methods

Ethics Statement

Approval for the study was obtained from the University of Sydney Human Research Ethics Committee (HREC). Recruitment of subjects occurred through a household doorknock survey. At the time of the home visit an introductory letter about the study was provided to an adult resident, the nature of the study was discussed, and consent to participate in the survey was sought. If consent was given, this adult was considered to be the "primary respondent" for the household, and was interviewed using a standardised questionnaire. Parents completed questionnaires for children less than 16 years, and children aged 16–18 years completed questionnaires with a parent/caregiver present. Self-completion questionnaires were left for other household adults not present at the time of the visit. Up to three follow-up phone calls were made for return of the self-completed questionnaires. Interviewers returned to households up to three times and left calling cards for households unable to be contacted. The household contact and interviewing process was documented using a standardised form. Participants in the panel sub-cohort provided separate and additional written consent. The recruitment process for the whole cohort and sub-cohort was approved by the HREC.

Subject Recruitment and Exposure Assignment

We recruited a cohort of 2,978 participants during June-December 2006 (pre-tunnel opening) and followed them up during the same time of year in 2007 and 2008 (post-tunnel opening), within two weeks of their initial participation date. Participants were recruited by door-knocking all households within four exposure zones (Figure 1), defined as follows:

a) a zone along the bypassed main road (predicted *'decreased exposure zone'*);

b) a zone around the tunnel feeder roads (predicted *'increased exposure zone'*);

c) a zone of 650 m radius around the tunnel's eastern ventilation stack (*'stack zone'*); and,

d) a control zone.

The first two zones were defined using NO_2 contours from Cal3qhc and Calmet/Calpuff dispersion modelling (see Supporting Information, Appendix S1). The zone around the eastern stack was included to address community concern about stack emissions, and the radius of 650 m was chosen to ensure a sufficient sample size. The control zone was in a nearby suburb, away from the

Table 1. Symptom prevalence and other characteristics of the study cohort by zone and year (n = 2978).

	Reduced exposure zone			Increased exposure zone[a]			Eastern stack zone[a]			Control zone		
	2006	**2007**	**2008**	**2006**	**2007**	**2008**	**2006**	**2007**	**2008**	**2006**	**2007**	**2008**
n	1024	756	608	589	422	302	645	500	394	768	537	424
(% of 2006)		(74)	(59)		(72)	(51)		(78)	(61)		(70)	(55)
%												
Adult (age ≥18 yrs)	72	70	68	74	72	71	75	72	72	72	72	72
Female	56	57	58	53	54	57	54	53	54	53	54	54
Diagnosed asthma	18	20	20	18	24	21	16	19	21	15	17	19
Current asthma	11	13	13	12	14	12	9	12	12	9	10	9
Wheeze	15	18	15	14	19	18	11	20	17	10	13	13
Asthma medication	11	14	15	11	14	14	10	14	13	9	10	11
Inhaled corticosteroids	6	8	9	4	8	7	4	7	6	5	5	7
Cough	40	49	50	33	52	48	36	52	53	34	47	48
Lower resp. symptoms[b]	39	47	47	35	51	50	34	48	52	33	43	43
Severe lower resp. symptoms[b]	10	11	12	12	14	13	8	17	12	8	9	10
Upper resp. symptoms[b]	17	24	22	10	25	28	12	27	25	12	22	20
Mouth symptoms	14	17	18	9	21	21	11	24	22	13	17	18
Smoker [c]	7	6	6	12	10	9	11	9	7	12	13	10
Home environmental factors												
Unflued gas heater	25	28	33	12	21	25	18	25	27	20	24	29
Gas cooktop or oven	52	58	59	25	52	51	44	56	56	42	46	48
ETS[d] at home	7	3	4	14	9	10	8	5	5	8	5	7
Educational status												
Tertiary educated	61	65	64	46	46	44	62	61	61	44	42	36
High school/diploma[e]	31	28	28	43	38	41	31	34	31	43	38	46
Up to middle school[f]	8	8	8	11	16	15	7	6	7	13	19	18
Work status												
Paid Work	74	76	77	66	70	70	75	76	78	71	70	73

[a]Participants in overlapping area of increased exposure zone and stack zone, entire cohort (2006 = 48; 2007 = 24; 2008 = 15) contributed data to both zones.
[b]Composite variables derived from questionnaire data (see Supporting information, Table S1).
[c]Subjects <18 years assumed to be non-smokers.
[d]Environmental tobacco smoke.
[e]Included Diploma/TAFE, and participants who responded "Other".
[f]Included participants who refused to respond.

influence of the tunnel, and not expected to experience any change in TRAP.

Eligibility criteria included: all residents within the zones; ages 2–75 years; being at home for ≥1 hour during rush hours (7:00 am–9.30 am or 4.30 pm–7:00 pm); and having sufficient English. Subjects were not told which zone they resided in. Participants were interviewed using the adult or children's version of the questionnaire.

A panel sub-cohort was recruited sequentially from the entire cohort to keep a nine week diary of symptoms and lung function. We originally intended recruiting only children to the sub-cohort, but, because of the scheduled opening of the tunnel and slower than expected recruitment, we decided, in late 2006, to also recruit adults. We recruited adults and children into two approximately equal strata until the sub-cohort quota was filled.

Questionnaire

Questionnaires collected information on respiratory symptoms and diagnoses (wheeze, cough and asthma), eye, nose and throat symptoms, and use of medications, and on potential confounders and effect modifiers including smoking, environmental tobacco smoke (ETS) exposure, and use of gas heating and gas cooking in the home. The respiratory health questions were adapted from the ISAAC [7] and Belmont studies [8], with questions about eye, nose and throat symptoms from the M5East study [2]. We asked about symptoms experienced within the preceding three months (in contrast to 12 months for ISAAC) so that the recall period for 2007 did not extend into the pre-tunnel period. Questions on mouth symptoms in adults, which we did not expect to be associated with air pollution exposure, were used to assess reporting bias. We used questionnaire responses to construct composite variables for three primary outcomes: lower respiratory tract symptoms (LRS), severe LRS (SLRS), and upper respiratory tract symptoms (URS) (see Supporting information, Table S1).

Diary Panel Sub-cohort

In panel sub-cohort participants (n = 380) we performed allergen skin prick tests to define atopic status in 2007 and

Table 2. Lung function measures, and other characteristics of the diary panel sub-cohort, by zone and year (n = 380)[a].

	Reduced exposure zone			Increased exposure zone[b]			Eastern stack zone[b]			Control zone		
	2006	**2007**	**2008**	**2006**	**2007**	**2008**	**2006**	**2007**	**2008**	**2006**	**2007**	**2008**
n [c] (% of 2006)	93	85 (91)	79 (85)	98	77 (78)	70 (71)	97	85 (88)	74 (76)	98	71(72)	61(62)
%												
Adult	46	42	41	48	44	43	45	42	46	52	54	56
Female	54	52	53	51	52	47	56	54	54	45	46	44
Atopic		58	60		57	53		50	48		58	58
Mean (SD)												
FEV_1 pre-bd[d] (L)	2.6 (0.8)	2.7 (0.8)	2.8 (0.8)	2.5 (0.9)	2.6(0.9)	2.8 (0.9)	2.8 (0.8)	2.9 (0.8)	3.1(0.8)	2.6 (0.9)	2.7 (0.8)	2.7 (0.7)
FEV_1 post-bd (L)	2.7 (0.8)	2.8 (0.8)	2.9 (0.8)	2.6 (0.9)	2.7(0.9)	2.8 (0.9)	2.9 (0.9)	3.0 (0.8)	3.1(0.8)	2.7 (0.9)	2.8 (0.8)	2.8 (0.8)
FVC pre-bd (L)	3.2 (1.1)	3.4 (1.1)	3.5(1.1)	3.1 (1.2)	3.3(1.2)	3.4 (1.1)	3.3 (1.1)	3.6 (1.1)	3.8(1.1)	3.2 (1.1)	3.4 (1.0)	3.4 (0.9)
FVC post-bd (L)	3.3 (1.1)	3.4 (1.1)	3.5 (1.1)	3.1 (1.2)	3.3(1.2)	3.4 (1.1)	3.4 (1.2)	3.6 (1.1)	3.8(1.1)	3.2 (1.1)	3.4 (1.0)	3.4 (1.0)
eNO (ppb)[e]	9.7 (1.8)	17.8(2.0)	8.9 (2.3)	8.9 (1.9)	18.7(1.8)	9.7 (2.0)	9.2 (2.0)	16.3(1.7)	8.0(2.0)	8.9 (1.8)	17.9(1.8)	9.1 (1.8)
n [f] (% of 2006)	86	76 (88)	70 (81)	82	57(70)	52 (63)	85	69 (81)	62(73)	81	57 (70)	46 (57)
Mean (SD)												
PEF am (L/min)[g]	453 (106)	456 (94)	447 (98)	372 (78)	377 (72)	351 (83)	429(137)	420(131)	459(125)	382(110)	377(103)	379 (86)
PEF pm (L/min)[g]	450 (108)	462 (90)	453 (97)	377 (83)	386 (79)	355 (80)	420(136)	429(126)	460 (136)	396 (103)	378 (97)	379 (87)

[a]Each subject contributed up to 63 days of data in the diary panel.
[b]Participants in overlapping area of increased exposure zone and stack zone: diary panel cohort (2006 = 6(4); 2007 = 5(2); 2008 = 5(3)) contributed data to both zones.
[c]Participants with spirometry and eNO.
[d]bd-bronchodilator.
[e]Geometric mean and SD.
[f]Participants with PEF/symptom diary.
[g]PEF recorded three times twice daily at morning and evening. Highest of three measurements used for each session.

measured spirometric lung function (forced expiratory volume in one second (FEV_1) and forced vital capacity (FVC)) and exhaled nitric oxide (eNO) once each year (2006–2008) at a home visit. We calculated the ratio of FEV_1 as a proportion of FVC (FEV_1/FVC) as a measure of airflow obstruction. We also asked participants to record symptoms, medication use and peak expiratory flows (PEF) twice daily for nine weeks each year at the same time of year (see Supporting information, Appendix S2).

Air Quality Monitoring

Two elevated air quality monitoring stations located near the ventilation stacks measured oxides of nitrogen (NOx), nitric oxide (NO), NO_2, particulate matter less than 2.5 μm and less than 10 μm in diameter respectively ($PM_{2.5}$ and PM_{10}) and carbon monoxide (CO). Monitoring methods were identical to those previously described for fixed site monitoring in the study area [9].

Statistical Analysis

We compared the study respondents' characteristics to the local population using Australian Census 2006 data [10].

Study hypotheses were tested using a regression model. Main study factors were study year, exposure zone and their interaction. We tested the hypotheses by fitting a contrast statement to this interaction term. This approach modelled changes in symptoms and lung function between pre-tunnel year (2006) and each post-tunnel year (2007 and 2008) by zone, adjusted for the change in the control zone. Coefficients for contrasts were reported as mean differences for the continuous variables, or odds ratios for binary or multinomial variables.

To adjust for autocorrelation we fitted random intercepts for subjects and households. An additional random intercept for study year was fitted for PEF data, due to repeated measures each year. Some models with both subject and household random effects did not converge, and we re-fitted the model with random intercepts for subject only.

We fitted both baseline models and models adjusted for covariates. All models, including baseline models, included adjustment for changes in the control zone. For adult questionnaire outcomes we fitted a model adjusting additionally for mouth symptoms to assess potential reporting bias. Covariates incorporated in adjusted models were specified *a priori* and included: asthma diagnosis at baseline; smoking status; ETS at home; use of gas cooking (stove or oven); use of unflued gas heaters; education; and employment status. Covariates were retained in these models if they contributed significantly ($P<0.05$) to the model.

Baseline models for FEV_1, FVC and PEF were adjusted for height, age, age^2, gender and their interaction, and models for FEV_1/FVC were adjusted for age and gender and their interaction, based on models used in predictive equations [11]. Baseline models for eNO were adjusted for age and gender. The same covariates as incorporated for the questionnaire analysis were used and all covariates were retained in these models.

Continuous variables were tested using linear regression. The distribution of eNO was skewed so its values were log-transformed. Other continuous variables were analysed without transformation. Questionnaire symptoms were analysed as binary outcomes using logistic regression. For the individual daily panel symptom data, we attempted to fit a general linear mixed model incorporating random effects for subject and day with a logit link. As these models did not converge, we used summarised data for each subject, converting daily

Table 3. Symptom prevalence of the diary panel sub-cohort, by zone and year (n = 380)[a].

	Reduced exposure zone			Increased exposure zone[b]			Eastern stack zone[b]			Control zone		
	2006	**2007**	**2008**	**2006**	**2007**	**2008**	**2006**	**2007**	**2008**	**2006**	**2007**	**2008**
N [c] (% of 2006)	86	76 (88)	70 (81)	82	57 (70)	52 (63)	85	69 (81)	62 (73)	81	57 (70)	46 (57)
%												
Symptom days [d]												
At least once	99	96	94	96	93	98	95	97	94	94	91	98
10% of days	93	84	79	89	70	77	82	78	89	84	72	76
>20% of days	69	67	60	73	53	62	62	64	66	63	60	61
Day-time symptoms [d]												
At least once	79	72	70	84	70	69	67	65	63	63	58	65
>10% of days	52	43	53	63	42	48	49	46	48	41	40	43
>20% of days	36	33	37	39	26	37	33	32	31	23	25	30
Night-time symptoms [d]												
At least once	64	61	59	73	44	50	66	59	65	54	42	52
>10% of days	36	30	27	38	25	29	34	32	40	30	26	28
>20% of days	19	13	13	21	12	15	18	22	21	15	19	17
Wheeze[d]												
At least once	28	24	20	35	21	23	22	28	35	35	28	33
>10% of days	9	7	9	17	4	10	11	9	5	15	14	7
>20% of days	3	7	3	10	4	4	6	4	0	11	9	4
Bronchodilator use [d]												
At least once	12	16	10	21	11	12	6	7	8	12	12	13
>10% of days	3	7	6	12	7	10	2	3	2	6	11	4
>20% of days	2	1	1	5	5	4	0	0	0	5	9	4
Cough [d]												
At least once	71	68	60	77	53	56	66	64	58	57	46	63
>10% of days	43	33	39	48	32	35	40	38	45	36	32	37
>20% of days	26	18	23	32	12	21	21	23	26	20	25	30
Upper resp. symptoms [d]												
At least once	50	50	53	60	44	44	48	49	53	40	44	50
>10% of days	19	17	29	24	11	21	18	23	18	9	14	26
>20% of days	5	8	13	12	7	10	6	9	3	5	11	15
Lower resp symptoms[e f]	37	44	33	26	57	51	38	32	44	31	38	39
Severe lower resp symp[e f]	6	7	8	12	16	13	5	7	6	9	7	12
Upper resp. symptoms [e f]	12	18	18	5	27	33	11	22	15	8	26	20
Mouth symptoms [f]	12	14	22	9	18	15	7	22	21	8	14	9

[a]Each subject contributed up to 63 days of data in the diary study.
[b]Participants in overlapping area of increased exposure zone and stack zone, diary panel cohort (2006 = 6(4); 2007 = 5(2); 2008 = 5(3)), contributed data to both zones. n = subjects in overlapping area with spirometry; (n) = subjects in overlapping area taking part in peak flow and symptom diary.
[c]Panel sub-cohort participants who took part in peak flow/symptom diary.
[d]Symptoms recorded by diary. % represents proportion of subjects experiencing those symptoms a) at least one day during the study period; b) >10% of days during the study period; c) >20% of days during the study period.
[e]Composite variables derived from variables measured in questionnaire (See Table S1, Supporting Information).
[f]Attributes measured by questionnaire.

symptom data to the proportion of days each year the subject reported symptoms, classified on an ordinal scale as: no days; 1–10%; 11–20%; and 21–100% of days. These data were analysed using ordinal regression with a cumulative logit link and a multinomial error distribution. The same covariates as specified previously were incorporated and retained in the diary symptom models.

Linear regression and multinomial regression models were fitted in Proc MIXED and Proc GLIMMIX, respectively using SAS 9.2. The logistic regression with two random effects used for the questionnaire analysis was fitted in MLwiN using the Markov Chain Monte Carlo (MCMC) estimation method, as these models did not converge when using numerical methods such as adaptive quadrature.

Table 4. Odds ratios for symptoms reported by questionnaire in 2007 and in 2008 relative to 2006 adjusted for change in the control zone.

Outcomes measured by questionnaire		2007 vs 2006		2008 vs 2006	
		Odds Ratio (95% CI)	P	Odds Ratio (95% CI)	P
Upper respiratory symptoms					
Reduced exp zone	Baseline model	0.7 (0.5, 1.2)	0.2	0.8 (0.5, 1.3)	0.3
	Adjusted	0.7 (0.4, 1.2)[a]	0.2	0.8 (0.5, 1.3)[a]	0.3
Increased exp zone	Baseline model	2.0 (1.2, 3.6)	0.02	3.2 (1.7, 6.1)	<0.001
	Adjusted	1.9 (1.0, 3.3)[a]	0.04	3.0 (1.6, 5.6)[a]	<0.001
Stack zone	Baseline model	1.8 (1.1, 3.2)	0.03	1.8 (1.0, 3.2)	0.05
	Adjusted	1.8 (1.1, 3.1)[a]	0.03	1.7 (1.0, 3.0)[a]	0.07
Lower respiratory symptoms					
Reduced exp zone	Baseline model	0.9 (0.6, 1.3)	0.7	0.9 (0.6, 1.4)	0.7
	Adjusted	0.9 (0.6, 1.3)[b]	0.6	0.9 (0.6, 1.4)[c]	0.6
Increased exp zone	Baseline model	1.5 (1.0, 2.2)	0.07	1.4 (0.8, 2.2)	0.2
	Adjusted	1.5 (1.0, 2.2)[b]	0.08	1.4 (0.8, 2.2)[c]	0.2
Stack zone	Baseline model	1.2 (0.8, 1.7)	0.5	1.6 (1.0, 2.4)	0.05
	Adjusted	1.2 (0.8, 1.7)[b]	0.4	1.6 (1.0, 2.5)[c]	0.04
Severe lower respiratory symptoms					
Reduced exp zone	Baseline model	1.1 (0.6, 2.0)	0.9	1.0 (0.5, 1.9)	1.0
	Adjusted	1.1 (0.6, 2.0)[d]	0.9	1.0 (0.5, 2.0)[d]	0.9
Increased exp zone	Baseline model	1.2 (0.6, 2.4)	0.7	1.0 (0.5, 2.1)	1.0
	Adjusted	1.1 (0.5, 2.2)[d]	0.8	1.0 (0.5, 2.0)[d]	0.9
Stack zone	Baseline model	3.0 (1.5, 6.0)	0.002	1.4 (0.7, 2.8)	0.4
	Adjusted	3.2 (1.6, 6.5)[d]	0.001	1.6 (0.8, 3.3)[d]	0.2
Wheeze in last 3 months					
Reduced exp zone	Baseline model	0.9 (0.5, 1.5)	0.7	0.9 (0.5, 1.5)	0.7
	Adjusted	0.9 (0.5, 1.6)[e]	0.8	0.7 (0.3, 1.2)[f]	0.2
Increased exp zone	Baseline model	1.1 (0.6, 2.0)	0.8	1.1 (0.6, 2.0)	0.8
	Adjusted	1.2 (0.6, 2.2)[e]	0.6	1.1 (0.5, 2.4)[f]	0.7
Stack zone	Baseline model	1.8 (1.0, 3.3)	0.06	1.8 (1.0, 3.3)	0.06
	Adjusted	2.0 (1.1, 3.6)[e]	0.03	1.5 (0.7, 3.0)[f]	0.3
Cough in last 3 months					
Reduced exp zone	Baseline model	0.8 (0.6, 1.1)	0.2	0.9 (0.6, 1.3)	0.4
	Adjusted	0.8 (0.6, 1.1)[g]	0.2	0.9 (0.6, 1.3)[h]	0.5
Increased exp zone	Baseline model	1.4 (0.9, 2.0)	0.1	1.1 (0.7, 1.8)	0.6
	Adjusted	1.5 (1.0, 2.2)[g]	0.06	1.2 (0.8, 1.9)[h]	0.4
Stack zone	Baseline model	1.1 (0.8, 1.7)	0.5	1.2 (0.8, 1.9)	0.3
	Adjusted	1.1 (0.8, 1.7)[g]	0.5	1.3 (0.8, 1.9)[h]	0.3

[a]Adjusted for: asthma, gas cooker or oven.
[b]Adjusted for: age, age^2, gender, asthma, smoker, unflued gas heater.
[c]Adjusted for: age, age^2, asthma, smoker, unflued gas heater.
[d]Adjusted for: age, age^2, asthma, smoker, ETS at home.
[e]Adjusted for: asthma, unflued gas heater.
[f]Adjusted for: age, age^2, asthma, unflued gas heater.
[g]Adjusted for: age, age^2, asthma, smoker, gas cooker or oven.
[h]Adjusted for: age, age^2, gender, asthma, smoker, gas cooker or oven.

We used an autoregressive model to test whether there was a difference in air quality at the elevated monitoring sites between 2006 and later years, adjusting for changes in regional air quality measured at three Sydney sites by the NSW Office of Environment and Heritage (OEH). Analysis was conducted in SAS 9.2 using the AUTOREG procedure.

Results

We identified 5,614 households in all study zones in 2006 (Figure 2). Overall household participation rate at baseline was 33%, varying from 25% in the stack zone to 38% in the control zone (see Supporting information, Table S2). 20% of households

Table 5. Differences in spirometry and peak flow and odds ratios for eNO, in 2007 and 2008 relative to 2006, adjusted for change in the control zone (n = 380).

Lung function outcomes		2007 vs 2006		2008 vs 2006	
		Difference (95% CI)	P	Difference (95% CI)	P
FEV1-pre bronchodilator (L)[a b]					
Reduced exp zone	Baseline model	−0.002 (−0.06, 0.06)	1.0	−0.01 (−0.08, 0.07)	0.9
	Adjusted	−0.002 (−0.06, 0.06)	1.0	−0.001 (−0.08, 0.08)	1.0
Increased exp zone	Baseline model	−0.02 (−0.08, 0.04)	0.5	−0.07 (−0.15, 0.00)	0.06
	Adjusted	−0.02 (−0.08, 0.04)	0.5	−0.07 (−0.15, 0.01)	0.09
Stack zone	Baseline model	0.01 (−0.05, 0.07)	0.8	−0.002 (−0.08, 0.07)	1.0
	Adjusted	0.01 (−0.04, 0.07)	0.6	0.004 (−0.07, 0.08)	0.9
FEV1-post bronchodilator (L)[a b]					
Reduced exp zone	Baseline model	−0.03 (−0.12, 0.05)	0.5	−0.05 (−0.15, 0.05)	0.4
	Adjusted	−0.03 (−0.12, 0.06)	0.5	−0.05 (−0.15, 0.06)	0.4
Increased exp zone	Baseline model	0.02 (−0.07, 0.10)	0.7	−0.06 (−0.17, 0.04)	0.2
	Adjusted	0.01 (−0.07, 0.10)	0.8	−0.06 (−0.17, 0.04)	0.2
Stack zone	Baseline model	−0.08 (0.17, 0.00)	0.06	−0.06 (−0.16, 0.05)	0.3
	Adjusted	−0.08 (−0.16, 0.01)	0.07	−0.05 (−0.16, 0.05)	0.3
FEV1/FVC-pre bronchodilator[b c]					
Reduced exp zone	Baseline model	0.01 (−0.01, 0.03)	0.2	0.006 (−0.01, 0.03)	0.6
	Adjusted	0.01 (−0.01, 0.03)	0.2	0.005 (−0.01, 0.02)	0.6
Increased exp zone	Baseline model	0.002 (−0.02, 0.02)	0.8	−0.002 (−0.02, 0.02)	0.8
	Adjusted	0.002 (−0.02, 0.02)	0.8	−0.003 (−0.02, 0.02)	0.8
Stack zone	Baseline model	0.004 (−0.01, 0.02)	0.7	0.004 (−0.01, 0.02)	0.7
	Adjusted	0.004 (−0.01, 0.02)	0.7	0.005 (−0.01, 0.02)	0.6
FEV1/FVC-post bronchodilator[b c]					
Reduced exp zone	Baseline model	0.009 (−0.00, 0.02)	0.2	0.003 (−0.01, 0.02)	0.6
	Adjusted	0.009 (−0.00, 0.02)	0.2	0.003 (−0.01, 0.02)	0.6
Increased exp zone	Baseline model	0.00 (−0.01, 0.01)	1.0	−0.009 (−0.02, 0.00)	0.2
	Adjusted	−0.001 (−0.01, 0.01)	0.9	−0.009 (−0.02, 0.00)	0.1
Stack zone	Baseline model	0.005 (−0.01, 0.02)	0.5	0.009 (−0.00, 0.02)	0.2
	Adjusted	0.005 (−0.01, 0.02)	0.5	0.009 (−0.00, 0.02)	0.1
FVC-pre bronchodilator (L)[a b]					
Reduced exp zone	Baseline model	−0.04 (0.12, 0.05)	0.4	−0.02 (−0.12, 0.08)	0.7
	Adjusted	−0.04 (−0.12, 0.05)	0.4	−0.02 (−0.12, 0.08)	0.8
Increased exp zone	Baseline model	−0.03 (−0.11, 0.05)	0.5	−0.09 (−0.19, 0.01)	0.09
	Adjusted	−0.03 (−0.11, 0.05)	0.5	−0.08 (−0.18, 0.02)	0.1
Stack zone	Baseline model	0.001 (−0.08, 0.08)	1.0	−0.02 (−0.11, 0.08)	0.8
	Adjusted	0.004 (−0.08, 0.08)	0.9	−0.01 (−0.11, 0.09)	0.8
FVC-post bronchodilator (L)[a b]					
Reduced exp zone	Baseline model	−0.05 (−0.15, 0.05)	0.4	−0.05 (−0.16, 0.07)	0.4
	Adjusted	−0.04 (−0.14, 0.06)	0.4	−0.05 (−0.16, 0.07)	0.4
Increased exp zone	Baseline model	0.01 (−0.09, 0.11)	0.8	−0.05 (−0.16, 0.06)	0.4
	Adjusted	0.01 (−0.09, 0.11)	0.8	−0.05 (−0.16, 0.07)	0.4
Stack zone	Baseline model	−0.11 (−0.21, −0.01)	0.03	−0.09 (−0.20, 0.02)	0.1
	Adjusted	−0.10 (−0.20, −0.01)	0.04	−0.09 (−0.20, 0.02)	0.1
Peak flow (L/min) morning[a b d]					
Reduced exp zone	Baseline model	−4.76 (−13.63, 4.11)	0.3	−5.74 (−16.90, 5.43)	0.3
	Adjusted	−5.20 (−14.40, 4.01)	0.3	−6.62 (−18.03, 4.79)	0.3
Increased exp zone	Baseline model	−2.56 (−11.85, 6.71)	0.6	−5.93 (−17.52, 5.65)	0.3
	Adjusted	−2.52 (−12.00, 6.95)	0.6	−4.56 (−16.46, 7.34)	0.5

Table 5. Cont.

Lung function outcomes		2007 vs 2006		2008 vs 2006	
		Difference (95% CI)	**P**	**Difference (95% CI)**	**P**
Stack zone	Baseline model	−9.69 (−18.66, −0.72)	0.03	4.31 (−6.83, 15.45)	0.5
	Adjusted	−9.77 (−18.94, −0.60)	0.04	4.64 (−6.70, 15.97)	0.4
Peak flow (L/min) evening [a b d]					
Reduced exp zone	Baseline model	−5.32 (−14.32, 3.67)	0.3	−6.93 (−18.39, 4.53)	0.2
	Adjusted	−5.29 (−14.61, 4.02)	0.3	−7.65 (−19.35, 4.06)	0.2
Increased exp zone	Baseline model	−0.83 (−10.24, 8.58)	0.9	−3.71 (−15.59, 8.18)	0.5
	Adjusted	−0.74 (−10.34, 8.85)	0.9	−1.82 (−14.02, 10.39)	0.8
Stack zone	Baseline model	−8.16 (−17.25, 0.94)	0.08	2.82 (−8.62, 14.25)	0.6
	Adjusted	−7.90 (−17.19, 1.38)	0.1	3.24 (−8.38, 14.86)	0.6
Exhaled nitric oxide (eNO) [b e]		**Ratio (95% CI)**		**Ratio (95% CI)**	
Reduced exp zone	Baseline model	0.94 (0.75, 1.19)	0.6	0.89 (0.71, 1.12)	0.3
	Adjusted	0.93 (0.74, 1.17)	0.5	0.92 (0.73, 1.14)	0.4
Increased exp zone	Baseline model	1.05 (0.83, 1.32)	0.7	1.08 (0.86, 1.35)	0.5
	Adjusted	1.00 (0.79, 1.26)	1.0	1.09 (0.87, 1.36)	0.4
Stack zone	Baseline model	0.90 (0.72, 1.13)	0.4	0.90 (0.73, 1.13)	0.4
	Adjusted	0.91 (0.73, 1.14)	0.4	0.93 (0.75, 1.15)	0.5

[a]Baseline models adjusted for: age; age^2; height; gender; gender*age; gender*age^2; gender*height.
[b]All adjusted models additionally adjusted for: diagnosed asthma at baseline; smoking; gas cooker or oven; unflued gas heating; ETS; education and employment status.
[c]Baseline models adjusted for: age, gender, gender*age.
[d]PEF recorded three times twice daily at morning and evening.
[e]Adjusted for age and gender.

refused to participate, 6% were ineligible and no contact could be made with 42% of households. Survey respondents had a similar distribution of demographic characteristics to the general population in the study area [10].

Of 2,978 cohort members who completed the 2006 questionnaire, 2,191 (74%) provided data in 2007 and 1,713 (58%) in 2008 (Figure 2). Loss to follow up was lower in the stack zone than the control zone (odds ratio (OR) 0.73, 95% CI 0.58–0.92 for 2007 and OR 0.82, 95% CI 0.67–1.01 for 2008) and higher in the reduced exposure zone than the control zone in 2008 (OR 1.28, 95% CI 1.04–1.57).

Baseline Data and Description of Study Group Over Time

Proportions of adults and females in the entire cohort did not vary substantially between zones or over the study period (Table 1). The prevalence of reporting wheeze, cough, diagnosed and current asthma, and use of asthma medication was similar across all zones at baseline (Table 1). The reduced exposure zone and the stack zone had more university-educated people and people in paid work, compared with the increased exposure zone and control zone. The proportion of participants using unflued gas heaters and gas cooktops/ovens increased in all zones over the study period. The prevalence of diagnosed asthma, cough, LRS, URS, and mouth symptoms increased in 2007 and 2008 in all zones.

Supporting information, Table S3, presents baseline data, reported in 2006, for the cohort that participated in each year of the study. Remaining members of the cohort in 2007 and in 2008 were similar in most baseline characteristics to the initial cohort, suggesting loss-to-follow-up is unlikely to have biased comparisons between years.

Of subjects asked to participate in the panel sub-cohort, 40% refused, 4% intended moving residence, and almost 6% were ineligible due to language or age constraints (Figure 2). The sub-cohort included more children than the entire cohort (Table 2), and the proportion of subjects with atopy was relatively high across all zones (50–58% in 2007). Spirometric measurements were similar across all zones in all years, although values for FEV_1 and FVC were slightly higher in subjects residing within the stack zone (Table 2). Participants in the stack zone and the reduced exposure zone had higher PEF readings in all years than participants from the other two zones. eNO readings were higher in 2007 than in 2006 and 2008, across all zones. Over the study period, participants in the stack zone reported increased prevalence of wheeze (at least once) and slightly more symptomatic days (>20% of days), whereas in other zones participants reported a decrease in these symptoms (Table 3). There was some year-to-year variation in the prevalence and frequency of diary symptoms, within each zone.

Supporting information, Table S4, presents baseline data, reported in 2006, for the panel sub-cohort that participated in each year. Remaining participants in 2007 and 2008 were similar in most baseline characteristics to the initial cohort except that, in all zones other than the increased exposure zone, those remaining had reported substantially more cough at baseline. Also, in the increased exposure zone and the stack zone, the remaining 2008 sub-cohort had reported slightly less "current asthma" and "asthma medication use" at baseline than the initial sub-cohort and those remaining in 2007. This suggests limited potential for selection bias due to loss-to-follow up in this sub-cohort.

Table 6. Odds ratios for symptoms reported in the diary panel sub-cohort in 2007 and in 2008 relative to 2006, adjusted for change in the control zone, (n = 380).

Symptoms measured by diary		2007 vs 2006		2008 vs 2006	
		Odds Ratio[a] (95% CI)	P	Odds Ratio[a] (95% CI)	P
Day symptoms					
Reduced exp zone	Baseline model	0.7 (0.2, 1.7)	0.4	0.6 (0.2, 1.6)	0.3
	Adjusted[b]	0.6 (0.2, 1.6)	0.3	0.5 (0.2, 1.4)	0.2
Increased exp zone	Baseline model	0.4 (0.1, 1.1)	0.06	0.4 (0.1, 1.3)	0.1
	Adjusted	0.3 (0.1, 0.9)	0.04	0.4 (0.1, 1.1)	0.09
Stack zone	Baseline model	1.1 (0.4, 2.9)	0.9	0.5 (0.2, 1.4)	0.2
	Adjusted	1.1 (0.4, 2.8)	0.9	0.4 (0.2, 1.3)	0.1
Night symptoms					
Reduced exp zone	Baseline model	1.0 (0.4, 2.7)	1.0	0.6 (0.2, 1.6)	0.3
	Adjusted	1.0 (0.4, 2.8)	1.0	0.5 (0.2, 1.4)	0.2
Increased exp zone	Baseline model	0.4 (0.2, 1.2)	0.1	0.5 (0.2, 1.4)	0.2
	Adjusted	0.5 (0.2, 1.3)	0.1	0.4 (0.1, 1.2)	0.1
Stack zone	Baseline model	1.4 (0.5, 3.8)	0.5	1.1 (0.4, 3.0)	0.9
	Adjusted	1.5 (0.5, 4.0)	0.5	1.0 (0.4, 2.8)	1.0
Symptom days					
Reduced exp zone	Baseline model	1.2 (0.4, 3.6)	0.7	0.6 (0.2, 2.0)	0.4
	Adjusted	1.1 (0.4, 3.4)	0.8	0.6 (0.2, 1.9)	0.4
Increased exp zone	Baseline model	0.5 (0.1, 1.4)	0.2	0.8 (0.3, 2.6)	0.7
	Adjusted	0.4 (0.1, 1.2)	0.1	0.6 (0.2, 2.0)	0.4
Stack zone	Baseline model	2.4 (0.8, 7.1)	0.1	1.8 (0.6, 5.6)	0.3
	Adjusted	2.5 (0.8, 7.5)	0.1	1.6 (0.5, 4.9)	0.4
Upper respiratory symptoms					
Reduced exp zone	Baseline model	0.6 (0.2, 1.6)	0.3	0.5 (0.2, 1.4)	0.2
	Adjusted	0.5 (0.2, 1.5)	0.2	0.4 (0.1, 1.2)	0.1
Increased exp zone	Baseline model	0.2 (0.07, 0.6)	0.005	0.1 (0.04, 0.4)	0.001
	Adjusted	0.2 (0.07, 0.6)	0.005	0.1 (0.04, 0.4)	<0.001
Stack zone	Baseline model	0.8 (0.3, 2.3)	0.7	0.3 (0.1, 0.9)	0.03
	Adjusted	0.8 (0.3, 2.2)	0.6	0.3 (0.1, 0.8)	0.02
Wheeze					
Reduced exp zone	Baseline model	1.2 (0.3, 4.0)	0.8	0.9 (0.3, 3.5)	0.9
	Adjusted	1.4 (0.4, 4.7)	0.6	1.1 (0.3, 4.0)	0.9
Increased exp zone	Baseline model	0.5 (0.1, 2.0)	0.3	1.2 (0.3, 4.6)	0.8
	Adjusted	0.5 (0.2, 1.9)	0.3	1.1 (0.3, 4.2)	0.9
Stack zone	Baseline model	2.2 (0.6, 7.5)	0.2	2.9 (0.8, 10.5)	0.1
	Adjusted	2.6 (0.8, 8.6)	0.1	3.3 (0.9, 11.6)	0.06
Cough					
Reduced exp zone	Baseline model	0.9 (0.3, 2.3)	0.8	0.4 (0.1, 1.1)	0.07
	Adjusted	0.8 (0.3, 2.2)	0.7	0.3 (0.1, 0.9)	0.03
Increased exp zone	Baseline model	0.4 (0.1, 1.1)	0.08	0.3 (0.1, 0.9)	0.03
	Adjusted	0.4 (0.1, 1.1)	0.06	0.2 (0.1, 0.7)	0.01
Stack zone	Baseline model	1.5 (0.5, 4.1)	0.4	0.6 (0.2, 1.8)	0.4
	Adjusted	1.5 (0.5, 4.1)	0.4	0.5 (0.2, 1.5)	0.2

[a]OR-odds of higher values on the ordinal scale (symptoms experienced: 0 days; 1–10% of days;10–20% of days; and 21–100% of days) compared with lower values, assuming proportional odds.

[b]Adjusted for: age; gender; diagnosis of asthma at baseline; smoker; ETS; gas stove or oven; unflued gas heating; education and employment status.

Table 7. Mean pollutant concentrations by year and difference in pollutant concentration from post-tunnel year compared to pre-tunnel year.

Site & pollutant	Mean pollutant levels (SD)		Difference in pollutant levels[a]	
	Pre-tunnel year[a]	First post-tunnel year[b]	Unadjusted estimate (95% CI)	Adjusted estimate (95% CI)[c]
Elevated site near eastern ventilation stack				
NO_2 ppb	16.7 (7.0)	15.7 (5.3)	−2.55 (−4.43, −0.66)	−1.00 (−2.21, 0.22)
NOx ppb	30.7 (17.8)	27.1 (14.1)	−6.33 (−10.97, −1.70)	−2.10 (−3.84, −0.37)
PM_{10} $\mu g/m^3$	20.2 (8.1)	16.1 (6.8)	−4.02 (−6.24, −1.80)	−1.74 (−3.04, −0.43)
$PM_{2.5}$ $\mu g/m^3$	6.4 (3.6)	5.3 (3.2)	−1.13 (−2.38, 0.12)	−0.26 (−0.65, 0.12)
Elevated site near western ventilation stack				
NO_2 ppb	13.4 (5.6)	12.1 (4.5)	−1.96 (−3.08, −0.84)	−1.07 (−1.68, −0.45)
NOx ppb	26.1 (19.9)	20.6 (14.6)	−6.05 (−10.47, −1.62)	−2.79 (−4.46, −1.11)
PM_{10} $\mu g/m^3$	18.7 (7.4)	16.2 (7.0)	−2.39 (−4.22, −0.56)	−0.46 (−1.94, 1.03)
$PM_{2.5}$ $\mu g/m^3$	6.2 (3.5)	5.3 (3.3)	−1.02 (−1.98, −0.06)	−0.19 (−0.65, 0.27)

[a]Difference between pre-tunnel and first post-tunnel years, estimated by autoregressive analysis.
[b]Pre-tunnel is 25 March 06–24 March 07 and first post-tunnel year is 25 March 2007–24 March 2008.
Data for only one post-tunnel year available.
[c]Adjusted for change in air quality at the three Sydney regional sites (averaged).

Questionnaire Symptom Analysis

Relative to 2006, the odds for URS in 2007 and in 2008 were significantly increased in the increased exposure zone and the stack zone, after adjusting for change in the control zone (Table 4). The odds for LRS and SLRS were significantly increased in 2008 and 2007 respectively, in the stack zone. The odds ratios for symptoms in the reduced exposure zone were generally at or below 1.0 in 2007 and 2008, although were not statistically significant. Adjusting for significant covariates did not change the estimates substantially (Table 4). Similarly, adjusting for mouth symptoms in the analysis of adult questionnaire responses ((model 2), made little difference to the effect estimates (model 1) (see Supporting information, Table S5).

Diary Analysis

In the panel sub-cohort, mean post-bronchodilator FVC and morning PEF were significantly reduced among stack zone participants in 2007, compared with 2006, after adjusting for change in the control zone (Table 5). This decrease was present, although not significant, for evening PEF. There was much less evidence of a similar change between 2006 and 2008. Although there were decreases in FEV_1 and FVC (pre-bronchodilator) in the increased exposure zone in 2008, they were not significant. There were no significant changes in lung function, airflow obstruction (FEV_1/FVC) or eNO in the other exposure zones.

In contrast to the questionnaire findings for the entire cohort, odds for symptoms recorded by panel sub-cohort participants were generally lower in all three exposure zones for both post-tunnel years, after adjustment for the control zone (Table 6). No significant increases in symptom prevalence were observed in any zone. Odds ratios for symptom days and wheeze in the stack zone were substantially greater than 1 but the confidence intervals (CIs) around these estimates were very broad. Adjustment for covariates did not materially alter the findings.

After adjusting for changes in regional air quality, monitors at the elevated sites near the ventilation stacks recorded significant decreases in PM_{10}, NOx and NO_2 in 2007 compared to 2006 (Table 7). Similarly, we previously showed using data from four additional ground level monitors that after adjusting for changes in regional air quality and meteorological conditions, PM_{10}, NOx and NO_2 concentrations decreased in the eastern part of the study area after the tunnel opened [9].

Discussion

We did not observe consistent evidence of an improvement in respiratory health among residents living near a major road which experienced a traffic volume reduction of 50% after opening of a road tunnel. Among residents living adjacent to tunnel feeder roads there was increased reporting of upper respiratory tract symptoms in the questionnaire survey but not in the sub-cohort who recorded symptoms in a diary. The most consistently observed change was in residents living within a 650 m radius of the eastern ventilation stack. After the tunnel opened these subjects reported more upper and lower respiratory tract symptoms and had evidence of lower spirometric function, particularly lower FVC. Significant adverse effects were seen only for the first year of tunnel use (2007) and not for the second year (2008), except for questionnaire reported lower respiratory symptoms.

The strengths of this study are the cohort intervention design, in which measurements were made in the same individuals before and after commencement of operation of the tunnel, the inclusion of both subjective and objective measures of respiratory health, adjustment for coincident trends in a control site and for time-varying confounders, and testing for reporting bias.

Apart from the stack zone findings, findings from the questionnaire survey of the entire cohort and the panel sub-cohort were not consistent. This is perplexing but points to limitations of questions and diary cards that rely on self-report. Using a summary annual measure of symptoms for each subject, rather than daily symptom scores, in the sub-cohort analysis resulted in reduced statistical power, possibly contributing to inconsistency in results, but does not explain odds ratios being in the opposite direction.

Recall or other reporting bias is a potential weakness of self-reporting, which we sought to address by asking adults about mouth symptoms. The reported prevalence of mouth symptoms increased in all zones and was greatest in the increased exposure zone and the stack zone, a pattern similar for wheeze and LRS, which might be suggestive of information bias. However, when we included mouth symptoms as a covariate in the analysis, the observed effects for other symptoms were not attenuated. Hence, we conclude that the observed associations are unlikely to be explained by reporting bias.

The previous health investigation on a different tunnel stack [2] did not find an association between modelled annual average NOx concentrations representing ground level exposures from stack emissions and self-reported eye, nose and throat symptoms assessed by telephone interview. That study did not assess other respiratory symptoms or include lung function measures, thus making it difficult to compare with findings from our study. A few traffic intervention studies have reported improvements in respiratory health outcomes including wheeze and peak flow [12], hospitalisations for asthma [13], emergency department presentations [14], eNO [15], and nose and eye symptoms [16], after redirection or closure of roads or implementation of traffic restrictions during major events. These health benefits were demonstrated despite somewhat modest reductions in air pollutant concentrations, aside from major reductions seen during the Beijing Olympic Games. However, other studies have not found an effect [17]. For this local traffic intervention we did not see substantial benefits in health outcomes along the bypassed main road.

The most likely explanation for the lack of change in respiratory outcomes where this was expected is that the traffic intervention did not result in substantial changes in TRAP for most subjects. Although NO_2 concentrations decreased adjacent to the bypassed main road, these changes were smaller than predicted, probably due to lower than predicted traffic volume through the tunnel [9]. Furthermore, changes in NO_2 decreased markedly with increasing distance from the affected roads [18], which is consistent with previous evidence of a rapid decay in NO_2 concentrations with distance from major roads [19,20].

The findings of this study suggest an adverse health effect associated with living in the stack zone. This is based on lower and upper respiratory tract symptoms and wheeze recorded by questionnaire in the entire cohort, lung function analysis of panel participants, and (albeit not significantly) diary symptom analysis. These findings related primarily, but not exclusively to 2007, the first year of tunnel operation. While the confidence intervals for the diary symptom scores were wide, the outcomes are consistent with other effects reported for the stack zone. These findings are unexpected, given that neither the modelled stack emissions nor measured NO_2, PM_{10} or $PM_{2.5}$ at the elevated monitors, showed any evidence of increased concentration after the tunnel commenced operation.

Bias of some kind as an explanation for positive associations cannot be ruled out in observational studies, although, as previously discussed, we believe reporting bias is an unlikely explanation. While bias due to differential loss to follow-up is possible, there was no evidence that the remaining cohort had a higher prevalence of diagnosed asthma, current asthma or wheeze at baseline compared to the complete cohort or that smoking, education or work status changed appreciably (see Supporting information, Tables S3 and S4). This suggests a lack of any major selective factors underlying loss to follow-up. Another possibility is that the findings are due to unmeasured confounding factors.

We considered whether the adverse effects in the stack zone could be due to exposure to tunnel portal emissions due to its location near the tunnel's eastern portal, an area with potential to experience an increase in pollutant concentrations [4]. However, the LCT was designed not to allow for portal emissions. Furthermore, the residents in the 'increased exposure zone', also located near the tunnel portal, did not show the same pattern of adverse health effects. The observed effects in the stack zone might have been due to TRAP associated with a highway which traverses the zone, however, highway traffic decreased after 2006 [9].

Another possible explanation for the stack zone findings is an alternative pollutant source located in the industrial area near the zone. While the local government authority indicated that land zoning did not change in the industrial estate during the study timeframe [21], it does not keep records of processes undertaken on individual properties. Thus we cannot rule out this possibility but also have no direct evidence for it.

The adverse findings in the stack zone are also not attributable to any measured pollutants emitted from the eastern ventilation stack, as discussed above. In addition, previous analysis of air quality data showed a decrease in NO_2, PM_{10} or $PM_{2.5}$ in the eastern section of the study area, over the study period [9]. Furthermore, the choice of a 650 m buffer for the stack zone was based on sample size considerations, rather than on any modelled or measured pollutant profiles.

It is possible, though, that the adverse findings are due to pollutants or size fractions other than those measured. Measurements of PM_{10} probably do not adequately reflect the particulate content of emissions in tunnels [4] and some studies show that PM_{10} and $PM_{2.5}$ do not represent TRAP as well as other pollutants such as NOx, NO_2, black carbon or elemental carbon [22,23]. Smaller size PM fractions that is, PM_1 or $PM_{0.1}$ (ultrafine particles (UFP)) are of lower mass, are present in far greater numbers, and, due to their oxidative potential, larger surface areas and ability to attach to other compounds, are considered to pose greater toxicity risks than PM_{10} [24–26]. Furthermore, elevated levels of UFP have been observed inside tunnels [27].

The finding of adverse respiratory health effects among residents living in the eastern stack zone is unexpected and difficult to explain, and as pollutant concentrations decreased in the immediate area, we are unable to attribute the effects to the ventilation stack. However, the findings imply the need to investigate the emission profile of the stack and other pollutant sources in the area that might explain the findings. The lack of improvement in respiratory health among people living along the bypassed main road is disappointing but probably reflects the short distance from the bypassed road within which exposure reduction might result in measurable health improvement.

Acknowledgments

We are grateful to the participants in this study. We acknowledge advice received from Professor Bert Brunekreef. We acknowledge the assistance of research staff: Kitty Ng, Angelica Ordonez, Paula Garai, Adrian Ferrero, and Jennifer Manning. The study was overseen by a Scientific Advisory Committee which provided advice on study design, methodology, analysis and interpretation, and consisted of Professor Bruce Armstrong, Professor Judy Simpson, Professor Bin Jalaludin, Dr Peter Lewis, and Dr Gavin Fisher.

Author Contributions

Conceived and designed the experiments: CTC VS ACW GBM. Performed the experiments: CTC WE ACW BGT GBM. Analyzed the data: CTC NR WX EB GBM. Wrote the paper: CTC NR WE WX ACW EB BGT VS GBM.

References

1. Bartonova A, Clench-Aas J, Gram F, Gronskei KE, Guerreiro C, et al. (1999) Air pollution exposure monitoring and estimation. Part V. Traffic exposure in adults. Journal of Environmental Monitoring 1: 337–340.
2. Capon A, Sheppeard V, Irvine K, Jalaludin B, Staff M, et al. (2008) Investigating health effects in a community surrounding a road tunnel stack - a cross sectional study. Environmental Health 7: 10.
3. Kuykendall JR, Shaw SL, Paustenbach D, Fehling K, Kacew S, et al. (2009) Chemicals present in automobile traffic tunnels and the possible community health hazards: A review of the literature. Inhalation Toxicology 21: 747–792.
4. National Health and Medical Research Council website. Air Quality In and Around Traffic Tunnels. Commonwealth of Australia 2008; ISBN online 1864964510 Available: http://wwwnhmrcgovau/guidelines/publications/eh42 Accessed 2011 Feb 25.
5. RTA (2001) Lane Cove Tunnel and Associated Road Improvements. Environmental Impact Statement. Working Paper Nine. Air Quality and Health Risk. Roads & Traffic Authority, NSW Government.
6. RTA (2010) Post Implementation Review. M7 Motorway, Cross City Tunnel and Lane Cove Tunnel. Roads & Traffic Authority (NSW), Motorways Project Branch.
7. ISAAC (1998) Worldwide variation in the prevalence of symptoms of asthma, allergic rhinoconjunctivitis, and atopic eczema: ISAAC. Lancet 351: 1225–1232.
8. Toelle B, Ng K, Belousova E, Salome C, Peat J, et al. (2004) The prevalence of asthma and allergy in schoolchildren in Belmont, Australia: three cross sectional surveys over 20 years. British Medical Journal 328: 386–387.
9. Cowie CT, Rose N, Gillett R, Walter S, Marks GB (2012) Redistribution of traffic related air pollution associated with a new road tunnel. Environmental Science & Technology 46: 2918–2927.
10. Australian Bureau of Statistics website. Census of Population and Housing, 2006. Available: http://www.abs.gov.au/census. Accessed 2008 Sep 17.
11. Hankinson J, Odencrantz J, Fedan K (1999) Spirometric Reference Values from a Sample of the General U.S. Population. American Journal of Respiratory & Critical Care Medicine 159: 179–187.
12. MacNeill SJ, Goddard F, Pitman R, Tharme S, Cullinan P (2009) Childhood peak flow and the Oxford Transport Strategy. Thorax 64: 651–656.
13. Lee JT, Son JY, Cho YS (2007) Benefits of mitigated ambient air quality due to transportation control on childhood asthma hospitalization during the 2002 summer Asian games in Busan, Korea. J Air Waste Manag Assoc 57: 968–973.
14. Li Y, Wang W, Kan H, Xu X, Chen B (2010) Air quality and outpatient visits for asthma in adults during the 2008 Summer Olympic Games in Beijing. Science of the Total Environment 408: 1226–1227.
15. Lin W, Huang W, Zhu T, Hu M, Brunekreef B, et al. (2011) Acute Respiratory Inflammation in Children and Black Carbon in Ambient Air before and during the 2008 Beijing Olympics. Environ Health Perspect 119: 1507–1512.
16. Burr ML, Karani G, Davies B, Holmes BA, Williams KL (2004) Effects on respiratory health of a reduction in air pollution from vehicle exhaust emissions. Occup Environ Med 61: 212–218.
17. Peel J, Klein M, Flanders W, Mulholland J, Tolbert P (2010) Impact of Improved Air Quality During the 1996 Summer Olympic Games in Atlanta on Multipple Cardiovascular and Respiratory Outcomes. HEI Research Report 148. Health Effects Institute, Boston, MA.
18. Rose N, Cowie C, Gillett R, Marks GB (2011) Validation of a Spatiotemporal Land Use Regression Model Incorporating Fixed Site Monitors. Environmental Science & Technology 45: 294–299.
19. Karner AA, Eisinger DS, Niemeier DA (2010) Near-Roadway Air Quality: Synthesizing the Findings from Real-World Data. Environmental Science & Technology 44: 5334–5344.
20. Zhou Y, Levy JI (2007) Factors influencing the spatial extent of mobile source air pollution impacts: a meta-analysis. BMC Public Health 7: 89.
21. WCC (2010) Willoughby, City Council. Personal communication, 2 December 2010.
22. Fischer PH, Hoek G, van Reeuwijk H, Briggs DJ, Lebret E, et al. (2000) Traffic-related differences in outdoor and indoor concentrations of particles and volatile organic compounds in Amsterdam. Atmospheric Environment 34: 3713–3722.
23. Roorda-Knape MC, Janssen NAH, De Hartog JJ, Van Vliet PHN, Harssema H, et al. (1998) Air pollution from traffic in city districts near major motorways. Atmospheric Environment 32: 1921–1930.
24. Lodovici M, Bigagli E (2011) Oxidative stress and air pollution exposure. Journal of Toxicology 2011, Article ID 487074, 9 pages.
25. Brook RD (2008) Cardiovascular effects of air pollution. Clinical Science 115: 175–187.
26. Mazzoli-Rocha F, Fernandes S, Einicker-Lamas M, Zin W (2010) Roles of oxidative stress in signaling and inflammation induced by particulate matter. Cell Biology and Toxicology 26: 481–498.
27. Knibbs LD, de Dear RJ, Morawska L, Mengersen KL (2009) On-road ultrafine particle concentration in the M5 East road tunnel, Sydney, Australia. Atmospheric Environment 43: 3510–3519.

3

Population Exposure to $PM_{2.5}$ in the Urban Area of Beijing

An Zhang, Qingwen Qi*, Lili Jiang, Fang Zhou, Jinfeng Wang

State Key Laboratory of Resources and Environmental Information System, Institute of Geographical Sciences and Natural Resources Research, Chinese Academy of Sciences, Beijing, China

Abstract

The air quality in Beijing, especially its $PM_{2.5}$ level, has become of increasing public concern because of its importance and sensitivity related to health risks. A set of monitored $PM_{2.5}$ data from 31 stations, released for the first time by the Beijing Environmental Protection Bureau, covering 37 days during autumn 2012, was processed using spatial interpolation and overlay analysis. Following analyses of these data, a distribution map of cumulative exceedance days of $PM_{2.5}$ and a temporal variation map of $PM_{2.5}$ for Beijing have been drawn. Computational and analytical results show periodic and directional trends of $PM_{2.5}$ spreading and congregating in space, which reveals the regulation of $PM_{2.5}$ overexposure on a discontinuous medium-term scale. With regard to the cumulative effect of $PM_{2.5}$ on the human body, the harm from lower intensity overexposure in the medium term, and higher overexposure in the short term, are both obvious. Therefore, data of population distribution were integrated into the aforementioned $PM_{2.5}$ spatial spectrum map. A spatial statistical analysis revealed the patterns of $PM_{2.5}$ gross exposure and exposure probability of residents in the Beijing urban area. The methods and conclusions of this research reveal relationships between long-term overexposure to $PM_{2.5}$ and people living in high-exposure areas of Beijing, during the autumn of 2012.

Editor: Qinghua Sun, The Ohio State University, United States of America

Funding: This study was supported by National Natural Science Foundation of China (41201412). The funders had no role in study design, data collection and analysis, decision to publish, or preparation of the manuscript.

Competing Interests: The authors have declared that no competing interests exist.

* E-mail: qiqw@igsnrr.ac.cn

Introduction

Beijing, the capital of China, has experienced a rapid increase in urban population, energy consumption, and vehicle numbers over the past several decades. As once occurred in notoriously foggy London, this development has been followed by more frequent episodes of haze in Beijing [1,2]. In the composition of the haze, particulate matter less than 10 microns in diameter (PM_{10}) is recognized as the main hazard linked to deleterious health effects, such as heart and lung disease [3]. Evidence has shown that when inhaled, fine particles ($PM_{2.5}$) are more toxic and more strongly associated with risks to human health than are coarse particles (between 2.5 and 10 microns in diameter) [4,5]. Therefore, overexposure to $PM_{2.5}$ poses a nontrivial risk to public health in Beijing.

A large number of case studies have associated the short-term exposure of limited individuals to $PM_{2.5}$ with aggravation of asthma, pulmonary dysfunction, lung cancer, heart disease, stroke, and other illnesses [6–8]. Some studies have suggested that long-term continuous $PM_{2.5}$ exposure is significantly related to deteriorating health conditions and increased morbidity [9–11]. Other studies have shown the accumulated health effects of $PM_{2.5}$ and have analyzed annual statistical data of long-term continuous $PM_{2.5}$ pollution, together with its impact on mortality, morbidity, and life expectancy. These studies are usually based on case studies and ignore temporal and spatial differences [12,13]. With regard to the health of a population, the importance of medium-scale and discontinuous exposure to $PM_{2.5}$ should be recognized, because this is connected to cumulative temporal effects and time variations associated with meteorological conditions. This approach is more likely to ascertain the spatiotemporal regularity in the relationship between public exposure and response, which is highly relevant to every person subject to urban particulate pollution.

To determine the spatiotemporal regularity and to monitor environmental health, high-resolution temporal and spatial measurement of atmospheric pollutants is of crucial importance [14]. Some investigators have monitored several sampling sites in Beijing in an attempt to analyze the spatial distribution and time-varying characteristics of $PM_{2.5}$ [15,16]. However, these analyses were based on discrete points and low time-resolution sampling, which affects the statistical average values. Furthermore, they did not address the population adequately; hence, exposures were not studied [15,16]. To investigate the risk from $PM_{2.5}$ for the total population of Beijing, a spatially continuous study with high temporal resolution sampling is necessary to detect hotspots, estimate vulnerability, and accurately assess population exposure.

In October 2012, the Beijing Environmental Protection Bureau (BJEPB) released hourly monitoring data from 35 sampling sites, covering a large area of the city. In this study, we collected data from 37 of those days. We also acquired and manipulated the latest census data of population distribution, based on appropriate population units. We performed a 37-day interpolation over the entire city, using the daily averages at these sites. We constructed time series charts of daily air quality index (AQI) levels for the

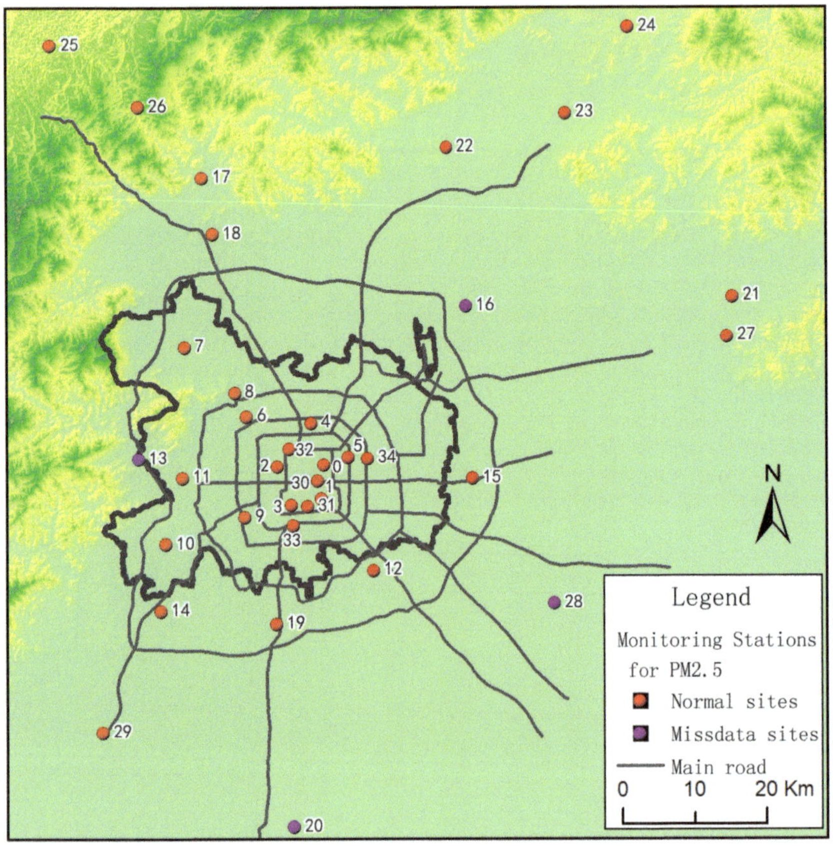

Code	Sitename
0	DongChengDongSi
1	DongChengTianTan
2	XiChengGuanYuan
3	XiChengWanShouXiGong
4	ChaoYangAoTiZhongXin
5	ChaoYangNongZhanGuan
6	HaiDianWanLiu
7	HaiDianBeiBuXinQu
8	HaiDianBeiJingZhiWuYuan
9	FengTaiHuaYuan
10	FengTaiYunGang
11	ShiJingShanGuCheng
12	YiZhuangKaiFaQu
13	MenTouGouLongQuanZhen
14	FangShanLiangXiang
15	TongZhouXinCheng
16	ShunYiXinCheng
18	ChangPingZhen
19	DaXingHuangCunZhen
21	PingGuZhen
22	HuaiRouZhen
23	MiYunZhen
25	YanQingZhen
17	ChangPingDingLing
20	JingNanYuZun
24	JingDongBeiMiYunShuiKu
26	JingXiBeiBaDaLing
27	JingDongDongGaoCun
28	JingDongNanYongLeDian
29	JingXiNanLiuLiHe
30	QianMenDongDaJie
31	YongDingMenNeiDaJie
32	XiZhiMenBeiDaJie
33	NanSanHuanXiLu
34	DongSiHuanBeiLu

Figure 1. Locations of monitoring stations for $PM_{2.5}$ in Beijing from BJEPB.

entire city, and related them to $PM_{2.5}$ concentration. To analyze population exposure to $PM_{2.5}$ in the Beijing urban area, especially in autumn, we used spatiotemporal statistical methods and made a distribution map of accumulated exceedance days, population exposure, and exposure probability exceeding the AQI standards.

Methods and Data

Source of $PM_{2.5}$ data

To strengthen ambient air pollution monitoring, 35 automatic monitoring stations for $PM_{2.5}$ fine particles were established in Beijing by BJEPB. At the end of September 2012, BJEPB gradually released these data in real time to the public. Figure 1 shows the locations of the BJEPB monitoring stations.

Each monitoring station automatically acquires hourly $PM_{2.5}$ concentration data. We collected the 37-day dataset from October 8 to November 13, 2012, for use as experimental data, from the Centre of the City Environmental Protection Monitoring Website Platform (www.bjmemc.com.cn), which is maintained by BJEPB. Daily 24-hour average concentration data were calculated based on the hourly real-time data. Thus, we obtained daily average $PM_{2.5}$ concentrations, between October 8 and November 13, 2012, for 35 sites in Beijing (see Table S1). The unit of concentration is $\mu g/m^3$ (micrograms per cubic meter). In addition, data on wind direction and scale for Beijing were also acquired for the same period from the Beijing Meteorological Bureau (see Table S2). Owing to equipment failure and transmission problems, some site data were incomplete or lost; thus, ultimately, we selected 31 of the 35 sites that had a complete record as our data source. The four omitted data sites are: Nos.13, 16, 20, and 28 (see Figure 1).

Census data

Beijing is the capital of the People's Republic of China, and has an area of 16,410 square kilometers. The latest Sixth National Population Census of the People's Republic of China [17], also known as the 2010 Chinese Census, was used in our analysis. This census revealed an official population for Beijing of 19,612,368. Its urban area includes six districts (county level): Xicheng, Dongcheng, Haidian, Chaoyang, Shijinshan, and Fengtai. The total urban area comprises 1378 square kilometers with a 2010 census population of 11,683,213.

To improve analysis precision, towns and sub-districts (street level) were chosen as the appropriate census geographic units. Using the Beijing Administrative Districts Atlas and Tabulation on the 2010 Population Census of Beijing Municipality (town and sub-district volume) [18,19], we matched the census data of the Beijing urban area to the administrative map. After data processing and merging, 129 census geographic units (street level)

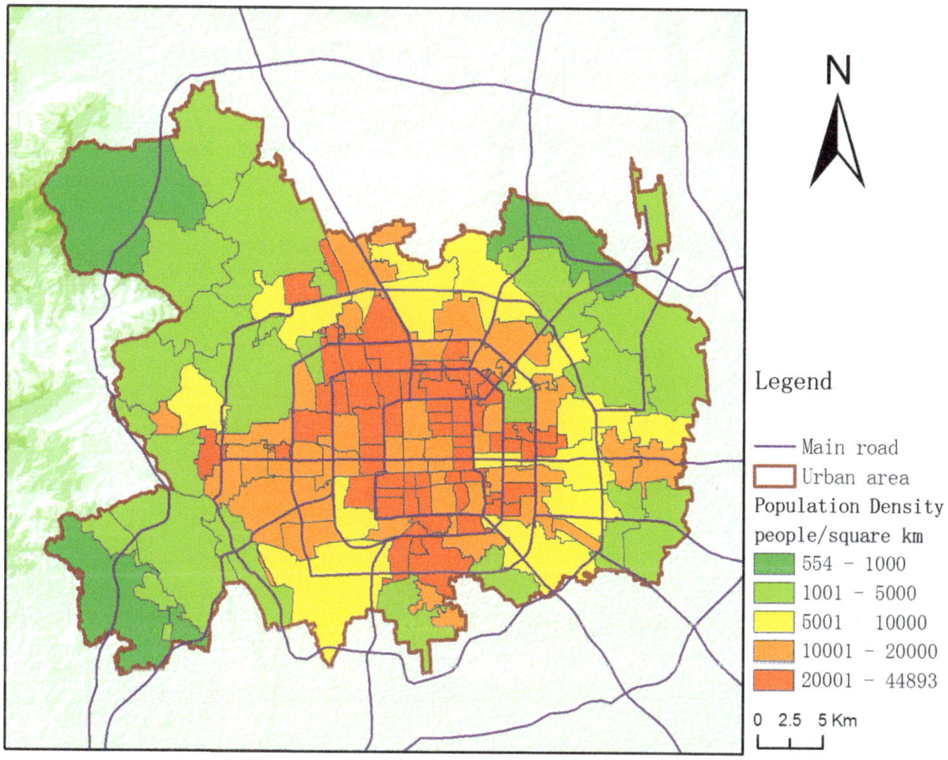

Figure 2. Census geographic unit population densities of Beijing urban area.

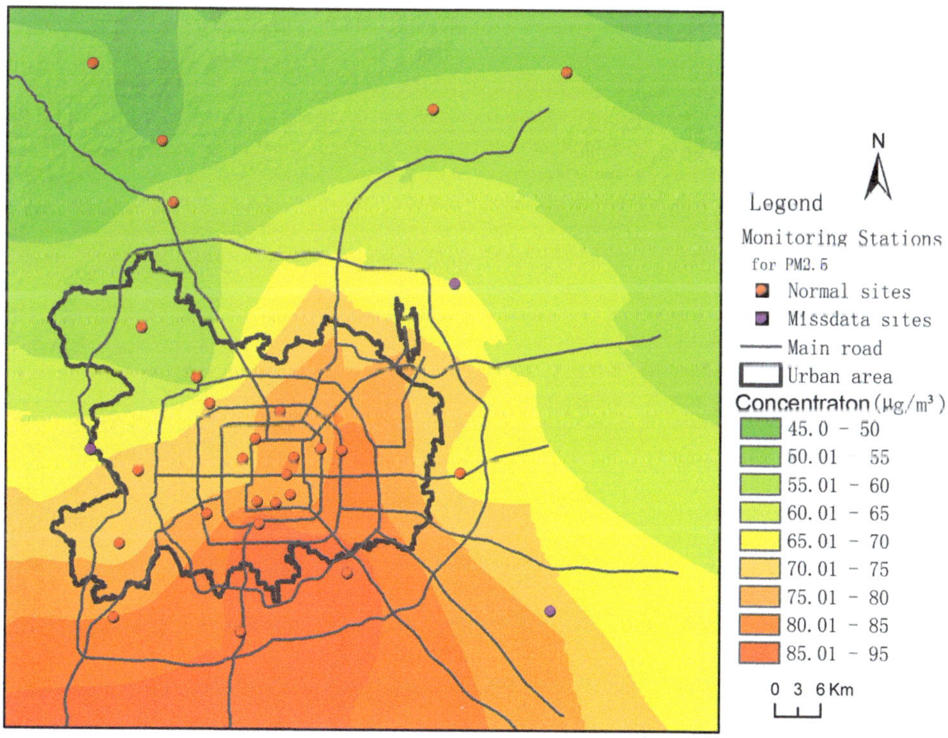

Figure 3. Sample of spatial interpolation for $PM_{2.5}$ concentrations (October 9, 2012).

Table 1. Individual Air Quality Index standards of $PM_{2.5}$*.

Individual Air Quality Index (AQI) Values	AQI Levels	Levels of Health Concern	24 Hour Average PM2.5 Concentrations Range (μg/m³)	Health Implications
0 to 50	Level 1	Good	0 to35	Air quality is considered satisfactory, and air pollution poses little or no risk.
51 to 100	Level 2	Moderate	35 to 75	Air quality is acceptable, but for some pollutants, there may be a moderate health concern for a very small number of people. People who are unusually sensitive to ozone may experience respiratory symptoms.
101 to 150	Level 3	Unhealthy for Sensitive Groups	75 to 115	Although the general public is unlikely to be affected in this AQI range, people with lung disease, older adults and children are at a greater risk from exposure to ozone; those with heart and lung disease, older adults and children are at greater risk from airborne particles.
151 to 200	Level 4	Unhealthy	115 to 150	Everyone may begin to experience adverse health effects, and members of the aforesaid sensitive groups may experience more serious effects.
201 to 300	Level 5	Very Unhealthy	150 to 250	This situation would trigger a health alert, signifying that everyone may experience more serious health effects.
301 to 500	Level 6	Hazardous	above 250	This would trigger a health warning of emergency conditions, and the entire population is more likely to be affected.

*From Ambient Air Quality Standards (GB 3095–2012) and Technical Regulation on Ambient Air Quality Index (HJ 633–2012, on trial) from the website of the Ministry of Environmental Protection, China (http://kjs.mep.gov.cn/hjbhbz/bzwb/dqhjbh/dqhjzlbz/201203/W020120410330232398521.pdf and http://kjs.mep.gov.cn/hjbhbz/bzwb/dqhjbh/jcgfffbz/201203/W020120410332725219541.pdf).

were formulated with the census data and area (see Table S3). Figure 2 shows the population density of these geographic units.

Spatial interpolation for $PM_{2.5}$ concentrations

The original $PM_{2.5}$ concentration data were at discrete points, i.e., at the automatic monitoring stations. We needed to interpolate the site data into surface data to characterize concentrations over the entire area. There are numerous spatial interpolation methods. Kriging, originated by Krige in 1951 and developed by Matheron [20], is one popular method used in various fields [21]. Kriging can interpolate the value of a random field at an unobserved location from observations of its value at nearby locations. Kriging has various forms, including: ordinary Kriging (OK), simple Kriging, indicator Kriging, universal Kriging, and intrinsic Kriging [22]. Of these, OK is the one used most frequently. It assumes a constant but unknown mean, which allows construction of an unbiased estimator that does not require prior knowledge of the stationary mean of the observed values [22]. Here, we use an OK method to interpolate spatially the $PM_{2.5}$ concentrations. Figure 3 shows a sample of one such spatial interpolation on October 9, 2012. The same method and parameters of interpolation were applied to all 37 days of daily site concentration data; ultimately, obtaining 37 daily averaged, areally interpolated concentrations.

Cumulative exposure hotspot detection for $PM_{2.5}$ air pollution

A series of epidemiology studies reported robust associations between short-term [6–8] and long-term exposure to $PM_{2.5}$ and adverse health effects, including cardiac and respiratory morbidity and mortality [23]. The focus of our study is on long-term cumulative effects. However, the estimation of the level of direct exposure of the population cannot be made from single-day concentrations. Harm from $PM_{2.5}$ concentration would not usually be immediately reflected, and there would be certain hysteresis and cumulative effects. Moreover, annual average concentration or periods of average concentration cannot indicate the risk level from $PM_{2.5}$. This is because average concentration does not reflect the difference between a normal concentration and its exceedance concentration. Furthermore, it is difficult to use average concentration to reveal spatiotemporal differences.

In addition, during the observation period, $PM_{2.5}$ pollution in Beijing showed a certain periodicity. By arrangement of the AQI-level diagrams in calendar order (Figure 5), each week appeared to have an independent and complete process of pollution. Except for October 8 (the first working day after the Chinese National Day holiday), the most serious pollution within the week always occurred on a Friday.

Therefore, we adopted a method for the evaluation of the cumulative effect. To calculate the level of exposure of human beings to particulate matter concentration, we must determine air pollution levels. The AQI helps in understanding the impact of local air quality on health. It is calculated from concentrations of the primary pollutant, which for Beijing in autumn is usually $PM_{2.5}$. According to the standard, and for ease of understanding, individual AQIs of $PM_{2.5}$ were divided into six categories by considering different concentration ranges (Table 1). Each category corresponds to a different level of health concern. We can convert the concentration of $PM_{2.5}$ to AQI levels with the help of the Ambient Air Quality Standards (GB 3095–2012) and Technical Regulation on Ambient Air Quality Index (HJ 633–2012, on trial) in Table 1.

We transformed the 37 daily averaged, areally interpolated concentrations to maps of AQI- level polygons, obtaining 37 AQI-level map layers. Through overlaying these map layers, an intersection layer was generated. Polygons in the intersection

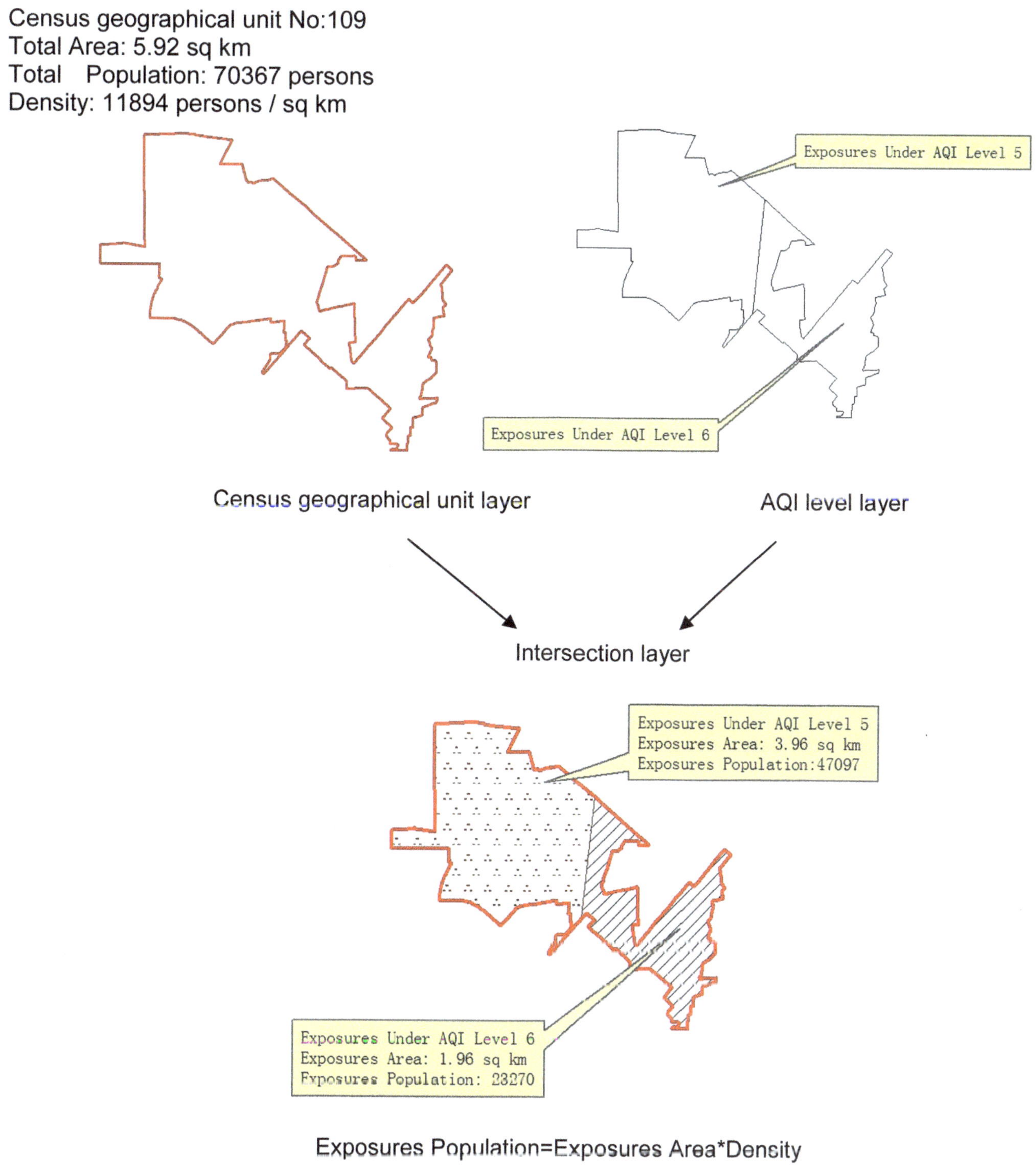

Figure 4. Population exposures estimation method.

layer are minimum exposure units. An AQI of $PM_{2.5}$ above the value of 100 (above level two) is considered as the urban air quality standard according to Ambient Air Quality Standards (GB 3095–2012), for which the 24-hour average $PM_{2.5}$ concentration exceeds 75 μg/m^3 [24,25]. Equation 1 is used to calculate the cumulative days exceeding the reference concentration of China's air quality guidelines in each minimum exposure unit:

$$C_i = \sum_{j=3}^{6} D_j, \tag{1}$$

where D_j denotes days at j air quality level in i minimum exposure units. C_i refers to the cumulative days under AQI levels from 3 to 6 exposure ($PM_{2.5}$ concentration exceeding 75 μg/m^3).

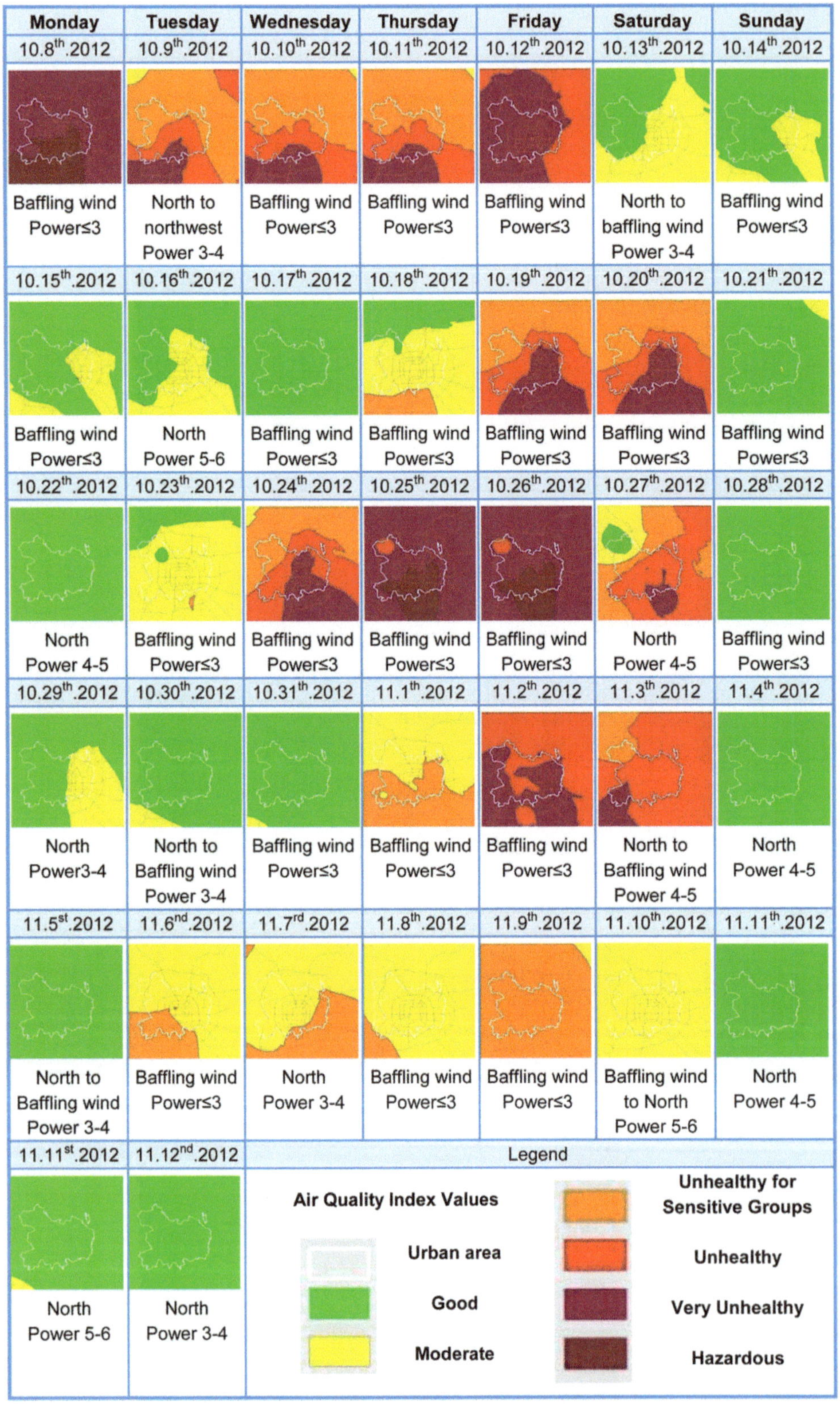

Figure 5. Thirty-seven daily Air Quality Index change in Beijing urban area.

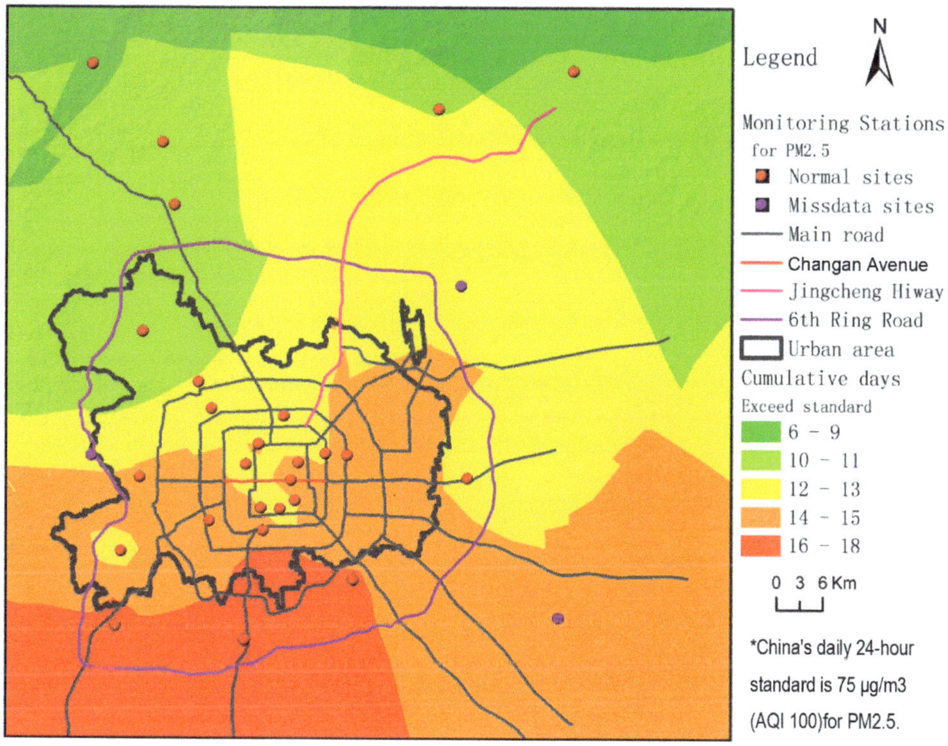

Figure 6. Number of cumulative exceedance days.

After calculation, the number of cumulative exceedance days in each minimum exposure unit is obtained. This facilitates the determination of areas of continued high exposure.

Area-weighting method estimating population exposure to $PM_{2.5}$

To estimate the level of exposure of the population to $PM_{2.5}$, we should ascertain which areas are exposed to high concentrations of $PM_{2.5}$ and the size of their populations. If there is no risk, then there is no exposure, and if there is no population, there is no exposure [9].

Based on a hotspot analysis, we can develop statistics regarding populations under continuous exposure. Previous studies used census geographic units as the smallest population exposure unit, but such a unit often has a different risk exposure level. To solve this problem, the area-weighting method was used to calculate the population under exposure. This method involves a straightforward algorithm for areal interpolation. Based on the assumption that population is evenly distributed within a source zone, the constant population density of each zone is initially estimated. Then, the size of each overlapping area between target and source

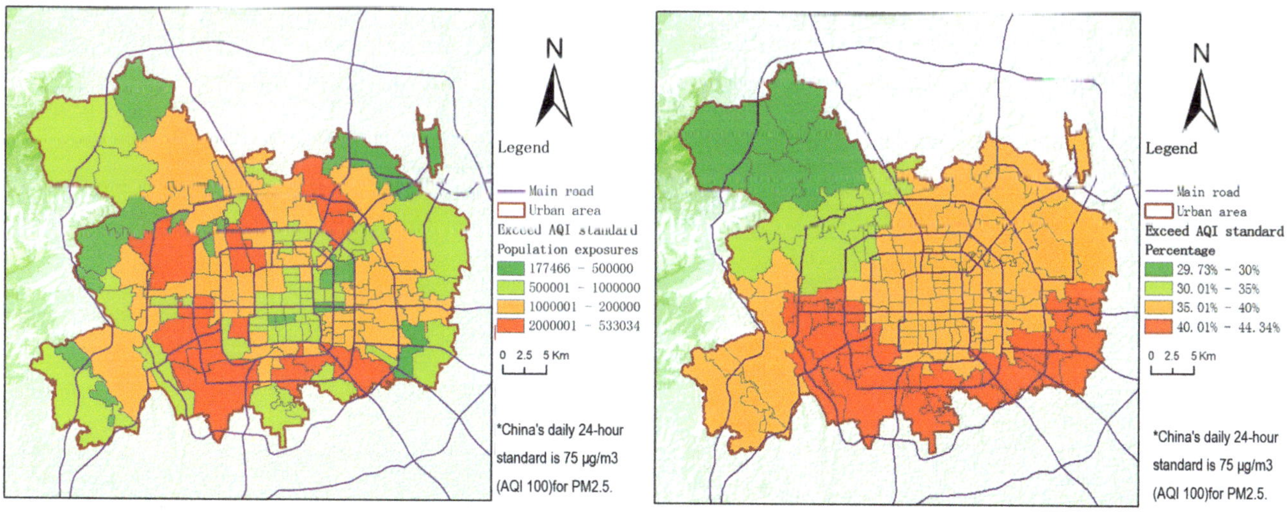

Figure 7. Population exposures (left) and percentages (right) exceeding AQI standard.

zones is used as a weight to estimate the population of the target zones [26].

To obtain cumulative population exposures exceeding the AQI standards, we assume that the population in urban internal areas is evenly distributed among the internal streets. The population density of every census geographic unit (street) was calculated separately using Equation 2:

$$d_k = P_k / A_k \, , \tag{2}$$

where d_k denotes census geographic unit k population density, P_k census geographic unit k population, and A_k census geographic unit k area.

Population exposures for each level in every census geographic unit are calculated using Equation 3:

$$E_{pop_{i,j}} = A_{level_i} \times d_k \, , \tag{3}$$

where $E_{pop_{i,j}}$ indicates census geographic unit j under AQI level i population exposure, A_{level_i} is the area under AQI level i, and d_k is census geographic unit k population density.

Then, the total population exposures under AQI level i are calculated with Equation 4.

$$E_{pop_i} = \sum_{j=1}^{n} E_{pop_{i,j}}, \tag{4}$$

where E_{pop_i} designates the total population exposures under AQI level i, and n is the number of census geographic units.

Total population exposures exceeding the AQI standard are calculated using Equation 5.

$$E_{pop} = \sum_{i=3}^{6} E_{pop_i}, \tag{5}$$

where E_{pop} represents the population exposures exceeding the AQI.

Then, the percentage of population exposures of the total population under AQI level i is calculated using Equation 6.

$$P_{level_i} = E_{pop_i} / E_{pop}, \tag{6}$$

where P_{level_i} denotes the percentage of the population exposures of the total population under AQI level i.

Figure 4 illustrates the estimated population exposure method for one census geographical unit.

Results

Figure 5 shows that during the 37 days, $PM_{2.5}$ pollution in Beijing had some spatiotemporal variations. The pollution process showed that southern Beijing began to accumulate $PM_{2.5}$ and exceed the standard, which was then followed gradually by northern Beijing. Following a change of wind direction, wind scale, humidity, other weather variables or related conditions, particulate pollution declined from north to south with the northern part of the city being the first to meet the levels of the standards. The third week of the time series charts depicts a complete, typical and long-lasting pollution process.

In our second result (Figure 6), it is not difficult to determine that Beijing suffers severe pollution of $PM_{2.5}$. The citywide cumulative number of exceedance days is generally high. Throughout the 37-day monitoring period, the number of exceedance days inside the 6th ring road was above ten. Meanwhile, the statistical result in this figure also displays an obvious spatial differentiation; the south and the southeast regions are hotspots of $PM_{2.5}$ pollution. Areas south of Chang An Avenue and east of the Beijing-Chengde expressway seem to be the high risk areas with up to two weeks or more of exceedance days. A large area south of the 4th ring road accumulated 16–18 days during this 37-day monitoring period. In other words, for half of the time or more, the people living there were exposed to excessive $PM_{2.5}$ pollution.

Statistically, over the duration of data collection, the cumulative exposed population is 160,561,842 (average daily exposed population is 4,339,509), representing 37.14% of Beijing's cumulative total urban population. The cumulative population exposed to hazardous levels of $PM_{2.5}$ (AQI level 6) is 12,315,860 (average daily exposed population is 332,861), or 2.85% of the total (see Table S3). As shown in Figure 7, the urban area between the second and fifth ring roads had generally high exposure. In contrast with the accumulated amount of exceedance days, the hotspots of accumulative exposed population were not just the areas to the south and southeast of Beijing; north to Chang An Avenue and west to the Beijing-Chengde expressway, there were also some census units in such hotspots of accumulated exposure.

Discussion

The spatiotemporal variations of $PM_{2.5}$ pollution in Beijing might be related to topography. West of Beijing is Xishan Mountain, part of the Taihang Mountains. North of the city is Jundu Mountain, part of the Yanshan Mountains. Surrounding the Beijing plain, these two mountains form a semicircular arc, extending to the southeast. The dominant wind direction during our monitoring period was northerly and northwesterly. Daily emissions from Beijing disperse towards the opening of the mountain arc, over the flat plain. When the wind loses strength or a weak southerly wind develops, atmospheric stability increases rapidly. Then, fine particulates accumulate from the south to north, as shown by the time series. Moreover, the provinces south of Beijing, such as Hebei, Henan and Shandong, receive large quantities of emissions every day. With wind transport from remote areas and increasing local background values, the pollutant accumulation and north-south spatial differences described above, become increasingly likely. Later in the pollution process, there is often a northerly wind of force 3 and above sweeping across the entire city, and with strong pollutant dispersal over the flat northeastern area, a pattern of reduction in levels of pollution appears from north to south.

The periodicity of $PM_{2.5}$ pollution in Beijing is probably linked to variations of meteorological conditions, as well as to periodic changes of transportation emissions caused by travel regulations in the city. This remains to be verified by our follow-up studies. Beijing is still expanding rapidly with tremendous daily emissions by traffic and other local sources. Combined with emissions from nearby provinces, it is easy for $PM_{2.5}$ pollution to exceed standards. Such conditions result in the high number of accumulated exceedance days. The spatial differences shown in Figure 6 might be explained by the spatiotemporal regularity of pollution processes. That is, the south is always the first area to exceed normal concentrations and the last to revert below them, causing a longer duration of pollution in that area compared with the north.

The northern part of the city saw fewer accumulative numbers of days of exceedance than the southern part. Nevertheless, given the spatial distribution of total population exposure and the dense population in northern areas, its 37-day cumulative exposure still easily reached the highest level. That is, the north may not have less population and economic harm from $PM_{2.5}$ pollution than the south. If the impact of population density is not considered, the exposure shows that the south still has a greater accumulated probability of exceedance days than the north.

Because we only collected $PM_{2.5}$ data for 37 days in autumn from the Beijing Environmental Protection Bureau, the above analysis and results may not be applicable to other seasons, because air pollution emissions and meteorological conditions change significantly between seasons in Beijing. We need to gather $PM_{2.5}$ data over longer periods to perform a long time series analysis.

Conclusions

Through spatiotemporal analysis of $PM_{2.5}$ data for 37 days in autumn, and by overlaying the population distribution of the Beijing urban area, we discovered hotspots in which cumulative days exceeded the AQI standard, and we calculated the population exposures exceeding that standard. The concentration levels of $PM_{2.5}$ have become a serious problem in Beijing that requires attention.

The research methods and conclusions here reveal relationships between long-term overexposure to $PM_{2.5}$ and people living in high-exposure areas during the autumn in Beijing. The research results will also support future studies, especially into the adverse health effects of long- and short-term exposure to $PM_{2.5}$.

Acknowledgments

We thank Mr. Hongsheng Li, Xiaolei Wang and Dong Li for their help.

Author Contributions

Conceived and designed the experiments: AZ QWQ. Performed the experiments: LLJ FZ. Analyzed the data: FZ. Contributed reagents/materials/analysis tools: JFW. Wrote the paper: QWQ AZ.

References

1. Sun Y, Zhuang G, Tang A, Wang Y, An Z (2006) Chemical characteristics of PM2. 5 and PM10 in haze-fog episodes in Beijing. Environmental science & technology 40: 3148–3155.
2. Wang JF, Hu MG, Xu CD, Christakos G, Zhao Y (2013) Estimation of Citywide Air Pollution in Beijing. PLOS ONE 8: e53400.
3. Kunii O, Kanagawa S, Yajima I, Hisamatsu Y, Yamamura S, et al. (2002) The 1997 haze disaster in Indonesia: its air quality and health effects. Archives of Environmental Health: An International Journal 57: 16–22.
4. Lippmann M (1998) The 1997 US EPA standards for particulate matter and ozone. Issues in environmental science and technology 10: 75–100.
5. Schwartz J, Neas LM (2000) Fine particles are more strongly associated than coarse particles with acute respiratory health effects in schoolchildren. Epidemiology 11: 6–10.
6. Gavett SH, Koren HS (2001) The role of particulate matter in exacerbation of atopic asthma. International archives of allergy and immunology 124: 109–112.
7. Haley VB, Talbot TO, Felton HD (2009) Surveillance of the short-term impact of fine particle air pollution on cardiovascular disease hospitalizations in New York State. Environmental Health 8: 42.
8. Vallejo M, Ruiz S, Hermosillo AG, Borja-Aburto VH, Cárdenas M (2005) Ambient fine particles modify heart rate variability in young healthy adults. Journal of Exposure Science and Environmental Epidemiology 16: 125–130.
9. Hao Y, Flowers H, Monti MM, Qualters JR (2012) US census unit population exposures to ambient air pollutants. International journal of health geographics 11: 3.
10. Jorgensen NW, Van Hee VC, Barr RG, O'Neill MS, Herrington DM, et al. (2012) Vascular Responses to Long-and Short-Term Exposure to Fine Particulate Matter. Journal of the American College of Cardiology 60.
11. Kan H, London SJ, Chen G, Zhang Y, Song G, et al. (2007) Differentiating the effects of fine and coarse particles on daily mortality in Shanghai, China. Environment international 33: 376.
12. Fann N, Lamson AD, Anenberg SC, Wesson K, Risley D, et al. (2011) Estimating the national public health burden associated with exposure to ambient PM2. 5 and ozone. Risk Analysis 32: 81–95.
13. Li P, Xin J, Wang Y, wang S, Pan X, et al. (2012) Research on the impact of Beijing atmospheric particulate pollution on population death rate. S7 Climatic and Environmental Changes and Human Health.
14. Urbat M, Lehndorff E, Schwark L (2004) Biomonitoring of air quality in the Cologne conurbation using pine needles as a passive sampler – Part I: magnetic properties. Atmospheric Environment 38: 3781–3792.
15. Yuan Y, Liu D, Che R, Dong X (2007) Reseach on the Characteristics of Beijing Atmospheric Particulates Pollution in Autumn. Ecology and Environment 16: 18–25.
16. Zhao W, Zhao W, Gong H, Gong Z (2012) Spatio-temporal Distribution Characteristics and Source Tracking of Particulate Matters in Beijing during Heating Period. Geographical Research 417–428.
17. Municipality OftSPCoB, Statistics BMBo, Beijing SOotNBoSi (2012) Tabulation on the 2010 Population Census of Beijing Municipality (Town and Sub-district Volume): China Statistics Press.
18. Bureau BCA, Mapping BIoSa (2005) Beijing Administrative Districts Atlas. Hunan Cartographic Publishing House.
19. Bureau BCA, Mapping BIoSa (2007) Beijing Administrative Districts Tuzhi. Beijing: China's tourism press.
20. Matheron G (1963) Principles of geostatistics. Economic geology 58: 1246–1266.
21. Bayraktar H, Turalioglu FS (2005) A Kriging-based approach for locating a sampling site In the assessment of air quality. Stochastic Environmental Research and Risk Assessment 19: 301–305.
22. Pang W, Christakos G, Wang JF (2010) Comparative spatiotemporal analysis of fine particulate matter pollution. Environmetrics 21: 305–317.
23. Du X, Kong Q, Ge W, Zhang S, Fu L (2010) Characterization of personal exposure concentration of fine particles for adults and children exposed to high ambient concentrations in Beijing, China. Journal of Environmental Sciences 22: 1757–1764.
24. China Ministry of Environmental Protection (2012) Technical Regulation on Ambient Air Quality Index (on trial). China Environmental Science Press.
25. China Ministry of Environmental Protection, General Administration of Quality Supervision, Inspection and Quarantine (2012) Ambient air quality standards. China Environmental Science Press.
26. Qiu F, Zhang C, Zhou Y (2012) The Development of an Areal Interpolation ArcGIS Extension and a Comparative Study. GIScience & Remote Sensing 49: 644–663.

4

Non-Specific Abdominal Pain and Air Pollution: A Novel Association

Gilaad G. Kaplan[1,2,3]⁹, Mieczyslaw Szyszkowicz[4*]⁹, Jakub Fichna[7], Brian H. Rowe[5], Eugeniusz Porada[4], Renaud Vincent[4], Karen Madsen[6], Subrata Ghosh[1,2], Martin Storr[1,2,8]

1 Division of Gastroenterology, University of Calgary, Calgary, Alberta, Canada, **2** Departments of Medicine, University of Calgary, Calgary, Alberta, Canada, **3** Community Health Sciences, University of Calgary, Calgary, Alberta, Canada, **4** Environmental Health Sciences and Research Bureau, Health Canada, Ottawa, Ontario, Canada, **5** Department of Emergency Medicine and School of Public Health, University of Alberta, Edmonton, Alberta, Canada, **6** Department of Medicine, Division of Gastroenterology, University of Alberta, Edmonton, Alberta, Canada, **7** Department of Biomolecular Chemistry, Medical University of Lodz, Lodz, Poland, **8** Department of Medicine, Ludwigs-Maximilians University, Munich, Germany

Abstract

Background: We studied whether short-term exposure to air pollution was associated with non-specific abdominal pain in epidemiologic and animal studies.

Methods: Patients visiting the emergency department with non-specific abdominal pain were identified in Edmonton (1992 to 2002, n = 95,173) and Montreal (1997 to 2002, n = 25,852). We calculated the daily concentrations for ozone (O_3), nitrogen dioxide (NO_2), sulfur dioxide (SO_2), carbon monoxide (CO), and particles <10 (PM_{10}) or <2.5 ($PM_{2.5}$) µm. A case crossover study design was used to estimate the odds ratio (OR) and 95% confidence interval (CI) associated with an increase in the interquartile range of the air pollutants. We investigated differential effects by age and sex. Mice were gavaged with urban particle extracts. In animal models, colonic motility was tested, and visceral abdominal pain was measured using a writhing test, and behavioral response to oil of mustard and neostigmine. Motility and pain was measured acutely (1.5 hours after gavage) and chronically (7-days and 21-days after gavage).

Results: Emergency department visits for non-specific abdominal pain were primarily by women between the ages of 15–24 years. Individuals aged 15 to 24 years were at increased risk of non-specific abdominal pain in Edmonton (same day CO: OR = 1.04, 95% CI = 1.02–1.06; and NO_2: OR = 1.06, 95% CI = 1.03–1.09). The risk of air pollution among 15–24 year olds in Montreal was significantly positive (same day CO: OR = 1.11, 95% CI = 1.05–1.17; NO_2: OR = 1.09, 95% CI = 1.01–1.16; SO_2: OR = 1.17, 95% CI = 1.10–1.25; $PM_{2.5}$: OR = 1.09, 95% CI = 1.04–1.15). Abdominal pain was increased by an acute gavage of pollution extract but not to chronic exposure to pollutants. Colonic transit was delayed following chronic but not acute exposure with the pollutants.

Conclusions: Epidemiological and animal data suggest that short-term exposure to air pollution may trigger non-specific abdominal pain in young individuals.

Editor: Devendra Amre, University of Montreal, Canada

Funding: Funding was provided by Health Canada and The Alberta IBD Consortium. The funders had no role in study design, data collection and analysis, decision to publish, or preparation of the manuscript.

Competing Interests: The authors have declared that no competing interests exist.

* E-mail: mietek.szyszkowicz@hc-sc.gc.ca

⁹ These authors contributed equally to this work.

Introduction

Abdominal pain is the second most common diagnosis (4.2%) claimed by emergency department physicians in the USA. In many cases patients are diagnosed with non-specific abdominal pain, which refers to abdominal pain that is not explained during acute assessment. [1,2] While many patients diagnosed with non-specific abdominal pain are asymptomatic at a later outpatient follow-up [3,4,5], a subset of these patients require recurrent admissions. [6,7,8] Consequently, non-specific abdominal pain is an important clinical presentation associated with impairment in quality of life for patients and significant costs to the health care system. [7,8]

The underlying cause of non-specific abdominal pain is likely multifactorial. non-specific abdominal pain may be due to gastrointestinal diseases such as Crohn's [9] and celiac [10] diseases that are diagnosed after the patient is discharged from the emergency department. Additionally, some patients experience a transient illness such as gastrointestinal infection. Cases of non-specific abdominal pain may also represent a functional disorder such as irritable bowel syndrome. [11] Despite extensive research into the causes of non-specific abdominal pain, risk factors are not known and thus, exploration of triggers of non-specific abdominal pain is warranted.

Air pollution directly effects pulmonary diseases (e.g. asthma) and has been associated with non-pulmonary diseases including cancer and strokes. [12,13,14] Recently, studies have indicated that air pollution may affect gastrointestinal disorders such as the inflammatory bowel diseases and appendicitis. [15,16] Air pollutants can cause systemic effects such as cardiovascular [17] or liver metabolic [18] changes after primary deposition and direct effects in the respiratory compartment. [19] However, most inhaled particles deposited in the nasopharyngeal compartment and airways are removed through mucociliary clearance and swallowed within a day [20] and thus, gastrointestinal effects may be due to direct effects of particles. [19]

Consequently, we investigated whether acute exposure to air pollutants was associated with emergency department visits for non-specific abdominal pain by: 1) evaluating the effect of short-term ambient air pollution exposure in a human population; 2) replicating this epidemiological study in a second population; and 3) exploring the biological plausibility in *in vitro and in vivo* models.

Methods

Phase 1: Epidemiological Studies

Study Population. As described previously [21], the administrative database study was approved by the individual research ethics boards at each participating institution and the data were transferred to the Health Canada following de-identification. No patient contact was made and patients could not be traced. The International Classification for Disease 9th revision was used to identify patients who were discharged from an emergency department with a primary diagnosis of non-specific abdominal pain (789.0×). [2] First, we identified a population-based sample of 95,173 individuals who visited one of 5 emergency department in Edmonton for non-specific abdominal pain from 1992–2002. Second, we identified a replication study population of 25,852 individuals who visited a downtown emergency department in Montreal for non-specific abdominal pain from 1997–2002. Three digit postal codes were used to restrict the study population to residents of the city of Edmonton or Montreal. While the study population from Edmonton was population-based and included all ages, pediatric emergency department visits were underrepresented in Montreal because of the immediate proximity of a Pediatric Hospital. Thus, for the study population from Montreal the analysis was restricted to individuals over the age of 15 years.

Air Pollution and Meteorological Data. The air pollutants measured included sulfur dioxide (SO_2), nitrogen dioxide (NO_2), carbon monoxide CO, ozone (O_3), and particulate matter characterized by an aerodynamic diameter of ≤10 μm (PM_{10}) and of ≤2.5 μm ($PM_{2.5}$). The data on air pollution were supplied by Environment Canada from fixed monitoring stations in Edmonton and Montreal. Environment Canada also supplied hourly data for temperature and relative humidity. For each air pollutant and weather variable, we obtained data on 24 measurements carried out at the monitoring station at hourly intervals. If more than 6 measurements were missing for a specific day, than the data for this monitor was considered as missing. An average was calculated for each monitor to represent daily mean value and average among monitors was used to represent ambient exposure across the city. [22] The distribution of air pollution and weather monitors for both cities can be found in Environment Canada's website. [23] A detailed description of air pollutant and weather measurements, monitor distribution, and correlations among monitors has been previously reported. [24]

Statistical Analysis. We applied a case crossover study design, which is an adaptation of the case–control study whereby cases serve as their own controls. [25] A time-stratified design was used to select referent periods, whereby the referent periods were selected from the same day of the week, month and year as the case's date of hospital visit. Because within-individual comparisons were made, confounding due to time-independent risk factors (e.g. genetics) was reduced. Similarly, the matching of control to case periods by day of week controlled for the influence of "day of week" effects. [25]

We used a conditional logistic regression (subroutine PHREG in SAS) to model the air pollutants and 6 components for natural splines for temperature and relative humidity. Estimates were expressed as the odds ratio (OR) with 95% confidence intervals (CI) associated with an increase in the interquartile range of the air pollutants (SO_2, NO_2, CO, O_3, PM_{10}, and $PM_{2.5}$) after adjusting for temperature and relative humidity. Control for potentially non-linear effects of weather was achieved by including natural spline functions of temperature and relative humidity. The interquartile range was calculated based on the daily mean concentration of each air pollutant during the study period (Table S1). Analyses were performed separately for Edmonton and Montreal. Additionally, the analysis was stratified by: age, categorized as <15 (only Edmonton), 15–24, 25 34, 35–44, 45–64, and >64 years. Different time lags for pollution and weather exposures were used in our analyses including same day, one day before admission (i.e. 1-day lag), and two days before admission (i.e. 2-day lag). Among individuals 15 to 24 years of age we examined for effect modification by sex by incorporating an interaction term (e.g. $PM_{2.5}$ X sex) into the model. All statistical analyses were conducted in SAS (version 8, SAS, Cary, North Carolina, US).

Phase 2: *In Vitro* & *In Vivo* Studies

Animals. Male Swiss albino mice (CD1, Charles River, Canada), weighing 24–26 g, were used. The animals were housed at a constant temperature (22°C) and maintained under a 12-h light/dark cycle with free access to laboratory chow and tap water. Animal use was approved by the University of Calgary Animal Care Committee.

Particulate Matter Preparation. The urban particles EHC-6802 consisted of a blend of total suspended particulate matter recovered from filters of the single-pass air purification system at the Environmental Health Centre (Tunney's Pasture, Ottawa, ON) in 1996, 1998, 2000, and 2002, as described previously. [26] Particles were mechanically sieved using a 36 μm mesh filter, and combined in equal proportions. EHC-6802 was recovered at the same site and manner as the urban particles EHC-93, and chemical characterization of the EHC-6802 material has confirmed that it was equivalent to the EHC-93 material. [26]

Experimental Design. Animals were randomly assigned to two groups with the same number of animals per treatment group: control (oral tap water, 200 μl/animal, once daily; n = 8) or EHC-6802 (oral 360 μg/200 μl tap water/animal, once daily; n = 8). Gavages were given once daily and experiments were performed in the following protocols: 1) 1.5 h after the gavage; 2) daily for 7 days; and 3) daily for 21 days.

Isolated smooth muscle strips. Full-thickness segments (1 cm) of ileum and distal colon were resected. The preparations were mounted between two platinum electrodes and placed longitudinally in separate organ baths as described previously. [27] One end of each preparation was attached to the bottom of the organ bath, while the other end was connected to a FT03 force displacement transducer (Grass Technologies, West Warwick, RI, USA). Changes in tension were amplified by a P11T amplifier

(Grass Technologies) and recorded on a computer using PolyView software (Polybytes, Cedar Rapids, Iowa).

Electrical field stimulation (EFS; 8 Hz; 60 V; pulse duration 0.5 ms; train duration 10 sec) was applied by a S88X stimulator (Grass Technologies). EFS of isolated smooth muscle strips caused twitch contractions, which were of cholinergic neuronal origin as they were virtually abolished by atropine (10^{-6} M) or TTX (10^{-6} M) (data not shown). In separate experiments the response of ileal and colonic muscle strips to direct muscular activation with stepwise addition of bethanechol (10^{-7} M–3×10^{-5} M) was recorded.

EHC-6802 urban ambient particles were added cumulatively into the organ baths and effects on the EFS-induced contractions were recorded. The mean amplitude of 4 successive twitch contractions was used as an internal control. Changes in contractions were reported as the percentage of the internal control. In control experiments the effects of the vehicle (tap water) were tested to control against possible effects on contractility by adding vehicle..

In separate experiments the response of ileal and colonic muscle strips to (i) neuronal electrical stimulation and (ii) direct muscular activation with stepwise addition of bethanechol (10^{-7} M–3×10^{-5} M) was recorded following 90 min, 7 days and 21 days following oral treatment with EHC-6802 particles.

Colonic expulsion test. Distal colonic expulsion was measured as reported previously. [27] 1.5 hours after receiving tap water (control) or EHC-6802 particle suspension (n = 5–10), a glass bead (2 mm) was inserted 2 cm into the distal colon and the time to bead expulsion was determined.

Behavioural Pain Responses. Behavioural pain responses were tested in 3 models used to study abdominal visceral pain responses (n = 5–10). The writhing test was performed as described by Gach et al. [28] Ninety minutes after oral administration of tap water (control group) or EHC-6802 particle suspension, mice received an intraperitoneal (i.p.) injection (10 ml/kg) of acetic acid solution (0.5%). The total number of writhes was counted 5 minutes after acetic acid injection, during three periods lasting 5 minutes each. The writhing response, considered as a nociceptive behaviour, was characterized by elongation of the body and the development of tension in the abdominal muscles and hind paws.

Behavioural responses to intracolonic (i.c.) oil of mustard (OM) and i.p. neostigmine were determined based on Laird et al. [29] and Eijkelkamp et al. (n = 5–10). [30] For the assessment of the OM-induced pain behaviours, 50 µl of OM (1% in 70% ethanol-30% saline) was administered i.c. under isoflurane anaesthesia. Petroleum jelly was applied to the perianal area to avoid stimulation of somatic areas. After 5 min of recovery, spontaneous behaviours were videotaped for analysis by an observer blinded to the experimental conditions.

In a separate experiment, neostigmine (2.5 µg/kg) or vehicle was administered i.p. and the pain-related behaviours were counted over a 10 minute period after the injection. Postures defined as pain-related behaviours: 1) licking of the abdomen, 2) stretching the abdomen, 3) squashing of lower abdomen against the floor, and 4) abdominal retractions, were each counted as 1.

Statistical Analysis. Statistical and curve-fitting analyses were performed using Prism 4.0 (GraphPad Software Inc., La Jolla, CA, USA). The data are expressed as means ± standard deviation. Student's t-test compared treatment means with control means. Two-sided P-values≤0.05 were statistically significant.

Results

Phase 1: Epidemiological Studies

Nearly 2/3 of emergency department visits for non-specific abdominal pain were by women and the majority of women presented in adolescence and early adulthood (Figure 1A and 1B; Table S1). In Edmonton, where pediatric emergency department visits were available, 61.4% of emergency department visits for non-specific abdominal pain occurred before the age 34 years. In Montreal, patients between the ages of 15 and 34 years accounted for 36.7% of emergency department visits (Table S1).

Table S2 describes the mean daily concentrations and variance of each air pollutant during the study periods. The effect of air pollution exposure was only observed when ages were stratified. In Edmonton, individuals aged 15 to 24 years were at elevated risk for presenting to the emergency department on the same day of higher concentrations of CO (OR = 1.04, 95% CI = 1.02–1.06), NO_2 (OR = 1.06, 95% CI = 1.03–1.09), SO_2 (OR = 1.02, 95% CI = 1.00–1.04), and $PM_{2.5}$ (OR = 1.03, 95% CI = 1.00–1.05). Similar effects were observed in Montreal for CO (OR = 1.11, 95% CI = 1.05–1.17), NO_2 (OR = 1.09, 95% CI = 1.01–1.16), SO_2 (OR = 1.17, 95% CI = 1.10–1.25), and $PM_{2.5}$ (OR = 1.09, 95% CI = 1.04–1.15) (Table 1). Among 15 to 24 year olds, O_3 was negatively associated with emergency department visits for non-

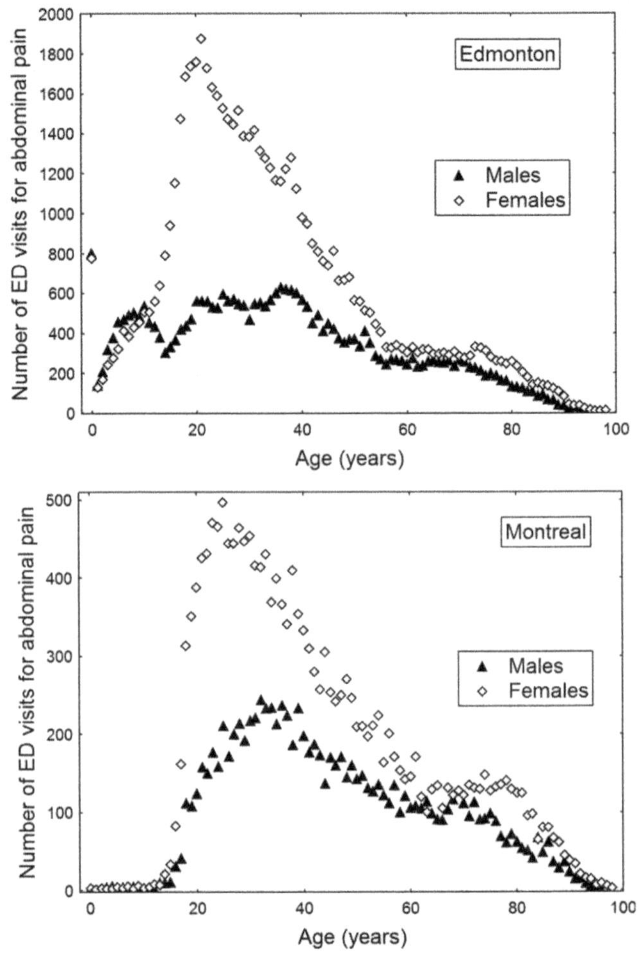

Figure 1. Age-specific frequency of admissions to emergency departments (ED) for non-specific abdominal pain stratified by sex.

specific abdominal pain in Edmonton (OR = 0.95, 95% CI = 0.91–0.99), but not in Montreal (OR = 1.03, 95% CI = 0.95–1.11) (Table 1). The effect of air pollutants on individuals aged 15 to 24 stratified by sex are presented in Table 2. Interaction terms were not significant between sexes (Table 2). The effect of air pollutants with 1-day and 2-day lags stratified by age and sex are presented in Tables S3 and S4, respectively.

Phase 2: *In Vitro & In Vivo* Studies

The course of body weight was not significantly different in animals treated with EHC-6802 urban particles by gavage compared to vehicle treated mice (Figure S1).

Gastrointestinal Motility. EHC-6802 (1:100000–1:1) produced a significant, concentration-dependent inhibitory effect on the EFS (8 Hz)-induced twitch contractions in the colon, but not the mouse ileum (Figure 2A). Electrically induced contractions were not changed at any time points (90 min, 7 days, 21 days), compared to bethanechol (10^{-6} M) treatment (Figure S2). Responses to direct muscular activation with bethanechol (10^{-7} M–3×10^{-5} M) were not changed by adding EHC-6802 particles to the organ bath (Figure S1C) at any time point (Figure S2).

In the colonic expulsion test a 7-day treatment with EHC 6802 particles significantly reduced the time to bead expulsion (Figure 2B). Neither acute (1.5 h) nor chronic (21 days) treatment with EHC-6802 significantly affected the colonic bead expulsion time.

Nociception. In our preliminary experiments, we confirmed that the i.p. injection of acetic acid solution, i.c. instillation of OM, and the i.p. injection of neostigmine induced pain behaviors in naïve mice (data not shown). The acute administration of EHC-6802 (1.5 h treatment) significantly increased the number of writhes compared to control group (Figure 3A). 7-day (Figure 3A) and 21-day (data not shown) treatment with EHC-6802 did not modify the response to acetic acid. EHC-6802 administered at 1.5 hour also showed a significantly higher number of spontaneous pain-related behaviours induced by OM (Figure 3B) and neostigmine (Figure 3C), compared to mice after 7- and 21-day treatment (21 days, data not shown).

Discussion

Non-specific abdominal pain is a common clinical presentation to the emergency department, typically involves multiple investigations, and may require admission to hospital. [1] A cause is often not discovered during emergency department visit and in most cases a subsequent organic diagnosis is not made during outpatient follow-up. [3] Although non-specific abdominal pain is a common disorder of the young [31,32], triggers of acute pain are not known. We found that young individuals had the highest prevalence of non-specific abdominal pain and were significantly more likely to visit the emergency department for non-specific abdominal pain when indicators of air pollution were elevated. Furthermore, ingestion of EHC-6802 urban ambient particles by mice led to an acute exacerbation of pain, which did not persist after chronic exposure of the pollutants to the animal. These observations suggest a novel association between air pollution exposure and exacerbation of abdominal pain.

In both Edmonton and Montreal, CO, $PM_{2.5}$, SO2, and NO_2 were associated with non-specific abdominal pain presentations in individuals aged 15 to 24 years. However, PM_{10} was positively associated with non-specific abdominal pain in Edmonton, but null in Montreal. Furthermore, ozone was negatively associated with non-specific abdominal pain in Edmonton, but not in Montreal. Additionally, the risk estimates were generally higher in Montreal than in Edmonton. The heterogeneity of findings between cities may be explained by residents living and/or working in downtown Montreal having greater exposure to air pollutants. In contrast, residents living throughout Edmonton would have been exposed to a wider range of air pollution concentrations. The greater patient-level variability to air pollution exposure experienced in Edmonton may have resulted in non-differential misclassification that would have the effect of biasing the risk estimates towards the null. Furthermore, the fixed monitoring stations are located to represent the background air pollution concentrations in a city; however, misclassifying the air pollution exposure may occur when multiple fixed monitoring sites are averaged into one daily value for the entire city. [22] Thus, future multi-city studies are required to confirm the associations observed in our study and to explore causes of heterogeneity.

Young women are known to have the highest prevalence of non-specific abdominal pain. [31,32] In both Edmonton and Montreal, women aged 15 to 24 years were more likely to visit an emergency department for non-specific abdominal pain. The reason for the disproportionate occurrence of non-specific abdominal pain in women as compared to men is not clear. Abdominal pain may be exacerbated during the menstrual cycle. [33] Furthermore, women have been shown to be more likely to seek medical attention for abdominal pain when compared to men. [34,35] We did not observe effect modification between men and women aged 15 to 24 years, suggesting that the effect of air pollution on non-specific abdominal pain was not sex-specific. However, because the risk estimates observed in our study were modest, we may have not been adequately powered to detect effect modification by sex.

The mechanism by which air pollutants may exacerbate abdominal pain is not known. In one study high concentrations of air pollutants (NO_2, CO, and O_3) impaired gastric basal contractility. [36] However, low-level exposures were not associated with gastric dysmotility. [36] Another study demonstrated that animals fed EHC-6802 had increased IL-8 secretion from the small bowel and experienced changes in the composition of colonic microflora. [37] Our *in vitro* studies demonstrated impaired colonic, but not ileal, contractility after the acute exposure of the ex-vivo tissue with the particulate matter. Neuronally mediated contractility was impaired whereas the direct smooth muscular reactivity was unchanged suggesting that the observed effects involved neuronal rather than muscular mechanisms. *In vivo* reduction of colonic motility was not observed acutely, but a significantly slowed colonic motility was recorded after 7 days, but not after 21 days of daily exposure to the EHC-6802 particles. Overall *in vitro* and *in vivo* effects on intestinal motility by acute and chronic exposure to EHC-6802 particles were negligible.

In contrast, acute exposure to EHC-6802 particles led to significantly increased pain response in mice and this observation was consistent in 3 different behavioral tests of visceral abdominal pain. Chronic exposure to ingested pollutants did not affect the pain response models suggesting that pain adaptation to the pollutants may have occurred over time. The cause for the acute hypersensitivity in mice is not known. The animal studies suggest that the pain response following acute exposure to the pollutants were not driven by intestinal dysmotility. Exposure to air pollution may exacerbate systemic inflammation [38] or lead to oxidative damage of colonic mucosa. [39] Furthermore, air pollution has been shown to mediate dysfunction of endothelial-dependent vasodilation [40] that may contribute to visceral hypersensitivity. Future studies will be needed to explore the exact mechanisms and

Table 1. Odds ratio for presenting to an emergency department with non-specific abdominal pain associated with an interquartile increase in an air pollutant concentration on the same day of ED visit stratified by age.

Pollutant	All Ages OR (95% CI)	<15 years OR (95% CI)	15–24 years OR (95% CI)	25–34 years OR (95% CI)	35–44 years OR (95% CI)	45–64 years OR (95% CI)	>64 years OR (95% CI)
Edmonton							
CO (ppm)	1.01(0.99–1.02)	0.99(0.97–1.02)	1.04(1.02–1.06)	0.99(0.97–1.01)	1.00(0.98–1.02)	1.01(0.99–1.03)	1.01(0.99–1.04)
NO_2 (ppb)	1.02(1.00–1.03)	1.00(0.97–1.04)	1.06(1.03–1.09)	0.99(0.96–1.01)	1.01(0.98–1.04)	1.03(0.99–1.06)	1.00(0.96–1.04)
SO_2 (ppb)	1.01(1.00–1.02)	1.00(0.98–1.03)	1.02(1.00–1.04)	0.99(0.97–1.02)	1.00(0.98–1.02)	1.00(0.97–1.02)	1.04(1.01–1.07)
O_3 (ppb)	0.99(0.97–1.01)	0.98(0.93–1.03)	0.95(0.91–0.99)	1.02(0.98–1.06)	1.02(0.98–1.07)	0.94 (0.90–0.98)	0.99(0.94–1.04)
PM_{10} (mg/m^3)	1.01(1.00–1.02)	1.01(0.98–1.04)	1.04(1.02–1.07)	1.01(0.98–1.03)	0.99(0.97–1.02)	1.00(0.97–1.02)	1.00(0.97–1.03)
$PM_{2.5}$ (mg/m^3)	1.00(0.99–1.01)	1.00(0.97–1.04)	1.03(1.00–1.05)	0.99(0.96–1.02)	0.99(0.96–1.02)	1.00(0.97–1.03)	0.99(0.96–1.03)
Montreal							
CO (ppm)	1.02(0.99–1.04)	NA	1.11(1.05–1.17)	1.04(1.00–1.09)	0.99(0.94–1.04)	1.08(1.04–1.13)	1.01(0.96–1.07)
NO_2 (ppb)	1.02(0.99–1.04)	NA	1.09(1.01–1.16)	1.06(1.00–1.12)	1.00(0.94–1.06)	1.14(1.08–1.20)	1.05(0.99–1.11)
SO_2 (ppb)	1.02(1.00–1.05)	NA	1.17(1.10–1.25)	1.08(1.03–1.14)	1.02(0.96–1.08)	1.13(1.08–1.19)	1.08(1.02–1.14)
O_3 (ppb)	0.99(0.97–1.02)	NA	1.03(0.95–1.11)	1.04(0.98–1.11)	1.09(1.02–1.16)	0.98(0.93–1.04)	1.03(0.96–1.09)
PM_{10} (mg/m^3)	1.01(0.98–1.04)	NA	1.03(0.95–1.11)	1.05(0.98–1.12)	0.97(0.91–1.04)	1.08(1.02–1.15)	0.98(0.91–1.05)
$PM_{2.5}$ (mg/m^3)	1.01(0.99–1.03)	NA	1.10(1.04–1.15)	1.06(1.01–1.10)	1.03(0.99–1.08)	1.08(1.04–1.12)	1.00(0.97–1.05)

NA – not available; CO- carbon monoxide; NO_2 – nitrogen dioxide; SO_2 - sulphur dioxide; O_3 - ozone; PM_{10} - particulate matter <10 microns; $PM_{2.5}$ - particulate matter <2.5 microns; PPM - parts per million; PPB - parts per billion; mg/m^3 - micrograms per meters cubed.

Table 2. Odds ratio for presenting to an emergency department with non-specific abdominal pain associated with an interquartile increase in an air pollutant concentration on the same day of ED visit stratified by gender among individuals aged 15–24 years.

	Edmonton			Montreal		
Pollutant (unit)	**Males OR (95% CI)**	**Females OR (95% CI)**	**p-Value for Interaction**	**Males OR (95% CI)**	**Females OR (95% CI)**	**p-Value for Interaction**
CO (ppm)	0.99(0.95–1.03)	1.05(1.03–1.08)	0.05	1.09(0.96–1.24)	1.10(1.03–1.18)	0.57
NO_2 (ppb)	1.00(0.94–1.06)	1.08(1.04–1.11)	0.07	1.06(0.91–1.22)	1.09(1.01–1.18)	0.10
SO_2 (ppb)	1.02(0.98–1.06)	1.03(1.00–1.05)	0.78	1.16(1.03–1.31)	1.14(1.07–1.23)	0.85
O_3 (ppb)	1.01(0.93–1.09)	0.93(0.89–0.98)	0.18	1.03(0.90–1.20)	1.02(0.93–1.12)	0.10
PM_{10} (mg/m^3)	1.04(1.00–1.09)	1.04(1.02–1.07)	0.66	1.00(0.83–1.21)	1.02(0.93–1.11)	0.26
$PM_{2.5}$ (mg/m^3)	1.06(1.01–1.11)	1.02(0.99–1.05)	0.07	1.09(0.98–1.21)	1.09(1.03–1.15)	0.62

OR – odds ratio; CI – confidence interval; CO- carbon monoxide; NO_2 – nitrogen dioxide; SO_2 - sulphur dioxide; O_3 - ozone; PM_{10} - particulate matter <10 microns; $PM_{2.5}$ - particulate matter <2.5 microns; PPM - parts per million; PPB - parts per billion; mg/m^3 - micrograms per meters cubed.

the associated neuronal circuits involved in triggering abdominal pain.

Several limitations should be considered. First, regional estimates of air pollutants from fixed monitoring stations may not reflect the variability of patient-level exposure to air pollutants. Though, the effects of air pollutants were likely underestimated as measurement errors were presumably non-differential. Second, we attempted to control for confounding through the use of a case crossover study design whereby cases served as their own control; however, residual confounding may have influenced our estimates.

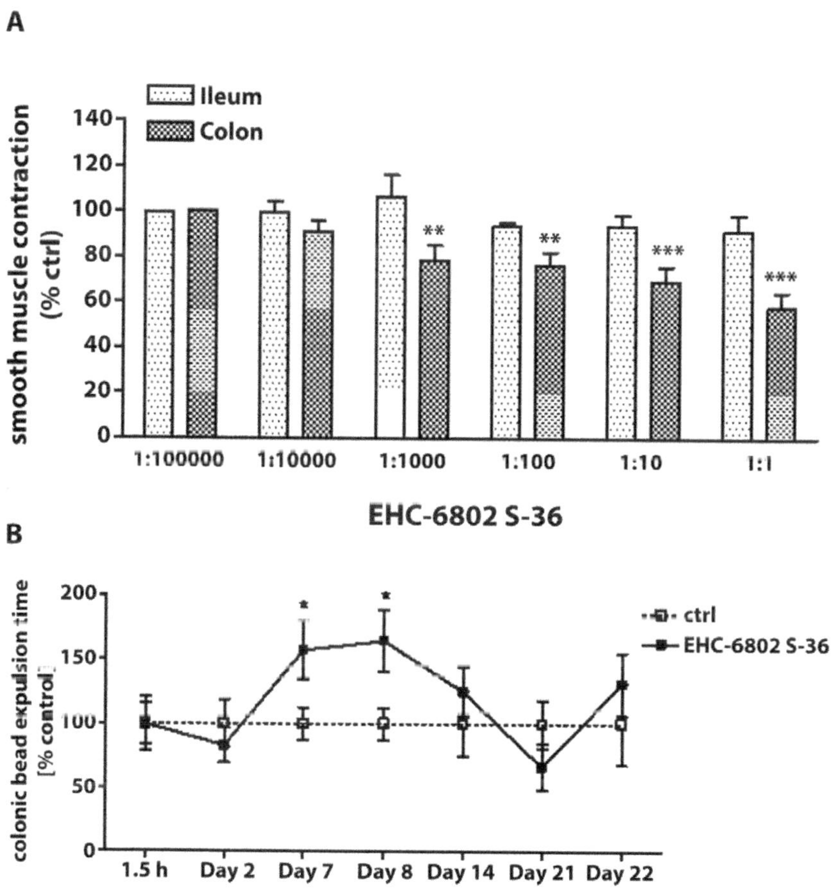

Figure 2. The effect of EHC-6802 on mouse GI motility. A) The effect of EHC-6802 (initial concentration 360 μg/200 μl regarded as 1:1 dilution) on EFS (8 Hz)-stimulated smooth muscle contractions in mouse ileum and colon. Note that EHC-6802 significantly, in a concentration-dependent manner inhibited twitch contractions in mouse colon. Data represent mean ± SEM for n = 8. *P<0.05, as compared with control. B) In vivo effects of EHC-6802 (360 μg/200 μl/animal, QD, p.o.) on colonic bead expulsion time in mice. The results are shown as mean ± SEM of n = 5–10 mice for each experimental group. *P<0.05, as compared with control (animals receiving tap water).

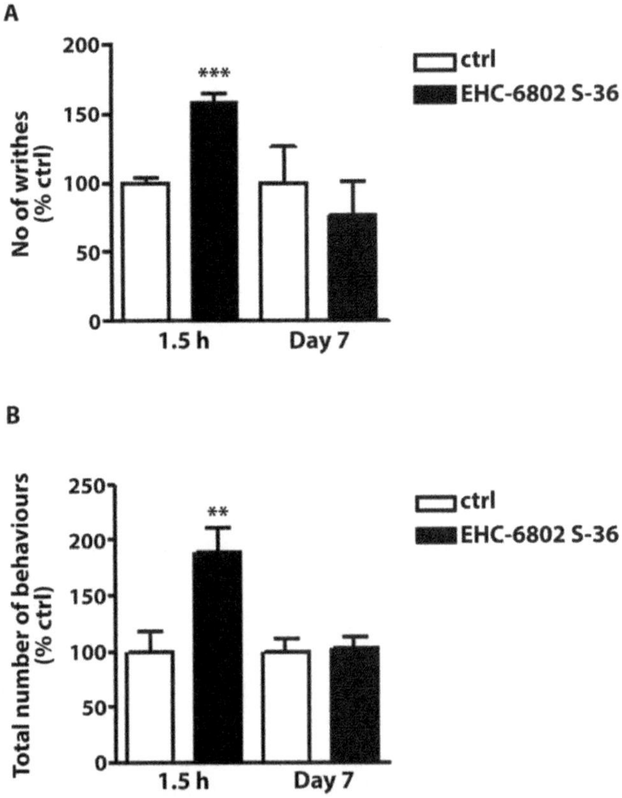

Figure 3. The effect of EHC-6802 (360 μg/200 μl/animal, p.o.) on behavioural pain responses in mice. A) The effect of EHC-6802 in the writhing test in mice. The number of writhes was determined 5 min after acetic acid injection (0.5%, 10 ml/kg, i.p.) over a period of 15 min. Note that a single 1.5 h- treatment with EHC-6802 significantly increased the total number of writhes. Data represent mean ± SEM of 5–10 mice per group. ***$P<0.001$, as compared to control (animals receiving tap water). B) The effect of EHC-6802 on the number of pain-related behaviours (licking of abdomen, stretching, squashing, abdominal contractions) evoked by i.c. administration of oil of mustard (1% in 70% EtOH - 30% saline), determined over a period of 20 min. Note that a single 1.5 h- treatment with EHC-6802 significantly increased the number of pain-related behaviours. Data represent mean ± SEM of 5–10 mice per group. **$P<0.01$, as compared to control (animals receiving tap water). C) The effect of EHC-6802 on the number of pain-related behaviours (licking of abdomen, stretching, squashing, abdominal contractions) evoked by the i.p. administration of neostigmine (2.5 μg/kg), determined over a period of 10 min. Note that a single 1.5 h- treatment with EHC-6802 significantly increased the number of pain-related behaviours. Data represent mean ± SEM of 5–10 mice per group. **$P<0.01$, as compared to control (animals receiving tap water).

For example aeroallergens, which may correlate with air pollution [41], were not evaluated in this study; however, aeroallergens are not known to trigger non-specific abdominal pain. Third, associations were explored in multiple pollutants using several lagged exposures, and stratified to investigate several factors (e.g. age groups). Thus, the probability of observing statistically significant findings by chance is increased. We attempted to minimize multiple comparison errors by including a replication study population and primarily reporting on associations observed in both populations. However, future studies are necessary to confirm our findings. Fourth, some of the non-specific abdominal pain patients may have had an unrecognized organic disease that was diagnosed after discharge from the emergency department. For example, NO_2 exposure (i.e., traffic pollutant) was associated with the development of appendicitis. [15] Furthermore, individuals less than 24 years and living in areas of higher concentration of NO_2 were at a two-fold increased risk of developing Crohn's disease. [16] Because our risk estimates were modest in size, our findings may reflect significant associations among a specific gastrointestinal disorder (e.g., Crohn's disease). Nonetheless, discovering a possible link between air pollution exposure and gastrointestinal disease is an important finding that should motivate further research. Finally, while EHC-6802 particles have been validated to represent daily city exposure of air pollutants [26], our findings should be interpreted cautiously because EHC-6802 particles were gathered in Ottawa and consists of multiple air pollutants.

In conclusion, individuals aged 15 to 24 years were most likely to present to emergency departments with non-specific abdominal pain and this occurred more frequently when the concentration of CO, $PM_{2.5}$, SO2, and NO_2 were elevated in the atmosphere. These findings were observed in a population-based study in Edmonton and replicated in an urban downtown emergency department in Montreal. Additionally, we demonstrated that acute exposure to ingested EHC-6802 particles exacerbated abdominal pain in mice, a response that led to adaptation after chronic exposure. The consistency of the epidemiological and animal studies suggests that air pollution may trigger non-specific abdominal pain in some cases. Thus, future patient-level and basic laboratory studies should be performed to substantiate the findings, further explore possible underlying mechanisms, and identify at risk populations.

Supporting Information

Figure S1 A) In vivo effects of EHC-6802 (360 μg/200 μl/animal, QD, p.o.) or tap water (200 μl/animal, QD, p.o.) during a 21 day treatment course. B) EFS (8 Hz)-stimulated smooth muscle contractions in mouse colon were not changed following 90 min, 7 days or 21 days of treatment with EHC-6802 S-36, as compared to vehicle treated mice. C) Bethanechol (10^{-7} M–3×10^{-5} M) stimulated smooth muscle contractions in mouse ileum (left) and colon (right) were not changed when EHC-6802 (360 μg/200 μl) was added to the organ bath.

Acknowledgments

Dr. Gilaad Kaplan is a Canadian Institutes of Health Research New Investigator and a Population Health Investigator from Alberta-Innovates Health Solutions. Dr. Rowe's research is supported by a Tier I Century Canada Research Chair in Evidence-Based Emergency Medicine from the Canadian Institutes for health Research through the Government of Canada (Ottawa, Ontario).

Author Contributions

Conceived and designed the experiments: M. Szyszkowicz GGK M. Storr BR. Performed the experiments: M. Szyszkowicz GGK JF EP M. Storr. Analyzed the data: M. Szyszkowicz GGK JF EP M. Storr. Contributed reagents/materials/analysis tools: RV. Wrote the paper: M. Szyszkowicz GGK M. Storr KM SG BR RV.

References

1. Sanders DS, Azmy IA, Hurlstone DP (2006) A new insight into non-specific abdominal pain. Ann R Coll Surg Engl 88: 92–94.
2. Nawar EW, Niska RW, Xu J (2007) National Hospital Ambulatory Medical Care Survey: 2005 emergency department summary. Adv Data: 1–32.
3. Jess P, Bjerregaard B, Brynitz S, Holst-Christensen J, Kalaja E, et al. (1982) Prognosis of acute nonspecific abdominal pain. A prospective study. Am J Surg 144: 338–340.
4. O'Byrne JM, Dempsey CB, O'Malley MK, O'Connell FX (1991) Non-specific abdominal pain in pre-menopausal women. Ir J Med Sci 160: 344–346.
5. Lukens TW, Emerman C, Effron D (1993) The natural history and clinical findings in undifferentiated abdominal pain. Ann Emerg Med 22: 690–696.
6. Decadt B, Sussman L, Lewis MP, Secker A, Cohen L, et al. (1999) Randomized clinical trial of early laparoscopy in the management of acute non-specific abdominal pain. Br J Surg 86: 1383–1386.
7. Raheja SK, McDonald PJ, Taylor I (1990) Non-specific abdominal pain–an expensive mystery. J R Soc Med 83: 10–11.
8. Sheridan WG, White AT, Havard T, Crosby DL (1992) Non-specific abdominal pain: the resource implications. Ann R Coll Surg Engl 74: 181–185.
9. Dubinsky MC, Johanson JF, Seidman EG, Ofman JJ (2002) Suspected inflammatory bowel disease–the clinical and economic impact of competing diagnostic strategies. Am J Gastroenterol 97: 2333–2342.
10. Sanders DS, Hopper AD, Azmy IA, Rahman N, Hurlstone DP, et al. (2005) Association of adult celiac disease with surgical abdominal pain: a case-control study in patients referred to secondary care. Ann Surg 242: 201–207.
11. Doshi M, Heaton KW (1994) Irritable bowel syndrome in patients discharged from surgical wards with non-specific abdominal pain. Br J Surg 81: 1216–1218.
12. Brunekreef B, Holgate ST (2002) Air pollution and health. Lancet 360: 1233–1242.
13. Gorham ED, Garland CF, Garland FC (1989) Acid haze air pollution and breast and colon cancer mortality in 20 Canadian cities. Can J Public Health 80: 96–100.
14. Villeneuve PJ, Chen L, Stieb D, Rowe BH (2006) Associations between outdoor air pollution and emergency department visits for stroke in Edmonton, Canada. Eur J Epidemiol 21: 689–700.
15. Kaplan GG, Dixon E, Panaccione R, Fong A, Chen L, et al. (2009) Effect of ambient air pollution on the incidence of appendicitis. Cmaj 181: 591–597.
16. Kaplan GG, Hubbard J, Korzenik J, Sands BE, Panaccione R, et al. (2010) The Inflammatory Bowel Diseases and Ambient Air Pollution: A Novel Association. Am J Gastroenterol.
17. Vincent R, Kumarathasan P, Goegan P, Bjarnason SG, Guenette J, et al. (2001) Inhalation toxicology of urban ambient particulate matter: acute cardiovascular effects in rats. Res Rep Health Eff Inst: 5–54; discussion 55–62.
18. Last JA, Gohil K, Mathrani VC, Kenyon NJ (2005) Systemic responses to inhaled ozone in mice: cachexia and down-regulation of liver xenobiotic metabolizing genes. Toxicol Appl Pharmacol 208: 117–126.
19. Beamish LA, Osornio-Vargas AR, Wine E (2011) Air pollution: An environmental factor contributing to intestinal disease. J Crohns Colitis 5: 279–286.
20. Lippmann M, Yeates DB, Albert RE (1980) Deposition, retention, and clearance of inhaled particles. Br J Ind Med 37: 337–362.
21. Stieb DM, Szyszkowicz M, Rowe BH, Leech JA (2009) Air pollution and emergency department visits for cardiac and respiratory conditions: a multi-city time-series analysis. Environ Health 8: 25.
22. Sajani SZ, Hanninen O, Marchesi S, Lauriola P (2010) Comparison of different exposure settings in a case-crossover study on air pollution and daily mortality: counterintuitive results. J Expo Sci Environ Epidemiol.
23. National Air Pollution Surveillance Program (NAPS) Website. Available: http://www.ec.gc.ca/rnspa-naps/. Accessed 2012 September 30.
24. Zemek R, Szyszkowicz M, Rowe BH (2010) Air pollution and emergency department visits for otitis media: a case-crossover study in Edmonton, Canada. Environ Health Perspect 118: 1631–1636.
25. Maclure M (1991) The case-crossover design: a method for studying transient effects on the risk of acute events. Am J Epidemiol 133: 144–153.
26. Thomson EM, Williams A, Yauk CL, Vincent R (2009) Toxicogenomic analysis of susceptibility to inhaled urban particulate matter in mice with chronic lung inflammation. Part Fibre Toxicol 6: 6.
27. Fichna J, Gach K, Perlikowska R, Cravezic A, Bonnet JJ, et al. (2010) Novel endomorphin analogues with antagonist activity at the mu-opioid receptor in the gastrointestinal tract. Regul Pept 162: 109–114.
28. Gach K, do-Rego JC, Fichna J, Storr M, Delbro D, et al. (2010) Synthesis and biological evaluation of novel peripherally active morphiceptin analogs. Peptides 31: 1617–1624.
29. Laird JM, Martinez-Caro L, Garcia-Nicas E, Cervero F (2001) A new model of visceral pain and referred hyperalgesia in the mouse. Pain 92: 335–342.
30. Eijkelkamp N, Kavelaars A, Elsenbruch S, Schedlowski M, Holtmann G, et al. (2007) Increased visceral sensitivity to capsaicin after DSS-induced colitis in mice: spinal cord c-Fos expression and behavior. Am J Physiol Gastrointest Liver Physiol 293: G749–757.
31. Guthery SL, Hutchings C, Dean JM, Hoff C (2004) National estimates of hospital utilization by children with gastrointestinal disorders: analysis of the 1997 kids' inpatient database. J Pediatr 144: 589–594.
32. Longstreth GF, Thompson WG, Chey WD, Houghton LA, Mearin F, et al. (2006) Functional bowel disorders. Gastroenterology 130: 1480–1491.
33. Heitkemper MM, Jarrett ME (2008) Update on irritable bowel syndrome and gender differences. Nutr Clin Pract 23: 275–283.
34. Ladwig KH, Marten-Mittag B, Formanek B, Dammann G (2000) Gender differences of symptom reporting and medical health care utilization in the German population. Eur J Epidemiol 16: 511–518.
35. Bertakis KD, Azari R, Helms LJ, Callahan EJ, Robbins JA (2000) Gender differences in the utilization of health care services. J Fam Pract 49: 147–152.
36. Roth RP, Tansy MF (1972) Effects of gaseous air pollutants on gastric secretomotor activities in the rat. J Air Pollut Control Assoc 22: 706–709.
37. Kish L, Hotte N, Cheng E, Rioux KP, Kaplan GG, et al. (2011) Orally Ingested Urban Particulate Matter Induces a Pro-Inflammatory Response and Decreases Microflora Diversity. Gastroenterology 140: S-46.
38. Sun Q, Wang A, Jin X, Natanzon A, Duquaine D, et al. (2005) Long-term air pollution exposure and acceleration of atherosclerosis and vascular inflammation in an animal model. Jama 294: 3003–3010.
39. Dybdahl M, Risom L, Moller P, Autrup H, Wallin H, et al. (2003) DNA adduct formation and oxidative stress in colon and liver of Big Blue rats after dietary exposure to diesel particles. Carcinogenesis 24: 1759–1766.
40. Tornqvist H, Mills NL, Gonzalez M, Miller MR, Robinson SD, et al. (2007) Persistent endothelial dysfunction in humans after diesel exhaust inhalation. Am J Respir Crit Care Med 176: 395–400.
41. Bartra J, Mullol J, del Cuvillo A, Davila I, Ferrer M, et al. (2007) Air pollution and allergens. J Investig Allergol Clin Immunol 17 Suppl 2: 3–8.

5

Role of Glutathione S Transferase Polymorphism in COPD with Special Reference to People Living in the Vicinity of the Open Cast Coal Mine of Assam

Tapan Dey[1], Kabita Gogoi[1], Bala Gopalan Unni[1*], Munmi Kalita[1], Moonmee Bharadwaz[1], Minakshi Bhattacharjee[2], Pranab Kumar Boruah[1], Thaneswar Bora[1], Dibyajyoti Ozah[1], Manoj Kalita[1]

1 Biotechnology Division, Council of Scientific & Industrial Research (CSIR) - North-East Institute of Science and Technology, Jorhat, Assam, India, **2** Biotechnology Department, University of Science and Technology Meghalaya (USTM), 9th Mile, Meghalaya, India

Abstract

Background: COPD may develop due to variation in the functioning of antioxidants along with smoking and environmental factors in genetically susceptible individuals. Since there are different views about the antioxidants responsible for detoxifying xenobiotic compound in the human body whose functional variation may lead to obstructive disease, this associative study has been taken up between GST gene polymorphism and COPD in populations exposed to coal dusts.

Methods: Genotypes of the 70 COPD patients and 85 non COPD patients were determined by PCR based methods followed by multiplex PCR of GSTT1 and GSTM1 genes taking albumin gene as a control. Suspended particulate analyses were determined through the Respirable Dust sampler along with the FTIR analysis of the dust samples from the glass microfiber filters.

Results: Dust sampling analysis reveals higher level of respirable suspended particulate matter, non respirable particulate matter, SO_2 and NO_2 present in air of the study site. FTIR analysis also suggests a higher concentration of organic silicone and aliphatic C-F compounds present in air of the study site and when spirometry was done, low lung function was observed among most of the subjects. GSTM1 null type was significantly associated with low lung function in smoker groups and the presence of at least one active allele (either GSTM1/GSTT1) seemed to have a protective role in the development of COPD.

Conclusions: GSTM1 (null genotype) appeared to be a risk factor for lower lung function in smokers living in the vicinity of coal mines. Apart from polluted environment and genetic susceptibility, mixed coal dust exposure rich in organic silicone and aliphatic C-F compounds also appears to be a factor for the low lung function.

Editor: Reiner Albert Veitia, Institut Jacques Monod, France

Funding: The project was entirely funded by Council of Scientific and Industrial Research, Govt. of India. The funders had no role in study design, data collection and analysis, decision to publish, or preparation of the manuscript

Competing Interests: The authors have declared that no competing interests exist.

* E-mail: bgunni@yahoo.com

Introduction

Globally, Chronic Obstructive Pulmonary Disease (COPD) is one of the leading causes of mortality and by 2020 it is expected to rise to the third position as a cause of death and in fifth position as the cause of disability adjusted life years (DALYs) as per projections made in the Global Burden of Disease study (GBDS) [1].

Toxic particles and gases that are present in the atmosphere are likely to be inhaled or often self-administered through cigarette smoke, causing lung injury. However, contamination of atmosphere from anthropogenic sources such as coal mining, industrial sources as well as local conditions generated either in the home or workplace makes a significant contribution to the development of COPD. The relative prevalence and severity of mining related occupational lung diseases are a function of the commodities mined, airborne hazard exposure levels, and co-existing illnesses or environmental conditions and lifestyle. Chronic Obstructive Pulmonary Disease (COPD) is thought to be the result of environmental triggered in genetically susceptible individuals. Alpha 1 Antitrypsin is the only known genetic cause of COPD. Bhattacharjee et al. earlier studied the polymorphism of α 1-antitrypsin gene in the population of the same area where we have taken up the study [2–3]. COPD is the consequence of an abnormal inflammatory response due to inhalation of noxious agents such as cigarette smoking, occupational or environmental exposure. In fact only a portion (10–20%) of heavy smokers develops a clinically detectable disease [4–5]. Antioxidants and other less well understood protective mechanism may also be important in preserving normal lung function in the face of a lifetime exposure to potentially injurious environmental factors. Oxidative injury may also play an important role in the pathogenesis of COPD, [6]. Such injury, resulting from an imbalance between free radicals and protective mechanisms, can alter the conformation of protease inhibitors and reparative enzymes, injure cell membranes, and may result in mutagenesis. Free radicals appear in the lungs through inhalation from the environment or by its release from inflammatory cells inside the body. Genetically controlled antioxidant defence systems may also play an important role in determining susceptibility, both to free radicals released by inflammatory cells and to oxidants inhaled from the environment. The lung possesses several enzymatic scavengers including glutathione which are under genetic control. The observation that the enzymatic antioxidants are under genetic control and the allelic variations of these antioxidants alter their abilities to reduce free radicals, [7–8] suggests that genetic factors may place some individuals at greater risk for oxidant injury. The glutathione system is the major antioxidant mechanism in the airways. The increased oxidative stress in the airways of COPD patients may play an important pathophysiological role in the disease development by amplifying the inflammatory response in COPD.

COPD is characterized by progressive development of airflow limitation that is not fully reversible. It encompasses chronic obstructive bronchitis, with obstruction of small airways, and emphysema with enlargement of air spaces and destruction of lung parenchyma, loss of lung elasticity and closure of small airways [9].

The Glutathione S Transferase's (GST's), a super family of enzymes consisting of alpha, mu, pi, theta, kappa, zeta, sigma, omega and delta families, are critical in the conversion of many reactive electrophilic compounds to less reactive metabolites which are excreted as glutathione conjugates. Several common variants of GST's have been well characterized and are associated with certain respiratory diseases, [10]. These genes are expressed in the respiratory tract and have common functional variant alleles that result in either a total absence or a substantial change in enzyme activity. Thus the study has taken up to evaluate the role of polymorphism of the detoxifying enzyme GST in the pathogenesis of COPD with respect to polluted environment where the population is staying.

Methods

The study was conducted at CSIR-North East Institute of Science and Technology, Jorhat, Assam after ethical clearance from the Institutional Ethics Committee, NEIST, Jorhat.

Air Analysis

Air sampling was done with the help of Respirable Dust Sampler (Envirotech Model APM 460 DXNL) and data's were calculated for Respirable suspended particulate matter (RSPM) and Suspended Particulate Matter (SPM). Determination of SO_2 concentration present in ambient air was done with the help of modified West and Gaeke Method and NO_2 concentration was determined by Sodium Arsenite Method (as per manufacturer's instructions by the Central Pollution Control Board, India). Air analysis was carried out during the period January 2012 to January 2013.

FTIR analysis of the dust samples

Organic extracts of dust samples of air were collected by the glass microfiber filters of size 1.6 µm. These extracts were then analysed through the FTIR system 2000 (Perkin Elmer, USA, ANC-1). The dust samples taken from the non-coal mine site i.e. around the research institute served as a control.

Survey at the study site

A survey was conducted in the vicinity of the open-cast coal mine area of Assam. Questions regarding smoking habits, age, gender, socioeconomic status, occupation, environment etc. were included in the questionnaire with special concern on the epidemiology of GST gene.

Among 575 subjects were surveyed from the study site and the overall disease symptoms reported were fever, frequent cough and cold, weakness, and high blood pressure. Among the 575 subjects surveyed only 155 subjects visited the nearest mini PHC at Ledo during the period January 2012 to January 2013. Of these, about 45.16% (70) were suffering from COPD related respiratory symptoms. Baseline characteristics of all the study subjects were shown in Table 1.

Definition of the subjects

The population consisted of 155 subjects (116 males and 39 females) living in the vicinity of open –cast coal mine area at Ledo, in Assam (Lat. 27°13′–27°23′N and Long. 95°35′–96′00E).

Diagnosis and inclusion criteria of subjects

COPD was diagnosed on the basis of medical history of exposure to coal dust, physical examination and spirometry data according to American Thoracic Society (ATS) guidelines. Inclusion criteria for COPD included the following symptoms: chronic airway symptoms and signs such as coughing, breathlessness, wheezing and chronic airway obstruction defined as the FEV_1/FVC ratio (Forced Expiratory Volume in one Second/ Forced Vital Capacity) of 70% and below and an FVC (Forced vital capacity) of 80% predicted value. Based on survey data and lung function analysis, the subjects were sorted out into the following categories: (a) Smokers with COPD (n = 38, female: male ratio = 0.31), (b) Non-smokers with COPD (n = 32, female: male ratio = 0. 52), (c) Healthy Non-smokers (n = 35, female: male ratio = 0. 43), (d) Smokers without COPD (n = 50, female: male ratio = 0.33).

Lung function analysis

Spirometry was conducted with Spirometer (Schiller) for each subject and the values of FVC (% predicted) and FEV1/FVC (% predicted) were recorded.

Blood Sample Collection

Blood samples were collected through health camps conducted in that area through Doctors of local PHC in collaboration with Doctors of Clinical Centre, NEIST, Jorhat. Each subject participated in the survey were examined by the doctors and recommended for blood sample collection after conducting spirometry. Blood samples were obtained only after getting subject's written consent.

DNA extraction and PCR amplification

Genomic DNA was extracted from whole blood using a GeNei^{TM} Whole Blood DNA extraction Kit, Bangalore Genei, India. Genomic DNA was amplified by PCR using a thermal cycler (Thermal Cycler Model 2720, Applied Biosystems, USA) in a 25 µl reaction mixture containing 1.5 mM MgCl2, 25 pM of each primer, 500 µM dNTPs, 0.1 U Taq Polymerase (MBI Fermentas, USA). The PCR conditions consisted of an initial cycle of 10 minutes at 95°C, 1 minute at 34°C and 10 seconds at 72°C. Specific GSTM1 and GSTT1 primers (Sigma) for the PCR based genotyping assays were synthesized as per the sequence given by Josef zidzik et.al. (2008)[11]. The products of multiplex PCR

Table 1. Baseline and clinical characteristics of the study subjects.

Variables Study population	**Smokers with COPD (n = 38)**	**Non-smokers with COPD (n = 32)**	**Healthy non Smokers (n = 35)**	**Smokers without COPD (n = 50)**
Male	29	21	24	42
Females	9	11	11	8
Sex ratio (Female/Male)	0.31	0.52	0.43	0.33
Mean age (Standard)				
Male	47±13.8	44.95±16.09	45±15.08	52±13.54
Females	57.7±15.94	37.80±11.83	38±11.83	60.62±7.42
FVC (% Predicted)				
Male	54±7.9	52±6.58	94±15.74	106 ±12.35
Female	52.6±6.20	50±6.97	113.9±25.20	91.5±8.34
FEV_1/FVC (% Predicted)				
Male	58±11.3	52.76±9.25	94±14.74	99±9.87
Female	53.5±4.92	51.45±10.68	86.9±20.94	98.37±8.61
Socio Economic Status	Poor	Poor	Poor	Poor

(GSTM1-215 bp; GSTT1-480 bp; Alb-360 bp) were separated using 3% Agarose gel (Biotechnology Grade, AMRESCO). For both GST genes, subjects were categorized as having null-type, non- null type, either null-type, both null-type, both non-null types based on the PCR amplification results.

Statistical Analysis

The frequency of alleles and genotypes prevalence and their association between COPD and Control subjects were assessed by Chi Square and Fishers Exact test. Odds Ratio (OR) and 95% Confidence Interval (CI) along with p value were calculated to describe the association. The associations for Smoker and Non Smoker between COPD and Control were analysed by binary logistic regression analysis to find out the potential confounders. A P value <0.05 was considered as statistically significant. SPSS v9 and GraphPad – Prism 5 software's were used for all the statistical analysis.

Results

Air sample analysis was done for three different periods of the year- December - March, April - July and August - November. The values for Respirable suspended particulate matter (RSPM), Suspended particulate matter (SPM), SO_2 and NO_2 were found to be significantly higher in the study site [Figure 1]. Air samples collected from 200 kms distant area (around the institute) served as control. The highest values for RSPM, SPM, SO_2 and NO_2 were observed during the period December - March, followed by April - July and August –November. The values of all the parameters were found to be higher than the standard value. RSPM (29, 10, 5), SPM (41, 41, 2), SO_2 (Trace) and NO_2 (14, 40, 2) for control samples were within the standard limits.

The organic extracts of the coal dust samples were analysed with the help of instrument software version 4.07. The IR spectra of the dust sample (A) of coal mine site shows peaks at 1023.5 cm^{-1} and 1190.1 cm^{-1} corresponding to allopathic flora

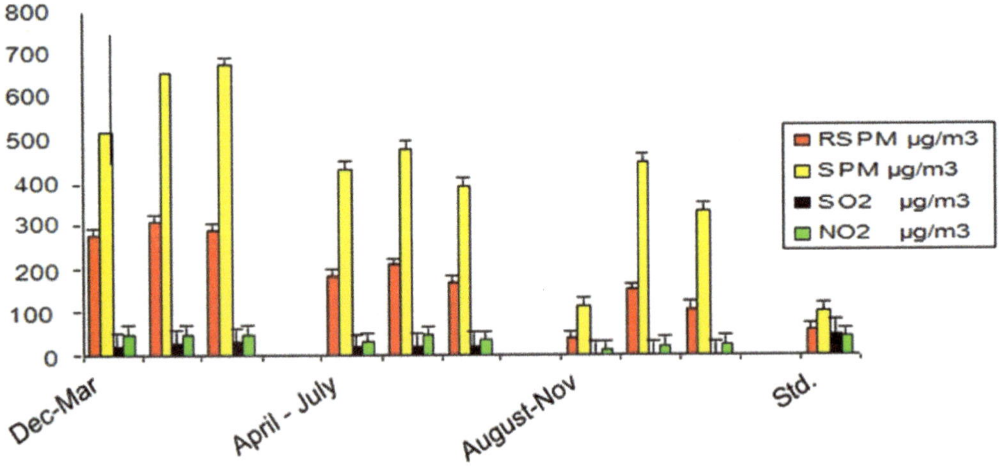

Figure 1. Comparative account of the concentration of respirable suspended particulate matter (RSPM), suspended particulate matter (SPM), SO_2 and NO_2 present in air of the coal mine site observed at different periods of the year (2012–2013) along with the standards [Variables are seasonal averages].

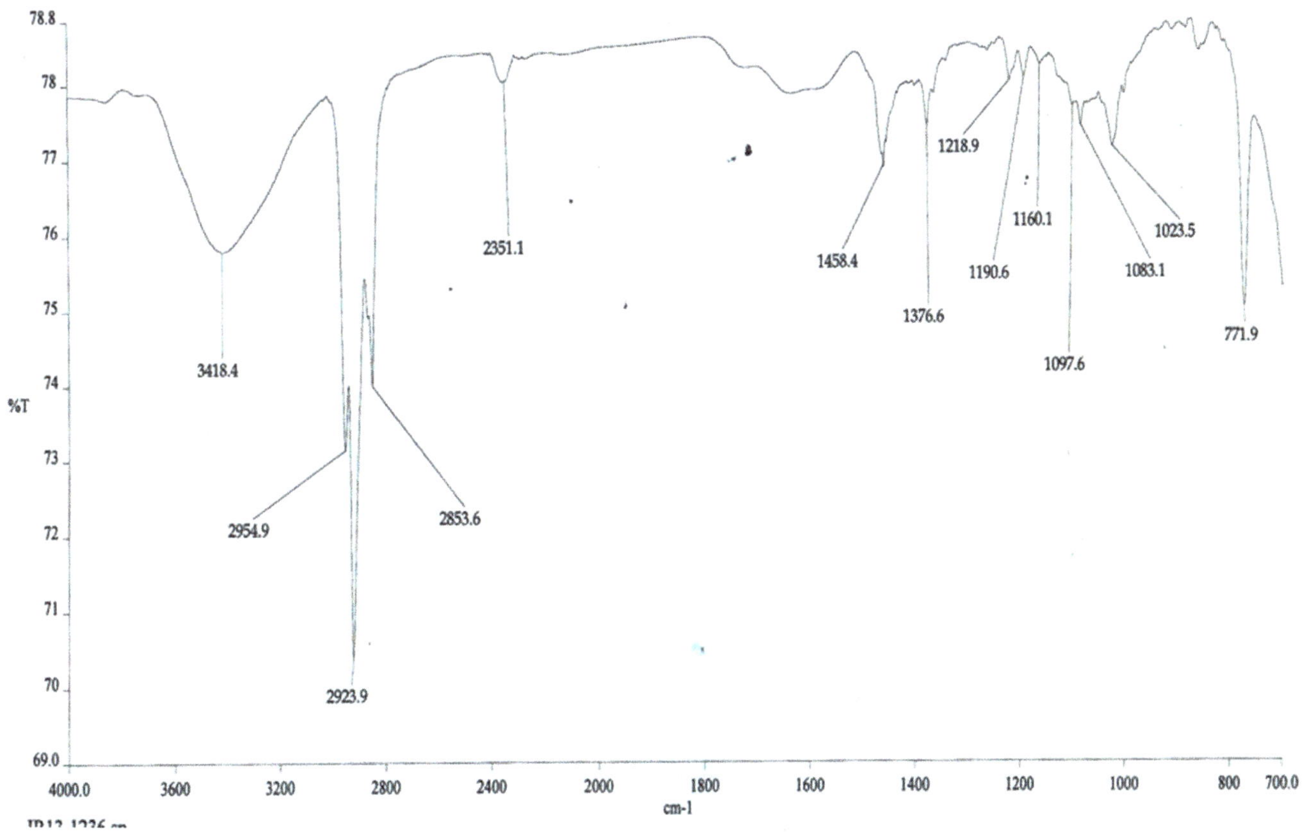

Figure 2. The overall FT-IR spectra of environmental dust particles present in air of coal mine site (A) collected with the help of Envirotech APM 460 DXNL respirable dust sampler, measured by Perkin Elmer 2000 FTIR spectrometer. The overall spectra from 1023.5 cm^{-1} to 1458 cm^{-1} illustrates the concentration of major air pollutants.

C-F stretch, 1083.1 cm^{-1} corresponding to organic silicone or silicone/aliphatic flora C-F stretch, 1217.4 cm^{-1} corresponding to aromatic C-H stretch in plane bend, whereas 1378.1 cm^{-1} and 1458.8 cm^{-1} corresponds to methyl C-H asymmetric/symmetric bending. On comparing the spectrum of dust samples of both the sites, it was observed that the sample (A) [Figure 2] shows very sharp peaks as compared to the spectrum of non - coal mine site dust sample (B) [Figure 3] which shows less intense broad peaks at 1021.2 cm^{-1}, 1082.2 cm^{-1},1217.0 cm^{-1},1377.3 cm^{-1}, 1400.0 cm^{-1}, 1461.0 cm^{-1}. The sharp peak at 1083.1 cm−1 clearly shows the higher organic silicone content in the coal mine site than the non -coal mine sites spectral peak at 1082.2 cm−1. Interpretations of the FTIR spectra were done as per John Coates [12].

For the assessment of lung function of each subject as per the spirometry results the Forced Vital Capacity (FVC) was calculated as % predicted. The smokers with COPD group and Non-smokers with COPD group showed an obstructive pattern of lung function. For smokers with COPD, FVC values were found to be 54±7.9 among males and 52.6±6.20 among females. For Non-smokers with COPD, among males the values found to be 52±6.51 and 50±6.97 among females. In case of healthy non-smokers and smokers without COPD, no lung function decline was observed. For healthy non-smokers, FVC % predicted values were 94±15.74 among males and 113.9±25.2 among females. The smokers without COPD the values were 106±12.35 among males and 91.5±8.34 among females. To find the ratio of (Forced expiratory volume in 1 second/Forced vital capacity) FEV_1/FVC, the values were calculated as % predicted. For male smokers with COPD and non-smokers with COPD, the values were observed as 58±11.3 and 52.76±9.25 respectively. For females, the values were found to be 53.5±4.92 and 51.45±10.68 respectively. Among the healthy non-smokers and smokers without COPD the values were found to be 94±15.74 and 106±12.35 in males, 113.9±25.20 and 91.5±8.34 in females respectively. So it can be said that smoking cessation remains the single most effective intervention to reduce the risk of COPD or to slow its progression.

The multiplex PCR products of both the genes were of expected size [Figure 4]. Among the studied individuals, the prevalence of the GSTM1 null genotype in COPD individuals (51.4%) was observed to be almost two times higher than that found in the Control (27.1%) [Table 2]. The study showed there was a significant link between GSTM1 null genotype and COPD with $p<0.05$ (OR = 2. 9, 95% CI = 1.5 5.6). However the prevalence of GSTT1 null genotype was not found statistically significant among COPD subjects (51.4%) and Controls (48.6%). When the proportion either null genotype of the two genes were considered we found two times higher ratio in COPD subjects compared to control. Statistical analysis by Chi square test and logistic regression analysis also reveals the same conclusion.

Discussion

The respirable mixed toxic dust particles generally inhaled may either be due to occupational exposure to polluted environment or self-administered as cigarette smoking. This form of inhalation injury causes a low grade inflammatory exudation of fluids into

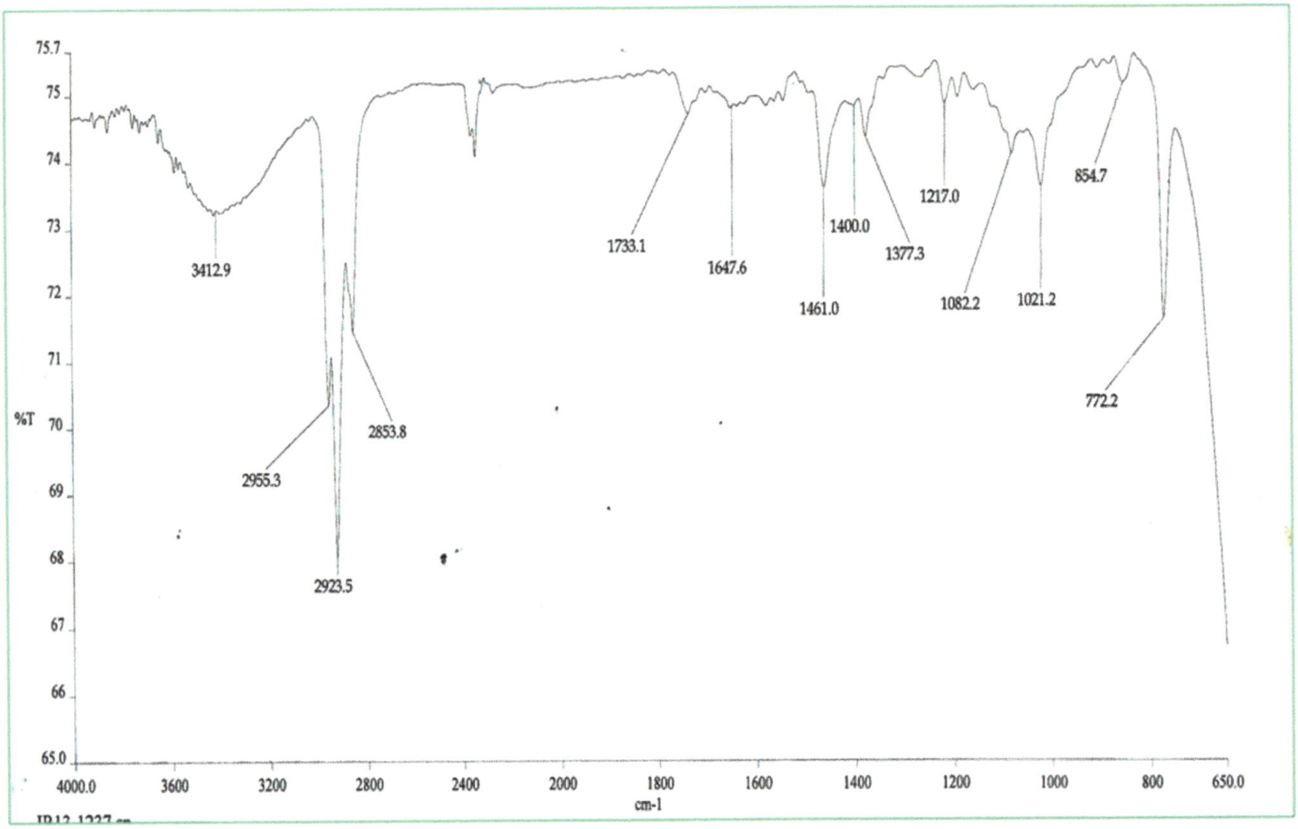

Figure 3. The overall FT-IR spectra of environmental dust particles present in air of the control site (B) collected with the help of Envirotech APM 460 DXNL respirable dust sampler, measured by Perkin Elmer 2000 FTIR spectrometer. The overall spectra from 1021.2 cm^{-1} to 1461.0 cm^{-1} illustrates the major air pollutants.

both large and small bronchi, bronchioles with minimal effects on lung function but in susceptible individuals, this normal inflammatory response is amplified. In our study, we have attempted to assess the effect of environmental determinants (smoking and air pollution) on pulmonary function related to some genetic determinants (GSTM1 and GSTT1).

The air analysis (done in three different periods of the year) showed highest RSPM, SPM, SO_2 and NO_2 during December - March and lowest during August - November. The overall annual data reveal that our study site is considerably polluted and the population in the area is exposed to recurrent episodes of acute air pollution. Among the surveyed population, 155 subjects attended the health camp. Spirometry was done for all the subjects to analyse the lung function in smokers with COPD and non-smokers with COPD. Healthy non-smokers and smokers without COPD showed normal lung functioning. The low lung function reveals an association of air pollution with smoking history. Other than smoking habit and genetic susceptibility our study population were

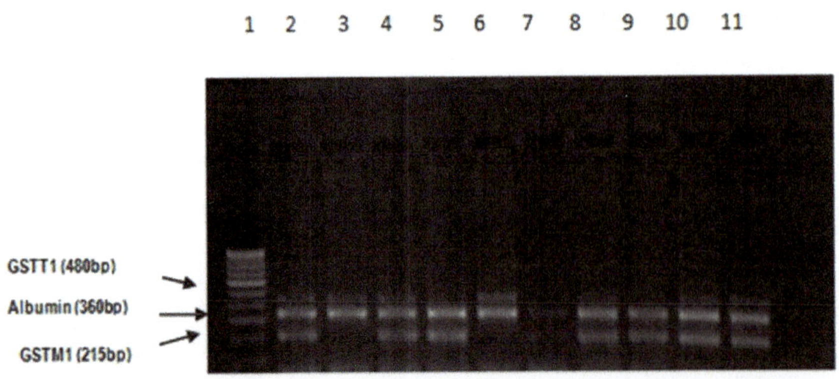

Figure 4. Polymorphism of GSTM1 and GSTT1 genes. Lane 1 is the molecular marker of 100 bps. Lane 2, 4,8,10 and 11 represents wild type for both GSTM1 and GSTT1 genes. Lane 3 represents a wild type of GSTM1 and a null type of GSTT1 gene. Lane 5 and 9 represents a wild type of GSTT1 gene and a null type of GSTM1 gene. Albumin gene is taken as a control.

Table 2. Frequency distribution chart of GSTT1 gene and GSTM1 gene polymorphism with respect to the risk of developing COPD (** - Statistically significant P<0.05).

Genotype	Controls (n = 85)	COPD (n = 70)	P value	OR (95% CI)
GSTM1				
NON NULL	62 (72.9%)	34 (48.6%)		1.0
NULL	23 (27.1%)	36 (51.4%)	.002**	2.9 (1.5–5.6)
GSTT1				
NON NULL	50 (58.8%)	34 (48.6%)		
NULL	35 (41.2%)	36 (51.4%)	0.203	1.5 (0.8–2.9)
GSTM1+GSTT1				
BOTH NON NULL	37 (43.5%)	12 (17.1%)		1.0
BOTH NULL	15 (17.6%)	12 (17.1%)	0.077	2.5 (1.0–6.7)
EITHER NULL	33 (38.8%)	46 (65.7%)	000**	4.3 (2.0–9.5)

also under significant exposure to respirable mixed coal dust. As suggested by Love et al. [13] our FTIR studies based on the principle of sharp peak in the spectra corresponds to higher concentration, also showed a higher concentration of organic silicone or silica dust and chloro flouro compounds (1083.1 cm^{-1}) present in air of coal mine site as compared to non-coal mine site (1082.2 cm^{-1}). It is generally known that the open cast coal mine areas are one of the most polluted areas, but to assess the compounds which are present in higher concentration in air surrounding the coal mine sites, we have studied the air samples through infrared spectroscopy. CFC as well as the organic silicone was found to cause many respiratory diseases including silicosis. Since we employed 1.6 μm filter papers for dust sampling, so certainly the size of these dust particles may be ≤1.6 μm. These tiny particles may get embedded in the alveolar sacs triggering the immune response to these xenobiotic compounds inside the lungs ultimately leading to cellular lesions. So the organic silicone may also play a role in the development of COPD.

Cigarette smoking is well known as the major risk factor for the development of COPD in the genetically susceptible subjects [14–16]. To confirm our hypothesis, the subjects were genotyped for GSTM1 and GSTT1gene. Enzymatic antioxidants are under

Table 3. Associative studies relating to smoking and non-smoking habit to the polymorphism of GSTM1 gene and GSTT1 gene by χ^2 test (** - Statistically significant P<0.05).

Habit	Genotype	Controls (n = 85)	COPD (n = 70)	P value	OR (95% CI)
NON SMOKER	GSTM1				
	NON NULL	24 (68.6%)	18 (56.3%)		1.0
	NULL	11 (31.4%)	14 (43.0%)	0.298	1.7 (0.6–4.6)
SMOKER	GSTM1				
	NON NULL	38 (76.0%)	16 (42.1%)		1.0
	NULL	12 (24.0%)	22 (57.9%)	0.001**	4.4 (1.7–10.9)
NON SMOKER	GSTT1				
	NON NULL	24 (28.6%)	14 (43.8%)		1.0
	NULL	11 (31.4%)	18 (56.3%)	0.041**	2.8 (1.0–7.6)
SMOKER	GSTT1				
	NON NULL	26 (52.0%)	20 (52.6%)		
	NULL	24 (48.0%)	18 (47.4%)	0.953	1.0 (0.4–2.3)
NON SMOKER	GSTM1+GSTT1				
	NON NULL	17 (48.6%)	4 (12.5%)		1.0
	EITHER NULL	12 (34.3%)	24 (75.0%)	000**	8.5 (2.3–30.9)
	BOTH NULL	6 (17.1%)	4 (12.5%)	0.381	2.8 (0.5–15.1)
SMOKER	GSTM1+GSTT1				
	NON NULL	20 (40.0%)	8 (21.1%)		1.0
	EITHER NULL	21 (42.0%)	22 (57.9%)	0.060	2.6 (1.0–7.2)
	BOTH NULL	9 (18.0%)	8 (21.1%)	0.209	2.2 (0.6–7.8)

Table 4. Associative studies relating to smoking and non-smoking habit with the polymorphism of GSTM1 and GSTT1 genes by binary logistic regression analysis (Variables taken were habit status and Null/Non Null genotype of both the genes) (** Statistically significant P<0.05).

Genotype	Controls (n = 85)	COPD (n = 70)	P value	OR (95% CI)
GSTM1				
NON SMOKER + NON NULL	24 (28.2%)	18 (25.7%)		1.0
SMOKER + NON NULL	38 (44.7%)	16 (22.9%)	0.181	0.6 (0.2–1.3)
NON SMOKER + NULL	11 (12.9%)	14 (20.0%)	0.299	1.7 (0.6–4.6)
SMOKER + NULL	12 (14.1%)	22 (31.4%)	0.060	2.4 (1.0–6.2)
GSTT1				
NON SMOKER + NON NULL	24 (28.2%)	14 (20.0%)		1.0
SMOKER + NON NULL	26 (30.6%)	20 (28.6%)	0.538	1.3 (0.5–3.2)
NON SMOKER + NULL	11 (12.9%)	18 (25.7%)	0.043**	2.8 (1.0–7.6)
SMOKER + NULL	24 (28.2%)	18 (25.7%)	0.584	1.3 (0.5–3.2)
GSTM1+GSTT1				
NON SMOKER+BOTH NON NULL	17 (20.0%)	4 (5.7%)		1.0
NON SMOKER+ EITHER NULL	12 (14.1%)	24 (34.3%)	.001**	8.5 (2.3–31.0)
NON SMOKER+ BOTH NULL	6 (7.1%)	4 (5.7%)	.221	2.8 (0.5–15.0)
SMOKER+ BOTH NON NULL	20 (23.5%)	8 (11.4%)	.446	1.7 (0.4–6.6)
SMOKER + EITHER NULL	21 (24.7%)	22 (31.4%)	.018**	4.5 (1.3–15.4)
SMOKER + BOTH NULL	9 (10.6%)	8 (11.4%)	.072	3.8 (0.9–16.1)

genetic control and allelic variation alters their abilities to reduce free radicals. The GSTs are important components of lung defence in response to oxidative stress and are highly polymorphic [9]. Studies in the role of polymorphism of genes regulating the GST enzyme including GSTT1, GSTM1 in reference to COPD has been conducted with controversial results in various ethnic groups and populations [16–18].

Several longitudinal, epidemiological and associative studies have established that acute episodes of atmospheric pollution were associated with a number of health complications [19–23]. A cross-sectional study in Merseyside conducted by Brabin et al. [24] in school children aged 5–11 years exposed to coal dust and air pollution established significant [Odds Ratio 1.55 (1.17–2.06) at 95% CI] increased prevalence of lung function decline irrespective of their parents being smokers or non-smokers. Leigh et al. [25] studied the relation between the emphysema and the lung coal content and found the extent of emphysema had a strong positive quantitative relation with coal content of the lungs ($p<0.0003$), age ($p<0.0001$), and smoking ($p<0.0001$). A Swiss study looked at the respiratory health of primary school children and noted that respiratory symptoms were highest in the area with the highest level of suspended particulates in the air and that they were independent of levels of nitrogen and sulphur oxides [26]. The coking works study from the north of England and the Munich/ Leipzig study from Germany also found similar results [27]. Such form of inhalational injury causes a low grade inflammatory exudation of fluid and cells into both large and small bronchi, bronchioles with minimal effects on lung function but in susceptible individuals, this normal inflammatory response is amplified. Studies related to the role of gene polymorphism regulating the GST enzymes including GSTT1, GSTM1 in reference to COPD has been conducted with controversial results in various ethnic groups and populations [28–30] in non-coal mine areas.

We made an attempt to evaluate the association of GSTM1, GSTT1 and for the combination of GSTM1 & GSTT1 genotypes among the individuals of Non Smokers and Smokers irrespective of age and sex. The prevalence of GSTM1 null genotype with smoking habit was found higher i.e. 57.9% of COPD individuals compared to 31.4% in Control and was found statistically significant with a fourfold risk among the individuals with a combination of GSTM1 null genotype and smoking habit ($p<0.\ 05$, OR $=4.\ 4$, CI $=1.7–10.9$). Whereas in GSTT1 genotype subjects with a combination of null genotype and no habit of smoking (Non Smoker) observed statistical significant ($p<0.\ 05$, OR $=2.8$, CI $=1.0–7.6$). It was also observed that subjects with GSTT1 null genotype and Non Smoker was with a frequency of 56.3% in COPD against 31.4% in Control. In the combination of GSTM1 & GSTT1 genotypes among the individuals of Non Smoker and Smoker it was found that the individuals with a combination with either null gene and Non-smoker was in eighth fold risk ($p<0.\ 05$, OR $=8.5$, CI $=2.3–30.9$).

From the regression analysis between the smoking status and polymorphism of both the genes revealed that the smokers with GSTM1 null genotype, non-smokers with either null genotype of the two genes found to be more susceptible to COPD [Table 3 & Table 4]. It is observed that Non Smoker with null genotype of GSTT1 genotype have a threefold risk compared to Non-smoker with Non Null GSTT1 genotype ($p<0.05$, OR $=2.8$ CI $=1.0–7.6$), whereas in case of GSTM1 and GSTT1 genotype both together, it was found that Smoker with either Null genotype have a four time risk ($p<0.\ 05$ OR $=4.\ 5$ CI $=1.3–15.4$).

From the study it is revealed that among the studied subjects prevalence of the GSTM1 null genotype and for combined GSTM1 & GSTT1 either null genotype in COPD individuals was higher than that found in Control. The study showed there was a significant link between GSTM1 null genotype and for combined GSTM1 & GSTT1 either null genotype as compared with COPD.

It is important to note that the study population were exposed to the respirable mixed coal dust. The analysis of the parameters was done during the period January 2012 to January 2013 and no seasonal variation was reported on the lung function of the subjects. There was no data available on the status of GST gene with reference to COPD subjects staying near the open-cast coal mine at Ledo, Assam. Further, the ethnicity (migrant labour) of the people of this region is very different from rest of the country. Barnes et al. [31] mentioned that differences in the prevalence of COPD in different ethnic groups are likely to be accounted by the different frequencies of genes relevant to pathogenesis, so that exploration of these differences at different molecular level may be informative. Although Yim et al. [32], He et al. [33], Cahkoglu et al. [34], Lakhdar et al. [35], have extensively studied the association of GSTM1 null genotype with the severity of COPD, but they have not found any correlation between the null genotype and COPD. In contrast, in the present study we have evaluated that the smokers with GSTM1 (null genotype) were more susceptible to COPD with additional contribution of respirable coal dust from the coal mine. However, Baranova et al. also suggested the homozygous deletion of GSTM1gene is associated with complete loss of enzyme production and is associated with severe chronic bronchitis in heavy smokers [32].

The limitations of the study were lack of age and sex matching between all the groups. We evaluated the effect of smoking history and GSTM1 (null) genotype in the pathogenesis of COPD.

Thus we can conclude that the GSTM1 (null genotype) appears to be a risk factor for the low lung function in smokers living in the vicinity of open cast coal mine area and the presence of at least one active allele (either GSTM1/GSTT1) seemed to have a protective role in the development of COPD. Further studies with the association of HMOX1 and MMP12 gene, will be necessary to elucidate the pathogenesis of COPD as a genetic disease in compliance with environment with larger sample size.

Acknowledgments

The authors are thankful to the Director, CSIR- North-East Institute of Science & Technology, Jorhat for his permission to carry out this work and CSIR, Govt. of India, New Delhi for financing the network project (BSC-0116).

Author Contributions

Conceived and designed the experiments: TD BGU. Performed the experiments: TD M. Bhattacharjee BGU KG DO. Analyzed the data: TD Manoj Kalita BGU PB TB DO. Contributed reagents/materials/analysis tools: BGU PB TB DO. Wrote the paper: TD BGU PB TB. Survey/health camp: M. Bharadwaz Munmi Kalita PB TB DO.

References

1. Jindal SK (2006) Emergence of Chronic Obstructive Pulmonary Disease as an epidemic in India. Indian J Med Res 124: 629–630
2. Bhattacharjee M, Unni BG, Das S, Baruah PK, Sharma P, et al. (2012) Alpha 1 antitrypsin gene: A case – control study in chronic obstructive pulmonary disease. Afr J Biotechnol 111: 207–215.
3. Bhattacharjee M, Unni BG, Das S, Deka M, Rao PG, (2011) Lung function decline: Screening of alpha -1 antitrypsin gene in a population exposed to coal dust. Int J Biotechnol Mol Biol Res 211:195–201.
4. Hogg JC (2004) Pathophysiology of airflow limitation in chronic obstructive pulmonary disease. Lancet 364:709–721.
5. Piqueras CMG, Cosio MG, (2001) Disease of the airways in Chronic Obstructive Pulmonary Disease. Eur Respir J 18 suppl 34: 41–49.
6. Repine JE, Bast A, Lankhorst I, (1997) Oxidative stress in Chronic Obstructive Pulmonary Disease. Am J Respir Crit Care Med 156: 341–357.
7. Koyama H, Geddea DM, (1998) Genes, Oxidative stress and the risk of Chronic Obstructive Pulmonary Disease Review. Thorax suppl 2:10–14.
8. Smith CA, Harrison DJ, (1997) Association between polymorphism and gene for microsomal epoxide hydrolase and susceptibility to emphysema Comments. Lancet 350:630–633.
9. Barnes PJ (2004) Small airways in COPD. N Engl J Med 350: 2635–2637.
10. Hayes JD, Strange RC, (2000) Glutathione S transferase polymorphisms and their biological consequences. *Pharmacol* 61:154–166.
11. Zidzik J, Slaba E, Joppa P, Kluchova Z, Dorkova Z, et al. (2008) Glutathione s-transferase and microsomal epoxide hydrolase gene polymorophisms and risk of chronic obstructive pulmonary disease in Slovak population. Croat Med J 492:182–191
12. Coates J (2000) Interpretation of Infrared Spectra: A pratical approach, Encyclopedia of analytical Chemistry. R.A.Meyers ed John Wiley & sons Ltd,Chichester, pp10815–10837
13. Love RG, Miller BG, Groat SK, Hagen S, Cowie HA, et al. (1997) Respiratory health effects of opencast coalmining: a cross sectional study of current workers. Occup Environ Med 54:416–423
14. Sanford AJ, Pane PD, (2000) Genetic risk factors for Chronic Obstructive Pulmonary Disease. Clin Chest Med 21: 633–643.
15. He JQ, Ruan J, Connett JE, Anthonisen NR, Pare PD, et al. (2002) Antioxidant Gene Polymorphism and Susceptibility to a rapid decline in lung function in smokers. Am J Respir Crit care Med 166: 323–328.
16. Cheng SL, Yu CJ, Chen CJ, Yang PC, (2004) Genetic polymorphism of epoxide hydrolase and Glutathione S Transferase in COPD. Eur Respir J 23:818–824.
17. Yim JJ, Yoo CG, Lee CT, Kim YW, Han SK, et al. (2002) Lack of association between Glutathione S Transferase polymorphism and COPD in Koreans. Lung 180: 119–125.
18. Lu B, He Q, (2002) Correlation between exon 5 polymorphism of Glutathione S-transferase P1 gene and susceptibility to Chronic Obstructive Pulmonary Disease in Northern Chinese population of Han nationality living in Beijing. Zhonghua nei Ke Za Zhi. 41:678–681.
19. Dockery DW, Pope CA, Xu X, Spengler JD, Ware JH, et al. (1993) An Association between air Pollution and Mortality in six US Cities. N Engl J Med 329: 1753–9.
20. Laden F, Schwartz J, Speizer FE, Dockery DW, (2006) Reduction in Fine Particulate Air Pollution and Mortality: extended follow-up of the Harvard Six Cities Study. Am J Respir Crit Care Med 17: 3667–72.
21. Brook RD, Franklin B, Cascio W, Hong Y, Howard G, et al. (2004) Air Pollution and Cardiovascular Disease: a statement for health-care professionals from the expert panel on Population and Prevention Science of the American Heart Association. Circulation 109: 2655–71.
22. Pope A, Dockery D, (2006) Health Effects of Fine Particulate Matter: lines that connect. J Air Waste Manag Assoc 56: 209–42.
23. Schwartz J (1997) Air Pollution and hospital admissions for cardiovascular disease in Tucson. Epidemiol 8:371–7
24. Brabin B, Smith M, Milligan P, Benjamin C, Dunne E, et al. (1994) Respiratory Morbidity in Merseyside schoolchildren exposed to coal dust and air pollution. Arch Dis Child 70: 305–312.
25. Leigh J, Driscoll TR, Cole BD, Beck RW, Hull BP, et al. (1994) Quantitative relation between emphysema and lung mineral content in coal workers. Occup Environ Med 516: 400–407.
26. Braun FC, Ackermann LU, Schwartz J, Gnehm HP, Rutishauser M, et al. (1992) Air Pollution and respiratory symptoms in preschool children. Am Rev Respir Dis 14: 542–47
27. Bhopal RS, Phillimore P, Moffatt S, Foy C, (1994) Is living near a coking works harmful to health? A study of industrial air pollution. J Epidemiol Community Health 48:237–247.
28. Schikowski T, Schaffner E, Meier F, Phuleria HC, Vierkotter A, et al. (2013) Improved air quality and attenuated lung function decline: modification by obesity in the sapaldia cohort, Environ. Health Perspect. 1219:1034–1039.
29. Cheng SL, Yu CJ, Chen CJ, Yang PC, (2004) Genetic polymorphism of epoxide hydrolase and glutathione S-transferase in COPD. Eur Respir J 23:818–824.
30. Lu B, He Q (2002) Correlation between exon 5 polymorphism of glutathione S-transferase P1 gene and susceptibility to chronic obstructive pulmonary disease in northern Chinese population of Han nationality living in Beijing. Zhonghua Nei Ke Za Zhi. 4:678–681.
31. Barnes PJ (1999) Molecular genetics of COPD. Thorax 53: 245–252.
32. Yim JJ, Park GY, Lee CT, Kim YW, Han SK, et al. (2000) Genetic susceptibility to chronic obstructive pulmonary disease in Koreans: combined analysis of polymorphic genotypes for microsomal epoxide hydrolase and glutathione s-transferase M1 and T1. Thorax 55:121–125.
33. Cahkoglu M, Tamer L, Aras NA, Karakas S, Ercan B, (2006) The association between polymorphic genotypes of glutathione s- Transferase and COPD in the Turkish population. Biochem Genet 447/8:307–319.
34. Lakhdar R, Denden S, Knani J, Leban N, Daimi H, et al. (2010) Association of GSTM1 and GSTT1 polymorphisms with Chronic obstructive pulmonary disease in a Tunisian population. Biochem Genet 48:647–657.
35. Baranova H, Perriot J, Albuisson E, Lvaschenko T, Baranov VS, et al. (1997) Peculiarities of the GSTM 0/0 genotype in French heavy smokers with various types of Chronic Bronchitis. Hum Genet 99: 822–826.

6

Comparison of the Effects of Air Pollution on Outpatient and Inpatient Visits for Asthma: A Population-Based Study in Taiwan

Hui-Hsien Pan[1,2¶], Chun-Tzu Chen[1,3¶], Hai-Lun Sun[1,2], Min-Sho Ku[1,2], Pei-Fen Liao[1], Ko-Hsiu Lu[2], Ji-Nan Sheu[1,2], Jing-Yang Huang[4], Jar-Yuan Pai[3], Ko-Huang Lue[1,2*]

1 Department of Pediatrics, Chung Shan Medical University Hospital, Taichung City, Taiwan R.O.C, **2** School of Medicine, Chung Shan Medical University, Taichung City, Taiwan R.O.C, **3** Department of Health Policy and Management, Chung Shan Medical University, Taichung City, Taiwan R.O.C, **4** Institute of Public Health, Department of Public Health, Chung Shan Medical University, Taichung City, Taiwan R.O.C

Abstract

Background: A nationwide asthma survey on the effects of air pollution is lacking in Taiwan. The purpose of this study was to evaluate the time trend and the relationship between air pollution and health care services for asthma in Taiwan.

Methods: Health care services for asthma and ambient air pollution data were obtained from the National Health Insurance Research database and Environmental Protection Administration from 2000 through 2009, respectively. Health care services, including those related to the outpatient and inpatient visits were compared according to the concentration of air pollutants.

Results: The number of asthma-patient visits to health-care facilities continue to increase in Taiwan. Relative to the respective lowest quartile of air pollutants, the adjusted relative risks (RRs) of the outpatient visits in the highest quartile were 1.10 (P-trend $=0.013$) for carbon monoxide (CO), 1.10 (P-trend $=0.015$) for nitrogen dioxide (NO_2), and 1.20 (P-trend $<$ 0.0001) for particulate matter with an aerodynamic diameter $\leqq 10\mu m$ (PM_{10}) in the child group (aged 0–18). For adults aged 19–44, the RRs of outpatient visits were 1.13 (P-trend $=0.078$) for CO, 1.17 (P-trend $=0.002$) for NO_2, and 1.13 (P-trend $<$ 0.0001) for PM_{10}. For adults aged 45–64, the RRs of outpatient visits were 1.15 (P-trend $=0.003$) for CO, 1.19 (P-trend $=0.0002$) for NO_2, and 1.10 (P-trend $=0.001$) for PM_{10}. For the elderly (aged$\geq$ 65), the RRs of outpatient visits in were 1.12 (P-trend $=0.003$) for NO_2 and 1.10 (P-trend $=0.006$) for PM_{10}. For inpatient visits, the RRs across quartiles of CO level were 1.00, 1.70, 1.92, and 1.86 (P-trend $=0.0001$) in the child group. There were no significant linear associations between inpatient visits and air pollutants in other groups.

Conclusions: There were positive associations between CO levels and childhood inpatient visits as well as NO_2, CO and PM_{10} and outpatient visits.

Editor: Stephania Ann Cormier, University of Tennessee Health Science Center, United States of America

Funding: The authors have no support or funding to report.

Competing Interests: The authors have declared that no competing interests exist.

* E-mail: cshy095@csh.org.tw

¶ These authors contributed equally to this work.

Introduction

Asthma is a common chronic inflammatory respiratory disease that affects 300 million people of all ages and all ethnic backgrounds and accounts for about 1 in every 250 deaths worldwide [1]. In Taiwan, the prevalence of asthma increased from 5.07% in 1985 to 11.9% in 2007 [2,3]. The risk factors for asthma include many external determinants such as mites, dust, mold, indoor and outdoor air pollution, and season variations [4–6]. Although air pollution has not been shown as the sole cause of respiratory illnesses, there is evidence that air pollution episodes lead to respiratory irritation, increased use of asthma medications and hospitalizations [7,8]. Traffic and industry-related pollutants, nitrogen dioxide (NO_2) and carbon monoxide (CO), were associated with asthma hospitalizations and outpatient visits [9,10]. Elevated levels of ozone (O_3), sulfur dioxide (SO_2) and particulate matter with an aerodynamic diameter $\leqq 10\mu m$ (PM_{10}) were reported to be related with increased asthma emergency room visits and admissions [11–13]. It has been reported that the rise in air pollution has increased respiratory and cardiovascular complications leading to elevated risk of death [14].

If air pollution is responsible for the observed enhanced respiratory complications and mortality, one would also expect to see an impact on clinic visits, outpatient visits, emergency department (ED) visits and hospitalization rates for asthma. However, there are no data regarding a population based survey with seasonal and air pollutants in asthma outpatient and inpatient visits in Taiwan. The objective of this study was to assess asthma-related outpatient and inpatient patterns of visits in different age groups based on the National Health Insurance Research Database (NHIRD) in Taiwan, and to compare the association with utilization of health care services and concentrations of air pollutants.

Materials and Methods

Database

The data were obtained from the NHIRD released by the National Health Research Institute (NHRI) in Taiwan. The National Health Insurance Program finances compulsory universal health care for 99% of all of residents of Taiwan [15]. The database contains demographic data, all health-care encounters, expenditure and dates of enrollment and withdrawal. To facilitate research, the NHRI randomly sampled a representative database of one million subjects enrolled in the National Health Insurance program in the year 2005 by a systematic sampling method. This one-million sample was validated to be representative of the entire insured population as reported by the NHRI. The identification numbers and personal information of all individuals in the NHRID were erased to protect the privacy of the individuals. This study was approved by the Institutional Review Board of the Chung-Shan Medical University Hospital, Taiwan.

Study Population

Cases of asthma were ascertained by the service claim for either outpatient or inpatient visit with a primary diagnosis of asthma (ICD-9-CM code 493.xx). Daily counts of clinic visits, outpatient visits, ED visits and hospital admissions for asthma were extracted from the medical insurance file for the period of 2000–2009. The outpatient visit was defined as a patient visit to a physician's office, clinic, or hospital outpatient department. The inpatient visits include ED visits and admissions. The analyses covered 54 municipalities, each with its own air quality monitoring station. We identified 306628 men and 315178 women who lived in municipalities with air quality monitoring stations during the study period. There were 33934 men and 34527 women with asthma. The patient's exact addresses were not available from the database. Therefore, we assumed that the municipality where a patient visit occurred was regarded as the same area where the patient was most likely exposed to air pollutants. Each asthma outpatient or inpatient visit was matched with the municipality's seasonal average pollutant concentrations. Each occurrence, limited to municipality with air quality monitoring stations, was counted as one visit. The outpatient and inpatient visits were further analyzed by seasons (spring as February, March and April; summer as May, June and July; autumn as August, September and October, winter as November, December and January) and four age groups (0–18, 19–44, 45–64 and ≥65 years).

Outdoor Air Pollution Monitoring

There were 76 air quality monitoring stations in Taiwan, established by the Environmental Protection Administration. Data from background and industrial air quality monitoring stations were excluded to avoid extreme levels of air pollutants. Finally, data from 54 air quality monitoring stations were enrolled (24 municipalities in northern, 9 municipalities in central, 19 municipalities in southern and 2 municipalities in eastern Taiwan). Complete monitoring data for the air pollutants included PM_{10}, SO_2, CO, O_3, and NO_2. The average seasonal concentration of each air pollutant from the Taiwan Environmental Protection Administration's air quality monitoring stations was calculated for further analysis from 2000 to 2009.

Statistical Analysis

All analyses were done by using the SAS ver. 9.3 software package (SAS Institute, Cary, NC, USA). Multivariate Poisson Regression was made in order to determine the relative risk (RR) of asthma inpatient and outpatient visits by sex, age, year and seasons. For use of health services in relation to air pollutants, the municipalities were defined as the observed units, and the level of air pollutants and counts of patient visits were collected in each municipality for each season from 2000 to 2009. As multiple visits by the same patient in a season in the municipality who lived, the count of the visits was as one visit per season. Generalized estimating equations (PROC GENMOD with repeated statement by the SAS Institute) were used to analyze of levels of pollutants by stratifying municipalities and to compare whether the inpatient and outpatient visits in those municipalities correlated with levels of pollutants. A p value <0.05 was considered statistically significant.

Results

The descriptive statistics for outpatient and inpatient visits and corresponding period and season data are shown in Table 1. Compared with women, the RRs for inpatient and outpatient visits in men were 1.31 ($p<0.0001$) and 1.15 ($p<0.0001$), respectively. Furthermore, when divided into 4 age groups, the highest RRs for inpatient and outpatient visits were among the elderly group (i.e. ≥65 years) followed by the child group (0 –18), the adults aged 45–64 group and the adult aged 19–44 group. For the period effect, there were increased outpatient visits (RR: 1.29, $P<0.0001$) since 2005 and inpatient visits (RR: 3.30, $P<0.0001$) since 2006 when compared with patient visits in 2000. The peak seasons for asthma inpatient and outpatient visits in Taiwan were spring (5.1/10,000 person-season) and winter (149.6/10,000 person-season), respectively.

Table 2 shows the RRs of outpatient health service use with respect to exposure to air pollutants in the four age groups following adjustments for sex, period and quartiles of air pollutants by the generalized estimating equations model. Relative to the respective lowest quartile of air pollutants, the adjusted RRs of the outpatient visits in the highest quartile were 1.10 (95% CI: 1.04, 1.16; P-trend = 0.013) for CO, 1.10 (95% CI: 1.01, 1.18; P-trend = 0.015) for NO_2, 0.94 (95% CI: 0.88, 0.99; P-trend = 0.016) for O_3, and 1.20 (95% CI:1.13, 1.27; P-trend <0.0001) for PM_{10} in the child group. For adults aged 19–44, the adjusted RRs of the outpatient visits were 1.13 (95% CI: 1.03, 1.23; P-trend = 0.078) for CO, 1.17 (95% CI: 1.05, 1.31; P-trend = 0.002) for NO_2, 0.88 (95% CI: 0.83, 0.94; P-trend<0.0001) for O_3, and 1.13 (95% CI: 1.05, 1.21; P-trend <0.0001) for PM_{10}. The adjusted RRs of the outpatient visits in the adults aged 45–64 were 1.15 (95% CI: 1.05, 1.26; P-trend = 0.003) for CO, 1.19 (95% CI: 1.08, 1.30; P-trend = 0.0002) for NO_2, 0.93 (95% CI: 0.89, 0.98; P-trend = 0.028) for O_3, and 1.10 (95% CI: 1.02, 1.18; P-trend = 0.001) for PM_{10}. The adjusted RRs for the elderly outpatient visits across quartiles of air pollutant level were 1.00, 1.02, 1.07, and 1.12 (P-trend = 0.003) for NO_2 and 1.00, 1.04, 1.06, 1.10 (P-trend = 0.006) for PM_{10}. There was no significant association between SO_2 and outpatient visits in any group.

The use of inpatient health services in relation to levels of air pollutants by different age groups is shown in Table 3. The adjusted RRs for the child inpatient visits across quartiles of CO level were 1.00, 1.70, 1.92, and 1.86 (P-trend = 0.0001) after multivariate adjustment by sex, year and other air pollutants. There were no significant linear associations between inpatient visits and air pollutants in other age groups.

Table 1. Inpatient and Outpatient Visits Stratified by Sex, Age, Year and Season in Taiwan.

		Inpatient visits[a]				Outpatient visits[b]			
	Person-seasons	Visits per season	Visit rate per 10000 person-season	RR	*P* value	Visits per season	Visit rate per 10000 person-season	RR	*P* value
Sex									
Female	8297518	2970	3.6	-		104358	125.7	-	
Male	7904684	3711	4.7	1.31	<0.0001	113915	144.0	1.15	<0.0001
Age									
0–18	3930291	2658	6.8	-		93580	237.9	-	
19–44	7373364	1083	1.5	0.22	<0.0001	40932	55.5	0.23	<0.0001
45–64	3674627	1317	3.6	0.53	<0.0001	46489	126.5	0.53	<0.0001
≥65	1223920	1623	13.3	1.96	<0.0001	37272	304.3	1.28	<0.0001
Year									
2000	1611185	369	2.3	-		18604	115.4	-	
2001	1611426	400	2.5	1.08	0.265	19932	123.6	1.07	<0.0001
2002	1628261	393	2.4	1.05	0.469	20624	126.6	1.10	<0.0001
2003	1635688	264	1.6	0.71	<0.0001	19501	119.2	1.03	0.002
2004	1692152	323	1.9	0.83	0.017	23061	136.2	1.18	<0.0001
2005	1719118	239	1.4	0.61	<0.0001	25529	148.4	1.29	<0.0001
2006	1699621	1284	7.6	3.30	<0.0001	24301	143.0	1.24	<0.0001
2007	1685165	1380	8.2	3.58	<0.0001	25208	149.6	1.30	<0.0001
2008	1672283	1212	7.3	3.17	<0.0001	23951	143.2	1.24	<0.0001
2009	1247303	817	6.6	2.86	<0.0001	17562	140.8	1.22	<0.0001
Season									
Spring (Feb-Apr)	4159829	2127	5.1	-		60648	145.7	-	
Summer(May-Jul)	4151745	1416	3.4	0.67	<0.0001	50591	121.8	0.84	<0.0001
Autumn(Aug-Oct)	4140026	1502	3.6	0.71	<0.0001	50887	122.9	0.84	<0.0001
Winter (Nov-Jan)	3750602	1636	4.4	0.85	<0.0001	56147	149.6	1.03	<0.0001

RR, relative risk.
[a]Inpatient visits include emergency department visits and hospitalizations.
[b]Outpatient visits include physician's office, clinic, or hospital outpatient department visits.

Table 2. Outpatient Visits in Relation to Air Pollutants by Different Age Groups.

	Outpatient visits[a]							
	Age group							
	0–18		**19–44**		**45–64**		**≥65**	
Variables	RR	95%CI	RR	95%CI	RR	95%CI	RR	95%CI
Men/women	**1.43**	1.37–1.48	**0.84**	0.79–0.90	**0.77**	0.72–0.82	**1.20**	1.12–1.29
Year	**1.06**	1.05–1.07	**1.03**	1.02–1.04	1.00	0.99–1.02	1.01	0.99–1.02
CO (ppm)								
0.19–0.43	-		-		-		-	
0.43–0.55	**1.07**	1.03–1.11	**1.13**	1.07–1.20	**1.06**	1.01–1.12	1.02	0.97–1.06
0.55–0.69	**1.06**	1.01–1.12	**1.12**	1.03–1.21	**1.11**	1.04–1.20	1.02	0.96–1.09
0.69–1.22	**1.10**	1.04–1.16	**1.13**	1.03–1.23	**1.15**	1.05–1.26	1.04	0.95–1.13
P-Trend	**0.013**		0.078		**0.003**		0.450	
NO_2 (ppm)								
4.45–14.01	-		-		-		-	
14.01–18.64	1.02	0.97–1.08	1.03	0.95–1.11	**1.07**	1.01–1.14	1.02	0.97–1.07
18.64–24.04	1.05	0.98–1.12	1.08	0.98–1.19	**1.14**	1.06–1.23	**1.07**	1.00–1.13
24.04–47.84	**1.10**	1.01–1.18	**1.17**	1.05–1.31	**1.19**	1.08–1.3	**1.12**	1.03–1.21
P-Trend	**0.015**		**0.002**		**0.0002**		**0.003**	
O_3 (ppm)								
14.23–23.83	-		-		-		-	
23.83–27.23	0.99	0.96–1.03	**0.96**	0.93–0.99	**0.96**	0.93–0.99	0.98	0.94–1.02
27.23–30.80	0.96	0.92–1.00	**0.94**	0.90–0.98	0.98	0.94–1.01	0.99	0.94–1.04
30.80–50.00	**0.94**	0.88–0.99	**0.88**	0.83–0.94	**0.93**	0.89–0.98	**0.95**	0.90–0.99
P-Trend	**0.016**		**<0.0001**		**0.028**		0.103	
PM_{10} ($\mu g/m^3$)								
23.00–43.67	-		-		-		-	
43.67–54.00	**1.06**	1.02–1.10	**1.07**	1.03–1.11	1.02	0.98–1.07	**1.04**	1.01–1.08
54.00–71.00	**1.11**	1.06–1.17	**1.14**	1.09–1.19	**1.09**	1.03–1.14	**1.06**	1.01–1.11
71.00–141.67	**1.20**	1.13–1.27	**1.13**	1.05–1.21	**1.10**	1.02–1.18	**1.10**	1.03–1.17
P-Trend	**<0.0001**		**<0.0001**		**0.001**		**0.006**	
SO_2 (ppm)								
0.20–2.80	-		-		-		-	
2.80–3.77	1.03	0.97–1.08	0.99	0.93–1.05	1.00	0.94–1.05	1.04	0.99–1.09
3.77–5.00	1.05	0.98–1.11	0.96	0.90–1.03	0.96	0.90–1.03	1.05	0.99–1.10
5.00–19.17	1.04	0.97–1.13	0.96	0.89–1.04	0.97	0.90–1.05	1.01	0.95–1.08
P-Trend	0.201		0.216		0.340		0.590	

Abbreviation: CO, carbon monoxide; NO_2, nitrogen dioxide; O_3, ozone; PM_{10}, particulate matter with an aerodynamic diameter $\leq 10\mu m$; RR, relative risk; SO_2, sulfur dioxide.
[a]Outpatient visits include physician's office, clinic, or hospital outpatient department visits.

Table 3. Inpatient Visits in Relation to Air Pollutants by Different Age Groups.

	Inpatient visits[a]							
	Age group							
	0–18		**19–44**		**45–64**		**≥65**	
Variables	RR	95%CI	RR	95%CI	RR	95%CI	RR	95%CI
Men/women	**1.86**	1.67–2.07	0.98	0.83–1.16	0.82	0.66–1.01	1.14	0.96–1.34
Year	**1.34**	1.28–1.39	**1.22**	1.17–1.27	**1.12**	1.08–1.15	**1.15**	1.10–1.20
CO (ppm)								
0.19–0.43	-		-		-		-	
0.43–0.55	**1.70**	1.40–2.06	1.08	0.83–1.40	1.17	0.92–1.49	1.28	0.98–1.66
0.55–0.69	**1.92**	1.53–2.40	1.07	0.81–1.42	1.18	0.90–1.55	1.35	0.93–1.97
0.69–1.22	**1.86**	1.39–2.48	0.94	0.64–1.38	1.21	0.87–1.68	1.42	0.94–2.15
P-Trend	**0.0001**		0.856		0.343		0.106	
NO_2 (ppm)								
4.45–14.01	-		-		-		-	
14.01–18.64	0.85	0.70–1.03	0.86	0.65–1.13	0.85	0.68–1.07	1.09	0.85–1.40
18.64–24.04	0.85	0.66–1.08	1.08	0.75–1.55	1.09	0.83–1.44	1.22	0.85–1.76
24.04–47.84	0.90	0.67–1.20	1.33	0.92–1.94	1.08	0.76–1.53	1.17	0.77–1.76
P-Trend	0.615		0.068		0.447		0.434	
O_3 (ppm)								
14.23–23.83	-		-		-		-	
23.83–27.23	**0.86**	0.74–0.99	1.02	0.90–1.16	1.00	0.81–1.23	1.14	0.99–1.31
27.23–30.80	0.87	0.74–1.02	1.08	0.87–1.32	1.13	0.88–1.44	**1.18**	1.01–1.39
30.80–50.00	0.83	0.66–1.05	1.04	0.79–1.35	1.06	0.80–1.40	1.20	0.99–1.45
P-Trend	0.136		0.548		0.450		0.070	
PM_{10} ($\mu g/m^3$)								
23.00–43.67	-		-		-		-	
43.67–54.00	**0.84**	0.73–0.97	1.04	0.83–1.30	**0.82**	0.69–0.97	0.89	0.74–1.07
54.00–71.00	0.97	0.83–1.13	1.25	0.96–1.63	1.01	0.82–1.22	1.00	0.81–1.23
71.00–141.67	0.89	0.71–1.12	1.05	0.77–1.44	1.04	0.78–1.37	0.93	0.70–1.23
P-Trend	0.757		0.336		0.393		0.900	
SO_2 (ppm)								
0.20–2.80	-		-		-		-	
2.80–3.77	0.95	0.77–1.18	1.13	0.83–1.52	0.90	0.69–1.17	0.95	0.72–1.26
3.77–5.00	1.03	0.79–1.33	1.26	0.96–1.65	0.91	0.70–1.19	1.00	0.76–1.31
5.00–19.17	0.82	0.63–1.07	0.96	0.71–1.29	0.81	0.61–1.08	0.78	0.57–1.05
P-Trend	0.369		0.882		0.204		0.223	

Abbreviation: CO, carbon monoxide; NO_2, nitrogen dioxide; O_3, ozone; PM_{10}, particulate matter with an aerodynamic diameter $\leqq 10\mu m$; RR, relative risk; SO_2, sulfur dioxide.
[a]Inpatient visits include emergency department visits and hospitalizations.

The use of inpatient health services increased with time as shown in Fig 1. In general, there were trends of increasing asthma outpatient visits in the children's group (Fig. 1A & B). In 2003, there was an outbreak of Severe Acute Respiratory Syndrome (SARS) in Hong Kong and nearly became a pandemic event [16]. During this period, inpatient visits dramatically decreased in all age groups. After the SARS outbreak, the rates of inpatient visits have increased with time since 2006, especially in the child and the elderly groups (Fig 1C & D).

Discussion

Among patients with asthma, air pollutant exposure causes increased asthma morbidity. Little is known about changes over time in air pollutant exposure among patients with asthma in a national sample. During the study period, the inpatient and outpatient visits by men were higher than in women. The peak seasons of asthma inpatient and outpatient visits for the total population were spring and winter, respectively. The inpatient and outpatient visits of asthmatics have not reached a plateau and have continued to increase. Inpatient visits for asthma increased with increased levels of CO in children but not for any pollutants in adults in the present study. Our study found that CO, NO_2 and PM_{10} had significant estimated associations on outpatient visits due to asthma and children are more susceptible than other age groups.

Health care visits only illustrate a small percentage and most severe inpatients of the total impacts of air pollution. Our study used the count of health care visits from NHRI databases as the measure of morbidity in the population. Clinic visits, outpatient visits, ED visits and admissions are types of health care service, but also possessed of potentially important disparities [17,18]. Because asthma is a chronic disease, patients with asthma were taught to deal with their symptoms and discomfort of an asthma attack [19,20]. When the concentrations of air pollutants rise, patients with asthma may have treated symptoms by themselves or visited neighborhood clinics and hospital outpatient departments for medical treatment. Subsequently, if patients did not receive any treatment or if the condition was deteriorated or ineffective after an outpatient visit, they would then visit hospital emergency departments for assistance. This may explain why there was no increased risk for inpatient visits for adults with increasing levels of air pollutants.

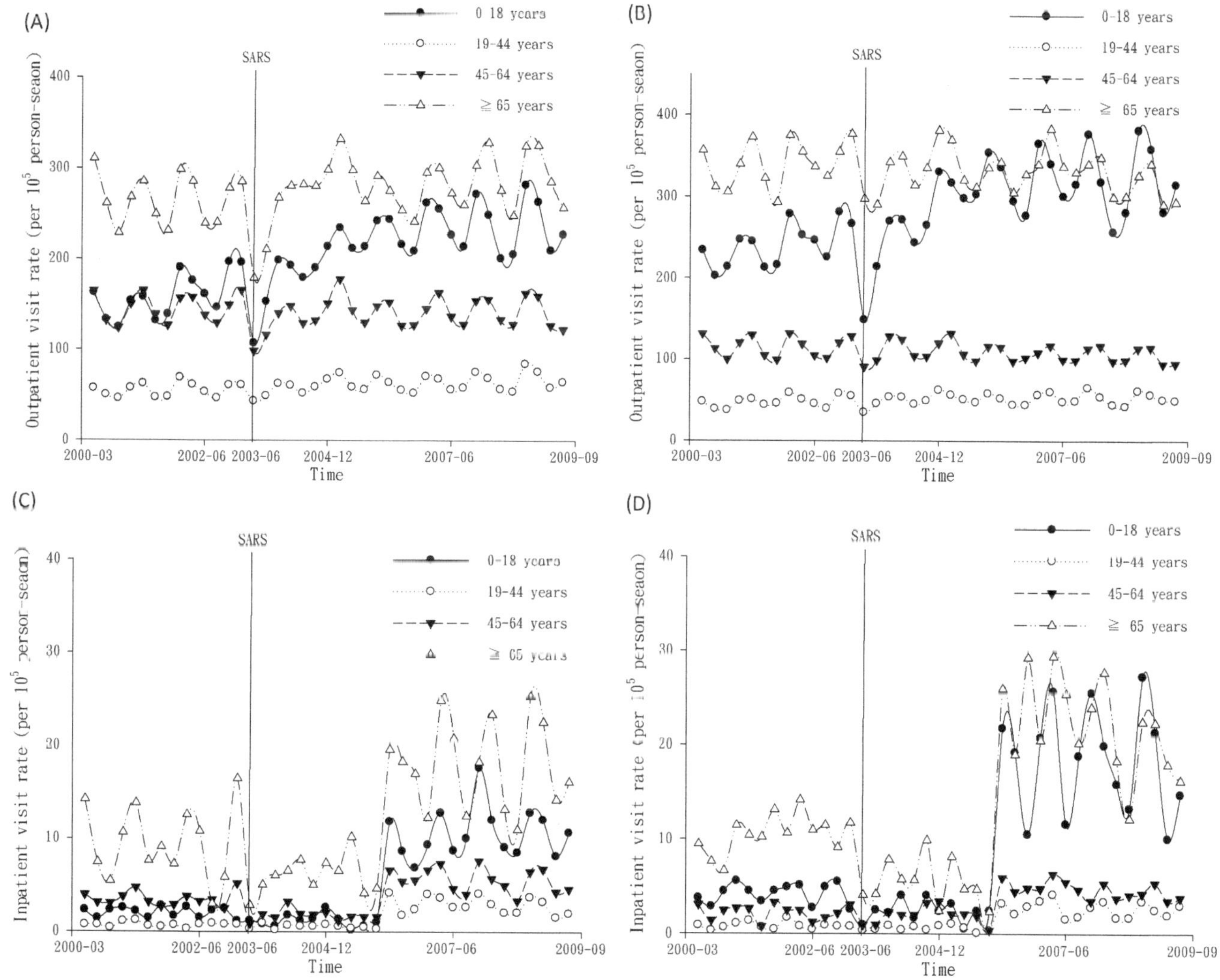

Figure 1. Outpatient visit rates for asthma in 4 age groups in women (A) and men (B), and inpatient visit rates in 4 age groups in women (C) and men (D) during 2000–2009. SARS, Severe Acute Respiratory Syndrome.

In response to the increasing prevalence, mortality rates, and medical cost of asthma, the Bureau of National Health Insurance initiated a Healthcare Quality Improvement Program for patients with asthma since November, 2001 [21]. The characteristics of the program were registry development, adherence to guidelines, patient education, and nursing care management. The Bureau offered financial incentives that motivated the nurses and physicians to change their practice patterns by following clinical guidelines; thereby, the asthma care team support from healthcare organizations was able to promote and enable patients to effectively self-manage their asthma with reduced healthcare resource utilization. During the SARS outbreak, people avoided hospital visits to prevent themselves from nasocomial infection and the inpatient visit rate for asthma decreased in all age groups [22]. However, there was an increase of inpatient and outpatient visits after the SARS outbreak. The exacerbation of air pollution probably plays a role in the rising rate of asthma visits due to growing populations, increased economic activity, rise in vehicular traffic, as well as the increasing intensity and occurrence of dust storms originating in Mongolia and China [23].

There have been associations between daily ambient air pollution levels and acute exacerbations of respiratory diseases in many time-series studies [11,12,24]. Urban air pollution constitutes a complex mixture of several compounds. The exposure of motor vehicle air pollutants, such as NO_2, CO, SO_2 and PM, increases the incidence and prevalence of asthma and bronchitic symptoms [25]. CO was reported to be associated with asthma admission and ED visits [9,26,27]. There were associations between short-term exposure to ambient CO and risk of cardiovascular disease hospitalizations, even at low ambient CO levels [28]. Sun *et al* conducted a one-year observation in central Taiwan that CO played a role in acute exacerbation of asthma in children and increased the number of childhood asthma ED visits, but not in adults [29]. Villeneuve et al reported that an increase in the interquartile range of the 5-day average for CO was associated with 48% increases in the risk of an asthma ED visit for children aged 2 to 4, but the associations were less pronounced in adult aged 15 to 44 [30]. In Rome, where air pollution comes mostly from combustion products of motor vehicles, CO was associated with most of the respiratory conditions in all ages, and it remained an independent predictor in multi-pollutant analysis for all respiratory admissions [31]. In London, there were significant associations between CO and daily consultations for asthma and other lower respiratory disease in children, whereas in adults the only consistent association was with PM_{10} [32]. In our study, CO was also significantly associated with asthma exacerbation and inpatient visits in children. However, a direct association between CO and asthma lacks a biologically plausible mechanism [33]. It is possible that CO might be a surrogate for other noxious incomplete combustion products [34]. Unlike with children, the other major confounders in adult asthmatic patients include occupational exposures, smoking, stress, emotional factors, and systemic diseases, which may also partly explain why outdoor air pollution was not associated with inpatient visits in the adults in our study.

Coal- and oil-fired power plants and diesel- and gasoline-powered motor vehicle engines are the main sources of ambient NOx emissions [35]. In a meta-analysis study, inhalation of NO_2 in the air significantly increased the development of childhood asthma and symptoms of wheezing [36]. A previous study in Taipei showed that most robust associations were found for NO_2 elevation and asthma admission rates [26]. NO_2 levels were associated with childhood asthma exacerbations and ED visits in Santa Clara, California [37]. In a spatiotemporal analysis of air pollution and asthma patient visits in Taipei, elevated levels of NO_2 had a positive association on outpatient visits [38]. In summary, there was a linear response in outpatient visits on days with elevated NO_2.

Previous studies have not yielded consistent results concerning associations between O_3 and SO_2 and asthma admissions. SO_2 was least frequently mentioned in the correlation with asthma hospitalization rate. Most previous studies have not shown a significant effect of SO_2 on asthma hospitalization rates supporting our findings [29,39]. O_3 is a highly reactive gas and induces bronchial inflammation, constriction of the airways and decreased lung function [40]. Long-term exposure to outdoor O_3 increases the prevalence of bronchitic symptoms among children [41]. O_3 levels have been previously reported to be associated with asthma admission and ED visits [11,24,38]. In contrast, asthma exacerbations were not associated with O_3 levels in North America [42]. Daily general-practice consultations for respiratory conditions were unrelated to O_3 in London [32] and Taiwan [43].

PM_{10} is a heterogeneous mixture of small solid or liquid particles with varying compositions in the atmosphere. There were no consistent results concerning associations between PM_{10} and asthma admissions. Some studies reported that PM_{10} was significantly associated with asthma admissions [8,39], other studies reported a lack of association between PM_{10} and asthma admission [27]. In our study, levels of PM_{10} were associated with outpatient visits for asthma, but not associated with admissions. Hwang and Chan used data obtained from clinic records and environmental monitoring stations in Taiwan and reported that PM_{10} had an impact on outpatient visits [43]. This was consistent with our findings.

Most of the previous studies are cross-sectional and have focused air pollutants on short-term, regional area and respiratory system diseases [8,29,38]. We conducted a nationwide asthma survey on the effects of air pollution in Taiwan and evaluated the association between different air pollutants and outpatient and inpatient visits. One of the strengths of the present study is the use of a computerized database, which is population-based and is highly representative. Because we enrolled all patients diagnosed with asthma from 2000 to 2009, we can rule out the possibility of selection bias. Since the data were obtained from a historical database, which collects all information, recall bias was avoided. Besides, we analyzed levels of pollutants at different municipalities and compared whether the inpatient and outpatient visits in those municipalities correlated with levels of pollutants. Thus this study directly associated the patient visits with the levels of pollutants in those municipalities. As multiple visits by the same patient would lead to misinterpretation of the data, we used the statistic method to reduce bias.

There were still several limitations of the present study. First, although we adjusted for several potential confounders in the statistical analysis, a number of possible confounding variables, including family history of atopy, dietary habits, physical activity, occupational exposures, smoking habits, stress and emotional factors, which are associated with asthma were not included in our database. Second, we were unable to ask patients for severity of asthma because of de-identifcation. Third, self-treatment with over-the-counter medications or alternative health services was not included in the database. These data also do not include the number of asthmatic subjects who had respiratory problems but did not search for health service. Therefore, the extent of the issue may have been considerably underestimated. Fourth, potentially inaccurate data in the records could lead to possible misclassification.

In conclusion, the present study provides evidence that exposure to the outdoor air pollutant, CO, exerted adverse effects on health and increases in the child admission. Our study also showed a linear association between NO_2, CO, and PM_{10} and outpatient visits. It is an important public health policy to monitor air quality and warn the public about atmospheric factors that could be associated with increased risks of asthma.

Acknowledgments

This study is based on data from the National Health Insurance Research Database provided by the Bureau of National Health Insurance, Department of Health and managed by National Health Research Institutes. The descriptions or conclusions herein do not represent the viewpoints of the Bureau of National Health Insurance, Department of Health or National Health Research Institutes.

Author Contributions

Conceived and designed the experiments: K.H. Lue HHP CTC. Performed the experiments: JYH JYP. Analyzed the data: K.H. Lue HLS JNS. Wrote the paper: HHP CTC. Reviewed the manuscript and gave input to the final version: HLS MSK PFL K.H. Lu JNS.

References

1. Masoli M, Fabian D, Holt S, Beasley R (2004) The global burden of asthma: executive summary of the GINA Dissemination Committee report. Allergy 59: 469–478.
2. Hsieh KH, Shen JJ (1988) Prevalence of childhood asthma in Taipei, Taiwan, and other Asian Pacific countries. J Asthma 25: 73–82.
3. Hwang CY, Chen YJ, Lin MW, Chen TJ, Chu SY, et al. (2010) Prevalence of atopic dermatitis, allergic rhinitis and asthma in Taiwan: a national study 2000 to 2007. Acta Derm Venereol 90: 589–594.
4. Han YY, Lee YL, Guo YL (2009) Indoor environmental risk factors and seasonal variation of childhood asthma. Pediatr Allergy Immunol 20: 748–756.
5. Chiang CH, Wu KM, Wu CP, Yan HC, Perng WC (2005) Evaluation of risk factors for asthma in Taipei City. J Chin Med Assoc 68: 204–209.
6. Guo Y, Jiang F, Peng L, Zhang J, Geng F, et al. (2012) The association between cold spells and pediatric outpatient visits for asthma in Shanghai, China. PLoS One 7: e42232.
7. Yeh KW, Chang CJ, Huang JL (2011) The association of seasonal variations of asthma hospitalization with air pollution among children in Taiwan. Asian Pac J Allergy Immunol 29: 34–41.
8. Kuo HW, Lai JS, Lee MC, Tai RC, Lee MC (2002) Respiratory effects of air pollutants among asthmatics in central Taiwan. Arch Environ Health 57: 194–200.
9. Delamater PL, Finley AO, Banerjee S (2012) An analysis of asthma hospitalizations, air pollution, and weather conditions in Los Angeles County, California. Sci Total Environ 425: 110–118.
10. Wang KY, Chau TT (2013) An association between air pollution and daily outpatient visits for respiratory disease in a heavy industry area. PLoS One 8: e75220.
11. Wilson AM, Wake CP, Kelly T, Salloway JC (2005) Air pollution, weather, and respiratory emergency room visits in two northern New England cities: an ecological time-series study. Environ Res 97: 312–321.
12. Qiu H, Yu IT, Tian L, Wang X, Tse LA, et al. (2012) Effects of coarse particulate matter on emergency hospital admissions for respiratory diseases: a time-series analysis in Hong Kong. Environ Health Perspect 120: 572–576.
13. Cadelis G, Tourres R, Molinie J (2014) Short-term effects of the particulate pollutants contained in Saharan dust on the visits of children to the emergency department due to asthmatic conditions in Guadeloupe (French Archipelago of the Caribbean). PLoS One 9: e91136.
14. Wong TW, Tam WS, Yu TS, Wong AH (2002) Associations between daily mortalities from respiratory and cardiovascular diseases and air pollution in Hong Kong, China. Occup Environ Med 59: 30–35.
15. Lu JF, Hsiao WC (2003) Does universal health insurance make health care unaffordable? Lessons from Taiwan. Health Aff (Millwood) 22: 77–88.
16. Tsang KW, Ho PL, Ooi GC, Yee WK, Wang T, et al. (2003) A cluster of cases of severe acute respiratory syndrome in Hong Kong. N Engl J Med 348: 1977–1985.
17. Gold LS, Thompson P, Salvi S, Faruqi RA, Sullivan SD (2014) Level of asthma control and health care utilization in Asia-Pacific countries. Respir Med 108: 271–277.
18. Sun HL, Kao YH, Lu TH, Chou MC, Lue KH (2007) Health-care utilization and costs in Taiwanese pediatric patients with asthma. Pediatr Int 49: 48–52.
19. Centers for Disease Control and Prevention (2011) Vital signs: asthma prevalence, disease characteristics, and self-management education: United States, 2001—2009. MMWR Morb Mortal Wkly Rep 60: 547–552.
20. Weng HC (2005) Impacts of a government-sponsored outpatient-based disease management program for patients with asthma: a preliminary analysis of national data from Taiwan. Dis Manag 8: 48–58.
21. Bureau of National Health Insurance (2001) A national comprehensive disease management program. Taipei: Bureau of National Health Insurance.
22. Huang YT, Lee YC, Hsiao CJ (2009) Hospitalization for ambulatory-care-sensitive conditions in Taiwan following the SARS outbreak: a population-based interrupted time series study. J Formos Med Assoc 108: 386–394.
23. Bell ML, Levy JK, Lin Z (2008) The effect of sandstorms and air pollution on cause-specific hospital admissions in Taipei, Taiwan. Occup Environ Med 65: 104–111.
24. Winquist A, Klein M, Tolbert P, Flanders WD, Hess J, et al. (2012) Comparison of emergency department and hospital admissions data for air pollution time-series studies. Environ Health 11: 70.
25. Gasana J, Dillikar D, Mendy A, Forno E, Ramos Vieira E (2012) Motor vehicle air pollution and asthma in children: a meta-analysis. Environ Res 117: 36–45.
26. Yang CY, Chen CC, Chen CY, Kuo HW (2007) Air pollution and hospital admissions for asthma in a subtropical city: Taipei, Taiwan. J Toxicol Environ Health A 70: 111–117.
27. Slaughter JC, Kim E, Sheppard L, Sullivan JH, Larson TV, et al. (2005) Association between particulate matter and emergency room visits, hospital admissions and mortality in Spokane, Washington. J Expo Anal Environ Epidemiol 15: 153–159.
28. Bell ML, Peng RD, Dominici F, Samet JM (2009) Emergency hospital admissions for cardiovascular diseases and ambient levels of carbon monoxide: results for 126 United States urban counties, 1999–2005. Circulation 120: 949–955.
29. Sun HL, Chou MC, Lue KH (2006) The relationship of air pollution to ED visits for asthma differ between children and adults. Am J Emerg Med 24: 709–713.
30. Villeneuve PJ, Chen L, Rowe BH, Coates F (2007) Outdoor air pollution and emergency department visits for asthma among children and adults: a case-crossover study in northern Alberta, Canada. Environ Health 6: 40.
31. Fusco D, Forastiere F, Michelozzi P, Spadea T, Ostro B, et al. (2001) Air pollution and hospital admissions for respiratory conditions in Rome, Italy. Eur Respir J 17: 1143–1150.
32. Hajat S, Haines A, Goubet SA, Atkinson RW, Anderson HR (1999) Association of air pollution with daily GP consultations for asthma and other lower respiratory conditions in London. Thorax 54: 597–605.
33. American Thoracic Society (1996) Health effects of outdoor air pollution. Committee of the Environmental and Occupational Health Assembly of the American Thoracic Society. Am J Respir Crit Care Med 153: 3–50.
34. Norris G, YoungPong SN, Koenig JQ, Larson TV, Sheppard L, et al. (1999) An association between fine particles and asthma emergency department visits for children in Seattle. Environ Health Perspect 107: 489–493.
35. Trasande L, Thurston GD (2005) The role of air pollution in asthma and other pediatric morbidities. J Allergy Clin Immunol 115: 689–699.
36. Takenoue Y, Kaneko T, Miyamae T, Mori M, Yokota S (2012) Influence of outdoor NO2 exposure on asthma in childhood: meta-analysis. Pediatr Int 54: 762–769.
37. Lipsett M, Hurley S, Ostro B (1997) Air pollution and emergency room visits for asthma in Santa Clara County, California. Environ Health Perspect 105: 216–222.
38. Chan TC, Chen ML, Lin IF, Lee CH, Chiang PH, et al. (2009) Spatiotemporal analysis of air pollution and asthma patient visits in Taipei, Taiwan. Int J Health Geogr 8: 26.
39. Lin M, Stieb DM, Chen Y (2005) Coarse particulate matter and hospitalization for respiratory infections in children younger than 15 years in Toronto: a case-crossover analysis. Pediatrics 116: e235–240.
40. Khatri SB, Holguin FC, Ryan PB, Mannino D, Erzurum SC, et al. (2009) Association of ambient ozone exposure with airway inflammation and allergy in adults with asthma. J Asthma 46: 777–785.
41. Hwang BF, Lee YL (2010) Air pollution and prevalence of bronchitic symptoms among children in Taiwan. Chest 138: 956–964.
42. Schildcrout JS, Sheppard L, Lumley T, Slaughter JC, Koenig JQ, et al. (2006) Ambient air pollution and asthma exacerbations in children: an eight-city analysis. Am J Epidemiol 164: 505–517.
43. Hwang JS, Chan CC (2002) Effects of air pollution on daily clinic visits for lower respiratory tract illness. Am J Epidemiol 155: 1–10.

7

A Novel Genetic Score Approach Using Instruments to Investigate Interactions between Pathways and Environment: Application to Air Pollution

Marie-Abele Bind[1,2]*, Brent Coull[2], Helen Suh[1], Robert Wright[1], Andrea Baccarelli[1,3], Pantel Vokonas[4], Joel Schwartz[1,3]

1 Department of Environmental Health, Harvard School of Public Health, Boston, Massachusetts, United States of America, **2** Department of Biostatistics, Harvard School of Public Health, Boston, Massachusetts, United States of America, **3** Department of Epidemiology, Harvard School of Public Health, Boston, Massachusetts, United States of America, **4** VA Normative Aging Study, Boston University School of Medicine, Boston, Massachusetts, United States of America

Abstract

Air pollution has been associated with increased systemic inflammation markers. We developed a new pathway analysis approach to investigate whether gene variants within relevant pathways (oxidative stress, endothelial function, and metal processing) modified the association between particulate air pollution and fibrinogen, C-reactive protein (CRP), intercellular adhesion molecule-1 (ICAM-1), and vascular cell adhesion molecule-1 (VCAM-1). Our study population consisted of 822 elderly participants of the Normative Aging Study (1999–2011). To investigate the role of biological mechanisms and to reduce the number of comparisons in the analysis, we created pathway-specific scores using gene variants related to each pathway. To select the most appropriate gene variants, we used the least absolute shrinkage and selection operator (Lasso) to relate independent outcomes representative of each pathway (8-hydroxydeoxyguanosine for oxidative stress, augmentation index for endothelial function, and patella lead for metal processing) to gene variants. A high genetic score corresponds to a higher allelic risk profile. We fit mixed-effects models to examine modification by the genetic score of the weekly air pollution association with the outcome. Among participants with higher genetic scores within the oxidative stress pathway, we observed significant associations between particle number and fibrinogen, while we did not find any association among participants with lower scores ($p_{interaction} = 0.04$). Compared to individuals with low genetic scores of metal processing gene variants, participants with higher scores had greater effects of particle number on fibrinogen ($p_{interaction} = 0.12$), CRP ($p_{interaction} = 0.02$), and ICAM-1 ($p_{interaction} = 0.08$). This two-stage penalization method is easy to implement and can be used for large-scale genetic applications.

Editor: Hugo ten Cate, Maastricht University Medical Center, Netherlands

Funding: This study was supported by the U.S. Environmental Protection Agency grants EPA RD-827353 and RD-832416; National Institute of Environmental Health Sciences (NIEHS) grants RO1-ES015172, ES014663, ES00002, PO1-ES008925, and Clean Air Act (CLARC) grant RD83479701. The VA Normative Aging Study (NAS) is supported by the Cooperative Studies Program/Epidemiology Research and Information Center of the U.S. Department of Veterans Affairs and is a component of the Massachusetts Veterans Epidemiology Research and Information Center, Boston, Massachusetts. The funders had no role in study design, data collection and analysis, decision to publish, or preparation of the manuscript.

Competing Interests: The authors have declared that no competing interests exist.

* E-mail: ma.bind@mail.harvard.edu

Introduction

To better understand molecular mechanisms, one can investigate the role of gene variants individually or collectively within a biological pathway. Pathway analysis may be more informative of the underlying biology. Mechanisms by which particulate air pollution is linked to exacerbation of cardiovascular disease (CVD) morbidity and mortality are not fully understood [1]. Previous studies have suggested biological changes after air pollution exposure such as thrombosis [2], systemic cytokine-mediated inflammation [3], and impaired endothelial function [4,5], especially among the elderly.

To better understand the molecular mechanisms linking air pollution and CVD, we chose to examine the role gene variants play in modifying the adverse air pollution effects by considering genetic pathway-air pollution interactions [6], since polymorphisms in pathways unrelated to how exposure produces response are unlikely to modify the exposure-response relation. Oxidative stress, endothelial dysfunction, and impaired metal processing are potential intermediate biological responses that may relate air pollution to CVD [6–9]. The oxidative stress pathway has been shown to be central to both the toxicology of air pollution and the pathogenesis of coronary atherosclerosis [10–12]. The endothelial dysfunction pathway has also been identified as a key player in air pollution molecular responses [8,13]. Metals in particles have been associated with increased oxidative stress and inflammation, and this was reduced in knockout mice deficient in metal transport mechanisms. In addition, exposure to metals on particles has been associated with decreases in heart rate variability [7,14,15] and whole-blood coagulation time [16–19]. We therefore hypothesized that gene variants related to metal processing play an important role in biological responses to air pollutants having metals attached to them.

This study builds upon previous research by investigating three potential pathways (oxidative stress, impaired endothelial function, and metal processing dysfunction) that may modify the association between air pollution and the outcomes fibrinogen, C-reactive protein, intercellular adhesion molecule-1 (ICAM-1) and vascular cell adhesion molecule-1 (VCAM-1), which have been related to coronary heart disease and atherosclerosis [20,21]. We developed a new allelic score method following an existing parsimonious approach that groups gene variants related to similar molecular mechanisms [22]. The study focused on the elderly, a potentially susceptible subgroup.

Materials

Study population

This study included participants from the Normative Aging Study (NAS), a prospective cohort of aging established in 1963, enrolling men from the Greater Boston area. Participants underwent medical examinations every three to five years. More detailed longitudinal characteristics on the study population can be found elsewhere [23]. We obtained one to five measurements of fibrinogen, C-reactive protein, ICAM-1, and VCAM-1 on 822 participants between 1999 and 2011.

Ethics statement

This study was approved by the Harvard School of Public Health and the Veteran Administration Institution Review Boards (IRB). Participants provided written informed consent to participate in this study, which was approved by the Veteran Administration IRB.

Air pollution assessment

We measured particulate concentrations at the Harvard supersite located 1 km from the examination site. We measured hourly particle number per cm^3 (0.007–3 μm) with a Condensation Particle Counter (TSI Inc, Model 3022A, Shoreview, MN), hourly ambient fine particle ($PM_{2.5}$) concentrations with a Tapered Element Oscillation Microbalance (Model 1400A, Rupprecht and Pastashnick, East Greenbush, NY), and hourly $PM_{2.5}$ black carbon concentrations using an Aethalometer (Magee Scientific Co., Model AE-16, Berkeley, CA).

The relevant exposure window for gene-air pollution interactions on coagulation, and inflammation is unclear. Results of previous studies suggested an association between air pollution and blood markers of inflammation and coagulation spread over several days and weeks prior to the time of assessment [5,23]. To limit the number of statistical comparisons, we *a priori* considered only concentrations averaged over the week preceding each participant's examination.

Genotyping assessment

An exhaustive list of the genes we considered can be found in Tables S1 to S3 in File S1. We related each gene to one of the three pathways (oxidative stress, impaired endothelial function, and metal processing dysfunction) based on their biological functionality provided *GeneCards* [24]. A brief description of each gene's function can be found in Tables S4 to S6 in File S1. We coded gene variants as follows: 0 for wild type, 1 for heterozygous, and 2 for homozygous; except for glutathione S-transferase mu 1 (GSTM1) and glutathione S-transferase theta 1 (GSTT1) that were coded 0 (deletion) or 1 (no deletion).

Assays were genotyped using the Sequenom MassArray MALDI-TOF Mass Spectrometer with semi-automated primer design and implementation of the short extension method (San Diego, CA). The MassARRAY System offers several features including very high sensitivity, capacity to analyze multiple classes of genetic markers, ability to quickly create and modify customized assay panels, flexible sample throughput (from few to thousands), and high quality data with low operating costs. Multiplexed polymerase chain reaction assays were designed using the Sequenom Spectro Designer software (San Diego, CA).

Table 1. Baseline characteristics of the 882 Normative Aging Study participants.

	Mean ± SD	5%	50%	95%
Age (years)	72.4±6.9	62	72	84
BMI (kg/m²)	28.3±4.1	22.5	27.8	35.2
Fibrinogen*	350±88.2	239	334	529
CRP**	3.1±5.7	0.3	1.7	9.0
ICAM-1***	298±101.8	188	288	433
VCAM-1***	1050±356.1	634	988	1622
	N (%)			
Diabetes	113 (14%)			
Smoking				
-Never	233 (28%)			
-Current	37 (67%)			
-Former	552 (5%)			
Statin use	296 (36%)			

*mg/dL.
**mg/L.
***ng/mL.

Table 2. Distribution and correlation between air pollutants for one-week averaged concentrations between 1999 and 2011.

Pollutants' distributions	Mean	25th percentile	Median	75th percentile
Particle Number (count/cm^3)	20238	12405	17388	27009
Black Carbon (µg/m^3)	0.96	0.65	0.87	1.18
$PM_{2.5}$ (µg/m^3)	10.83	7.99	10.12	12.78
Spearman correlations between pollutants pairs				
	Particle number	Black carbon	$PM_{2.5}$	
Particle number	1	0.04	0.12*	
Black carbon		1	0.72*	
$PM_{2.5}$			1	

* p-value <0.05.

Blood marker outcomes assessment

At each visit, we measured concentrations of plasma fibrinogen using MDA Fibriquick (Trinity Biotech, Bray, Ireland), a bovine thrombin reagent (Clauss, reagent number 17). We quantified C-reactive protein levels using an immunoturbidimetric assay on the Hitachi 917 analyzer (Roche Diagnostics Indianapolis, IN). We assessed plasma ICAM-1 and VCAM-1 concentrations with an enzyme-linked immunoabsorbent assay method (R&D Systems, Minneapolis, MN). This technique has been exhaustively tested for superior quality and reproducibility.

Methods

1st-stage: Gene variants selection using the least absolute shrinkage and selection operator (Lasso)

We related gene variants to each pathway based on their functionality. Because of the large number of gene variants for each gene and within each pathway, we chose to penalize them using the Lasso method, which simultaneously performs variables selection and regression coefficients estimation [25]. To select the relevant gene variants, we considered the following outcomes (not being examined for air pollution) for each pathway: 8-hydroxydeoxyguanosine (8-ohdg) for oxidative stress, augmentation index for endothelial function, and patella lead for metal processing. We included gene variants following dominant models ({0} versus {1 or 2}), recessive ({0 or 1} versus {2}), deletions ({0} versus {1}), and adjusted for age. The regression model was:

$$
\begin{aligned}
E(Y_i) = \beta_0 + \sum_{p=1}^{P_1} \beta_p \, I(Variant_{i,p} = \{1 \text{ or } 2\}) \\
+ \sum_{p=P_1+1}^{P_1+P_2} \beta_p \, I(Variant_{i,p} \\
= \{2\}) + \sum_{p=P_1+P_2 1+1}^{P_1+P_2+P_3} \beta_p \, I(Variant_{i,p} \\
= \{1\}) + \beta_{P_1+P_2+P_3+1} Age_i
\end{aligned}
$$

$$
\text{subject to : } \sum_{p=1}^{P_1+P_2+P_3+1} |\beta_p| \le c
$$

where Yi corresponds to levels of 8-ohdg, augmentation index, and patella lead levels for each participant i and c is a non-negative tuning parameter. P_1 and P_2 represent the numbers of gene variants that could follow a dominant and recessive model, respectively. P_3 corresponds to the number of gene variants that were deletions. The Lasso problem can be rewritten in the equivalent Lagrangian form in which the regression coefficients solved:

$$\hat{\beta} = \arg\min_{\beta}\left\{\frac{1}{2}\sum_{i=1}^{n}\left[Y_i - \beta_0 - \sum_{p=1}^{P_1}\beta_p\, I(Variant_{i,p} = \{1\ or\ 2\}) - \sum_{p=P_1+1}^{P_1+P_2}\beta_p\, I(Variant_{i,p} = \{2\}) - \sum_{p=P_1+P_21+1}^{P_1+P_2+P_3}\beta_p\, I(Variant_{i,p} = \{1\}) - \beta_{P_1+P_2+P_3+1}\, Age_i\right]^2 + \lambda \sum_{p=1}^{P_1+P_2+P_3+1}|\beta_p|\right\}$$

where λ is a non-negative parameter that controls the amount of shrinkage: the larger the value of λ, the greater the amount of shrinkage. There is a one-to-one correspondence between λ and c. The latter constraint ($\sum_{p=1}^{P_1+P_2+1}|\beta_p|$) makes the solutions nonlinear in the Y_i and computing the Lasso solution is a quadratic programming problem. We chose the penalty term of the Lasso regression (λ) based on the lowest Akaike Information Criterion (AIC). This approach created a reduced set of candidate gene variants related to each pathway. We adjusted for age at this stage to explain some variability of the instrumental outcomes, and thus to help the selection process to be more accurate. We conducted the analysis using the Least Angle Regression ("lars") R package [26].

2nd-stage: Genetic-score construction

Using the signs of the non-zero coefficients of the Lasso penalization, we summed up gene variants related to each pathway to construct three score variables representing the allelic profile of each participant. The score for each individual i was:

$$Score_i = \sum_{p=1}^{P_1+P_2+P_3}(-1)^{I(\beta_p<0)}\, I(Variant_{i,p} = k)\, I(|\beta_p| > 0),$$

where $k = \{1 or 2\} or \{2\} or \{1\}$

3rd-stage: Genetic-pathway-air pollution interactions

We then dichotomized these score variables using the median of each score's distribution and used these resulted binary variables to investigate effect modification. A high score corresponds to a higher allelic risk profile. Another score approach has been used successfully before to examine gene-environment by pathway interactions [22]. Fibrinogen, C-reactive protein, ICAM-1, and VCAM-1 were log-transformed to generate a more normal distribution in the residuals. Because many participants had their blood markers measured more than once, we fit separate linear mixed-effects models with a random intercept for each participant. Regression models included the air pollutant, seasonal sine and cosine, temperature, relative humidity, age, body mass index, diabetes, smoking status, statin use, the binary genetic score, and the interaction between the binary genetic score and the air pollutant. At this stage, we still need to control for age because it is a still potential confounding variable.

Sensitivity analysis

Because our analysis involved 9 tests ($n_{Tests} = 3$ exposures *3 pathways), we conducted some multiple testing corrections using Bonferroni adjustment.

Results

Descriptive statistics

At baseline, the median age of the study participants was 72 years (Table 1). About 14% were diabetics and 37% were statin users. More detailed longitudinal characteristics of the study population can be found elsewhere [23]. Some summary statistics of the air pollutant concentrations are presented in Table 2. $PM_{2.5}$

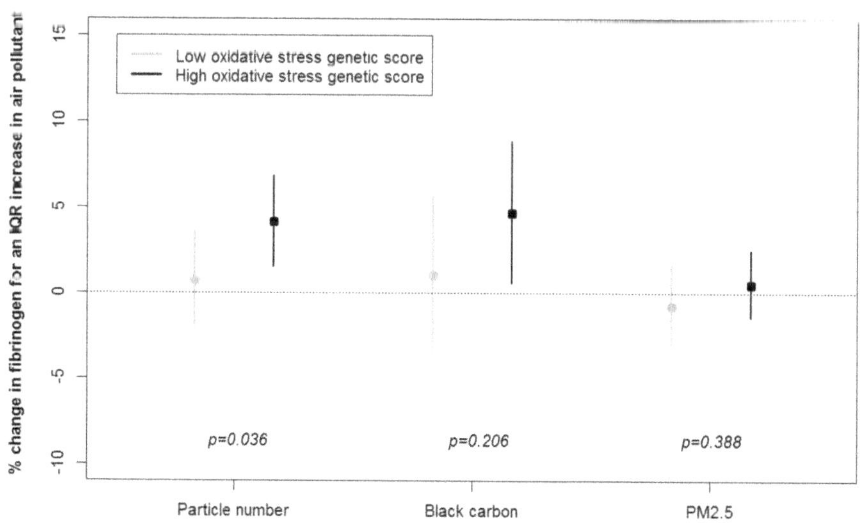

Figure 1. Percent change in fibrinogen for one interquartile range increase in air pollutant according to the oxidative stress genetic score (low versus high).

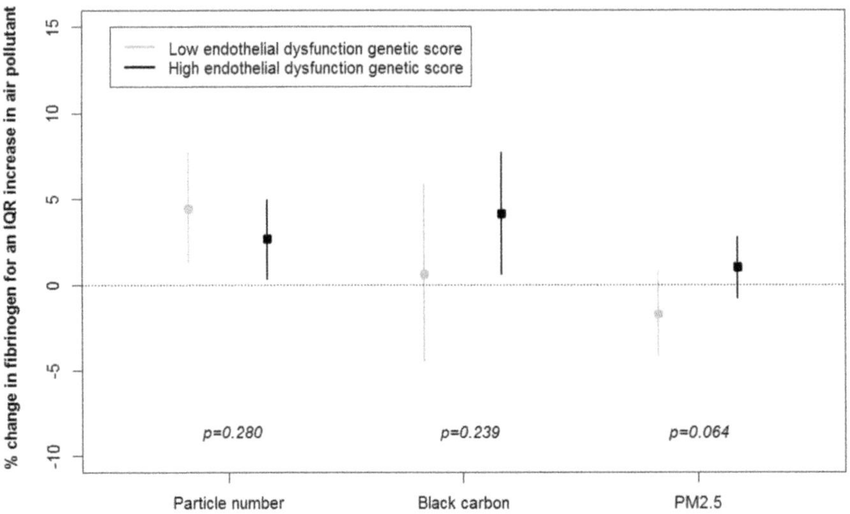

Figure 2. Percent change in fibrinogen for one interquartile range increase in air pollutant according to the endothelial dysfunction genetic score (low versus high).

and black carbon were significantly correlated (correlation coefficient of 0.72).

Genetic scores

Among the 18 gene variants candidates for the oxidative stress pathway, seven were chosen by the Lasso procedure: *rs1001179* (+), *rs480575* (−), *rs2071746* (+), *rs5995098* (+), *rs1800566* (+), *rs2282679* (−), and *rs3170633* (−). (+) and (−) represent the sign we used when we added these seven variants up to create the genetic score. Among the nine gene variants candidates for the endothelial function pathway, the Lasso approach chose four of them: *rs12944039* (+), *rs2072324* (+), *rs2255929* (+), and *rs1137933* (−). Similarly, among the nine gene variants candidates for the metal processing pathway, five were chosen by the Lasso penalization *rs224572* (−), *rs422982* (+), *rs12227734* (−), *rs1005559* (−), and *rs1799945* (−). Tables S1 to S3 in File S1 present the *rs* numbers and their corresponding genes.

Gene-air pollution interactions by pathways

We observed significant genetic pathway-air pollution interactions on fibrinogen, C-reactive protein, and ICAM-1. We found that the effect of particle number on fibrinogen was modified by the oxidative stress genetic score ($p_{interaction} = 0.04$) (Figure 1). An interquartile range increase in particle number was associated with a 0.7% (95% CI: −2.0% to 3.5%) and a 4.1% (95% CI: 1.5% to 6.8%) increase in fibrinogen, among participants with low and high oxidative stress scores, respectively. Among participants classified with high endothelial dysfunction genetic scores, we found a greater $PM_{2.5}$ association with fibrinogen than among individuals with lower endothelial dysfunction genetic scores ($p_{interaction} = 0.06$) (Figure 2). We also found some effect modification by the allelic profile related to metal processing. Compared to individuals classified with lower metal processing genetic scores, among participants with higher scores we found greater associations between particle number and fibrinogen ($p_{interaction} = 0.12$), C-reactive protein ($p_{interaction} = 0.02$), and ICAM-1 ($p_{interaction} = 0.08$) (Figure 3). For instance, an interquartile range increase in particle number was associated with a 1.9% (95% CI: −10.0% to 15.5%) and a 19.8% (95% CI: 7.3% to 33.9%) increase in C-reactive protein, among participants classified with low and high metal processing genetic scores, respectively.

In addition, we observed significant associations between black carbon and fibrinogen only among those with high genetic score for oxidative stress, endothelial dysfunction, and impaired metal processing (Figures 1 to 3). Numerical values of the regression coefficients and their associated 95% confidence intervals are presented in Tables S7 to S9 in File S1.

Sensitivity analysis

After adjusting for multiple comparisons using the Bonferroni correction, none of the interactions remained statistically significant.

Discussion

We developed a new method to investigate the role of air pollution interacting with genetic variation within relevant pathways. Our genetic score approach suggests oxidative stress, endothelial function, and metal processing as biological pathways that play a role in enhancing air pollution effects on intermediary CVD-related blood markers. We observed significant greater associations between one-week exposure to particulate air pollution and fibrinogen, C-reactive protein, and ICAM-1 among elderly participants with higher allelic risk profiles, as compared to the same associations among individuals with lower allelic risk profiles.

Our results support previous findings that showed similar intermediate-term associations between air pollutants (main effects and gene-environment interactions) and markers of coagulation and inflammation [3,5,27]. In the same cohort, previous studies found that levels of fibrinogen, C-reactive protein, ICAM-1, and VCAM-1 were higher after days or weeks of exposure to primary and secondary pollutants [4,23]. Peters et al. reported that the association between 5-day averages of PM_{10} and fibrinogen was modified by single-nucleotide polymorphisms (SNPs) within the fibrinogen genes [28]. Another study found some significant interactions between air pollution exposure and fibrinogen polymorphisms [27]. Madrigano et al. also observed stronger associations between black carbon exposure and VCAM-1 among participants who were GSTM1 null homozygous [5].

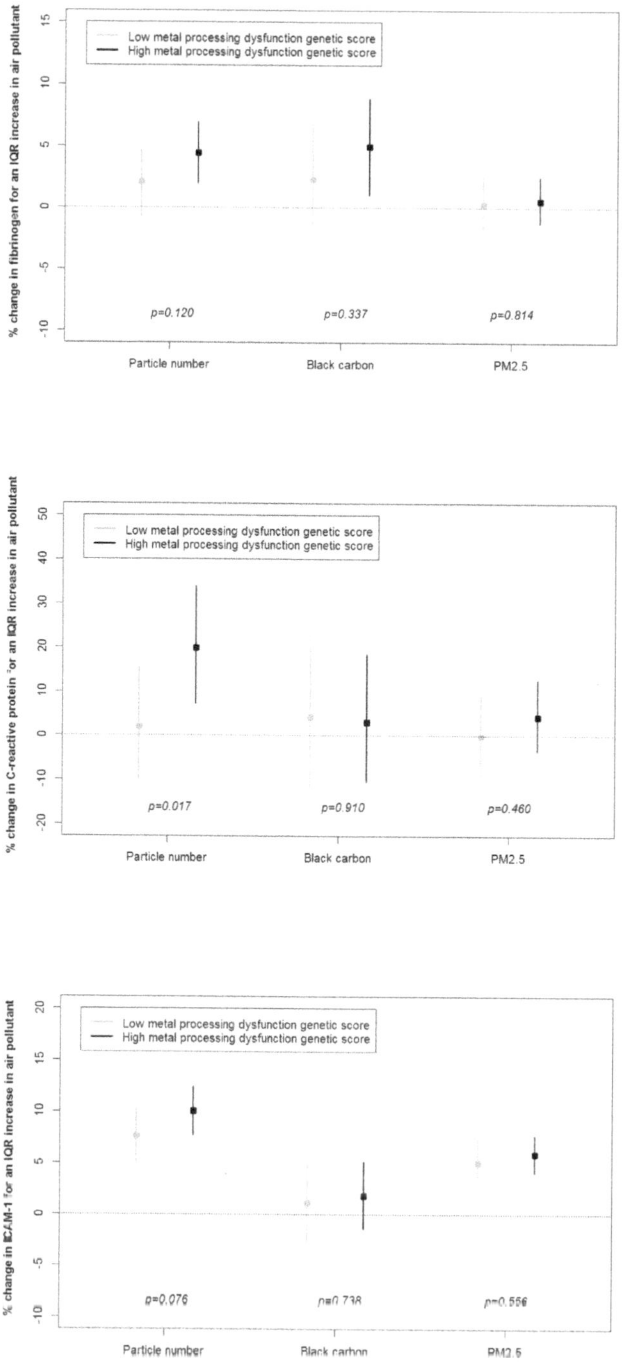

Figure 3. Percent change in blood markers (fibrinogen, C-reactive protein, and ICAM-1) for one interquartile range increase in air pollutant according to metal processing dysfunction genetic score (low versus high).

The greater increases in coagulation and inflammatory markers observed among subgroups with higher allelic risk profiles happened after one-week exposure to particle number and black carbon, which are traffic-related air pollutants. In Boston, black carbon originates mostly from local emissions during the morning and afternoon rush hours, as well as from transported traffic emissions [29]. We observed significant associations between black carbon and fibrinogen among participants with higher genetic score related to the three pathways of interest. This finding suggests that genetic variations related to oxidative stress, endothelial function, and metal processing play an important role in enhancing coagulation responses occurring after black carbon exposures.

In addition, while particle number was not associated with C-reactive protein in the overall sample [23], we found a positive association among the sensitive subgroup with a higher allelic risk profile related to metal processing dysfunction. Another interesting finding in this study is that the metal processing genetic score modified the association between particle number and fibrinogen, C-reactive protein, and ICAM-1, but not the effect of fine particles. Particle number concentrations are surrogates for ultrafine particles and come mostly from fresh local traffic emissions. In Boston, although secondary organic aerosols, sulfate particles, and metals constitute a large fraction of $PM_{2.5}$ mass [30], metals also are constituents of ultrafine particles. A previous study has suggested that transition metals bound to ambient particles and their related oxidative stress response may play an important role in cardiac toxicity of particles [15]. A study suggested that either particles or metals may translocate into the bloodstream, resulting in direct toxic effects on the atherosclerotic plaque stability, vascular endothelium function, and thrombosis [16]. Our results seem plausible and consistent with this last study and add more evidence on health effects due to metal toxicity. Our findings suggest that increased coagulation and systemic inflammatory responses may occur after metals exposure and that ultrafine particles may carry more adverse metals than fine particles.

Our results also support previous studies suggesting oxidative stress as an intermediary pathway between air pollution and cardiovascular related events [6,7,11,12,31,32]. Oxidative stress can arise from an imbalance between the production of reactive nitrogen or oxygen species (O_2^-, H_2O_2, and $ONOO^-$) and a biological system's ability to promptly detoxify intermediary reactive products or simply repair the resulting damage [33]. Disturbances in the normal redox state can cause many adverse biological effects through the production of peroxides and free radicals which damage all components of the cell. For instance, oxidative stress can lead to lipid, protein, and DNA oxidation, leading to the formation of adducts which disrupt normal cellular function. Traffic emissions have also been associated with an increase in reactive oxygen species [34], as well as a diminution of antioxidant enzyme activity [35]. Oxidative stress can also initiate proinflammatory cascades, which in turn lead to activation of antioxidant proteins and substances such as catalase and glutathione. In addition, participants with genetic variants related to oxidative stress, such as heme oxygenase 1, had a higher risk of recurrent venous thromboembolism [32]. The associations we observed between traffic-related air pollutants and fibrinogen among those with higher oxidative stress genetic score support this last study. Variants within genes that mediate oxidative stress can convey susceptibility to traffic-related air pollution and enhance coagulation responses, atherosclerosis, and aging [36].

In our study, we observed a positive association between black carbon and fibrinogen only among participants with higher genetic risk profile related to endothelial function. There is some evidence that oxidative stress and impaired endothelial function are interrelated. Many of the genes implicated in diseases associated with vascular dysfunction, such as atherosclerosis, are oxidative stress-sensitive, suggesting that an imbalance in cellular oxidative stress is a critical underlying factor [37]. The most prevailing theory of impaired endothelial function is an increase in reactive oxygen species, which can impair nitric oxide (NO) production and activity. For example, Miller et al. suggested that exposure to diesel exhaust particulates leads to oxidative stress through the generation of oxygen-centered free radicals that reduce the bioavailability of endothelium-derived NO without prior interaction with the lung or vascular tissue [38]. These two interconnected mechanisms could explain why we observed associations between black carbon and fibrinogen only among individuals with high genetic risk profiles related to oxidative stress and endothelial function. The endothelium is the layer of epithelial cells that lines the cavities of the heart, blood, lymph vessels, and the serous cavities of the body. Impaired endothelial function leads to impaired control of blood pressure, inflammation, formation of new blood vessels, and blood clotting [39]. Endothelial dysfunction is considered as a key early event in the development of atherosclerosis [1]. A high endothelial dysfunction score, representing a higher allelic risk profile, may therefore identify a subset of participants having stronger coagulation responses to traffic-related air pollution.

Limitations and strengths

This novel score approach we developed requires relevant instrumental outcomes to perform the first selection stage. The Lasso procedure returns regression coefficients that can be used to create the score. Since our outcome instruments were specific pathway-related measurements but did not represent the entire pathway, we chose to use weights equal to 1 to construct the score from the selected gene variants. Researchers can decide which weights they prefer to use based on their instrumental outcomes. We assumed equal weights for each of the gene variants chosen by the Lasso penalization. This assumption may also be a limitation, as some of the genes polymorphisms might have different functional effects and magnitude. Our defined biological pathways were not inclusive of all genes which regulate oxidative stress, endothelial dysfunction, or impaired metal processing. Therefore, we cannot definitely rule out false negative findings. We found some strong evidence of interactions between genetic-pathway and traffic-related air pollutants, which have been measured with ambient local monitors. Although this personal urban exposure approximation is likely to induce some Berkson measurement error [40], correction for measurement error may yield less biased estimates. The air pollution exposure assessment also did not take into account individual daily activities, especially whether some participants spend more time indoors versus outdoors. Traffic emission sources of organic chemicals have been associated with increased systemic inflammation [41]. Organic carbon monitoring was only available for the last few years of our study and we therefore could not address this issue. To minimize issues related to multiple testing, we focused on only one moving average and three air pollutants. Different exposure windows and pollutants may also be relevant. Our results did not remain significant after Bonferroni's correction. Even though this adjustment is very conservative, we cannot rule out false positive results. We considered a limited number of gene variants, especially for the endothelial function and metal processing pathways. Considering a larger number of variants at the first selection stage may yield more accurate scores.

This new statistical approach extends our knowledge on known and potential genetic pathways between air pollution and cardiovascular outcomes by prospectively investigating genetic pathway-environment interactions. Our score approach, selecting the gene variants by pathway and penalizing them by a Lasso procedure, enables us to reduce the set of candidate gene variants within pathway, to test for gene-air pollution interactions by pathway, and to reduce multiple comparisons since the significance examination of the interaction is based on a one degree of freedom test. Pathway analysis borrows information from correlated genes that belong to the same pathway and provides results with increased power. Our results demonstrated that the association between air pollutants and cardiovascular-related biomarkers are likely to differ according to the genetic background.

Acknowledgments

We thank Tania Kotlov.

Author Contributions

Conceived and designed the experiments: JS PV. Performed the experiments: HS RW AB. Analyzed the data: MAB BC JS. Wrote the paper: MAB.

References

1. Brook RD, Rajagopalan S, Pope CA 3rd, Brook JR, Bhatnagar A, et al. (2010) Particulate matter air pollution and cardiovascular disease: An update to the scientific statement from the american heart association. Circulation 121: 2331–2378.
2. Nemmar A, Hoylaerts MF, Hoet PH, Dinsdale D, Smith T, et al. (2002) Ultrafine particles affect experimental thrombosis in an in vivo hamster model. Am J Respir Crit Care Med 166: 998–1004.
3. Ruckerl R, Ibald-Mulli A, Koenig W, Schneider A, Woelke G, et al. (2006) Air pollution and markers of inflammation and coagulation in patients with coronary heart disease. Am J Respir Crit Care Med 173: 432–441.
4. Alexeeff SE, Coull BA, Gryparis A, Suh H, Sparrow D, et al. (2011) Medium-term exposure to traffic-related air pollution and markers of inflammation and endothelial function. Environ Health Perspect 119: 481–486.
5. Madrigano J, Baccarelli A, Wright RO, Suh H, Sparrow D, et al. (2010) Air pollution, obesity, genes and cellular adhesion molecules. Occup Environ Med 67: 312–317.
6. Zanobetti A, Baccarelli A, Schwartz J (2011) Gene-air pollution interaction and cardiovascular disease: A review. Prog Cardiovasc Dis 53: 344–352.
7. Chahine T, Baccarelli A, Litonjua A, Wright RO, Suh H, et al. (2007) Particulate air pollution, oxidative stress genes, and heart rate variability in an elderly cohort. Environ Health Perspect 115: 1617–1622.
8. Ren C, Baccarelli A, Wilker E, Suh H, Sparrow D, et al. (2010) Lipid and endothelium-related genes, ambient particulate matter, and heart rate variability – the VA normative aging study. J Epidemiol Community Health 64: 49–56.
9. Van Hee VC, Adar SD, Szpiro AA, Barr RG, Roux AD, et al. (2010) Common genetic variation, residential proximity to traffic exposure, and left ventricular mass: The multi-ethnic study of atherosclerosis. Environ Health Perspect 118: 962–969.
10. Lee S, Park Y, Zuidema MY, Hannink M, Zhang C (2011) Effects of interventions on oxidative stress and inflammation of cardiovascular diseases. World J Cardiol 3: 18–24.
11. Mazzoli-Rocha F, Fernandes S, Einicker-Lamas M, Zin WA (2010) Roles of oxidative stress in signaling and inflammation induced by particulate matter. Cell Biol Toxicol 26: 481–498.
12. Ren C, Park SK, Vokonas PS, Sparrow D, Wilker E, et al. (2010) Air pollution and homocysteine: More evidence that oxidative stress-related genes modify effects of particulate air pollution. Epidemiology 21: 198–206.
13. Thomson EM, Kumarathasan P, Calderon-Garciduenas L, Vincent R (2007) Air pollution alters brain and pituitary endothelin-1 and inducible nitric oxide synthase gene expression. Environ Res 105: 224–233.
14. Lim S, Chung HU, Paek D (2010) Low dose mercury and heart rate variability among community residents nearby to an industrial complex in korea. Neurotoxicology 31: 10–16.
15. Park SK, O'Neill MS, Wright RO, Hu H, Vokonas PS, et al. (2006) HFE genotype, particulate air pollution, and heart rate variability: A gene-environment interaction. Circulation 114: 2798–2805.
16. Sangani RG, Soukup JM, Ghio AJ (2010) Metals in air pollution particles decrease whole-blood coagulation time. Inhal Toxicol 22: 621–626.

17. Ghio AJ, Richards JH, Dittrich KL, Samet JM (1998) Metal storage and transport proteins increase after exposure of the rat lung to an air pollution particle. Toxicol Pathol 26: 388–394.
18. Ghio AJ, Piantadosi CA, Wang X, Dailey LA, Stonehuerner JD, et al. (2005) Divalent metal transporter-1 decreases metal-related injury in the lung. Am J Physiol Lung Cell Mol Physiol 289: L460–7.
19. Molinelli AR, Madden MC, McGee JK, Stonehuerner JG, Ghio AJ. (2002) Effect of metal removal on the toxicity of airborne particulate matter from the utah valley. Inhal Toxicol 14: 1069–1086.
20. Danesh J, Collins R, Appleby P, Peto R (1998) Association of fibrinogen, C-reactive protein, albumin, or leukocyte count with coronary heart disease: Meta-analyses of prospective studies. JAMA 279: 1477–1482.
21. Hwang SJ, Ballantyne CM, Sharrett AR, Smith LC, Davis CE, et al. (1997) Circulating adhesion molecules VCAM-1, ICAM-1, and E-selectin in carotid atherosclerosis and incident coronary heart disease cases: The atherosclerosis risk in communities (ARIC) study. Circulation 96: 4219–4225.
22. Baja ES, Schwartz JD, Wellenius GA, Coull BA, Zanobetti A, et al. (2010) Traffic-related air pollution and QT interval: Modification by diabetes, obesity, and oxidative stress gene polymorphisms in the normative aging study. Environ Health Perspect 118: 840–846.
23. Bind MA, Baccarelli A, Zanobetti A, Tarantini L, Suh H, et al. (2012) Air pollution and markers of coagulation, inflammation, and endothelial function: Associations and epigene-environment interactions in an elderly cohort. Epidemiology 23: 332–340.
24. Safran M, Dalah I, Alexander J, Rosen N, Iny Stein T, et al. (2010) GeneCards version 3: The human gene integrator. Database (Oxford) 2010: baq020.
25. Tibshirani R (1996) Regression shrinkage and selection via the lasso. Journal of the Royal Statistical Society 58: 267–288.
26. Efron B, Hastie T, Johnstone I, Tibshirani R (2004) Least angle regression. The Annals of Statistics 32: 407.
27. Ljungman P, Bellander T, Schneider A, Breitner S, Forastiere F, et al. (2009) Modification of the interleukin-6 response to air pollution by interleukin 6 and fibrinogen polymorphisms. Environ Health Perspect 117: 1373–1379.
28. Peters A, Greven S, Heid IM, Baldari F, Breitner S, et al. (2009) Fibrinogen genes modify the fibrinogen response to ambient particulate matter. Am J Respir Crit Care Med 179: 484–491.
29. Park SK, O'Neill MS, Vokonas PS, Sparrow D, Schwartz J (2005) Effects of air pollution on heart rate variability: The VA normative aging study. Environ Health Perspect 113: 304–309.
30. Kang CM, Koutrakis P, Suh HH (2010) Hourly measurements of fine particulate sulfate and carbon aerosols at the harvard-U.S. environmental protection agency supersite in boston. J Air Waste Manag Assoc 60: 1327–1334.
31. Mordukhovich I, Wilker E, Suh H, Wright R, Sparrow D, et al. (2009) Black carbon exposure, oxidative stress genes, and blood pressure in a repeated-measures study. Environ Health Perspect 117: 1767–1772.
32. Mustafa S, Weltermann A, Fritsche R, Marsik C, Wagner O, et al. (2008) Genetic variation in heme oxygenase 1 (HMOX1) and the risk of recurrent venous thromboembolism. J Vasc Surg 47: 566–570.
33. Agarwal A, Saleh RA, Bedaiwy MA (2003) Role of reactive oxygen species in the pathophysiology of human reproduction. Fertil Steril 79: 829–843.
34. Delfino RJ, Staimer N, Tjoa T, Polidori A, Arhami M, et al. (2008) Circulating biomarkers of inflammation, antioxidant activity, and platelet activation are associated with primary combustion aerosols in subjects with coronary artery disease. Environ Health Perspect 116: 898–906.
35. Delfino RJ, Staimer N, Tjoa T, Gillen DL, Polidori A, et al. (2009) Air pollution exposures and circulating biomarkers of effect in a susceptible population: Clues to potential causal component mixtures and mechanisms. Environ Health Perspect 117: 1232–1238.
36. Runge MS, Molnar K, Madamanchi NR (2010) "Old" hearts and arteries: The role of oxidative stress. Trans Am Clin Climatol Assoc 121: 52–60.
37. Hennig B, Oesterling E, Toborek M (2007) Environmental toxicity, nutrition, and gene interactions in the development of atherosclerosis. Nutr Metab Cardiovasc Dis 17: 162–169.
38. Miller MR, Borthwick SJ, Shaw CA, McLean SG, McClure D, et al. (2009) Direct impairment of vascular function by diesel exhaust particulate through reduced bioavailability of endothelium-derived nitric oxide induced by superoxide free radicals. Environ Health Perspect 117: 611–616.
39. Spanier AJ, Kahn RS, Hornung RW, Wang N, Sun G, et al. (2009) Environmental exposures, nitric oxide synthase genes, and exhaled nitric oxide in asthmatic children. Pediatr Pulmonol 44: 812–819.
40. Zeger SL, Thomas D, Dominici F, Samet JM, Schwartz J, et al. (2000) Exposure measurement error in time-series studies of air pollution: Concepts and consequences. Environ Health Perspect 108: 419–426.
41. Delfino RJ, Staimer N, Tjoa T, Arhami M, Polidori A, et al. (2010) Association of biomarkers of systemic inflammation with organic components and source tracers in quasi-ultrafine particles. Environ Health Perspect 118: 756–762.

8

Allostatic Load Model Associated with Indoor Environmental Quality and Sick Building Syndrome among Office Workers

Chien-Cheng Jung[1], Hsiu-Hao Liang[1], Hui-Ling Lee[2], Nai-Yun Hsu[1], Huey-Jen Su[1]*

1 Department of Environmental and Occupational Health, College of Medicine, National Cheng Kung University, Tainan city, Taiwan, **2** Department of Chemistry, College of Science and Engineering, Fu Jen Catholic University, New Taipei city, Taiwan

Abstract

This study investigates whether indoor environmental quality (IEQ) influences allostatic load (AL) and whether AL can be a predictor for sick building syndrome (SBS). We also assessed and compared the associations between AL and SBS versus 8-hydroxydeoxyguanosine (8-OHdG) and SBS. A total of 115 office workers from 21 offices completed self-reported SBS questionnaires, and provided 11 biomarkers for their AL. Multiple linear regressions and logistic regression analysis were applied to examine the correlations between IEQ and AL or 8-OHdG and between AL or 8-OHdG and SBS, respectively. Our data revealed that the neuroendocrine system was correlated with CO_2, the difference between indoor and outdoor CO_2 levels (dCO_2), and the indoor-outdoor ratio of CO_2 (CO_2 I/O). Metabolic system effects were associated with illumination. The relationships between illumination, CO_2, dCO_2, CO_2 I/O and 8-OHdG were consistent with those and AL in specific systems. Furthermore, we found that risks for SBS syndromes were related with neuroendocrine and metabolic system of the AL. 8-OHdG was associated with eye dryness or irritation, eye tiredness and vomiting. We conclude that IEQ significantly influences AL and that AL can be a predictor for reporting SBS with information on system-specific effects.

Editor: Yinping Zhang, Tsinghua University, China

Funding: The project is funded by Taiwan Environmental Protection Administration (EPA-100-FA11-03-A018, EPA-101-FA11-03-D059). The funders had no role in study design, data collection and analysis, decision to publish, or preparation of the manuscript.

Competing Interests: The authors have declared that no competing interests exist.

* E-mail: hjsu@mail.ncku.edu.tw

Introduction

Sick building syndrome (SBS) refers to situations in which building occupants suffer from discomforts of the nose, skin or eyes, and the symptoms are relieved when the occupants leave the building. Both personal and environmental factors may be determinants for SBS [1]. Environmental factors include ventilation rate, levels of total volatile organic compounds (TVOCs) or water dampness, etc. [2–5], while personal factors include stress from job or socioeconomic status, life quality, job satisfaction, and others [6–10].

Currently, researchers have investigated biomarkers that may be associated with SBS, one of which is 8-OHdG. The oxidized nucleoside 8-OHdG is the most frequently detected DNA lesion resulting from the action of reactive oxygen species, and upon DNA repair, this molecule is excreted in the urine [11–13]. Previous studies have found that urinary 8-OHdG in smokers and subjects of frequent exposure to traffic pollution is higher than that of non-smokers and subjects with low exposure to traffic pollution [14,15]. Therefore, 8-OHdG can be adopted as an effect marker for exposure to pollution. Furthermore, urinary 8-OHdG is associated with both indoor pollutant exposure and self-reported SBS [16,17]. Although these studies suggest that 8-OHdG is a potential marker for reporting SBS, the etiology of SBS has yet to be fully understood, and a comprehensive approach for examining SBS is not available.

Allostasis is a process that reestablishes homeostasis after the human body experiences neuroendocrine stress response [18]. However, continued or chronic stress can result in overactive stress responses or exhaustion, leading to wear and tear on the body, referred to as allostatic load (AL). In other words, frequent activation of stress response can damage the body over the long run and increase risks of cardiovascular diseases, diabetes, depression, or neurodegenerative diseases [19–21]. The occurrence of AL is often associated with stress from work, socioeconomic status [22–27], adverse childhood experiences, or chronic stress in childhood [28–31]. Given that AL is a physiological consequence of stress and that stress is one risk factor for SBS, understanding the relationship between SBS and AL may provide further insights for the etiology of SBS.

Recent studies indicate that environmental factors may derive various levels of biomarkers or physiologic parameters that are adopted to quantify AL. Systolic and diastolic pressure, heart rate, triacylglycerol, and hemoglobin A1c were found to be significantly increased upon exposure to increased $PM_{2.5}$ and PM_{10} [32,33]. A1c level was also significantly associated with ozone exposure [32]. In addition, exposure to TVOC, fungi or β-1,3-D Glucan caused elevated interleukin (IL)-6, tumor necrosis factor (TNF)-α and C-reactive protein (CRP) levels [34–37]. Perry and co-workers found that illumination was associated with creatinine (Perry et al., 1979). These studies suggest that AL could be affected by environmental factors, although there has been no investigation for whether environmental factors influence AL.

Our hypothesis is that indoor environmental quality may affect AL level. We therefore aimed to investigate the association between AL and indoor environmental quality in office workers and their corresponding office spaces, respectively. We further hypothesized that AL may account for the reporting of SBS symptoms, and therefore the relationship between AL and SBS symptoms was assessed. In view of the existing literature, we also measured 8-OHdG levels and investigated the relationships among 8-OHdG, IEQ and SBS, respectively.

Materials and Methods

Study design and study subjects

A cross-sectional study was carried out from July 2011 to December 2012. We randomly selected four organizations in Kaohsiung and Tainan, Taiwan, who agreed to participate in the study. Twenty-one office spaces and 143 office workers were randomly chosen from the four organizations mentioned above, and subjects younger than 20 years, part-time employees, and subjects with mental diseases were excluded from further analyses [38,39]. Another 28 subjects were further eliminated either for the lack of their urine and saliva samples, or for missing information on the questionnaires. A total of 115 subjects were included in the final assessment. Indoor and outdoor environmental quality of these offices was assessed during office hours (09:00–17:00). Physiological measurements, urine and saliva samples were collected from participants at the end of the workday. All participants were also asked to complete a self-reported questionnaire including questions regarding their personal information and assessment of SBS. The study was approved (ER-99-040) by Institutional Review Board of National Cheng Kung University Hospital, Tainan, Taiwan. All participants signed and provided their consent form to study group.

Indoor and outdoor environmental quality measurement

Indoor and outdoor levels of carbon monoxide (CO), carbon dioxide (CO_2), total volatile organic compounds (TVOC), particulate matter diameter below 2.5 micrometers ($PM_{2.5}$), fungi, bacteria and illumination were monitored simultaneously during office hours. CO and CO_2 were examined by the Q-TRAK Indoor Air Quality Monitor (Model 7575, TSI Corporation, Shoreview, USA). A real-time RAE/PGM-730 instrument was used to measure the levels of TVOC (Model ppbRAE 3000, ProRAE Corporation, San Jose, USA). The level of $PM_{2.5}$ was determined by a DUST-TRAK Aerosol Monitor (Model 8520, TSI Corporation, Shoreview, USA). Fungi and bacteria were sampled every three hours by a signal-stage Anderson bioaerosol sampler with the flow rate of 28.3 l/min and sampling time of one min. Concentrations of bacteria and fungi in the air were obtained by two culturing media and conditions. Dichloran Glycerol (DG-18) at 25±1°C for five days and Tryptic Soy Agar (TSA) at 30±2°C for two days. Illumination was measured by an Illumination instrument (CL2000, Minolta Co., Japan), and these data were measured on each subject' desk.

Biomarkers and the measure

The AL model based on multi-systemic physiological measurements was used to estimate the AL [40]. Each participant received eleven non-invasive measurements categorized in five systems: (1) cardiovascular and respiratory system, including systolic blood pressure (SBP), diastolic blood pressure (DBP) and heart rate (Terumo, Japan); (2) anthropometric system, including body mass index (BMI) and body fat (OMRON, Japan); (3) neuroendocrine system, including epinephrine and norepinephrine in urine (liquid chromatography with tandem mass spectrometry, LC-MS/MS), and cortisol (enzyme-linked immunosorbent assay (ELISA), BD Biosciences, USA); (4) metabolic system, including creatinine in urine (liquid chromatography); and (5) immune system, including IL-6 and TNF-α in saliva (ELISA kit, BD Biosciences, USA). Epinephrine, norepinephrine and cortisol were expressed as micrograms per gram creatinine. Saliva and urine samples were centrifuged at 3000 rpm for 10 min at 4°C, and the supernatant was drawn and stored at −80°C until analysis. To stabilize epinephrine and norepinephrine in urine, 0.5 ml H_2SO_4 (6 N) was added to urine samples before storage. A simple count score was used as a measure of AL. The cut-off point was set to the 75th percentile, and the number of biomarkers for which the individual falls into the highest risk quartile of each measure's sample distribution was taken as the AL score. Urinary 8-OHdG was determined by liquid chromatography with tandem mass spectrometry (LC-MS/MS).

Questionnaire

The questionnaire contained questions for personal information and an SBS (SBS standard questionnaire, [41]). Queries for personal information included age, gender, working years, weekly working hours, education, monthly income, marital status, smoking behavior and exercise frequency. For SBS symptoms, participants self-assessed if during the past four weeks inside the building they had experienced any of the following conditions: eye dryness or irritation; stridor; headache; dry throat; tiredness; chest discomfort; nose itch; cough; eye tiredness; nervous feeling; pain in neck, shoulders, or back; sneeze; concentration difficulty; dizziness; depression; breathing difficulty; vomit; or skin dryness or itch.

Statistical analysis

Chi-square was used to test the difference of SBS prevalence between males and females and the difference of personal characteristics. A multiple linear regression model was applied to investigate the association between indoor environmental quality and AL score or 8-OHdG level, and a logistic regression model was used to calculate odd ratio (OR). Data analysis was performed by SAS 9.2 software (SAS Institute Inc, Cary, USA). The statistical significance was set to $p<0.05$.

Results

Study population characteristics

The study recruited 115 office workers from 21 offices to assess the relationship between indoor environmental quality, allostatic load and SBS risk. Characteristics of the subjects are summarized in Table 1. Of the study population, 83 were male and 32 female, and the mean age was 34.2 years (SD = 5.7). The majority of participants (86.1%) were non-smokers, and approximately half (51.4%) habitually exercised. Other parameters are shown in Table 1.

Based on the allostatic load model proposed by Juster et al., we took multi-parameter measurements from 115 participants, and defined the AL score for each individual as the number of measures that fell outside of normal range [40]. The cut-off point was set to be the 75th percentile of each measure's sample distribution in our study, as shown in Table 2. From a total of 11 biomarker measurements, the average total AL score was 2.6 (SD = 1.7) (Table 3), with the highest 7 and the lowest 0.

All subjects were required to provide self-reported SBS symptoms according to a standard SBS questionnaire developed

Table 1. Summary statistics for characteristics of subjects.

Variables	Mean ±S.D. or n (%)
Age (years)	34.2±5.7
Males	83 (72.2%)
Working years	4.4±4.3
Weekly working hours	42.5±14.2
Education (years)	
≤12 (High school)	3 (2.6%)
13–16 (College)	66 (57.4%)
≥17 (Master or higher)	46 (40.0%)
Monthly income (NTD)*	
<20,000	2 (0.9%)
20,000–40,000	36 (31.3%)
40,000–60,000	28 (24.3%)
60,000–80,000	29 (25.2%)
80,000–100,000	8 (7.0%)
≥100,000	12 (10.4%)
Marital status (n, %)	
Unmarried or single	59 (51.3%)
Married or live together	54 (47.0%)
Others	2 (1.7%)
Smoking behavior (n, %)	
Non-smoker	99 (86.1%)
Smoker	16 (13.9%)
Exercise frequency (≥45 min per week. n, %)**	
Always	5 (4.5%)
Usually	
Often	16 (14.5%)
Sometime	57 (51.4%)
Never	11 (9.9%)

S.D. Standard deviation NTD: New Taiwanese Dollars.
*Missing 1 samples **Missing 4 samples.

by the World Health Organization. We analyzed the prevalence of SBS symptoms, and found the most common symptom be eye tiredness while the least to be stridor (Table 4). In addition to eye tiredness, the 8 most common discomforts of more than 30% prevalence included eye dryness or irritation, nose itching, cough, nervous, pain in neck, shoulders or back, sneeze, concentration difficulty, and depression. We also discovered a generally higher prevalence rate of SBS in females than in males, and thus performed chi-square test to evaluate the significance of the observation. In support of this observation, several symptoms were significantly more prevalent ($p<0.05$) in women, including eye dryness or irritation, dry throat, eye tiredness, sneeze, concentration difficulty, dizziness, depression, vomit, and skin dryness or itch, as shown in Table 4.

Indoor environmental quality

The indoor environmental indicators levels in the 21 offices are summarized in Table 5. The average $PM_{2.5}$ level in our study (37.9 ± 31.3 μg/m^3) was higher than the Taiwan indoor air quality standards (35 μg/m^3) while the average level of the other pollutants were all below standards for each substance. However, we detected an extraordinary TVOC level of 1130 ppb in one office, which was substantially higher than the standard of 560 ppb, and we found that this office was located next to a clean room for organic solvent usage. Furthermore, the highest $PM_{2.5}$ and total fungi levels detected in this study of 143.7 μg/m^3 and 3180.2 CFU/m^3, respectively, were significantly greater than the standards of 35 μg/m^3 and 1000 CFU/m^3, respectively. Onsite investigation revealed that the office with high $PM_{2.5}$ was naturally ventilated and located in the vicinity of roads with heavy traffic, and the indoor-to-outdoor ratio of $PM_{2.5}$ was 0.80. On the other hand, the office with an extremely high level of fungi was located near areas of vegetation with natural ventilation, and the indoor-to-outdoor ratio of fungi was 0.90. Taken together, these findings indicate that types of ventilation and characteristics of external environment play an important role in indoor air quality.

Table 2. Cutoff for each biomarker of allostatic load.

Biomarkers	Cutoff point
BMI (kg/m^2)	≥26.8
Fat (%)	≥29.4
SBP (mmHg)	≥132.5
DBP (mmHg)	≥86.5
Heart rate (times/min)	≥83.0
IL-6 (pg/ml)	≥15.4
TNF-α (pg/ml)	≥10.7
Cortisol (μg/g creatinine)	≥20.8
Epinephrine (μg/g creatinine)	≥17.7
Norepinephrine (μg/g creatinine)	≥42.9
Creatinine (mg/dl)	≥109.3

Relationship between indoor environmental quality and AL score

As shown in Table 6, the relationship between indoor environmental quality and AL was analyzed by a multiple linear regression model. Age, gender, smoking behavior, education attainment, monthly income, marital status and exercise frequency were found to can affect AL condition [23,25–28,30]. In this study, we adjusted these variables to avoid causing effects. We found no association between total AL score and any of the indoor environmental indicators in our study. However, the AL score

Table 3. AL score and 8-OHdG level.

Biomarkers	Mean ±S.D.	Median	Min-Max
AL scores			
Cardio. and Resp. system	0.8±0.9	0	0–3
Anthropometric system	0.5±0.8	0	0–2
Neuroendocrine system	0.7±0.8	1	0–3
Immune system	0.4±0.7	0	0–2
Metabolic system	0.3±0.4	0	0–1
Total allostatic scores	2.6±1.7	2	0–7
8-OHdG (mg/g creatinine)	4.1±2.2	3.5	0.8–12.9

S.D. Standard deviation.
Cardio. and Resp. (Cardiovascular and respiratory).

Table 4. Prevalence of SBS in males and females.

Symptoms	All (%, n)	Male (%, n)	Female (%, n)	*p-value**
Eye dryness or irritation	35.7 (41)	28.9 (24)	53.1 (17)	<0.05
Stridor	2.6 (3)	1.2 (1)	6.3 (2)	0.18
Headache	29.6 (34)	26.5 (22)	37.5 (12)	0.25
Dry throat	27.8 (32)	19.3 (16)	50.0 (16)	<0.05
Tiredness	27.8 (31)	26.5 (22)	28.1 (9)	0.81
Chest discomfort	13.9 (16)	10.8 (9)	21.9 (7)	0.14
Nose itch	31.2 (37)	27.7 (23)	43.8 (14)	0.12
Cough	36.5 (42)	33.7 (28)	43.8(14)	0.439
Eye tiredness	63.5 (73)	54.2 (45)	87.5 (28)	<0.05
Nervous feeling	33.9 (39)	30.1 (25)	43.8 (14)	0.19
Pain in neck, shoulders or back	55.7 (64)	50.6 (42)	68.8 (22)	0.10
Sneeze	38.3 (44)	28.9 (24)	62.5 (20)	<0.05
Concentration difficulty	45.2 (52)	38.6 (32)	62.5 (20)	<0.05
Dizziness	23.5 (27)	18.1 (15)	37.5 (12)	<0.05
Depression	33.0 (38)	24.1 (20)	56.3 (18)	<0.05
Breathing difficulty	12.2 (14)	8.4 (7)	21.9 (7)	0.06
Vomit	13.4 (15)	7.2 (6)	28.1 (9)	<0.05
Skin dryness or itch	26.1 (30)	18.1 (15)	46.9 (15)	<0.05

*Statistical significance at $p<0.05$.

Table 5. Indoor environmental quality in 21 office units.

Environmental indexes	Mean ±S.D.	Median	Range
Illumination (Lux)*	495.2±392.0	384.5	57.6–1485.0
CO (ppm)†	0.73±0.10	0.74	0.54–0.90
CO_2 (ppm)†	742.0±212.3	664.2	464.0–1193.6
TVOC (ppb)†	539.1±348.8	380.6	1.4–1330.0
$PM_{2.5}$ (μg/m³)	37.9±31.3	27.3	6.1–126.5
Total fungi (CFU/m³)	507.2±751.5	254.4	115.0–3180.2
Total bacterial (CFU/m³)	385.2±222.4	333.3	123.7–927.3
dCO_2 (ppm)	285.5±190.3	264.2	31.2–701.9
CO_2 I/O	1.62±0.38	1.65	1.07–2.43

S.D. Standard deviation.
*Illumination was measured from each subject's desk.
†CO, CO_2 and TVOC missing 2, 2 and 1 samples, respectively.

on individual body systems was associated with specific environmental indexes. CO_2, the concentration difference of indoor and outdoor CO_2 (dCO_2) and the ratio of indoor and outdoor CO_2 (CO_2 I/O ratio) levels which were a marker of ventilation rate, were associated with AL score on the neuroendocrine system ($p<0.05$). Furthermore, there was a significantly negative relationship between illumination and AL score on the metabolic system. These results suggest that indoor environmental quality influences homeostasis of the body, and each indicator may target specific body systems.

Relationship between AL score and SBS

Logistic regression model was applied to calculate the odds ratio (OR) of SBS with AL score. The control variables included age, gender, smoking status, educational level, monthly income, marital status, exercise frequency, lux, CO, TVOC, $PM_{2.5}$, bacteria, fungi CO_2, dCO_2 and CO_2 I/O ratio, and results of the analysis were shown in Table 7. AL score on the neuroendocrine system was associated with reported cough and pain in neck, shoulders or back ($p<0.05$). AL scores on the metabolic system were significantly associated with depression ($p<0.05$). Collectively, our analysis suggests that the AL model could be a potential predictive marker for SBS.

Relationships between 8-OHdG and indoor environmental quality and SBS

The average 8-OHdG level was 4.1±2.2 mg/g creatinine (table 4) and it was associated with illumination, CO_2, dCO_2 levels and CO_2 I/O ratio ($p<0.05$) (table 6). Moreover, logistic regression model analysis shows that 8-OHdG level was correlated with likelihood of reporting eye dryness or irritation, eye tiredness and vomit ($p<0.05$) (table 7), which are part of the SBS.

Discussion

The current study reveals that allostatic load can be influenced by indoor environmental quality in addition to previously proposed personal risk factors such as job stress, socioeconomic status, and hunger [25,28,29]. More remarkably, we also demonstrate for the first time that AL scores are associated with the risk of SBS. Based on these findings, we propose that AL may potentially serve as a predictive index for SBS, and compared to 8-OHdG as a general indicator for DNA lesions, the analysis of AL allows further investigation of specific physical systems to help determine major risk factors of SBS.

The study measured allostatic load by examining 11 physiologic parameters on 115 subjects. The average AL score obtained in this study is 2.6±1.7, which is close to previous studies, which ranged from 2.8 to 3.4 [23,42–46]. Notably, in earlier studies blood samples were required for analysis of IL-6 and TNF-α, both of which were parameters for assessing the immunological system of AL [47], while our study measured these two parameters from the saliva samples. From the standpoint of feasibility of field assessment, saliva sample collection is non-invasive, easy to collect, and requires much lower cost than the collection and analysis of blood samples. More importantly, since they are produced upon systemic or local inflammation, IL-6 and TNF-α are secreted into both the circulatory system and the saliva, and salivary IL-6 and TNF-α are well correlated with plasma IL-6 and TNF-α, respectively [48–50]. In addition, recent studies found that immunological indexes in saliva can be used to predict effects of stress [51,52]. Together, these studies suggest that saliva is a valid tool for assessing immunological system of AL, and that non-invasive sampling may be sufficient to estimate specific systems of AL.

Several biomarkers that compose the multi-system index for AL were shown to be influenced by air pollutants. For example, exposure to $PM_{2.5}$ or PM_{10} can decrease lung function and increase blood pressure and heart rate in the cardiovascular and respiratory system [33,53–55]. Exposure to volatile organic compounds was associated with the levels of IL-6 and TNF-α in the immune system [36,37]. In addition, higher concentration of ozone exposure was associated with elevated glycated hemoglobin and blood pressure in the metabolic and cardiovascular and respiratory system, respectively [32]. Therefore, air pollution may be one risk factor for AL. The current study found that CO_2 concentration, dCO_2 and CO_2 I/O ratio are associated with AL in the neuroendocrine system ($p<0.05$, Table 6). Additionally, the

Table 6. β relating to indoor environmental quality, AL score and 8-OHdG level (Standard error).

Environmental factors	Total AL score	System AL score					8-OHdG
		Cardio. and Resp.	Anthropometric	Neuroendocrine	Immune	Metabolic	
Illumination (Lux)	0.000 (0.001)	0.000 (0.000)	0.000 (0.000)	0.000 (0.000)	−0.000 (0.000)	−0.000 (0.000)*	0.002 (0.000)*
CO_2 (ppm)	0.001 (0.001)	0.001 (0.000)	−0.000 (0.000)	0.001 (0.001)*	−0.001 (0.000)	−0.000 (0.000)	0.000 (0.000)*
CO (ppm)	−0.515 (0.489)	−0.332 (0.248)	−0.065 (0.215)	−0.224 (0.234)	0.487 (0.210)	−0.124 (0.123)	−0.636 (0.536)
TVOC (ppm)	−0.000 (0.001)	0.000 (0.000)	−0.000 (0.000)	0.000 (0.000)	−0.000 (0.000)	−0.000 (0.000)	−0.000 (0.001)
$PM_{2.5}$ (μg/m^3)	−0.014 (0.006)	−0.002 (0.003)	−0.002 (0.003)	−0.006 (0.003)	−0.005 (0.003)	−0.002 (0.002)	0.006 (0.008)
Total bacteria (CFU/m^3)	0.000 (0.001)	−0.001 (0.000)	0.001 (0.000)	0.000 (0.000)	−0.000 (0.000)	−0.000 (0.000)	0.002 (0.001)
Total fungi (CFU/m^3)	−0.000 (0.000)	0.000 (0.000)	−0.000 (0.000)	−0.000 (0.000)	−0.000 (0.000)	−0.000 (0.000)	0.000 (0.000)
dCO_2 (ppm)	0.001 (0.001)	0.001 (0.001)	−0.000 (0.000)	0.001 (0.001)*	−0.001 (0.001)	−0.000 (0.000)	0.004 (0.001)*
CO_2 I/O ratio	0.619 (0.592)	0.348 (0.304)	0.019 (0.260)	0.561 (0.263)*	−0.229 (0.255)	−0.278 (0.145)	1.978 (0.621)*

Adjusted for age, gender, smoking behavior, education attainment, monthly income, marital status and exercise frequency.
*Statistical significance at $p<0.05$.

study found that intensity of illumination has a negative relationship with AL on the metabolic system ($p<0.05$, Table 6), consistent with an earlier study that serum creatine level is negatively related to light exposure and light intensity [56]. Together, our research has shown that air pollutants influence the levels of AL, and each pollutant may impact only specific bodily systems.

Sick building syndrome refers to discomforts resulted from synergistic or independent effects of personal and environmental factors, with symptoms including headache, dry throat and concentration difficulty [1]. Among 115 subjects in this study, the most common symptoms are eye tiredness (63.5%), pain in neck, shoulders or back (55.7%), and concentration difficulty (45.2%) (Table 4). Other symptoms of more than 30% prevalence include eye dryness, nose itch, cough, nervous, sneeze and depression, which are similar to those in previous studies [57,58]. Moreover, our study reveals that female subjects are more prone than males to suffer from eye dryness or irritation, dry throat, eye tiredness, sneeze, concentration difficulty, dizziness, depression, vomit and skin dryness or itch. Consistently, earlier reports have indicated that female prevalence of eye irritation, tiredness, dryness, pain in throat, and skin dryness are higher than

Table 7. Odds ratio of SBS associated with AL scores and 8-OHdG level (Odds ratio 95% CI).

Symptoms	Total AL score	System AL score					8-OHdG
		Cardio. and Resp.	Anthropometric	Neuroendocrine	Immune	Metabolic	
Eye dryness or irritation	1.22(0.88–1.69)	1.26(0.66–2.38)	1.74(0.84–3.61)	1.28(0.62–2.66)	0.82(0.35–1.89)	1.00(0.25–4.02)	1.47(1.01–2.15)*
Headache	1.07(0.76–1.52)	1.03(0.52–2.03)	0.93(0.42–2.06)	1.05(0.48–2.31)	1.24(0.51–3.03)	1.43(0.31–6.51)	1.38(0.97–1.95)
Dry throat	1.25(0.90–1.74)	1.89(0.87–3.94)	1.77(0.82–3.83)	1.13(0.52–2.44)	1.12(0.48–2.63)	0.51(0.12–2.23)	1.38(0.97–1.97)
Tiredness	1.16(0.83–1.62)	1.38(0.70–2.74)	1.45(0.69–3.06)	1.02(0.47–2.20)	1.85(0.74–4.65)	0.20(0.04–1.01)	1.33(0.94–1.88)
Chest Discomfort	1.21(0.75–1.96)	1.06(0.41–2.77)	0.62(0.20–1.94)	1.95(0.73–5.23)	1.45(0.48–4.34)	4.84(0.47–50.30)	1.33(0.85–2.07)
Nose itch	0.82(0.58–1.16)	0.41(0.16–1.06)	0.57(0.24–1.32)	1.96(0.89–4.31)	0.68(0.25–1.84)	0.72(0.14–3.71)	1.39(0.97–1.97)
Cough	1.13(0.84–1.53)	0.97(0.51–1.85)	0.95(0.46–1.95)	3.31(1.36–8.06)**	0.97(0.46–2.08)	0.60(0.15–2.36)	1.27(0.91–1.76)
Eye tiredness	0.96(0.71–1.32)	1.18(0.62–2.26)	1.04(0.50–2.19)	1.20(0.59–2.43)	0.70(0.33–1.51)	0.29(0.07–1.28)	1.83(1.11–3.02)*
Nervousness	0.95(0.69–1.29)	0.74(0.36–1.51)	0.70(0.32–1.51)	1.02(0.50–2.08)	0.87(0.36–2.09)	3.34(0.81–13.84)	1.10(0.89–1.52)
Pain in neck, shoulders or back	1.13(0.85–1.49)	1.26(0.70–2.28)	0.96(0.51–1.81)	1.95(1.01–3.77)*	0.73(0.35–1.49)	0.93 (0.26–3.38)	1.07(0.79–1.46)
Sneeze	0.66(0.44–0.96)	0.34(0.13–0.91)	0.83(0.38–1.79)	1.22(0.55–2.73)	0.37(0.12–1.12)	0.18(0.03–1.05)	1.19(0.82–1.71)
Concentration difficulty	1.14(0.83–1.58)	1.15(0.59–2.23)	1.20(0.59–2.42)	1.76(0.85–3.62)	1.06 (0.48–2.32)	0.31(0.07–1.42)	1.45(0.96–2.19)
Dizziness	0.94(0.65–1.37)	0.57(0.24–1.36)	0.98(0.39–2.45)	1.39(0.57–3.41)	1.60(0.50–5.11)	0.64(0.13–3.32)	1.41(0.95–2.10)
Depression	1.16(0.82–1.64)	0.83 (0.41–1.67)	0.92(0.43–1.94)	1.41(0.66–3.01)	1.57(0.65–3.80)	6.15(1.29–29.39)*	1.24(0.86–1.80)
Breathing difficulty	0.86(0.46–1.61)	0.52(0.12–2.36)	0.76 (0.20–2.92)	2.57(0.61–10.80)	0.57(0.09–3.65)	–	1.49(0.74–2.98)
Vomit	0.90(0.59–1.37)	0.41(0.11–1.57)	0.36(0.10–1.24)	2.34(0.75–7.29)	1.38(0.45–4.23)	0.88(0.13–6.25)	1.91(1.11–3.27)*
Skin dryness or itch	1.36 (0.92–2.01)	0.43(0.75–3.97)	2.33 (0.90–6.06)	1.84(0.76–4.49)	0.64(0.22–1.86)	0.94(0.18–4.97)	1.01(0.71–1.45)

Adjusted for age, gender, smoking behavior, education attainment, monthly income, marital status, exercise frequency, lux, CO, CO_2, TVOC, $PM_{2.5}$, bacteria, fungi, CO_2 indoor and outdoor ratio, different of indoor and outdoor CO_2.
*Statistical significance $p<0.05$ ** Statistical significance $p<0.01$.

in males [57,58]. These results imply that females may be more susceptible than males to SBS.

The focus of most previous studies on SBS was either on environmental risks or on personal factors, leaving synergistic effects unexplored and potential risk factors for SBS undiscovered. For example, work stress, one of the personal factors, not only influences the occurrence of AL [25,28,29], but also increases the risk of SBS [7–10]; but the relationship between AL and SBS has not been investigated previously. This study found that air pollutant affects the expression of AL, suggesting that AL is a combined effect of air pollutants and personal factors. More remarkably, our analysis indicates that AL on the neuroendocrine system is associated with cough and pain in the neck, shoulders or back ($p<0.05$, Table 7). Together with a previous reports finding that pressure levels represented by stress hormones were linked to the prevalence of SBS [7,10], these data support the hypothesis that AL on the neuroendocrine system can influence the incidence of SBS. Our study identified that CO_2, dCO_2 and CO_2 I/O ratio were positively associated with AL on the neuroendocrine system. Since high prevalence of SBS is known to be associated with low levels of indoor ventilation [2,3], our results provide a possible mechanism that the neuroendocrine system mediates the effect of ventilation rate on the prevalence of SBS. In addition, the prevalence of depression is positively correlated with AL on the metabolic system in our study ($p<0.05$, Table 7). This is the first discovery of the relationship between depression and metabolic system and deserves further study in the future. Taken together, in the light of our research, AL appears to be a combinational result of air pollutants and personal factors, and AL score may be considered an assessment approach for identifying SBS.

Production of 8-OHdG is known to be induced by exposure to industrial pollutants such as benzene, engine oil or chromium [59–61], and its correlation with non-industrial pollutant exposure was firstly revealed by Lu and co-workers [16]. Subsequently, 8-OHdG was considered a likely predictive marker for SBS [17]. The current study further compares the feasibility and effectiveness of AL and 8-OHdG in predicting risk of reporting SBS, and analyzes the relationship between 8-OHdG, IEQ and SBS in the same group. The average concentration of 8-OHdG in our study was comparable with Lu's work, the SBS questionnaire was the standard WHO questionnaire, and some variables with personal and environmental factors were adjusted. Table 6 shows that selected indoor air quality indicators were associated with 8-OHdG level. The relationships between illumination, CO_2, dCO_2 and CO_2 I/O ratio and 8-OHdG were consistent with those and AL in specific system. In particular, visible light is known to irradiate iron ion-binding molecules such as haem or myoglobin, generating reactive oxygen species OH., H_2O_2 or O_2^- that in turn cause DNA damage and produce 8-OHdG [12,62,63]. Therefore, our findings of correlation analysis are consistent with previous mechanistic studies. Furthermore, our finding of an association between CO_2 and 8-OHdG agrees with Lu's work, [16], yet we found no significant association between 8-OHdG and TVOC, as suggested in that work. It is possible that ppbRAE measures only total VOC levels and there are often different compositions of VOCs in different office spaces. Furthermore, we found that 8-OHdG was associated with reported eye dryness or irritation, eye tiredness and vomiting (table 7), and AL was also associated with SBS symptoms.

Different SBS symptoms are physiologically related to specific human body systems. AL, however, is profiled by summing outcomes of 5 individual body system and therefore can be assessed with different specific SBS. In this study, we found that the association between CO_2 and "reported pain in the neck and lack of concentration" is explained, with biological plausibility, by the neuroendocrine system identified through the AL analysis. Furthermore, AL is shown to serve as surrogate outcomes responding to both environmental and personal stress [25,27,28], though no study to date has demonstrated any potential association between 8-OHdG levels and the characterization of personal factors. Therefore, AL may be considered far more attractive than 8-OHdG for appropriate application in research on the etiology of SBS.

Finally, we review a few limitations of this study. First, the sampling was carried out during working hours, and so the actual exposure to environmental factors could not be precisely quantified if the subjects were required to leave their seats for a substantial period of time due to increased work demands. Secondly, work stress was not exactly determined and thus is not included as an adjustable variable in our analysis. Furthermore, since only non-invasive sampling was used, there is no information on plasma C-reactive protein (CRP), total cholesterol (TC) and high-density lipoprotein (HDL) that could be used for further analysis of the association between AL and indoor air quality and SBS. Finally, although gender and exercise frequency were included as adjustable variables in assessing correlations, creatinine was the only marker used in our study to represent metabolic system of AL.

Conclusions

AL is influenced by environmental pollutants and is positively associated with SBS, and compared to the single measure of 8-OHdG, multi-systemic AL investigation may provide significant insights for the etiology of SBS.

Acknowledgments

The authors wish to thank Hsin-Wen Chang, Yi-Ho Lee and Wen-Chi Wang for assistance in the field measurements and experiments. Dr. Yi-Jen Chiu provided editorial comments.

Author Contributions

Conceived and designed the experiments: CCJ HHL NYH HLL HJS. Performed the experiments: CCJ HHL. Analyzed the data: CCJ HHL. Contributed reagents/materials/analysis tools: HLL HJS. Wrote the paper: CCJ HHL HJS.

References

1. Groes L (1995) A Statistical Analysis of Indoor Environmental Factors: Laboratory of Heating and Air Conditioning, Technical University of Denmark. The European IAQ-Audit Project.
2. Fisk WJ, Mirer AG, Mendell MJ (2009) Quantitative relationship of sick building syndrome symptoms with ventilation rates. Indoor Air 19: 159–165.
3. Norback D, Nordstrom K (2008) Sick building syndrome in relation to air exchange rate, CO(2), room temperature and relative air humidity in university computer classrooms: an experimental study. International Archives of Occupational and Environmental Health 82: 21–30.
4. Takeda M, Saijo Y, Yuasa M, Kanazawa A, Araki A, et al. (2009) Relationship between sick building syndrome and indoor environmental factors in newly built Japanese dwellings. Int Arch Occup Environ Health 82: 583–593.
5. Wang BL, Takigawa T, Yamasaki Y, Sakano N, Wang DH, et al. (2008) Symptom definitions for SBS (sick building syndrome) in residential dwellings. Int J Hyg Environ Health 211: 114–120.
6. Gomzi M, Bobic J, Radosevic-Vidacek B, Macan J, Varnai VM, et al. (2007) Sick building syndrome: psychological, somatic, and environmental determinants. Arch Environ Occup Health 62: 147–155.

7. Hansen AM, Meyer HW, Gyntelberg F (2008) Building-related symptoms and stress indicators. Indoor Air 18: 440–446.
8. Kinman G, Griffin M (2008) Psychosocial factors and gender as predictors of symptoms associated with sick building syndrome. Stress and Health 24: 165–171.
9. Runeson R, Wahlstedt K, Wieslander G, Norback D (2006) Personal and psychosocial factors and symptoms compatible with sick building syndrome in the Swedish workforce. Indoor Air 16: 445–453.
10. Wahlstedt K, Lindgren T, Norback D, Wieslander G, Runeson R (2010) Psychosocial work environment and medical symptoms among Swedish commercial airline cabin crew. Am J Ind Med 53: 716–723.
11. Chuang CY, Lee CC, Chang YK, Sung FC (2003) Oxidative DNA damage estimated by urinary 8-hydroxydeoxyguanosine: influence of taxi driving, smoking and areca chewing. Chemosphere 52: 1163–1171.
12. Floyd RA, West MS, Eneff KL, Schneider JE (1990) Mediation of 8-Hydroxy-Guanine Formation in DNA by Thiazin Dyes Plus Light. Free Radical Biology and Medicine 8: 327–330.
13. Staessen JA, Nawrot T, Den Hond E, Thijs L, Fagard R, et al. (2001) Renal function, cytogenetic measurements, and sexual development in adolescents in relation to environmental pollutants: a feasibility study of biomarkers. Lancet 357: 1660–1669.
14. Lai CH, Liou SH, Lin HC, Shih TS, Tsai PJ, et al. (2005) Exposure to traffic exhausts and oxidative DNA damage. Occupational and Environmental Medicine 62: 216–222.
15. Lodovici M, Casalini C, Cariaggi R, Michelucci L, Dolara P (2000) Levels of 8-hydroxydeoxyguanosine as a marker of DNA damage in human leukocytes. Free Radical Biology and Medicine 28: 13–17.
16. Lu CY, Ma YC, Lin JM, Chuang CY, Sung FC (2007) Oxidative DNA damage estimated by urinary 8-hydroxydeoxyguanosine and indoor air pollution among non-smoking office employees. Environ Res 103: 331–337.
17. Lu CY, Ma YC, Lin JM, Li CY, Lin RS, et al. (2007) Oxidative stress associated with indoor air pollution and sick building syndrome-related symptoms among office workers in Taiwan. Inhal Toxicol 19: 57–65.
18. McEwen BS (1998) Protective and damaging effects of stress mediators. N Engl J Med 338: 171–179.
19. Goldstein DS (2012) Stress, Allostatic Load, Catecholamines, and Other Neurotransmitters in Neurodegenerative Diseases. Cell Mol Neurobiol.
20. Korte SM, Koolhaas JM, Wingfield JC, McEwen BS (2005) The Darwinian concept of stress: benefits of allostasis and costs of allostatic load and the trade-offs in health and disease. Neurosci Biobehav Rev 29: 3–38.
21. Swaab DF, Bao AM, Lucassen PJ (2005) The stress system in the human brain in depression and neurodegeneration. Ageing Res Rev 4: 141–194.
22. Clark MS, Bond MJ, Hecker JR (2007) Environmental stress, psychological stress and allostatic load. Psychol Health Med 12: 18–30.
23. Doan SN, Evans GW (2011) Maternal responsiveness moderates the relationship between allostatic load and working memory. Dev Psychopathol 23: 873–880.
24. Dowd JB, Simanek AM, Aiello AE (2009) Socio-economic status, cortisol and allostatic load: a review of the literature. Int J Epidemiol 38: 1297–1309.
25. Gallo LC, Jimenez JA, Shivpuri S, Espinosa de los Monteros K, Mills PJ (2011) Domains of chronic stress, lifestyle factors, and allostatic load in middle-aged Mexican-American women. Ann Behav Med 41: 21–31.
26. Hu P, Wagle N, Goldman N, Weinstein M, Seeman TE (2007) The associations between socioeconomic status, allostatic load and measures of health in older Taiwanese persons: Taiwan social environment and biomarkers of aging study. J Biosoc Sci 39: 545–556.
27. Sun J, Wang S, Zhang JQ, Li W (2007) Assessing the cumulative effects of stress: The association between job stress and allostatic load in a large sample of Chinese employees. Work and Stress 21: 333–347.
28. Gustafsson PE, Janlert U, Theorell T, Westerlund H, Hammarstrom A (2011) Socioeconomic status over the life course and allostatic load in adulthood: results from the Northern Swedish Cohort. J Epidemiol Community Health 65: 986–992.
29. Rogosch FA, Dackis MN, Cicchetti D (2011) Child maltreatment and allostatic load: consequences for physical and mental health in children from low-income families. Dev Psychopathol 23: 1107–1124.
30. Suglia SF, Duarte CS, Sandel MT, Wright RJ (2010) Social and environmental stressors in the home and childhood asthma. J Epidemiol Community Health 64: 636–642.
31. Von Korff M, Alonso J, Ormel J, Angermeyer M, Bruffaerts R, et al. (2009) Childhood psychosocial stressors and adult onset arthritis: broad spectrum risk factors and allostatic load. Pain 143: 76–83.
32. Chuang KJ, Yan YH, Cheng TJ (2010) Effect of air pollution on blood pressure, blood lipids, and blood sugar: a population-based approach. J Occup Environ Med 52: 258–262.
33. Lin LY, Chen HW, Su TL, Hong GB, Huang LC, et al. (2011) The effects of indoor particle exposure on blood pressure and heart rate among young adults: An air filtration-based intervention study. Atmospheric Environment 45: 5540–5544.
34. Gross O, Gewies A, Finger K, Schafer M, Sparwasser T, et al. (2006) Card9 controls a non-TLR signalling pathway for innate anti-fungal immunity. Nature 442: 651–656.
35. Kumar H, Kumagai Y, Tsuchida T, Koenig PA, Satoh T, et al. (2009) Involvement of the NLRP3 inflammasome in innate and humoral adaptive immune responses to fungal beta-glucan. J Immunol 183: 8061–8067.
36. Ma CM, Lin LY, Chen HW, Huang LC, Li JF, et al. (2010) Volatile organic compounds exposure and cardiovascular effects in hair salons. Occupational Medicine-Oxford 60: 624–630.
37. Sundblad BM, von Scheele I, Palmberg L, Olsson M, Larsson K (2009) Repeated exposure to organic material alters inflammatory and physiological airway responses. European Respiratory Journal 34: 80–88.
38. Lopez-Duran NL, Kovacs M, George CJ (2009) Hypothalamic-pituitary-adrenal axis dysregulation in depressed children and adolescents: A meta-analysis. Psychoneuroendocrinology 34: 1272–1283.
39. Seeman TE, Singer BH, Rowe JW, Horwitz RI, McEwen BS (1997) Price of adaptation - Allostatic load and its health consequences - MacArthur studies of successful aging. Archives of Internal Medicine 157: 2259–2268.
40. Juster R-P, McEwen BS, Lupien SJ (2010) Allostatic load biomarkers of chronic stress and impact on health and cognition. Neuroscience & Biobehavioral Reviews 35: 2–16.
41. WHO (1986) Indoor air quality research. Report on a WHO meeting. Stockholm 27–31 August 1984. EURO Rep Stud: 1–64.
42. Bellingrath S, Weigl T, Kudielka BM (2009) Chronic work stress and exhaustion is associated with higher allostastic load in female school teachers. Stress 12: 37–48.
43. Crews DE, Harada H, Aoyagi K, Maeda T, Alfarano A, et al. (2012) A Pilot Study of Allostatic Load among Elderly Japanese Living on Hizen-Oshima Island. J Physiol Anthropol 31: 18.
44. Hasson D, Von Thiele Schwarz U, Lindfors P (2009) Self-rated health and allostatic load in women working in two occupational sectors. J Health Psychol 14: 568–577.
45. Mattei J, Demissie S, Falcon LM, Ordovas JM, Tucker K (2010) Allostatic load is associated with chronic conditions in the Boston Puerto Rican Health Study. Social Science & Medicine 70: 1988–1996.
46. Peek MK, Cutchin MP, Salinas JJ, Sheffield KM, Eschbach K, et al. (2010) Allostatic Load Among Non-Hispanic Whites, Non-Hispanic Blacks, and People of Mexican Origin: Effects of Ethnicity, Nativity, and Acculturation. Am J Public Health 100: 940–946.
47. Mair CA, Cutchin MP, Peek MK (2011) Allostatic load in an environmental riskscape: The role of stressors and gender. Health & Place 17: 978–987.
48. Liu JY, Duan YX (2012) Saliva: A potential media for disease diagnostics and monitoring. Oral Oncology 48: 569–577.
49. Pfaffe T, Cooper-White J, Beyerlein P, Kostner K, Punyadeera C (2011) Diagnostic Potential of Saliva: Current State and Future Applications. Clinical Chemistry 57: 675–687.
50. Zhang YY, Lin M, Zhang ST, Wang Z, Jiang L, et al. (2008) NF-kappa B-dependent cytokines in saliva and serum from patients with oral lichen planus: A study in an ethnic Chinese population. Cytokine 41: 144–149.
51. Lester SR, Brown JR, Aycock JE, Grubbs SL, Johnson RB (2010) Use of Saliva for Assessment of Stress and Its Effect on the Immune System Prior to Gross Anatomy Practical Examinations. Anatomical Sciences Education 3: 160–167.
52. Sjogren E, Leanderson P, Kristenson A, Ernerudh J (2006) Interleukin-6 levels in relation to psychosocial factors: Studies on serum, saliva, and in vitro production by blood mononuclear cells. Brain Behavior and Immunity 20: 270–278.
53. Barraza-Villarreal A, Escamilla-Nunez MC, Hernandez-Cadena L, Texcalac-Sangrador JL, Sienra-Monge JJ, et al. (2011) Elemental carbon exposure and lung function in schoolchildren from Mexico City. European Respiratory Journal 38: 548–552.
54. Flouris AD, Metsios GS, Carrillo AE, Jamurtas AZ, Gourgoulianis K, et al. (2009) Acute and short-term effects of secondhand smoke on lung function and cytokine production. Am J Respir Crit Care Med 179: 1029–1033.
55. Hoffmann B, Luttmann-Gibson H, Cohen A, Zanobetti A, de Souza C, et al. (2012) Opposing Effects of Particle Pollution, Ozone, and Ambient Temperature on Arterial Blood Pressure. Environ Health Perspect 120: 241–246.
56. Perry B, Doumas B, Jendrzejczak B (1979) Effect of Light and Temperature on the Stability of Creatine-Kinase in Human-Sera and Controls. Clinical Chemistry 25: 625–628.
57. Rios JLD, Boechat JL, Gioda A, dos Santos CY, Neto FRD, et al. (2009) Symptoms prevalence among office workers of a sealed versus a non-sealed building: Associations to indoor air quality. Environ Int 35: 1136–1141.
58. Tsai DH, Lin JS, Chan CC (2012) Office workers' sick building syndrome and indoor carbon dioxide concentrations. J Occup Environ Hyg 9: 345–351.
59. Kuo HW, Chang SF, Wu KY, Wu FY (2003) Chromium(VI) induced oxidative damage to DNA: increase of urinary 8-hydroxydeoxyguanosine concentrations (8-OHdG) among electroplating workers. Occupational and Environmental Medicine 60: 590–594.
60. Lagorio S, Tagesson C, Forastiere F, Iavarone I, Axelson O, et al. (1994) Exposure to Benzene and Urinary Concentrations of 8-Hydroxydeoxyguanosine, a Biological Marker of Oxidative Damage to DNA. Occupational and Environmental Medicine 51: 739–743.
61. Nilsson R, Nordlinder R, Moen BE, Ovrebo S, Bleie K, et al. (2004) Increased urinary excretion of 8-hydroxydeoxyguanosine in engine room personnel exposed to polycyclic aromatic hydrocarbons. Occupational and Environmental Medicine 61: 692–696.
62. Huang YP, Li J, Ma WH, Cheng MM, Zhao JC, et al. (2004) Efficient H2O2 oxidation of organic pollutants catalyzed by supported iron sulfophenylporphyrin under visible light irradiation. Journal of Physical Chemistry B 108: 7263–7270.
63. Wang XX, Gu Y, Fang YF, Huang YP (2012) Mechanism of oxidative damage to DNA by Fe-loaded MCM-41 irradiated with visible light. Chinese Science Bulletin 57: 1504–1509.

Acrolein and Asthma Attack Prevalence in a Representative Sample of the United States Adult Population 2000 – 2009

B. Rey deCastro*

Centers for Disease Control and Prevention, National Center for Environmental Health, Division of Laboratory Sciences, Atlanta, Georgia, United States

Abstract

Background: Acrolein is an air toxic and highly potent respiratory irritant. There is little epidemiology available, but US EPA estimates that outdoor acrolein is responsible for about 75 percent of non-cancer respiratory health effects attributable to air toxics in the United States, based on the Agency's 2005 NATA (National-Scale Air Toxics Assessment) and acrolein's comparatively potent inhalation reference concentration of 0.02 μg/m^3.

Objectives: Assess the association between estimated outdoor acrolein exposure and asthma attack reported by a representative cross-sectional sample of the adult United States population.

Methods: NATA 2005 chronic outdoor acrolein exposure estimates at the census tract were linked with residences oif adults (≥18 years old) in the NHIS (National Health Interview Survey) 2000 – 2009 (n = 271,348 subjects). A sample-weighted logistic regression model characterized the association between the prevalence of reporting at least one asthma attack in the 12 months prior to survey interview and quintiles of exposure to outdoor acrolein, controlling for potential confounders.

Results: In the highest quintile of outdoor acrolein exposure (0.05 – 0.46 μg/m^3), there was a marginally significant increase in the asthma attack pOR (prevalence-odds ratio [95% CI] = 1.08 [0.98:1.19]) relative to the lowest quintile. The highest quintile was also associated with a marginally significant increase in prevalence-odds (1.13 [0.98:1.29]) in a model limited to never smokers (n = 153,820).

Conclusions: Chronic exposure to outdoor acrolein of 0.05 – 0.46 μg/m^3 appears to increase the prevalence-odds of having at least one asthma attack in the previous year by 8 percent in a representative cross-sectional sample of the adult United States population.

Editor: Yinping Zhang, Tsinghua University, China

Funding: The author received no external funding for the study.

Competing Interests: The author has declared that no competing interests exist.

* E-mail: rdecastro@cdc.gov

Introduction

Acrolein is among the 189 pollutants designated under the 1990 Clean Air Act [1] to be a hazardous air pollutant (or air toxic) known to or suspected of causing cancer or other serious health problems. A simple aldehyde, acrolein (CAS 107-0-28; IUPAC prop-2-enal) is a potent irritant of the respiratory tract, including the nose and throat [2–5]. Acrolein can also irritate the eyes and skin [6,7]. Acrolein was deployed as a chemical warfare agent (tear gas) at least once during World War I [8–11].

Acrolein is used mainly as a precursor in the manufacture of acrylic acid and as a biocide. Acrolein is ubiquitous in the environment, and since it is formed through the combustion of petroleum (especially diesel fuel), acrolein is a concern for mobile sources, aircraft, and industrial boilers [12]. Another major component of mobile emissions — 1,3-butadiene — oxidizes in the atmosphere to acrolein, further augmenting acrolein's presence in the environment. US EPA (United States Environmental Protection Agency) estimates that approximately three-quarters of ambient acrolein exposure originates from mobile sources [13]. Acrolein exists in the atmosphere in the vapor phase, and is subject to photochemical degradation by hydroxyl radicals, ozone, and nitrate radicals. Acrolein's half-life in the atmosphere based on observed gas-phase photolysis indicates a half-life of 10.9 days [14], but calculated half-life estimates range from 3.4 hours to 28 days depending on the type of photolytic reaction [15–20].

Acrolein is also a major component of the vapor phase of tobacco smoke, as well as smoke from cooking, forest fires, and biomass combustion. Small amounts of acrolein have also been detected in some foods, such as fried foods, cooking oils, and roasted coffee [6,20–22]. Acrolein is also formed endogenously as part of physiological oxidative stress response and polyamine metabolism [23].

Although direct evidence of acrolein's carcinogenicity in humans or experimental animals is considered inadequate [24], acrolein is known to induce DNA damage [25] and to form DNA adducts relevant to lung cancer and inhibition of tumor suppression [26,27]. Acrolein has also been shown to interact with a prominent carcinogenic constituent of tobacco smoke — benzo[*a*]pyrene — to inhibit p53 tumor suppressor activity, which suggests a role for acrolein in lung cancer initiation [28].

Attention has focused recently on the potential role of endogenous acrolein — produced as part of oxidative stress response — in a variety of neurologic disorders, such as Alzheimer's disease, Parkinson's disease, and amyotrophic lateral sclerosis [29–32]. Endogenous acrolein has also been observed in connection to spinal cord injury [33], as well as myelin damage [34]. The possibly adverse neurological effects of endogenous acrolein have prompted concern about environmental exposure to acrolein, particularly through air pollutant emissions and tobacco smoke.

Acrolein's role as a respiratory toxicant is well established [6,7]. Because of its high reactivity with human tissue, inhalation of acrolein has been hypothesized to induce or exacerbate acute lung injury and chronic obstructive pulmonary disease [21]. A risk assessment for human lung function extrapolated from rat data suggested that ambient concentrations of acrolein in the United States may be associated with reduced respiratory function [35]. In a comprehensive review considering the exposure prevalence and toxic potency of hazardous air pollutants, acrolein was recommended for further research into its role in the initiation and exacerbation of asthma [36]. Other risk assessments of air toxics have singled out acrolein as a prominent noncancer respiratory health risk in the United States [37,38] and in Pittsburgh, PA [39].

Asthma is a chronic inflammatory disorder of the small respiratory airways. Acute episodes, or asthma attacks, in susceptible individuals are associated with airflow obstruction characterized by wheezing, breathlessness, chest tightness, and coughing. These episodes result from a combination of airway inflammation and elevated bronchial hyperresponsiveness to a variety of triggers. Important environmental triggers include ETS (environmental tobacco smoke), dust mites, cockroach allergen, outdoor air pollution, wood smoke, pets, and mold [40,41]. As a potent respiratory irritant, acrolein may have a prominent role as an environmental trigger of asthma attacks. Asthma attacks are usually reversible, either spontaneously or with medication, and while there is no cure for asthma, symptoms can be managed through medication and the avoidance of environmental triggers [42–44].

There are standard US EPA methods for measuring ambient acrolein (Method TO-11a, based on a canister sampler, and TO-15, a cartridge method), with temporal resolution of hours to days [45,46]. Negative bias has been reported with Method TO-11a [47]. A real-time quantum cascade laser infrared absorption device is also commercially available [48]. Because acrolein is highly reactive, ambient measurement remains notoriously difficult, which increases uncertainty about the consistency and reliability of acrolein monitoring data [49,50]. Improving measurement of ambient acrolein is the subject of ongoing research [51,52].

With the difficulties in measuring ambient acrolein, it is not surprising that there is scant epidemiologic research into the health effects of acrolein in general populations. In striking contrast to this lack of information, US EPA estimates that acrolein is responsible for about 75 percent of non-cancer respiratory health effects attributable to air toxics in the United States, based on the Agency's NATA (National-Scale Air Toxics Assessment) for 2005 [13]. Acrolein's environmental ubiquity notwithstanding, its prominence in risk assessments like NATA and others [35–39] is driven primarily by its distinctively high potency for respiratory health effects compared to other air toxics, with a chronic inhalation RfC (reference concentration) of 0.02 $\mu g/m^3$ [53].

The aim of this study was to provide empirical evidence of the association between acrolein exposure and respiratory irritation, as denoted by the prevalence of self-reported asthma attacks among adults in the general population. This was achieved by geographically linking NATA 2005 acrolein exposure concentration estimates at the census tract level with residences of participants in NHIS (National Health Interview Survey) 2000 – 2009 [54]. NHIS data have been geographically linked to US EPA ambient air monitoring data to examine the associated prevalence of childhood asthma [55] and respiratory allergies [56,57], but this is the first time that NHIS data have been linked to NATA exposure estimates. Geographically linked NATA data has also been used to study air toxics exposure and asthma in a children's birth cohort [58], cancer [59], and autism spectrum disorder [60,61].

Materials and Methods

National-Scale Air Toxics Assessment 2005

NATA 2005 estimates chronic inhalation exposure risks for 179 air toxics, and is a tool for prioritizing geographic areas, pollutants, and emission sources for further evaluation of public health impact [13]. US EPA developed NATA to augment data from the Agency's nationwide air toxics monitoring network by taking a model-based approach to quantifying the potential health risks from air toxics. NATA provides risk estimates for each of the approximately 66,000 census tracts comprising the United States. Census tracts are delineated by the US Census Bureau, and each typically contains about 4,000 residents and ranges from 1,500 to 8,000 residents. Tracts also vary in geographic size, with urban tracts usually smaller than two square miles and rural tracts being much larger [62].

NATA encompasses outdoor emissions of air toxics from stationary point sources, non-point sources, and mobile sources (on-road and off-road), as well as estimates of background concentration and secondary formation of acrolein from the decay of 1,3-butadiene. In 2011, US EPA released NATA 2005 [13] based on 2005 National Emissions Inventory data [63], which were the most complete and up-to-date inventory data at the time of the analysis. NATA 2005 combines data from the National Emissions Inventory and US EPA's air monitoring system as inputs to air dispersion and photochemical models (American Meteorological Society/EPA Regulatory Model [AERMOD], Assessment System for Population Exposure Nationwide Model [ASPEN], and Community Multiscale Air Quality Model [CMAQ]) to estimate outdoor air toxics concentrations. The air dispersion models account for the physical topography and population density in each census tract. The outdoor concentration estimates are then used as the basis for estimating chronic air toxics exposure concentrations from a screening-level inhalation exposure model (Hazardous Air Pollutant Exposure Model, Version 5: HAPEM5). These exposure concentrations account for differences in exposure resulting from demographics, human activities, commuting patterns, climate data, and indoor-to-outdoor variability. The NATA process concludes with estimation of cancer and non-cancer public health risks [64], but for the analyses reported here only the acrolein inhalation exposure concentrations estimated at the census tract level were used. And, unlike NATA health risk estimates, NATA exposure concentration estimates do not utilize unit health risk factors like the RfC.

National Health Interview Survey 2000 – 2009

NHIS is conducted by NCHS (National Center for Health Statistics) of the CDC (Centers for Disease Control and Prevention) and is a principal source of information on the health conditions, behaviors, and access to health services of the civilian, non-institutionalized population of the United States. Information is collected through a comprehensive interview of a nationally representative cross-section of households [65]. NHIS is designed to produce national estimates of disease prevalence, and provides reliable data for estimating asthma prevalence across the United States, over time, and across subgroups [66]. Sampling is conducted annually following a multistage area probability design that permits the representative sampling of households and noninstitutional group quarters (e.g., college dormitories). This study evaluated the cross-sectional prevalence of having at least one self-reported asthma attack in the 12 months prior to each subject's interview, which represented affirmative answers to both NHIS questions "Have you ever been told by a doctor or other health professional that you had asthma?" and "During the past 12 months, have you had an episode of asthma or an asthma attack?" This definition identifies subjects with symptomatic asthma and is the standard CDC definition for evaluating asthma attack prevalence [55].

Additional NHIS data on potential confounders were evaluated in the logistic regression models: sex, age, race/ethnicity, poverty status, education, health insurance coverage, access to health care, whether the subject's residence was located in an urban or rural area, as well as NHIS interview year and calendar quarter. Acrolein is a component of tobacco smoke, so smoking status (never smoked, former smoker, and current smoker) was evaluated as a potential confounder and separate models were fit for never smokers and never & former smokers. Information for potential confounders was reported by the participant or an adult in the participant's household, except for poverty and insurance status, urban residence, survey year, and calendar quarter. Poverty status of each subject's family was determined according to a calculated PIR (poverty income ratio): the ratio of family income to the US Census poverty threshold adjusted for family size. Families with a PIR <1.00 were categorized as poor. The uninsured were subjects who did not have health insurance at the time of the interview under private health insurance, Medicare, Medicaid, State Children's Health Insurance Program, a state-sponsored health plan, other government programs, or military health plan. Insurance status for the subject may have been reported by an adult in the subject's family. This definition of uninsured matches that used in Health United States [67]. Access to health care was assessed through the NHIS question "Is there a place that you usually go to when you are sick or need advice about your health?" Relevant places include hospital emergency rooms and outpatient departments. Residence in an urban or rural area followed US Census Bureau classification and is available from restricted-use NHIS data files.

NATA-NHIS Linked Data and Statistical Models

One adult 18 years of age or older was randomly selected from each participating NHIS family to provide detailed information about their health, including symptomatic asthma and smoking status. This study compiled ten years of data (2000 – 2009) regarding these sample adults, numbering 287,530 subjects [54]. Of these, 9,524 subjects were missing census tract information, and another 6,658 subjects were missing data on the outcome or predictors. This left 271,348 subjects (94.4 percent) who were included in the logistic regression models reported here. The Census 2000 tract identifiers for NHIS subjects's residences are available from restricted-use NHIS data files, which enabled assignment of subjects to their respective census tracts. Using Census 2000 tract delineations, NATA acrolein exposure data at the census tract were geographically linked with health outcome data for individual NHIS subjects.

Since NHIS is a complex survey in which subjects are obtained through a multistage sampling design involving stratification, clustering, and oversampling of specific population subgroups, the complex design must be accounted for in order to ensure properly estimated variances of regression coefficients obtained from statistical models of NHIS data [66,68]. Robust estimation of these variances may be accomplished through a generalized estimating equations approach incorporating Taylor series linearization and applying sampling weights to each survey subject. This approach was used as it was implemented by the CROSSTAB, DESCRIPT, and RLOGIST subroutines of SUDAAN version 10.0.1 [69] called from the SAS version 9.3 statistical software application [70]. To parameterize the variance estimation procedure, variables indicating the NHIS sampling stratum and primary sampling unit were used. Pursuant to NCHS guidance [71,72], sampling strata were renumbered to distinguish strata used in different NHIS sample design periods: 2000 – 2005 and 2006 – 2009. In addition, the final annual sample weights from the NHIS 2000 – 2009 sample adult files (WTFA_SA) for subjects 18 years old and older were adjusted for incomplete data (see below). To yield annualized totals, sample weights were divided by ten, which accounts for combining data across ten NHIS sampling years [71,72].

Sample-weighted logistic regression models were fit to the linked NATA-NHIS data where the dichotomous outcome was whether an NHIS subject answered affirmatively that they had both ever been diagnosed with asthma and had at least one asthma attack in the previous 12 months. The reference outcome group comprised NHIS subjects who: a) answered affirmatively that they had ever been diagnosed with asthma, but answered negatively that they had had at least one asthma attack in the previous 12 months; and b) answered negatively that they had ever been diagnosed with asthma. The predictor of interest was the quintile of NATA 2005 chronic inhalation exposure concentration to acrolein for the census tract where the NHIS subject resided. Additional predictors representing potential confounders (described above) were evaluated in the regression models.

Acrolein is a prominent component of tobacco smoke, so it is important to assess acrolein's influence on asthma attack prevalence among subjects who do not smoke. This was evaluated in two sample-weighted logistic regression models for NATA-NHIS 2000 – 2009 subjects limited to those reporting: a) that they never smoked ("never smokers"; n = 153,820); and b) that they either never smoked or were former smokers ("never & former smokers"; n = 212,537). No subjects in these strata were current smokers, but the second, larger stratum enabled us to evaluate the potential influence of discontinued smoking. The predictors in both stratified models were identical to the unstratified, except that the stratified models did not control for smoking status. Statistical significance was set at $\alpha \le 0.05$, and marginal significance was set at $0.05 < \alpha \le 0.15$.

Among all NHIS 2000 – 2009 sample adults, 24.0 percent were missing information on PIR, which derives from the historic difficulty of obtaining self-reported income data. To address this problem, NCHS provides multiply imputed data on income and PIR. Specifically, for each subject, NCHS provides public-use PIR data that have been imputed five times [54], which were included in the statistical analyses reported here. Statistical analysis of multiply imputed data yields multiple results that can be combined to obtain unified variance estimates that reflect the additional variation due to the multiple imputation procedure [73]. Earlier evaluation of the performance of multiply imputed NHIS income data indicated that it corrects biases that would otherwise occur without multiple imputation and that statistical efficiency is improved [74]. Consistent with NCHS guidance [75], this approach was adopted through the use of the MI_VAR option in the CROSSTAB and RLOGIST subroutines of SAS-callable SUDAAN version 10.0.1. All results reported here and the sample size described above are based on multiply imputed PIR data.

Regression models require complete information on all dependent and predictor data for each subject, which were available for 271,348 of 287,530 subjects (94.4 percent) in the NATA-NHIS linked data. Since subjects with incomplete data were not part of the statistical models reported here, model estimates would have been biased to the degree there was non-randomness in the pattern of missing data. To ameliorate this potential bias, the original NHIS survey sample weights (WTFA_SA) were adjusted prior to statistical analysis using the WTADJUST subroutine of SAS-callable SUDAAN version 10.0.1, which treats those subjects with complete information as a random subsample of all NHIS subjects. In this approach, the probability of an NHIS subject being selected into this subsample was assumed to be a logistic function of covariates. The parameters of this logistic selection model were estimated implicitly by WTADJUST in adjusting the sample weights. This approach is pursuant to current NCHS guidance [76], and because all models reported here used sample weights adjusted in this manner, results may be considered statistically representative of the United States civilian, noninstitutionalized adults comprising the NHIS study population [77].

Ethics Statement

The NCHS Research ERB (Ethics Review Board) protected the rights and welfare of participants in the NHIS. In accordance with Federal regulations, the NCHS ERB reviewed and approved NHIS protocols and any changes made to them. This process ensured the ethical treatment of NHIS participants. Documented signed informed consent was obtained from each participant.

Results

Figure 1 displays the NATA 2005 acrolein exposure concentration quintile for each census tract in the contiguous United States overlaid with locations of 102 facilities that reported to the 2010 Toxic Release Inventory [78] (accessed through TOXMAP [79]) that they released acrolein to the air on-site categorized by the quintile of the amount released. This map demonstrates how acrolein exposure is more widespread than suggested by the locations of emitting facilities, which is consistent with the prominence of mobile sources for acrolein exposure.

Table 1 displays the sample-weighted proportion of subjects across acrolein exposure quintiles and across demographic characteristics among the 271,348 adult subjects comprising the NATA-NHIS data used in the logistic regression models. The national sample-weighted prevalence of 12-month asthma attacks was 3.8 (standard error: 0.1) percent among adults during 2000 – 2009. Because assignment of subjects to acrolein exposure quintiles did not account for sample weighting, the sample-weighted distribution of subjects among acrolein quintiles reported in Table 1 do not appear even. Table 2 summarizes the distribution of outdoor acrolein exposure concentrations: the sample-weighted median acrolein chronic inhalation exposure concentration was 2.5E-2 (IQR: 3.2E-2) $\mu g/m^3$, with a range from 1.4E-4 to 4.6E-1 $\mu g/m^3$.

Based on the sample-weighted logistic regression model for NATA-NHIS 2000 – 2009 subjects reported in Table 3 (comprising the same data described in Table 1), there was a marginally significant (p-value = 0.10) increase in the 12-month asthma attack pOR (prevalence-odds ratio) of 1.08 [95% CI (95 percent confidence interval): 0.98:1.19] at the highest exposure quintile (5.5E-2 to 4.6E-1 $\mu g/m^3$) relative to the lowest quintile (1.4E-4 to 1.1E-2 $\mu g/m^3$), controlling for sex, age, race/ethnicity, poverty status, education, smoking status, access to health care, whether the subject's residence was located in an urban or rural area, and NHIS interview year. The reference group for this model was 35 – 44 year old, non-Hispanic white, non-smoking males, with one or more places to go for medical care or advice, and educational attainment of a high school diploma or more, who were interviewed in 2005, and were living in an urban household with a poverty index ≥1.00. Health insurance coverage and NHIS interview quarter were initially included in the model, but since they were not statistically significant they were dropped. For the second through fourth acrolein exposure quintiles (1.1E-2 to 5.5E-2 $\mu g/m^3$), pORs ranged from 1.01 to 1.02 and were statistically indistinguishable from asthma attack prevalence in the lowest quintile of acrolein exposure.

Among adult never smokers and never & former smokers (Table 3), the prevalence-odds of having at least one asthma attack in the previous 12 months was 13 percent [1.13, 95% CI: 0.98:1.29] and 9 percent [1.09, 95% CI: 0.98:1.22] higher at the highest acrolein exposure quintile compared to the lowest quintile. These elevated prevalence-odds, however, had marginally significant p-values of 0.08 and 0.13. Additionally, there was a marginally significant 11 percent increase in asthma attack prevalence-odds in the third acrolein exposure quintile compared to the lowest (p-value = 0.11), but this pattern was confined to never smokers.

Discussion

In a representative cross-sectional sample of the adult civilian, non-institutionalized population of the United States for 2000 – 2009, chronic residential exposure to outdoor acrolein (estimated for 2005) at the highest quintile of 0.05 – 0.46 $\mu g/m^3$ was found to increase the prevalence-odds of having at least one asthma attack in the previous 12 months relative to the lowest quintile by about 8 percent at a marginally significant level, as estimated in a sample-weighted logistic regression model and controlling for potential confounders. Since the NHIS data were pooled over 2000 – 2009, statistical estimates are interpretable as being averaged over the time interval of the pooled data [71,72].

These marginally significant results among adults are worth comparing to a recent study of a nationally representative cohort of United States children followed since infancy (n = 6,950) [58]. Adopting a somewhat different approach to using the NATA estimates, this study geographically linked by zip code air toxics respiratory health risk estimates from NATA 2002 to cohort subjects in order to evaluate associated changes in asthma prevalence. This study found no evidence of significantly increased prevalence of asthma associated with total air toxics respiratory health risk. The important differences between the designs of Stoner, Anderson and Buckley [58] and the present study notwithstanding, it is notable that although both utilized long-term averages derived from NATA, neither respiratory risk exposure comprising all air toxics [58] nor ambient concentration exposure for acrolein was strongly associated with increased asthma prevalence. This may point to the larger difficulty of associating short-term, acute health episodes with long-term

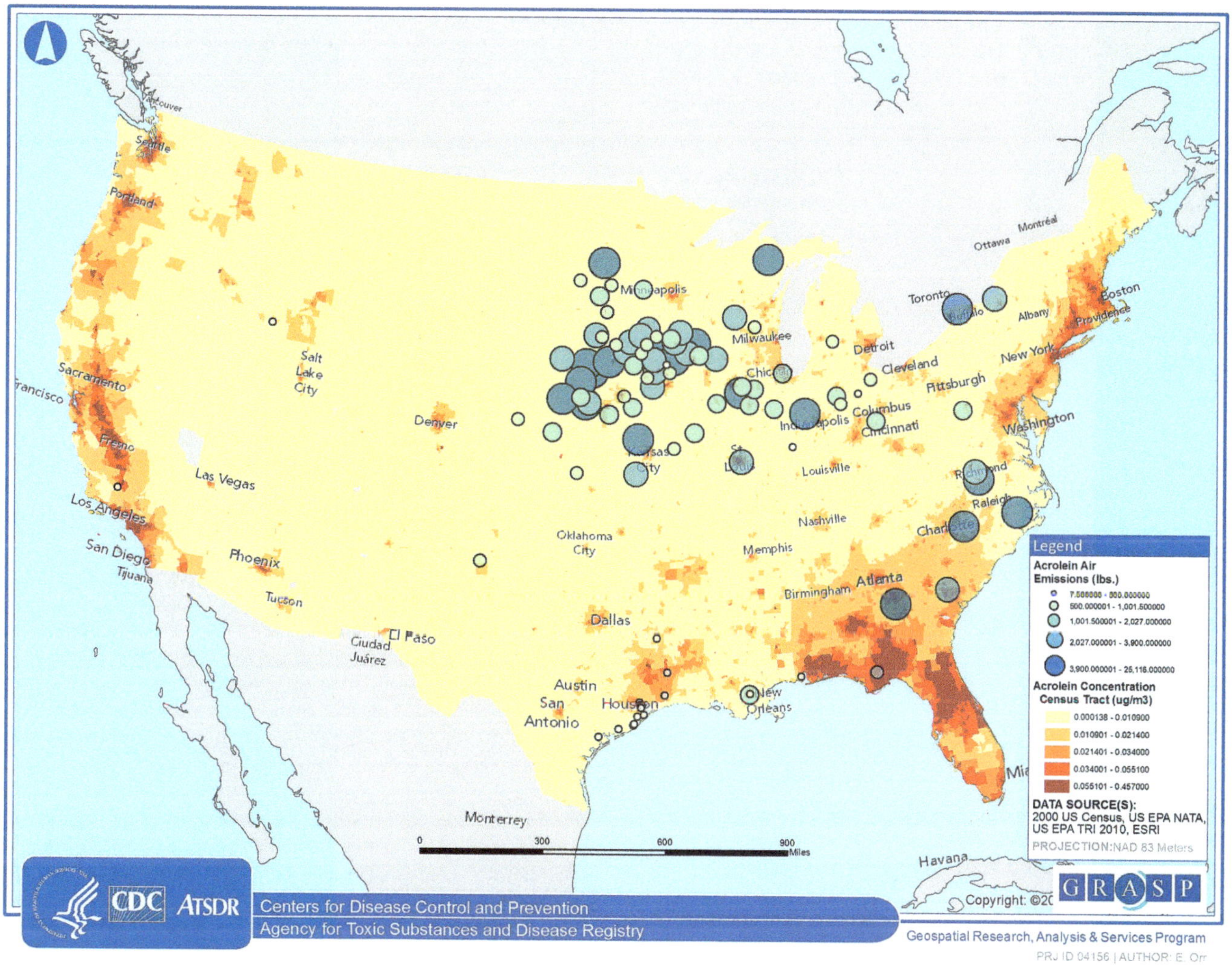

Figure 1. Map of the contiguous United States indicating the NATA 2005 acrolein exposure concentration quintile [µg/m3] for each census tract (shaded areas) overlaid with locations of 102 facilities reporting on-site air releases of acrolein [lbs] to the 2010 Toxic Release Inventory (bubble size corresponds to quintile of acrolein released).

average exposure rather than, say, coincident real-time exposure measurements that captures highly resolved temporal spikes in exposure more relevant to the health outcome.

The highest quintile of acrolein exposure concentration was more than double the chronic inhalation RfC of 0.02 µg/m^3 [53] and encompassed California OEHHA's (Office of Environmental Health Hazard Assessment) chronic REL (reference exposure level) of 0.35 µg/m^3 [80]. The acrolein exposure concentration quintile encompassing the RfC — the second quintile of 0.01 – 0.02 µg/m^3 — was not associated with significantly increased asthma attack prevalence-odds. The importance of these comparisons should be viewed in light of the uncertainty factor for acrolein's RfC, which is 1000 [53], and REL, which is 200 [80].

Elevated rates of adverse respiratory health effects, including asthma, have been widely observed among residents, especially children, living near heavily trafficked roadways [81–87]. Adverse respiratory health effects have been also correlated with traffic-related air pollutants like particulates, nitrogen dioxide, and sulfur dioxide [88–91]. Acrolein is a constituent of the complex mixture of traffic-related pollutants that are implicated in short-term, respiratory (non-cancer) health effects like asthma attacks. Since traffic density is correlated with urbanization, competing effects from other traffic-related air pollutants were accounted for by including urban residence as a model predictor, but this statistical adjustment is likely incomplete. While acknowledging these considerations, acrolein is distinguished among air toxics because of its high potency for respiratory (non-cancer) health effects. Among the 85 substances that have inhalation RfCs listed on US EPA's Integrated Risk Information System [53], only two have more toxic RfCs: chromium (VI; aerosol) and 1,6-hexamethylene diisocyanate. Twenty-seven substances have RfCs that are within two orders of magnitude less toxic than acrolein's, including 1,3-butadiene, acrylonitrile, and chromium (VI; particulate), but these three air toxics are of concern because of their cancer health risks [37], contrasted with the non-cancer health risk posed by acrolein. And while some studies have been put forward to support a proposal to relax acrolein's RfC from 0.02 to 0.62 µg/m^3 (0.27 ppb) [92–94], acrolein would still be a major concern among air toxics at this proposed level.

Acrolein is a component of tobacco smoke, but both stratified non-smoker models suggest that outdoor acrolein's adverse effect on asthma attacks does not appear to be driven solely by acrolein inhaled directly from smoking tobacco.

Table 1. Characteristics of subjects comprising NATA-NHIS 2000 – 2009 data: 271,348 of 287,530 subjects (94.4 percent) with 3.8 (0.1) percent (standard error) prevalence of at least one asthma attack in previous 12 months (sample-weighted).

Predictor	Sample-Weighted Percent (SE)
Acrolein Exposure Quintile	
1st: 1.4E-4 –1.1E-2 μg/m3	21.7 (0.5)
2nd: 1.1E-2 – 2.1E-2	21.9 (0.5)
3rd: 2.1E-2 – 3.4E-2	20.7 (0.5)
4th: 3.4E-2 – 5.5E-2	18.8 (0.4)
5th: 5.5E-2 – 4.6E-1	16.8 (0.4)
Sex	
Female	51.9 (0.1)
Male	48.1 (0.1)
Age	
18 – 24 years-old	13.0 (0.2)
25 – 34	18.0 (0.1)
35 – 44	20.0 (0.1)
45 – 54	19.2 (0.1)
55 – 64	13.6 (0.1)
65 – 74	8.7 (0.1)
75 – 84	5.8 (0.1)
≥85	1.8 (0.04)
Race/Ethnicity	
Hispanic	12.4 (0.2)
Non-Hispanic Black/African American only	11.3 (0.2)
Non-Hispanic Other	5.5 (0.1)
Non-Hispanic White only	70.8 (0.3)
Smoking Status	
Current Smoker	21.7 (0.1)
Former Smoker	21.9 (0.1)
Never Smoked	56.5 (0.2)
Education	
High School Graduate or Higher	83.2 (0.2)
Less than High School Graduate	16.8 (0.2)
Poverty Income Ratio (Imputed)	
1.00 or Higher	89.2 (0.1)
<1.00 (Poor)	10.8 (0.1)
Place to Go When Sick	
No place	14.7 (0.1)
One or more places	85.3 (0.1)
Residence	
Rural	25.0 (0.5)
Urban	75.0 (0.5)
NHIS Interview Year	
2000	9.4 (0.1)
2001	9.5 (0.1)
2002	9.6 (0.1)
2003	9.9 (0.1)
2004	10.0 (0.1)
2005	10.1 (0.1)
2006	10.2 (0.1)
2007	10.4 (0.1)
2008	10.5 (0.1)
2009	10.6 (0.1)

Although it would have been desirable to also evaluate the influence of ETS, during 2000 – 2009 NHIS only asked about ETS in 2005. Restricting analysis to a single year of data would have curtailed statistical power so severely as to hinder proper evaluation of ETS as a potential predictor of asthma attack prevalence.

Median outdoor 24-hour measurements of acrolein at downtown urban sites (Pittsburgh, PA) ranged from 0.11 – 0.14 μg/m^3 [39]. NATA focuses on outdoor sources of acrolein, and so ignores influences of indoor origin, such as cooking and building materials. Acrolein levels in indoor air, however, have been observed to be higher than outdoors. In a comprehensive quantitative literature review of indoor hazardous air pollutants in United States residences [95], the median concentration for acrolein was estimated to be 0.84 μg/m^3, nearly double the maximum estimated outdoor exposure concentration in NATA 2005. Logue, McKone, Sherman and Singer [95] also concluded that acrolein should be considered a priority chronic and acute hazard among chemical air contaminants. Using a mist chamber technique, indoor acrolein measured in nine California homes in the morning averaged 4.0 (standard deviation: 2.2) μg/m^3, while simultaneous outdoor measurements averaged 0.58 (0.64) μg/m^3 [96]. In the same study, substantial diurnal fluctuation in indoor acrolein ranging up to a factor of 2.5 was correlated with cooking and ambient temperature. In a model-based exposure assessment of the United States population, nonsmokers exposed to indoor ETS by living with a smoker were estimated to be exposed to acrolein in a range from 1.6 to 3.6 μg/m^3 [97]. Evaluated with respect to the marginally significant increase in asthma attack prevalence-odds reported here in the acrolein range of 0.05 – 0.46 μg/m^3, the higher indoor acrolein levels described above suggest that indoor exposure may further increase asthma attack prevalence in the general population.

In order to link NATA acrolein exposure estimates with NHIS asthma data, NHIS subjects living in the same census tract were assigned the same NATA acrolein exposure concentration. This approach, however, ignores: a) intra-tract exposure variation; b) exposure subjects may have obtained in other Census tracts; c) differences in time subjects resided in a census tract. In the temporal domain, NHIS subjects were assigned the same 2005 acrolein exposure estimate across all NHIS surveys from 2000 to 2009, which ignores temporal variation in acrolein exposure. Short-term, transient emissions profiles — such as startups, shutdowns, malfunctions, and upsets — are not represented, so NATA's chronic exposure estimates for acrolein would tend to underestimate total and acute exposure. In addition, estimates from NATA 2005 do not reflect long-term trends in air toxics exposure, as would be expected as air pollution regulations for mobile and fixed sources are implemented and as older facilities are retired.

Table 2. Sample-weighted statistics of outdoor acrolein exposure concentrations for NATA-NHIS 2000 – 2009 data.

Statistic	Quantity [μg/m³]
Mean (SE)	3.4E-2 (3.6E-4)
Geometric Mean (geometric standard error)	2.3E-2 (2.3E-4)
Median (interquartile range)	2.5E-2 (3.2E-2)
5th Percentile	4.9E-3
10th	6.6E-3
25th	1.2E-2
75th	4.4E-2
90th	6.9E-2
95th	8.7E-2

Representative of the United States civilian, noninstitutionalized adult (>=18 years) population.

NATA comprises a rigorous and complex process in environmental health risk assessment encompassing the entire United States. US EPA has emphasized, however, that NATA is a tool primarily for prioritizing air toxics, emissions sources, and geographic areas based on health risks [64]. This prioritizing information, rather than being the final assessment of risk, instead enables regulators to target resources for additional, more refined studies. The NATA process itself is also continually refined, and although NATA 2005 incorporated improvements in data and modeling, US EPA has noted that further improvements are possible in several areas to reduce uncertainty: emissions data quality, monitoring data quality, representativeness of meteorologic data, representativeness of background air toxics concentrations, and atmospheric model formulation [64]. In addition, in a quantitative comparison of model-based NATA 2005 estimates with ambient monitoring data [49], NATA 2005 accurately predicted ambient concentrations for several air toxics, but there appears to be a tendency for NATA to underestimate concentrations for others, which may be attributable to emissions sources missing from the National Emissions Inventory, under-estimation of emission rates, bias in monitoring techniques, and difficulties in estimating background concentrations. Still, other studies indicate that NATA estimates are reasonably reliable estimates of personal exposure to ambient air toxics [98,99]. The limitations of NATA estimates described above would contribute to exposure misclassification, adding a degree of uncertainty to the significance of the estimated association between outdoor acrolein exposure and asthma attack prevalence.

A distinct feature of NHIS is that it is conducted as a serial cross-sectional survey, hence a causative relationship between the dependent variable and the predictors cannot be concluded based on the models alone, but must be evaluated in context of dispositive epidemiologic and toxicologic evidence [100]. A particular strength of this study is the large sample size of the NATA-NHIS data, as well as its national representativeness and comprehensive detail on subject characteristics, which enabled good control of potential confounders for asthma attack prevalence in the logistic models, while at the same time examining the effect of acrolein exposure.

This is the first report suggesting an association between outdoor acrolein exposure and adverse effects on health in a general population. This was achieved through the availability of a rigorous model-based and geographically linked estimate of exposure that accounts for human activity and microenvironments. These exposure estimates also have the advantage of being at a high spatial resolution that account for variation resulting from local population density and topography. Model-based exposure estimates provide one avenue of exposure assessment, and although this report demonstrates the feasibility of conducting national epidemiologic analysis for air toxics with modeled

Table 3. Adjusted prevalence-odds ratios for having at least one asthma attack in the previous 12 months among a representative cross-sectional sample of adult (≥18 years-old) United States population.

	All Subjects		Never Smokers		Never & Former Smokers	
	n=271,348		n=153,820		n=212,537	
Acrolein Exposure Quintile [μg/m3]	pOR (95% CI)	p-Value	pOR (95% CI)	p-Value	pOR (95% CI)	p-Value
1st: 1.4E-4 – ≤1.1E-2	Reference		Reference		Reference	
2nd: >1.1E-2 – ≤2.1E-2	1.02 (0.94, 1.12)	0.61	1.07 (0.93, 1.22)	0.35	1.03 (0.92, 1.15)	0.60
3rd: >2.1E-2 – ≤3.4E-2	1.02 (0.93, 1.12)	0.69	1.11 (0.97, 1.27)	0.11	1.04 (0.94, 1.16)	0.46
4th: >3.4E-2 – ≤5.5E-2	1.01 (0.92, 1.11)	0.83	0.98 (0.86, 1.12)	0.77	0.97 (0.87, 1.08)	0.60
5th: >5.5E-2 – ≤4.6E-1	1.08 (0.98, 1.19)	0.10	1.13 (0.98, 1.29)	0.08	1.09 (0.98, 1.22)	0.13

Adjusted for sex, age, race/ethnicity, poverty income ratio (imputed five times), education, access to health care, whether a subject's residence was located in an urban or rural area, and NHIS interview year. The unstratified model for all subjects also controlled for smoking status, while the stratified models for never smokers and never & former smokers did not.

exposure estimates, there continues to be a need for epidemiologic studies to further establish acrolein's risks in the general population. This need will be addressed as methods for measuring acrolein in ambient air continue to progress [51,52], and as recent advances in measuring acrolein metabolites in human samples become broadly applied [101].

Conclusion

Environmental exposure to acrolein arising from outdoor emissions in the range of 0.05 – 0.46 $\mu g/m^3$ was associated with a marginally significant 8 percent increase in the prevalence-odds of having at least one asthma attack in the past 12 months among a representative cross-sectional sample of the 2000 – 2009 United States adult civilian, non-institutionalized population, as estimated in a sample-weighted logistic regression model controlling for potential confounders. Outdoor acrolein exposure in this range was also associated with marginally significant increases in asthma attack prevalence-odds among adults who do not directly smoke tobacco, a major source of acrolein exposure.

Acknowledgments

Disclaimer: The findings and conclusions in this article are those of the author and do not necessarily represent the views of the Centers for Disease Control and Prevention. The author wishes to thank Jennifer D. Parker (CDC): critical contributions to the development of this research, including advice on statistical analysis and data linkage; Lara Akinbami (CDC): helpful discussions about United States asthma prevalence and trends; Phillip Kott (RTI International): helpful discussions about survey sample weight adjustment; Nataliya Kravets (CDC): data linkage and geocoding; Ted Palma (US EPA): helpful discussions about acrolein exposure modeling in NATA; and the CDC Guest Researcher Program. In addition, the author wishes to thank the staff of the CDC Research Data Center, especially Ajay Yesupriya and Stephanie Robinson.

Author Contributions

Conceived and designed the experiments: BRdC. Analyzed the data: BRdC. Wrote the paper: BRdC. Critically revising: BRdC. Final approval: BRdC.

References

1. US EPA (2010) The Clean Air Act Amendments of 1990 List of Hazardous Air Pollutants. Available: http://www.epa.gov/ttn/atw/orig189.html. Accessed 2012.
2. Alarie Y (1973) Sensory irritation by airborne chemicals. CRC Crit Rev Toxicol 2: 299–363.
3. Grafstrom RC, Dypbukt JM, Willey JC, Sundqvist K, Edman C, et al. (1988) Pathobiological effects of acrolein in cultured human bronchial epithelial cells. Cancer Res 48: 1717–1721.
4. Kehrer JP, Biswal SS (2000) The molecular effects of acrolein. Toxicol Sci 57: 6–15.
5. Monteil C, Le Prieur E, Buisson S, Morin JP, Guerbet M, et al. (1999) Acrolein toxicity: comparative in vitro study with lung slices and pneumocytes type II cell line from rats. Toxicology 133: 129–138.
6. ATSDR (2007) Toxicological Profile for Acrolein. Available: http://www.atsdr.cdc.gov/toxprofiles/tp124.pdf. Accessed 2012.
7. Faroon O, Roney N, Taylor J, Ashizawa A, Lumpkin MH, et al. (2008) Acrolein health effects. Toxicol Ind Health 24: 447–490.
8. Beswick FW (1983) Chemical agents used in riot control and warfare. Hum Toxicol 2: 247–256.
9. Cowell EM (1939) Chemical Warfare and the Doctor-I. Br Med J 2: 736–738.
10. Hilmas CJ, Smart JK, Hill BA (2008) History of chemical warfare. In: SD . Tuorinsky, editor editors. Medical aspects of chemical warfare. Washington, DC: US Army Office of the Surgeon General. pp. 9–76.
11. Tuorinsky SD, Sciuto AM (2008) Toxic inhalation injury and toxic industrial chemicals. In: SD . Tuorinsky, editor editors. Medical aspects of chemical warfare. Washington, DC: US Army Office of the Surgeon General. pp. 339–370.
12. Faroon O, Roney N, Taylor J, Ashizawa A, Lumpkin MH, et al. (2008) Acrolein environmental levels and potential for human exposure. Toxicol Ind Health 24: 543–564.
13. US EPA (2011) 2005 National-scale Air Toxics Assessment Results. Available: http://www.epa.gov/ttn/atw/nata2005/. Accessed 2012.
14. Haag WR, Yao CD, Pettit T, Mill T (1988) Estimation of photolysis rate constants for acrolein (Magnacide H Herbicide and Magnacide B Microbiocide) in the environment. 50032-020.
15. Atkinson R (1985) Kinetics and mechanisms of the gas-phase reactions of hydroxyl radicals with organic compounds under atmospheric conditions. Chemical Reviews 85: 69–201.
16. GDCh Advisory Committee on Existing Chemicals of Environmental Relevance (BUA) (1994) Acrolein. BUA Report No. 157. Stuttgart.
17. Edney EO, Kleindienst TE, Corse EW (1986) Room-temperature rate constants for the reaction of OH with selected chlorinated and oxygenated hydrocarbons. International Journal of Chemical Kinetics 18: 1355–1371.
18. Howard PH (1989) Handbook of Environmental Fate and Exposure Data for Organic Chemicals: Large Production and Priority Pollutants. Lewis Publishers.
19. MacKay D, Mackay D, Shiu WY, Ma KC (1995) Illustrated Handbook of Physical-Chemical Properties and Environmental Fate for Organic Chemicals. Boca Raton, FL: Lewis Publishers.
20. US EPA (2003) Toxicological Review of Acrolein. Available: http://www.epa.gov/iris/toxreviews/0364tr.pdf. Accessed 2012.
21. Bein K, Leikauf GD (2011) Acrolein - a pulmonary hazard. Mol Nutr Food Res 55: 1342–1360.
22. Health Effects Institute Air Toxics Review Panel (2007) Mobile-Source Air Toxics: A Critical Review of the Literature on Exposure and Health Effects. Boston, MA: Health Effects Institute.
23. Abraham K, Andres S, Palavinskas R, Berg K, Appel KE, et al. (2011) Toxicology and risk assessment of acrolein in food. Mol Nutr Food Res 55: 1277–1290.
24. International Agency for Research on Cancer (1985) Dry Cleaning, Some Chlorinated Solvents and Other Industrial Chemicals. Available: http://monographs.iarc.fr/ENG/Monographs/vol63/mono63-14.pdf. Accessed 2012.
25. Voulgaridou GP, Anestopoulos I, Franco R, Panayiotidis MI, Pappa A (2011) DNA damage induced by endogenous aldehydes: current state of knowledge. Mutat Res 711: 13–27.
26. Feng Z, Hu W, Hu Y, Tang MS (2006) Acrolein is a major cigarette-related lung cancer agent: Preferential binding at p53 mutational hotspots and inhibition of DNA repair. Proc Natl Acad Sci U S A 103: 15404–15409.
27. Hecht SS (2006) Smoking and lung cancer-a new role for an old toxicant? Proc Natl Acad Sci U S A 103: 15725–15726.
28. Biswal S, Maxwell T, Rangasamy T, Kehrer JP (2003) Modulation of benzo[a]pyrene-induced p53 DNA activity by acrolein. Carcinogenesis 24: 1401–1406.
29. Jomova K, Vondrakova D, Lawson M, Valko M (2010) Metals, oxidative stress and neurodegenerative disorders. Mol Cell Biochem 345: 91–104.
30. Lopachin RM, Gavin T, Barber DS (2008) Type-2 alkenes mediate synaptotoxicity in neurodegenerative diseases. Neurotoxicology 29: 871–882.
31. Singh M, Dang TN, Arseneault M, Ramassamy C (2010) Role of by-products of lipid oxidation in Alzheimer's disease brain: a focus on acrolein. J Alzheimers Dis 21: 741–756.
32. Tanuma N, Miyata R, Hayashi M, Uchiyama A, Kurata K (2008) Oxidative stress as a biomarker of respiratory disturbance in patients with severe motor and intellectual disabilities. Brain Dev 30: 402–409.
33. Shi R, Rickett T, Sun W (2011) Acrolein-mediated injury in nervous system trauma and diseases. Mol Nutr Food Res 55: 1320–1331.
34. Leung G, Sun W, Zheng L, Brookes S, Tully M, et al. (2011) Anti-acrolein treatment improves behavioral outcome and alleviates myelin damage in experimental autoimmune encephalomyelitis mouse. Neuroscience 173: 150–155.
35. Woodruff TJ, Wells EM, Holt EW, Burgin DE, Axelrad DA (2007) Estimating risk from ambient concentrations of acrolein across the United States. Environ Health Perspect 115: 410–415.
36. Leikauf GD (2002) Hazardous air pollutants and asthma. Environ Health Perspect 110 Suppl 4: 505–526.
37. McCarthy MC, O'Brien TE, Charrier JG, Hather HR (2009) Characterization of the Chronic Risk and Hazard of Hazardous Air Pollutants in the United States Using Ambient Monitoring Data. Environ Health Perspect 117: 790–796.
38. Woodruff TJ, Axelrad DA, Caldwell J, Morello-Frosch R, Rosenbaum A (1998) Public health implications of 1990 air toxics concentrations across the United States. Environ Health Perspect 106: 245–251.
39. Logue JM, Small MJ, Stern D, Maranche J, Robinson AL (2010) Spatial Variation in Ambient Air Toxics Concentrations and Health Risks between Industrial-Influenced, Urban, and Rural Sites. Journal of the Air & Waste Management Association 60: 271–286.
40. CDC (2012) Asthma webpage. Available: http://www.cdc.gov/asthma/default.htm. Accessed 2012.

41. US EPA (2012) Asthma webpage. Available: http://www.epa.gov/asthma/. Accessed 2012.
42. National Heart Lung and Blood Institute (1995) Global Initiative for Asthma. 95–3659. Bethesda, MD.
43. National Heart Lung and Blood Institute (2007) Guidelines for the Diagnosis and Management of Asthma. Available: http://www.nhlbi.nih.gov/guidelines/asthma/asthgdln.pdf. Accessed 2012.
44. National Heart Lung and Blood Institute (2012) What is asthma. Available: http://www.nhlbi.nih.gov/health/health-topics/topics/asthma/. Accessed 2012.
45. US EPA (1999) Compendium Method TO-15, Determination of Volatile Organic Compounds (VOCs) In Air Collected In Specially-Prepared Canisters and Analyzed by Gas Chromatography/Mass Spectrometry (GC/MS). Available: http://www.epa.gov/ttnamti1/files/ambient/airtox/to-15r.pdf. Accessed 2012.
46. US EPA (1999) Compendium Method TO-11A, Determination of Formaldehyde in Ambient Air Using Adsorbent Cartridge Followed by High Performance Liquid Chromatography (HPLC). Available: http://www.epa.gov/ttnamti1/files/ambient/airtox/to-11ar.pdf. Accessed 2012.
47. Swift JL, Howell M, Tedder D, Merrill R (2006) Collection and Analysis of Acrolein Using Compendium Method TO-15. 22nd Annual National Environmental Monitoring Conference. Arlington, VA.
48. Shorter JH (2012) Acrolein Monitor Using Quantum Cascade Laser Infrared Absorption. Aerodyne Research, Inc.
49. Eastern Research Group (2010) Results of the 2005 NATA Model-to-Monitor Comparison, Final Report. Available: http://www.epa.gov/ttn/atw/nata2005/05pdf/nata2005_model2monitor.pdf. Accessed 2012.
50. US EPA (2010) EPA Schools Monitoring Initiative Fact Sheet: Acrolein Update. Available: http://epa.gov/schoolair/pdfs/acroleinupdate.pdf. Accessed 2012.
51. Cahill TM, Charles MJ, Seaman VY and Committee HEIHR (2010) Development and application of a sensitive method to determine concentrations of acrolein and other carbonyls in ambient air. Res Rep Health Eff Inst: 3–46.
52. Mason JB, Fujita EM, Campbell DE, Zielinska B (2011) Evaluation of passive samplers for assessment of community exposure to toxic air contaminants and related pollutants. Environ Sci Technol 45: 2243–2249.
53. US EPA (2003) Integrated Risk Information System (IRIS) on Acrolein. Available: http://www.epa.gov/iris/subst/0364.htm. Accessed 2012.
54. CDC (2000–2009) National Health Interview Survey. Available: http://www.cdc.gov/nchs/nhis.htm. Accessed 2012.
55. Akinbami LJ, Lynch CD, Parker JD, Woodruff TJ (2010) The association between childhood asthma prevalence and monitored air pollutants in metropolitan areas, United States, 2001–2004. Environ Res 110: 294–301.
56. Parker JD, Akinbami LJ, Woodruff TJ (2009) Air pollution and childhood respiratory allergies in the United States. Environ Health Perspect 117: 140–147.
57. Parker JD, Woodruff TJ, Akinbami LJ and Kravets N (2008) Linkage of the US National Health Interview Survey to air monitoring data: an evaluation of different strategies. Environ Res 106: 384–392.
58. Stoner AM, Anderson SE, Buckley TJ (2013) Ambient air toxics and asthma prevalence among a representative sample of US kindergarten-age children. PLoS One 8: e75176.
59. Linder SH, Marko D, Sexton K (2008) Cumulative cancer risk from air pollution in Houston: disparities in risk burden and social disadvantage. Environ Sci Technol 42: 4312–4322.
60. Kalkbrenner AE, Daniels JL, Chen JC, Poole C, Emch M, et al. (2010) Perinatal exposure to hazardous air pollutants and autism spectrum disorders at age 8. Epidemiology 21: 631–641.
61. Windham GC, Zhang L, Gunier R, Croen LA, Grether JK (2006) Autism spectrum disorders in relation to distribution of hazardous air pollutants in the san francisco bay area. Environ Health Perspect 114: 1438–1444.
62. US Census Bureau (2012) Census Tracts and Block Numbering Areas. Available: http://www.census.gov/geo/www/cen_tract.html. Accessed 2012.
63. US EPA (2010) 2005 National Emissions Inventory Data & Documentation. Available: http://www.epa.gov/ttn/chief/net/2005inventory.html. Accessed 2012.
64. US EPA (2011) An Overview of Methods for EPA's National-Scale Air Toxics Assessment. Available: http://www.epa.gov/ttn/atw/nata2005/05pdf/nata_tmd.pdf. Accessed 2012.
65. CDC (2010) NHIS Survey Description. Available: ftp://ftp.cdc.gov/pub/Health_Statistics/NCHS/Dataset_Documentation/NHIS/2009/srvydesc.pdf. Accessed 2012.
66. Botman SL, Moore TF, Moriarity CL, Parsons VL (2000) Design and estimation for the National Health Interview Survey, 1995–2004. Vital Health Stat 2.
67. CDC (2010) Health, United States, 2010. Available: http://www.cdc.gov/nchs/hus.htm. Accessed 2012.
68. Binder DA (1983) On the Variances of Asymptotically Normal Estimators from Complex Surveys. International Statistical Review 51: 279–292.
69. Research Triangle Institute (2008) SAS-Callable SUDAAN. Research Triangle Park, NC.
70. SAS Institute (2010) SAS. Cary, NC.
71. CDC (2009) Variance Estimation and Other Analytic Issues in the 1997–2005 NHIS. Available: http://www.cdc.gov/nchs/data/nhis/9705var.pdf. Accessed 2012.
72. CDC (2011) Variance Estimation and Other Analytic Issues, NHIS 2006–2010. Available: http://www.cdc.gov/nchs/data/nhis/2006var.pdf. Accessed 2012.
73. Rubin DB (2004) Multiple Imputation for Nonresponse in Surveys. Wiley-Interscience.
74. Schenker N, Raghunathan TE, Chiu PL, Makuc DM, Zhang GY, et al. (2006) Multiple imputation of missing income data in the National Health Interview Survey. Journal of the American Statistical Association 101: 924–933.
75. CDC (2010) Multiple Imputation of Family Income and Personal Earnings in the National Health Interview Survey: Methods and Examples. Available: http://www.cdc.gov/nchs/data/nhis/tecdoc4.pdf. Accessed 2012.
76. CDC (2011) Use of Survey Weights for Linked Data Files – Preliminary Guidance. Available: http://www.cdc.gov/nchs/data/datalinkage/use_of_survey_weights_for_linked_data_files.pdf. Accessed 2012.
77. Folsom RE, Singh AC (2009) The generalized exponential model for sampling weight calibration for extreme values, nonresponse, and poststratification. In: Proceedings of the Joint Statistical Meetings 2000, Section on Survey Research Methods, 2000, 598-603.
78. US EPA (2010) Toxics Release Inventory. Available: http://www.epa.gov/tri/). Accessed 2012.
79. National Library of Medicine (2010) TOXMAP. Available: http://toxmap.nlm.nih.gov. Accessed 2012.
80. CA Office of Environmental Health Hazard Assessment (2008) Air Toxics Hot Spots Risk Assessment Guidelines: Technical Support Document For the Derivation of Noncancer Reference Exposure Levels. Available: http://www.oehha.ca.gov/air/hot spots/2008/NoncancerTSD_final.pdf. Accessed 2012.
81. Baumann LM, Robinson CL, Combe JM, Gomez A, Romero K, et al. (2011) Effects of distance from a heavily transited avenue on asthma and atopy in a periurban shantytown in Lima, Peru. J Allergy Clin Immunol 127: 875–882.
82. Holguin F, Flores S, Ross Z, Cortez M, Molina M, et al. (2007) Traffic-related exposures, airway function, inflammation, and respiratory symptoms in children. Am J Respir Crit Care Med 176: 1236–1242.
83. Kan H, Heiss G, Rose KM, Whitsel E, Lurmann F, et al. (2007) Traffic exposure and lung function in adults: the Atherosclerosis Risk in Communities study. Thorax 62: 873–879.
84. Kim JJ, Huen K, Adams S, Smorodinsky S, Hoats A, et al. (2008) Residential traffic and children's respiratory health. Environ Health Perspect 116: 1274–1279.
85. Morgenstern V, Zutavern A, Cyrys J, Brockow I, Koletzko S, et al. (2008) Atopic diseases, allergic sensitization, and exposure to traffic-related air pollution in children. Am J Respir Crit Care Med 177: 1331–1337.
86. Pujades-Rodriguez M, Lewis S, McKeever T, Britton J, Venn A (2009) Effect of living close to a main road on asthma, allergy, lung function and chronic obstructive pulmonary disease. Occup Environ Med 66: 679–684.
87. Venn AJ, Lewis SA, Cooper M, Hubbard R, Britton J (2001) Living near a main road and the risk of wheezing illness in children. Am J Respir Crit Care Med 164: 2177–2180.
88. Brauer M, Hoek G, Van Vliet P, Meliefste K, Fischer PH, et al. (2002) Air pollution from traffic and the development of respiratory infections and asthmatic and allergic symptoms in children. Am J Respir Crit Care Med 166: 1092–1098.
89. Kelly FJ, Fussell JC (2011) Air pollution and airway disease. Clin Exp Allergy 41: 1059–1071.
90. Salam MT, Islam T and Gilliland FD (2008) Recent evidence for adverse effects of residential proximity to traffic sources on asthma. Curr Opin Pulm Med 14: 3–8.
91. Stone V (2000) Environmental air pollution. Am J Respir Crit Care Med 162: S44–47.
92. Dorman DC, Struve MF, Wong BA, Marshall MW, Gross EA, et al. (2008) Respiratory tract responses in male rats following subchronic acrolein inhalation. Inhal Toxicol 20: 205–216.
93. Schroeter JD, Kimbell JS, Gross EA, Willson GA, Dorman DC, et al. (2008) Application of physiological computational fluid dynamics models to predict interspecies nasal dosimetry of inhaled acrolein. Inhal Toxicol 20: 227–243
94. Struve MF, Wong VA, Marshall MW, Kimbell JS, Schroeter JD, et al. (2008) Nasal uptake of inhaled acrolein in rats. Inhal Toxicol 20: 217–225.
95. Logue JM, McKone TE, Sherman MH, Singer BC (2011) Hazard assessment of chemical air contaminants measured in residences. Indoor Air 21: 92–109.
96. Seaman VY, Bennett DH, Cahill TM (2007) Origin, occurrence, and source emission rate of acrolein in residential indoor air. Environ Sci Technol 41: 6940–6946.
97. Nazaroff WW, Singer BC (2004) Inhalation of hazardous air pollutants from environmental tobacco smoke in US residences. J Expo Anal Environ Epidemiol 14 Suppl 1: S71–77.
98. Payne-Sturges DC, Burke TA, Breysse P, Diener-West M, Buckley TJ (2004) Personal exposure meets risk assessment: a comparison of measured and modeled exposures and risks in an urban community. Environ Health Perspect 112: 589–598.
99. Rosenbaum AS, Axelrad DA, Woodruff TJ, Wei YH, Ligocki MP, et al. (1999) National estimates of outdoor air toxics concentrations. J Air Waste Manag Assoc 49: 1138–1152.

100. Hill AB (1965) The Environment and Disease: Association or Causation? Proc R Soc Med 58: 295–300.
101. Alwis KU, Blount BC, Britt AS, Patel D, Ashley DL (2012) Simultaneous analysis of 28 urinary VOC metabolites using ultra high performance liquid chromatography coupled with electro spray tandem mass spectrometry (UPLC-ESI/MSMS) (accepted). Anal Chim Acta.

DNA Damage in Buccal Mucosa Cells of Pre-School Children Exposed to High Levels of Urban Air Pollutants

Elisabetta Ceretti[1], Donatella Feretti[1]*, Gaia C. V. Viola[1], Ilaria Zerbini[1], Rosa M. Limina[1], Claudia Zani[1], Michela Capelli[2], Rossella Lamera[2], Francesco Donato[1], Umberto Gelatti[1]

1 Unit of Hygiene, Epidemiology and Public Health, Department of Medical and Surgical Specialities, Radiological Sciences and Public Health, University of Brescia, Brescia, Italy, **2** Post-Graduate School of Public Health, University of Brescia, Brescia, Italy

Abstract

Air pollution has been recognized as a human carcinogen. Children living in urban areas are a high-risk group, because genetic damage occurring early in life is considered able to increase the risk of carcinogenesis in adulthood. This study aimed to investigate micronuclei (MN) frequency, as a biomarker of DNA damage, in exfoliated buccal cells of pre-school children living in a town with high levels of air pollution. A sample of healthy 3-6-year-old children living in Brescia, Northern Italy, was investigated. A sample of the children's buccal mucosa cells was collected during the winter months in 2012 and 2013. DNA damage was investigated using the MN test. Children's exposure to urban air pollution was evaluated by means of a questionnaire filled in by their parents that included items on various possible sources of indoor and outdoor pollution, and the concentration of fine particulate matter (PM10, PM2.5) and NO_2 in the 1–3 weeks preceding biological sample collection. 181 children (mean age±SD: 4.3±0.9 years) were investigated. The mean±SD MN frequency was 0.29±0.13%. A weak, though statistically significant, association of MN with concentration of air pollutants (PM10, PM2.5 and NO_2) was found, whereas no association was apparent between MN frequency and the indoor and outdoor exposure variables investigated via the questionnaire. This study showed a high MN frequency in children living in a town with heavy air pollution in winter, higher than usually found among children living in areas with low or medium-high levels of air pollution.

Editor: Susanna Esposito, Fondazione IRCCS Ca' Granda Ospedale Maggiore Policlinico, Università degli Studi di Milano, Italy

Funding: The study was funded by the Q-TECH Research and Study Centre (Quality and Technology Assessment, Governance and Communication Strategies in Health Systems), University of Brescia. The Lombardy Regional Authority provided a three-year research grant under agreements with the university for the promotion of research in the Lombardy region. The funders had no role in study design, data collection and analysis, decision to publish, or preparation of the manuscript.

Competing Interests: The authors have declared that no competing interests exist.

* E-mail: feretti@med.unibs.it

Introduction

Air pollution is a global health problem, particularly in urban areas. Close to 90% of European citizens residing in urban areas are exposed to air pollution exceeding EU limit values (Air Quality Directive 2008/50/EC) and WHO guideline levels [1–3].

Several epidemiologic studies have demonstrated the association between air pollution exposure, especially to particulate matter, and mortality and morbidity in humans [4–8].

The finest fractions of particulate matter (PM2.5 and less) play a major role in causing chronic diseases because they are retained in the alveolar regions of the lungs and diffuse into the blood stream, inducing inflammation, oxidative stress, and blood coagulation [9,10]. Extracts of urban air particles can induce cancer in animals [11,12] and are mutagenic in bacteria, plant and mammalian cells in *in-vitro* tests [13–17]. *In-vivo* mutagenicity tests have been performed on humans as well. The micronuclei (MN) formation has been used as an indicator of chromosome damage, induced by substances that cause chromosome breakage (clastogens) as well as by agents that affect the spindle apparatus (aneugens) [18]. Other nuclear anomalies were investigated as biomarkers of DNA damage and cell death, which may be useful for a more comprehensive assessment of genotoxic damage [19–22].

An increase in cancer risk has been observed in the presence of a high level of chromosomal aberrations and micronuclei in several studies [23–26]. This biomarker can be investigated in various organs, tissues and body fluids, such as leukocytes or lymphocytes in peripheral blood, though cells derived from target tissues are considered more appropriate. In particular, exfoliated buccal and nasal cells have been used in the biological monitoring of people exposed to airborne pollutants as they are representative of epithelial respiratory tract cells and are easier to collect than those of other respiratory organs [21,27–29].

An increase of micronuclei in leukocytes in peripheral blood has been observed in people exposed to urban air pollutants [30–36] and a strong correlation of MN frequency in buccal exfoliated cells and peripheral lymphocytes has been found [37,38].

Children are a high-risk group in terms of the health effects of air pollution [2,5,39–42]. Some studies suggest that early exposure during childhood can play an important role in the development of chronic diseases in adulthood: the earlier the exposure, the greater the risk of chronic disease, including cancer [43].

Few studies have considered genetic damage in mucosa buccal cells as MN frequency in cells of children exposed to air pollution, and they only involved a small number of subjects showing cytogenetic damage in children or young adults living in polluted

Table 1. Micronuclei (MN) frequencies observed in buccal mucosa cells of children according to socio-demographic features and exposure variables (N = 181).

Subjects and demographic features	N (%)	% MN (Mean±SD)
Sex		
M	103 (56.9)	0.30±0.13
F	78 (43.1)	0.29±0.13
Children's age		
3 years	34 (18.8)	0.31±0.16
4 years	59 (32.6)	0.28±0.13
5–6 years	88 (48.6)	0.29±0.12
Parents' education (at least one parent)		
Primary school or less	19 (10.5)	0.28±0.12
Secondary school	53 (29.3)	0.32±0.16
College or university	109 (60.2)	0.28±0.12
Home characteristics		
Traffic in the area		
Heavy	97 (53.6)	0.31±0.14
Moderate	59 (32.6)	0.28±0.13
Very light	25 (13.8)	0.25±0.08
Truck traffic in the area		
Heavy	31 (17.1)	0.30±0.12
Moderate	67 (37.0)	0.31±0.17
Very light	82 (45.3)	0.27±0.09
Indoor exposure		
Gas stove in home	61 (33.7)	0.28±0.11
Fireplace in home	37 (20.4)	0.28±0.11
Presence of smokers in home	28 (15.5)	0.30±0.18
School characteristics		
Traffic in the area		
Heavy	102 (56.4)	0.31±0.15
Moderate	68 (37.6)	0.27±0.09
Very light	10 (5.5)	0.29±0.15
Truck traffic in the area		
Heavy	27 (14.9)	0.33±0.19
Moderate	80 (44.2)	0.29±0.12
Very light	72 (39.8)	0.28±0.12
Child's habits		
Plays outdoors		
Less than 1 hour	78 (43.1)	0.31±0.16
More than 1 hour but less than 3 hours	67 (37.0)	0.28±0.10
3 hours or more	35 (19.3)	0.28±0.10
Remains in the kitchen while meals are cooked		
Never	24 (13.3)	0.29±0.13
Sometimes	119 (65.8)	0.30±0.14
Often/always	37 (20.4)	0.27±0.09
Consumes fried/grilled/smoked food		
Never	5 (2.8)	0.36±0.17
More than once per month	123 (68.0)	0.28±0.14
Parents' smoking habits		
Neither parent smoke	118 (65.2)	0.29±0.12
Mother smoked during pregnancy	40 (22.1)	0.29±0.14
Mother smokes	30 (16.6)	0.29±0.15

Table 1. Cont.

Subjects and demographic features	N (%)	% MN (Mean±SD)
Father smoke	49 (27.1)	0.30±0.17
Both parents smokers	42 (23.2)	0.29±0.15

areas with a high concentration of PM or oxidant pollutants [33,44–47].

The aim of this study was to investigate MN frequency, as a biomarker of DNA damage, in exfoliated buccal cells of pre-school children living in a town with high levels of air pollution during the winter season, when the highest levels of particulate matter and other pollutants are usually found.

Materials and Methods

Study design

This study is part of the RESPIRA study (Italian acronym for Rischio ESPosizione Inquinamento aRia Atmosferica), a molecular epidemiology cross-sectional study aiming to assess the presence of MN frequency in pre-school children living in Brescia, a highly polluted town in Northern Italy, located in the Po Valley, one of the most highly polluted areas of Europe. The children were recruited in 6 schools located in different areas of the town. The study enrolled children aged 3–6 years, born in Italy to European parents, without malignant tumours, who had not undergone radiotherapy or chemotherapy in the previous 12 months or X-rays in the previous 3 months.

The presence of MN and other nuclear anomalies were investigated in buccal mucosa cells taken from the children.

The biological samples were collected during two consecutive winter seasons (2012 and 2013) since the highest values of PM10 and PM2.5 in Brescia are usually observed in the winter months. The project was approved by the Ethics Committee of Local Unit Health of Brescia (Comitato Etico dell'ASL -Azienda Sanitaria Locale- di Brescia). The children's parents provided their written informed consent to participate in this study. All the data collected were treated confidentially in accordance with current Italian legislation (privacy law).

Questionnaire

The children's parents were interviewed using an *ad hoc* questionnaire designed to gather information on exposure to air pollutants from both indoor and outdoor sources, including some characteristics of the area of residence (e.g. traffic, factories), parents' smoking habits, and children's respiratory diseases and drug consumption.

Collection of air pollution data

Chemical data regarding daily concentration of the most commonly measured air pollutants (CO, NO_2, SO_2, benzene, O_3, PM10 and PM2.5) were retrieved from the freely available ARPA (Regional Agency for Environmental Protection) database to characterize urban air quality.

Collection of biological samples

All the biological samples were taken during or after a series of days with high levels of PM10, PM2.5 and NO_2.

For the collection of buccal mucosa cells, the children rinsed their mouths twice with mineral water. Interdental brushes were used to collect epithelial buccal cells for the micronucleus test, by gently scraping the inside of both cheeks and dipping them into tubes containing 15 ml of PBS (phosphate buffered saline solution) [21]. This method is simple and non-invasive and therefore easily acceptable by both children and parents.

Micronucleus test

After shaking and removing the brush, the epithelial buccal cells in PBS were centrifuged for 10 minutes at 1100 g at 4°C and re-suspended in warm PBS (37°C). In order to determine whether enough cells had been collected to perform the test, 10 µl of cell suspension was applied in a Burker chamber on which the number of cells was scored. The PBS cell suspension was then centrifuged for 4 minutes at 8700 g and the pellet was re-suspended in 700 µl of warm hypotonic solution (KCl 0.56%, 37°C). After 1 minute, 700 µl of cold methanol/acetic acid (14:1, −20°C) was added. This fixed suspension was centrifuged for 4 minutes at 8700 g, and the pellet was re-suspended in 40 µl of warm PBS and dropped onto two frosted slides, which were dried and stained with Giemsa dye (5 minutes at room temperature). The slides were then washed with distilled water, dried and mounted with Eukitt.

The slides were examined at 1000X magnification for microscope analysis. Before MN frequency was assessed, cells were divided into two categories: "normal" cells and ones that are considered "abnormal" based on their cytobiological and nuclear features, which are indicative of DNA damage (MN and nuclear buds), cytokinetic failure (binucleated cells), proliferative potential (basal cells) or cell death (condensed chromatin, karyorrhexis, pyknotic, karyolitic and without nucleus cells), according to the Buccal Micronucleus Cytome (BMCyt) assay [22]. To assess MN, nuclear buds, binucleated and basal cells frequency, at least 2000 "normal" cells per slide (two slides per subject, 4000 cells per subject) were scored by two expert operators with duplicate reading. The results are given as the percentage of cells with MN and buds, and binucleated and basal cells [19,22]. Moreover, condensed chromatin, karyorrhexis, pyknotic, karyolitic, and without nucleus cells were evaluated scoring 2000 total cells per subject, and expressed as percentages.

Statistical analysis

All the data were processed to investigate the associations between air pollution parameters and MN frequency. As buccal cells have a short life, no more than 3 weeks [22], the associations between frequency of MN and other biomarkers frequency and air pollutants concentration (CO, NO_2, SO_2, benzene, O_3, PM10 and PM2.5) were analysed using air pollutants mean levels at 0 (sampling day), 1, 2 and 3 weeks before sampling. Both univariate analysis and multivariate analysis (multiple regression, logistic regression) were performed to assess the association investigated, adjusting for confounding factors. Particularly, linear regression with biological parameters as dependent variables and the concentration of air pollutants as independent variables were fitted. In order to improve the interpretation of these findings the coefficients of the linear regression were computed on a 10 µg/m^3

Table 2. Frequency of micronuclei (MN) and other nuclear anomalies in all children and according to sex and age.

	N° children	% MN	% Nuclear buds	% Binucleated cells	% Basal cells	% Condensed chromatin cells	% Karyorrhetic cells	% Pyknotic cells	% Karyolytic cells	% Without nucleus cells
		(Mean±SD)	(Mean±SD)	(Mean±SD)	(Mean±SD)	(Mean±SD)	(Mean±SD)	(Mean±SD)	(Mean±SD)	(Mean±SD)
All children	181	0.29±0.13	0.02±0.04	0.11±0.08	0.72±1.15	17.20±6.13	7.44±4.29	0.57±0.48	3.72±3.37	1.42±1.28
Sex										
Males	103	0.30±0.13	0.02±0.04	0.11±0.09	0.84±1.38	17.10±6.26	7.88±4.72	0.61±0.50	3.96±3.97	1.42±1.18
Females	78	0.29±0.13	0.02±0.03	0.10±0.08	0.56±0.73	17.33±6.00	6.86±3.60	0.52±0.46	3.42±2.34	1.43±1.40
Children's age										
3 years	34	0.31±0.16	0.02±0.06	0.10±0.09	0.56±0.80	16.70±4.52	7.48±4.84	0.65±0.53	4.07±2.91	1.65±1.27
4 years	59	0.28±0.13	0.01±0.02	0.13±0.10	0.64±0.69	16.58±6.18	7.96±4.41	0.52±0.49	3.21±2.31	1.23±1.15
5–6 years	88	0.29±0.12	0.02±0.03	0.09±0.06	0.83±1.46	17.80±6.62	7.07±3.99	0.57±0.46	3.94±4.05	1.47±1.35

scale of measures of air pollutants. Two-tailed statistical tests were performed, with 0.05 p-value as the threshold for rejecting the null hypothesis. All the analyses were performed using the Stata TM 12.0 statistical package (Stata Statistical Software Release 12.0, 2012; Stata Corporation, College Station, Texas, USA).

Results

A total of 222 children were recruited, of whom 181 (mean age ±SD: 4.35±0.84 years; 56.9% males) were examined because 41 samples were not eligible for MN analysis due to an insufficient number of cells collected.

The mean±SD MN frequency expressed as a percentage was 0.29±0.13 (median 0.28 and range 0.085–0.990).

Table 1 shows the results of the MN analysis, according to children's socio-demographic characteristics and habits, and indoor and outdoor exposure data. No variable was associated with MN frequency in exfoliated buccal cells.

Table 2 sets out all the biomarkers evaluated in the buccal mucosa cells of children according to children's sex and age. No statistically significant difference was observed according to sex and age for each parameter.

The daily levels of PM10, PM2.5 and NO_2 from January to March 2012 and 2013, and biological sampling days are shown in Figure 1. During these months, the concentrations of PM10 and PM2.5 were almost always over the EU limit values for daily means (50 and 25 μg/m^3, respectively). Likewise, the annual EU limit for NO_2 (40μg/m3) was exceeded on all the days. On the contrary, the concentration of CO, ozone, SO_2 and benzene remained low throughout the period considered and were always below the EU limit values (data not shown).

The mean MN frequency according to concentrations of PM10, PM2.5 and NO_2 on the day of sampling and one, two and three weeks before sampling are shown in Table 3. A fair variation (from 0.21% to 0.62%) was found in MN frequency from one day to another, without a clear relationship with concentration of air pollutants.

The coefficients of linear regressions of the biological markers on the concentration of PM10, PM2.5 and NO_2 on the same day and at 1, 2 and 3 weeks preceding biological sampling computed for 10 μg/m^3 units of increase are shown in Table 4. A modest, though statistically significant, increase of the frequency of nuclear anomalies (MN, buds and binucleated cells) and of basal cells for an increase of PM10, PM2.5 and NO_2 was found. No clear pattern was evident for the other parameters.

Discussion

The main finding of this study was a surprisingly high level of MN frequency in exfoliated buccal cells of pre-school children living in Brescia.

The MN frequency (mean: 0.29%) observed in our study was higher than that observed in healthy children "without important exposure" (0.108%) as a result of a pooled analysis of 321 children aged up to 9 years [48]. Furthermore, the value observed in children living in Brescia was higher than that shown in adolescents or young adults working in an engine repair workshop (mean age: 15.5 years, MN frequency: 0.07%) [49] or exposed to ozone air pollution (university students, MN frequency: 0.12%) [46]. However, these studies are not comparable to ours due to the different age of the people investigated. Age is in fact one of the most important factors affecting MN data in both lymphocytes and oral cells, with a progressively increase of MN with age [33,36,48,50]. However, the MN frequency observed in our study was higher than that found in children (aged 6–17) living in the

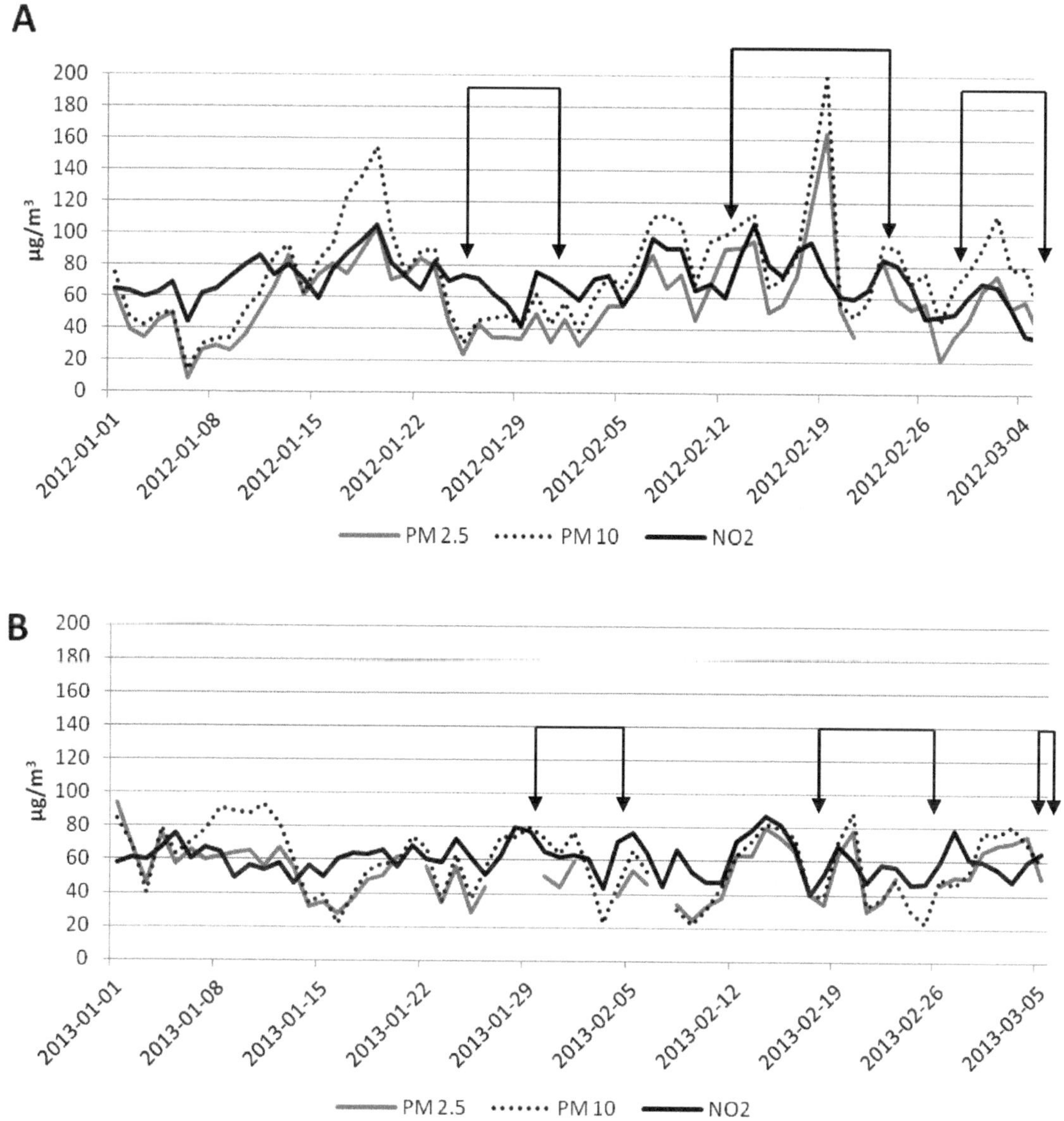

Figure 1. PM10, PM2.5 and NO_2 concentration in 2012 (A) and 2013 (B). The arrows indicate the days on which biological sampling was performed. The dotted lines refer to the days without measurement of air pollutants.

urban area of Calcutta [44], which was higher than that found in those living in rural area (0.22% vs 0.17%). Children of a similar age to those in our study were included in a study on genotoxicity of air pollutants generated by biomass burning [47], which found a significant difference in MN frequency in oral cells between children exposed to high PM2.5 concentration and those living in a control area with a lower PM2.5 concentration (0.12% vs 0.02% in children under 7 years of age). A MN frequency similar to, or higher than, that observed in our study was reported in some studies carried out in Eastern Europe, as revised by Holland et al. [36].

According to a wide dataset which includes adults and children, a MN frequency interval of between 0.03% and 0.17% can be considered as a range of spontaneous MN frequency in buccal exfoliated cells of "unexposed" children [48]. Other authors have reported a range of 0.05–0.08% as the baseline MN frequency in exfoliated cells in healthy people [46]. Therefore, the mean MN frequency of 0.29% found in our study is about two-three-fold higher than that considered as a "reference" value for children of this age.

The MN frequency did not vary according to sex, age, parental education and all the variables investigated through the questionnaire. A weak, though statistically significant, association of MN frequency with concentration of air pollutants in the week preceding the buccal mucosa cell collection, but not on the same day and in the second or third week before sampling, was found. These results are partially in agreement with findings from previous studies comparing people at different exposure levels to air pollutants, which showed that industrial or urban air pollutants had a genotoxic effect on mucosa cells [44,47,49]. It should be pointed out, however, that the concentration of PM10, PM2.5 and NO_2 was high throughout the study period, and always above the EU proposed limits for daily levels. The range of variation of pollutant concentration may therefore have been too narrow to determine a substantial change in MN frequency in the children. Alternatively, a threshold instead of a dose-response mechanism

Table 3. Micronuclei (MN) frequency and PM10, PM2.5 and NO_2 concentrations on the day of sampling and in the 1–3 weeks preceding biological sampling.

Day of sampling	N° children	% MN (Mean±SD)	Same day PM10 (μg/m³)	Same day PM2.5 (μg/m³)	Same day NO_2 (μg/m³)	1 week PM10 (μg/m³)	1 week PM2.5 (μg/m³)	1 week NO_2 (μg/m³)	2 weeks PM10 (μg/m³)	2 weeks PM2.5 (μg/m³)	2 weeks NO_2 (μg/m³)	3 weeks PM10 (μg/m³)	3 weeks PM2.5 (μg/m³)	3 weeks NO_2 (μg/m³)
2012-01-24	8	0.62±0.23	52.0	44	70.3	110.2	82.6	84.5	92.9	73.8	80.0	73.9	59.6	74.1
2012-01-25	4	0.27±0.10	31	24	73.9	99.8	78.1	82.0	93.0	74.4	79.3	74.4	60.0	74.6
2012-01-26	4	0.30±0.14	45.5	43	71.8	84.7	68.7	78.9	90.8	72.4	78.4	73.5	59.0	75.1
2012-01-30	4	0.18±0.03	62.0	50	76.1	50.9	42.0	65.4	80.7	62.4	74.6	76.2	60.7	74.7
2012-01-31	7	0.27±0.10	43.0	32	71.9	46.8	37.9	64.5	78.5	60.2	74.5	77.5	61.8	74.9
2012-02-13	12	0.26±0.15	106.5	92	87.2	97.1	72.0	77.4	77.1	58.1	72.5	68.4	52.8	70.2
2012-02-14	12	0.34±0.08	113.0	96	107.1	100.1	75.0	80.1	80.3	61.1	73.3	69.1	53.4	70.4
2012-02-20	12	0.25±0.09	57.0	53	61.3	112.3	93.1	87.0	104.7	82.6	82.2	88.9	69.8	77.4
2012-02-22	11	0.29±0.04	57.5	nd	65.4	95.9	79.0	76.6	98.2	67.3	79.0	88.9	70.1	76.1
2012-02-23	10	0.23±0.05	93.5	83	84.4	94.6	71.7	74.4	94.3	72.9	77.2	88.9	68.0	76.1
2012-02-28	11	0.24±0.09	67.5	36	50.5	68.7	44.3	65.0	87.0	65.9	74.2	91.4	69.0	76.1
2012-02-29	12	0.30±0.13	77.5	46	60.8	71.5	44.3	63.7	83.7	61.6	70.1	89.3	66.5	73.9
2012-03-01	12	0.24±0.07	92.0	65	69.7	74.4	50.9	63.3	84.5	61.3	68.7	87.7	65.6	72.5
2012-03-05	6	0.25±0.04	51.0	38	34.5	78.8	50.7	55.6	74.6	49.8	61.2	87.2	64.2	69.8
2013-01-29	6	0.40±0.19	80	nd	77.4	57.9	36.7	63.8	53.6	40.1	62.5	61.3	46.4	60.0
2013-01-30	3	0.21±0.03	73.5	51	65.5	59.8	32.8	66.2	56.5	40.5	64.5	60.8	45.6	60.6
2013-02-04	1	0.32±0.00	45.0	39	71.4	63.7	38.5	64.7	60.8	41.6	63.4	55.0	41.7	62.1
2013-02-18	9	0.31±0.05	39.0	34	52.2	65.2	60.6	67.9	50.3	46.8	64.3	54.8	44.9	64.4
2013-02-19	5	0.33±0.06	71.0	64	69.5	64.4	60.0	68.6	49.9	46.4	62.9	53.1	46.8	63.1
2013-02-25	7	0.41±0.15	23	nd	46.9	58.5	48.7	55.8	55.7	55.1	61.8	50.1	47.8	61.4
2013-03-04	16	0.25±0.07	70.5	75	60.1	58.0	59.0	58.6	53.9	53.8	57.2	57.7	56.3	60.7
2013-03-05	9	0.21±0.04	58.0	50	65.0	64.8	61.3	60.5	56.1	57.3	57.8	58.9	58.3	61.4

Table 4. Coefficients of linear regression of the biological markers for 10 $\mu g/m^3$ units of increase of PM10, PM2.5 and NO_2 concentration.

	Same day			Weeks preceding biological sampling								
				1 week			2 weeks			3 weeks		
	PM 10	PM 2.5	NO_2	PM 10	PM 2.5	NO_2	PM 10	PM 2.5	NO_2	PM 10	PM 2.5	NO_2
MN	−0.01	−0.00	0.00	0.01*	0.01	0.03**	0.00	0.00	0.02	−0.01	−0.03**	−0.01
Nuclear buds	−0.00	0.00	0.08*	0.01***	0.01**	0.01***	0.00	0.00	0.01**	−0.00	−0.00	0.00
Binucleated cells	−0.00	−0.00	0.0[illegible]	0.01**	0.01*	0.02**	0.01**	0.02**	0.03***	0.00	0.00	0.03**
Basal cells	0.02	0.00	0.10	0.20***	0.15**	0.40***	0.15**	0.26**	0.42***	0.06	0.03	0.40**
Condensed chromatin cells	0.20	0.06	1.07***	1.13***	1.06***	2.93***	1.26***	1.94***	3.61***	0.74*	0.71	3.79***
Karyorrhetic cells	0.04	−0.36**	−0.26	1.73	−0.29	0.20	0.66***	0.60	1.22**	1.05***	1.32**	2.22***
Pyknotic cells	0.01	0.04**	0.02	0.02	0.03	0.03	−0.03	−0.01	−0.05	−0.07**	−0.11**	−0.12*
Karyolytic cells	−0.04	0.02	−0.25	−0.48***	−0.36**	−0.93***	−0.61***	−0.93***	−1.40***	−0.61**	−0.71*	−2.05***
Without nucleus cells	0.09*	0.07	0.12	−0.01	−0.08	0.05	−0.01	−0.04	0.03	−0.01	−0.12	−0.04

statistically significant: *$p<0.05$;
**$p<0.01$;
***$p<0.001$.

may be proposed, so similar genotoxic effects on children's buccal mucosa cells may have been produced by air pollutant values above a defined level. Lastly, the role of confounding factors, such as children's diet, physical activity and others, which was not investigated in our study, cannot be excluded. Indeed, some studies have observed an association between MN frequency in buccal mucosa cells and diet (e.g. fruit consumption, supplementation with B vitamins or antioxidants) and some specific behaviours, such as smoking habit, intake of alcohol, and betel quid chewing [21,37,48,51,52].

Other nuclear anomalies apart from MN were associated with the PM10, PM2.5 and NO_2 concentration, particularly during the week preceding exfoliated cells sampling confirming findings of MN analysis on the possible effects of air pollutant exposure. However, the biological significance of these biomarkers in exfoliated oral cells is still unclear. Some nuclear anomalies, as buds, could be the result of genotoxic damage but these events are also associated with natural degenerative processes in these short-lived cells [53,54]. The frequency of binucleated cells was related to air pollutants concentration too. This parameter is primarily an indicator of cytotoxicity because of failures in cell division. Although these findings should be considered with caution because of paucity of data on the relationship between these biomarkers and air pollutants levels, overall they suggest that air pollution exposure may induce both genotoxic and cytotoxic damage in buccal mucosa cells of children. This observation strengthens the need to investigate these additional biomarkers of DNA damage together with MN for a more comprehensive evaluation of these issues.

This study has various strengths. First, the number of subjects recruited (181 children) is relatively high compared to previous studies of MN frequency in oral cells, particularly in this age group.

Second, it is unlikely that other important factors biased our results: there is no major industrial exposure in the area and few children could have been exposed to indoor pollution sources, including passive smoking at home, according to data collected via the questionnaire. Furthermore, the inclusion and exclusion criteria, e.g. residence in town, born to European parents, no malignant tumours, no radiotherapy or chemotherapy in the previous 12 months, no X-ray exposure in the three months before buccal cell collection, allowed us to rule out a possible role of other mutagenic factors.

Third, the collection of exfoliated buccal cells during two consecutive winter seasons (2012 and 2013) provided information on two periods characterized by similar environmental conditions and it showed consistent results, with no substantial difference in MN frequency in the two years.

Lastly, 4000 cells per child were scored for assessing MN frequency, representing a very high number of observations, higher than normally performed in current practice, in order to reduce the variability of the estimate, as suggested by Ceppi et al. [37].

This study has some limitations, however, mainly the lack of a control group of subjects living in a less polluted area or another collection of biological samples from the same or similar children in a period with lower airborne pollutant levels, for evaluating the role of spatial or seasonal differences in influencing MN frequency in buccal mucosa cells. Differences between more and less polluted areas have been found in some research [47] and seasonal differences in air pollution composition could be relevant for genotoxic damage, such as the summer increase of ozone concentration [46]. Nevertheless, this study was designed to assess MN frequency in a sufficiently large sample of children regularly exposed to high levels of air genotoxic pollutants (PM and PAH), and not seasonal differences. We are, however, planning to extend the research to summer months and to children living in less polluted areas.

Only one sampling of oral cells was performed for each child, so the intra-individual variability of this early effect biomarker was not evaluated. The use of only one effect measure does not cause an important bias, however, because it reflects the mean exposure of the three preceding weeks, when the climate situation had not substantially changed. On the other hand, previous studies which found an association between this biomarker and air pollution also used a single measure of effect [31,44,47,49,50,55,56].

MN are a marker of early biological effect able to detect both clastogens and aneuploidy-inducing chemicals [18]. They are formed from acentric chromosomal fragments or whole chromosomes that are not included in the main daughter nuclei during nuclear division and their induction therefore reflects clastogenic and/or aneugenic events. They represent stable cytogenetic alterations which are the result of recent exposure of buccal mucosa cells, in a rapidly dividing tissue [18,28]. These cells are short-lived and they are the first barrier for substances introduced into the body by inhalation or ingestion and may be an excellent target tissue for detecting early genotoxic effects induced by mutagenic airborne compounds. Their use can, therefore, be proposed for assessing exposure to airborne mutagens, especially in the paediatric population as they are easy to collect, considering also the strong correlation of MN frequency in these cells and in lymphocytes, which have been shown to be related to the subsequent risk of developing cancer [37,48].

Brescia is a highly industrialized area with a high level of motor vehicle traffic. It is located in the Po Valley, one of the most highly polluted areas of Europe, where the concentrations of PM10, PM2.5 and NOx are usually above the EU reference values for many days of the year, as in the two years of this study (2012 and 2013), similar to those found in other towns and cities in the Po Valley [3].

In conclusion, this study shows that children living in a town with high levels of air pollutants in a Western country have a high level of MN in buccal mucosa cells, confirming previous findings of a mutagenic effect of urban air pollution on human beings.

Acknowledgments

We wish to thank the Brescia Municipality for its logistic support and the school personnel and all children and parents involved in the research.

Author Contributions

Conceived and designed the experiments: EC DF FD UG. Performed the experiments: EC DF GCVV IZ CZ. Analyzed the data: RML MC RL FD. Wrote the paper: EC DF FD UG.

References

1. WHO (2006) Air Quality Guidelines. Global Update 2005. World Health Organization Office for Europe, Copenhagen, Denmark.
2. WHO (2007) Children's health and the environment in Europe: a baseline assessment. World Health Organization Regional Office for Europe, Copenhagen, Denmark.
3. EAA (2012) Air Quality in Europe 2011. European Environment Agency, Technical report No 4/2012.
4. Pope CA 3rd, Burnett RT, Thun MJ, Calle EE, Krewski D, et al. (2002) Lung cancer, cardiopulmonary mortality, and long-term exposure to fine particulate air pollution. JAMA 287(9):1132–1141.
5. ERS (2010) Air quality and health. European Respiratory Society. Lausanne, Switzerland.
6. Janssen NA, Fischer P, Marra M, Ameling C, Cassee FR (2013) Short-term effects of PM2.5, PM10 and PM2.5-10 on daily mortality in the Netherlands. Sci Total Environ 463–464: 20–26.
7. Raaschou-Nielsen O, Andersen ZJ, Beelen R, Samoli E, Stafoggia M, et al. (2013) Air pollution and lung cancer incidence in 17 European cohorts: prospective analyses from the European Study of Cohorts for Air Pollution Effects (ESCAPE). Lancet Oncol 14: 813–822.
8. Shah ASV, Langrish JP, Nair H, McAllister DA, Hunter AL, et al. (2013) Global association of air pollution and heart failure: a systematic review and meta-analysis. Lancet 382: 1039–1048.
9. Sørensen M, Autrup H, Møller P, Hertel O, Jensen SS, et al. (2003) Linking exposure to environmental pollutants with biological effects. Mutat Res 544: 255–271.
10. Lewtas J (2007) Air pollution combustion emissions: characterization of causative agents and mechanisms associated with cancer, reproductive, and cardiovascular effects. Mutat Res 636: 95–133.
11. Claxton LD, Woodall GM Jr (2007) A review of the mutagenicity and rodent carcinogenicity of ambient air. Mutat Res 636: 36–94.
12. Møller P, Folkmann JK, Forchhammer L, Bräuner EV, Danielsen PH, et al. (2008) Air pollution, oxidative damage to DNA, and carcinogenesis. Cancer Lett 266: 84–97.
13. Monarca S, Crebelli R, Feretti D, Zanardini A, Fuselli S, et al. (1997) Mutagens and carcinogens in size-classified air particulates of a northern Italian town. Sci Total Environ 205: 137–144.
14. Monarca S, Feretti D, Zanardini A, Falistocco E, Nardi G (1999) Monitoring of mutagens in urban air samples. Mutat Res 426: 189–192.
15. Claxton LD, Matthews PP, Warren SH (2004) The genotoxicity of ambient outdoor air, a review: Salmonella mutagenicity. Mutat Res 567: 347–399.
16. Traversi D, Degan R, De Marco R, Gilli G, Pignata C, et al. (2009) Mutagenic properties of PM2.5 urban pollution in the northern Italy: the nitro-compounds contribution. Environ Int 35: 905–910.
17. de Brito KC, de Lemos CT, Rocha JA, Mielli AC, Matzenbacher C, et al. (2013) Comparative genotoxicity of airborne particulate matter (PM2.5) using Salmonella, plants and mammalian cells. Ecotoxicol Environ Saf 94: 14–20.
18. Kirsch-Volders M, Plas G, Elhajouji A, Lukamowicz M, Gonzalez L, et al. (2011) The in vitro MN assay in 2011: origin and fate, biological significance, protocols, high throughput methodologies and toxicological relevance. Arch Toxicol 85: 873–899. doi:10.1007/s00204-011-0691-4
19. Tolbert PE, Shy CM, Allen JW (1992) Micronuclei and other nuclear anomalies in buccal smears: methods development. Mutat Res 271: 69–77.
20. Fenech M, Crott JW (2002) Micronuclei, nucleoplasmic bridges and nuclear buds induced in folic acid deficient human lymphocytes – evidence for breakage – fusion-bridge cycles in the cytokinesis-block micronucleus assay. Mutat Res 504: 131–136.
21. Holland N, Bolognesi C, Kirsch-Volders M, Bonassi S, Zeiger E, et al. (2008) The micronucleus assay in human buccal cells as a tool for biomonitoring DNA damage: the HUMN project perspective on current status and knowledge gaps. Mutat Res 659: 93–108.
22. Thomas P, Holland N, Bolognesi C, Kirsch-Volders M, Bonassi S, et al. (2009) Buccal micronucleus cytome assay. Nat Protoc 4(6):825–837.
23. Hagmar L, Brøgger A, Hansteen IL, Heim S, Högstedt B, et al. (1994) Cancer risk in humans predicted by increased levels of chromosomal aberrations in lymphocytes: Nordic study group on the health risk of chromosome damage. Cancer Res 4(11):2919–2922.
24. Bonassi S, Abbondandolo A, Camurri L, Dal Prá L, De Ferrari M, et al. (1995) Are chromosome aberrations in circulating lymphocytes predictive of future cancer onset in humans? Preliminary results of an Italian cohort study. Cancer Genet Cytogenet 79(2):133–135.
25. Bonassi S, Znaor A, Ceppi M, Lando C, Chang WP, et al. (2007) An increased micronucleus frequency in peripheral blood lymphocytes predicts the risk of cancer in humans. Carcinogenesis 28(3):625–631.
26. Bonassi S, El-Zein R, Bolognesi C, Fenech M (2011) Micronuclei frequency in peripheral blood lymphocytes and cancer risk: evidence from human studies. Mutagenesis 26(1):93–100.
27. Coronas MV, Pereira TS, Rocha JA, Lemos AT, Fachel JM, et al. (2009) Genetic biomonitoring of an urban population exposed to mutagenic airborne pollutants. Environ Int 35(7):1023–1029.
28. Kashyap B, Reddy PS (2012) Micronuclei assay of exfoliated oral buccal cells: means to assess the nuclear abnormalities in different diseases. J Cancer Res Ther 8(2):184–191.
29. Samanta S, Dey P (2012) Micronucleus and its applications. Diagn Cytopathol 40: 84–90.
30. Bolognesi C, Merlo F, Rabboni R, Valerio F, Abbondandolo A (1997) Cytogenetic biomonitoring in traffic police workers: micronucleus test in peripheral blood lymphocytes. Environ Mol Mutagen 30: 396–402.
31. Maffei F, Hrelia P, Angelini S, Carbone F, Cantelli Forti G, et al. (2005) Effects of environmental benzene: Micronucleus frequencies and haematological values in traffic police working in an urban area. Mutat Res 583: 1–11.
32. Cavallo D, Ursini CL, Carelli G, Iavicoli I, Ciervo A, et al. (2006) Occupational exposure in airport personnel: characterization and evaluation of genotoxic and oxidative effects. Toxicology 223: 26–35.
33. Huen K, Gunn L, Duramad P, Jeng M, Scalf R, et al. (2006) Application of a geographic information system to explore associations between air pollution and

micronucleus frequencies in African American children and adults. Environ Mol Mutagen 47(4):236–246.
34. Pedersen M, Vinzents P, Petersen JH, Kleinjans JC, Plas G, et al. (2006) Cytogenetic effects in children and mothers exposed to air pollution assessed by the frequency of micronuclei and fluorescence in situ hybridization (FISH): a family pilot study in the Czech Republic. Mutat Res 608: 112–120.
35. Tovalin H, Valverde M, Morandi MT, Blanco S, Whitehead L, et al. (2006) DNA damage in outdoor workers occupationally exposed to environmental air pollutants. Occup Environ Med 63: 230–236.
36. Holland N, Fucic A, Merlo DF, Sram R, Kirsch-Volders M (2011) Micronuclei in neonates and children: effects of environmental, genetic, demographic and disease variables. Mutagenesis 26: 51–56.
37. Ceppi M, Biasotti B, Fenech M, Bonassi S (2010) Human population studies with the exfoliated buccal micronucleus assay: statistical and epidemiological issues. Mutat Res 705(1):11–19.
38. Desai SS, Ghaisas SD, Jakhi SD, Bhide SV (1996) Cytogenetic damage in exfoliated oral mucosal cells and circulating lymphocytes of patients suffering from precancerous oral lesions. Cancer Lett 109: 9–14.
39. Landrigan PJ, Kimmel CA, Correa A, Eskenazi B (2004) Children's health and the environment: public health issues and challenges for risk assessment. Environ Health Perspect 112(2):257–265.
40. WHO (2005) Effects of air pollution on children's health and development. World Health Organization Regional Office for Europe, Copenhagen, Denmark.
41. Bateson TF, Schwartz J (2008) Children's response to air pollutants. J Toxicol Environ Health A 71(3):238–243.
42. Grigg J (2009) Particulate matter exposure in children: relevance to chronic obstructive pulmonary disease. Proc Am Thorac Soc 6(7):564–569.
43. Wild CP, Kleinjans J (2003) Children and increased susceptibility to environmental carcinogens: evidence or empathy? Cancer Epidemiol Biomarkers Prev 12(12):1389–1394.
44. Lahiri T, Roy S, Basu C, Ganguly S, Ray MR, et al. (2000) Air pollution in Calcutta elicits adverse pulmonary reaction in children. Indian J Med Res 112: 21–26.
45. Montero R, Serrano L, Dávila V, Segura Y, Arrieta A, et al. (2003) Metabolic polymorphisms and the micronucleus frequency in buccal epithelium of adolescents living in an urban environment. Environ Mol Mutagen 42: 216–222.
46. Chen C, Arjomandi M, Qin H, Balmes J, Tager I, et al. (2006) Cytogenetic damage in buccal epithelia and peripheral lymphocytes of young healthy individuals exposed to ozone. Mutagenesis 21(2):131–137.
47. Sisenando HA, Batistuzzo de Medeiros SR, Artaxo P, Saldiva PH, Hacon Sde S (2012) Micronucleus frequency in children exposed to biomass burning in the Brazilian Legal Amazon region: a control case study. BMC Oral Health 12: 6. doi:10.1186/1472-6831-12-6
48. Bonassi S, Coskun E, Ceppi M, Lando C, Bolognesi C, et al. (2011) The HUman MicroNucleus project on eXfoLiated buccal cells (HUMN(XL)): the role of life-style, host factors, occupational exposures, health status, and assay protocol. Mutat Res 728(3):88–97.
49. Karahalil B, Karakaya AE, Burgaz S (1999) The micronucleus assay in exfoliated buccal cells: application to occupational exposure to polycyclic aromatic hydrocarbons. Mutat Res 442(1):29–35.
50. Rossnerova A, Spatova M, Pastorkova A, Tabashidze N, Veleminsky M Jr, et al. (2011) Micronuclei levels in mothers and their newborns from regions with different types of air pollution. Mutat Res 715: 72–78.
51. Mondal NK, Ghosh S, Ray MR (2011) Micronucleus formation and DNA damage in buccal epithelial cells of Indian street boys addicted to gasp 'Golden glue'. Mutat Res 721: 178–183.
52. Thomas P, Wu J, Dhillon V, Fenech M (2011) Effect of dietary intervention on human micronucleus frequency in lymphocytes and buccal cells. Mutagenesis 26: 69–76.
53. Nersesyan AK (2005) Nuclear buds in exfoliated human cells. Letter to the Editor. Mutat Res 588: 64–68.
54. Cerqueira EMM, Gomes-Filho IS, Trindade S, Lopes MA, Passos JS, et al. (2004) Genetic damage in exfoliated cells from oral mucosa of individuals exposed to X-rays during panoramic dental radiographies. Mutat Res 562: 111–117.
55. Neri M, Ugolini D, Bonassi S, Fucic A, Holland N, et al. (2006) Children's exposure to environmental pollutants and biomarkers of genetic damage. II. Results of a comprehensive literature search and meta-analysis. Mutat Res 612: 14–39.
56. Pedersen M, Wichmann J, Autrup H, Dang DA, Decordier I, et al. (2009) Increased micronuclei and bulky DNA adducts in cord blood after maternal exposures to traffic-related air pollution. Environ Res 109: 1012–1020.

11

Prevalence of Subjective Olfactory Dysfunction and Its Risk Factors: Korean National Health and Nutrition Examination Survey

Woo Hyun Lee[1], Jee Hye Wee[1], Dong-Kyu Kim[1], Chae-Seo Rhee[1], Chul Hee Lee[1], Soyeon Ahn[2], Ju Hyun Lee[2], Yang-Sun Cho[3], Kun Hee Lee[4], Kyung Soo Kim[5], Si Whan Kim[6], Ari Lee[7], Jeong-Whun Kim[1]*

1 Department of Otorhinolaryngology, Seoul National University Bundang Hospital, Seongnam; Seoul National University College of Medicine, Seoul, South Korea, **2** Medical Research Collaborating Center, Seoul National University Bundang Hospital, Seongnam, South Korea, **3** Department of Otorhinolaryngology-Head and Neck Surgery, Sungkyunkwan University School of Medicine, Seoul, South Korea, **4** Department of Otorhinolaryngology-Head and Neck Surgery, Kyung Hee University School of Medicine, Seoul, South Korea, **5** Department of Otorhinolaryngology-Head and Neck Surgery, Yonsei University College of Medicine, Seoul, South Korea, **6** Department of Otorhinolaryngology-Head and Neck Surgery, Hallym University College of Medicine, Anyang, South Korea, **7** Division of Chronic Disease Surveillance, Korea Centers for Disease Control & Prevention, Seoul, South Korea

Abstract

Background: Population-based studies for olfactory dysfunction are lacking. The aim of this study is to evaluate the prevalence of subjective olfactory dysfunction and its risk factors in the Korean general population.

Methods: The data were obtained from the 2009 Korea National Health and Nutrition Examination Survey (KNHANES), which was a cross-sectional survey of non-institutionalized population all around the country (n = 10,533). All interviewees underwent medical interviews, physical examinations, endoscopic examination and blood/urine tests. Whether sense of smell has been normal or abnormal during the last 3 months was asked. Complete olfaction data were obtained from 7,306 participants and the participants were divided into normosmic and hyposmic group. Multivariate logistic regression analyses were performed to identify its risk factors.

Results: The weighted prevalence of subjective olfactory dysfunction was 4.5%. Its increased prevalence was significantly associated with the increasing age for both men and women. In the multivariate analyses, low income (adjusted odds ratio [OR] = 1.43, 95% Confidence Interval [CI] = 1.01–2.03), habitual exposure to air pollutants (adjusted OR = 2.18, CI = 1.33–3.55), a history of hepatitis B (adjusted OR = 3.10, CI = 1.25–7.68), rhinitis (adjusted OR = 1.78, CI = 1.26–2.51) and chronic sinusitis (adjusted OR = 14.55, CI = 10.06–21.05) were risk factors of olfactory dysfunction.

Conclusion: Our population-based study showed that olfactory dysfunction was quite prevalent and several risk factors were associated with impaired sense of smell. Given its prevalence, further researches for its prevention and management are required.

Editor: Yiqing Song, Brigham & Women's Hospital, and Harvard Medical School, United States of America

Funding: The authors have no support or funding to report.

Competing Interests: The authors have declared that no competing interests exist.

* E-mail: kimemail@snu.ac.kr

Introduction

The loss of sense of smell decreases quality of life and may contribute to the failure in recognizing hazardous substances or unpleasant odors. [1] This problem can develop changes in appetite and this may make a nutritional problem especially in the elderly. Persons who have olfactory dysfunction are at an increased risk of danger from leaking gas, smoke, food spoilage, and pollution. [2] Although it is usually not a life-threatening or highly morbid condition, [3] it is important to recognize undiagnosed olfactory dysfunction and its etiology in order to prevent these possible complications. So far, population-based studies for olfactory dysfunction are lacking. The National Health Interview Survey (NHIS) of the United States including about 42,000 households showed that the prevalence of self-reported olfactory problems was 1.4%. [1] There were another two population-based studies in the Western countries, which were based on psychophysical tests. In the Epidemiology of Hearing Loss Study (EHLS) of individuals aged 53 to 97 years in Beaver Dam, 24.5% were found to have impaired sense of smell, [4] and the Skovde population-based study of the general Swedish adult population aged ≥20 years, 19.1% were found to have olfactory dysfunction. [5] However, there has been no large group study for olfactory impairment in the Asian population.

Although olfactory dysfunction needs to be objectively diagnosed and managed by clinicians, it is also important to know the prevalence of self-reported olfactory dysfunction which may determine whether they are likely to seek medical attention. [6]

The current investigation was conducted to determine the prevalence of self-reported olfactory dysfunction in the large Korean population who were randomly selected. This study is also aimed to identify the risk factors of olfactory dysfunction through analyses of systematic medical questionnaires, endoscopic examinations and blood/urine tests.

Materials and Methods

Population

The Korea Centers for Disease Control and Prevention and the Korean Society of Otorhinolaryngology-Head and Neck Surgery have collaboratively collected medical history and clinical measurement data from a representative sample of the Korean population in the Korean National Health and Nutrition Examination Survey (KNHANES). The KNHANES 2009 data set was used, which included information on the presence of subjected olfactory dysfunction and endoscopic findings of the ear, nose and throat (ENT). A total of 12,722 individuals of 4,000 households were invited to participate in the KNHANES 2009. These participants were randomly selected from a panel to represent the population of South Korea. The panel was extracted by using a multistage clustered and stratified random sampling method that was based on the National Census Data. Among 12,722 selected individuals, 10,533 (82.8%) agreed to participate in the clinical examination. Analyses were conducted for the data from 7,306 participants who answered to the questionnaire for the presence of olfactory dysfunction. However, the other general characteristics of the subjects included were comparable with those of the subjects excluded. The average age of the 7,306 participants was 49.3±16.5 years (range, 20–95 years) and the ratio of male to female was 1:1.32. Written informed consents were obtained from all the participants prior to the survey and approval for this research was obtained from the Institutional Review Board of the Seoul National University Bundang Hospital.

ENT Evaluation, Medical History and Clinical Examination

The endoscopic ENT examinations and interview for otolaryngologic medical history were performed by trained ENT residents according to the standardized protocols. A total of 108 surveys were conducted by four survey teams during a time span of 27 weeks all over the country. Each survey team consisted of one ENT resident, three nurses, two interviewers, and one coordinator. Each survey was performed in a mobile examination unit vehicle at the pre-assigned places. A single visit was required for each participant to the examination unit vehicle. All the questionnaires, ENT examinations and blood/urine samplings were performed during the single visit. A total of 87 ENT residents from 45 teaching hospitals were recruited for performing ENT examinations. The Epidemiology Committee of the Korean Society of Otorhinolaryngology-Head and Neck Surgery verified the quality control of the survey.

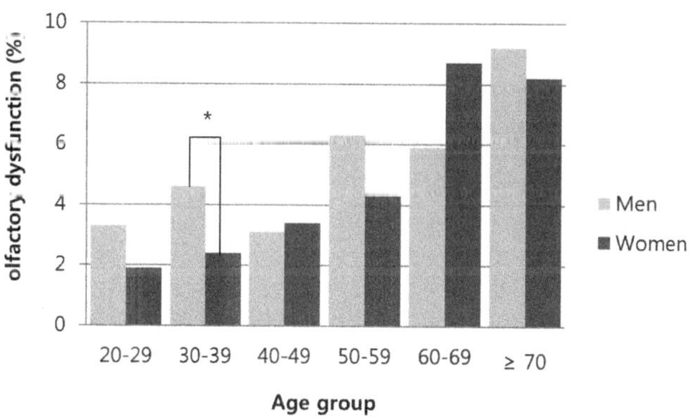

Figure 1. Prevalence of subjective olfactory dysfunction by age groups and genders. There were increasing tendency of olfactory dysfunction prevalence with increasing age in both genders, asterisk; $p<0.05$.

The olfactory questionnaire was asked about whether the participants have had problems with the sense of smell during the past three months. The participants with a positive response were considered hyposmic and those with a negative response normosmic. Chronic rhinosinusitis was ascertained when nasal polyps were observed or more than one of the symptoms such as anterior/posterior nasal drip, nasal obstruction, facial pain/pressure, and olfactory dysfunction were present for more than 3 months (anterior/posterior nasal drip or nasal obstruction should be included as a presenting symptom). Otitis media was defined when there was a perforation in the tympanic membrane, cholesteatoma including a retraction pocket, or inflammation in the tympanic membrane with effusion. The diagnosis of rhinitis was made when they have had a history of rhinorrhea, sneezing, itching, and nasal obstruction for the last one year without cold symptoms. The deviation of the nasal septum was defined when an asymmetric displacement to one or both sides of the nasal cavities was observed after vasoconstriction of the nasal cavities.

Other medical histories were obtained by trained interviewers of the Korea Center of Disease Control and Prevention. A set of structured questionnaires were asked. The socioeconomic status (income, education, residency and occupation), smoking, drinking, a history of habitual exposure to air pollutants or chemical substances were asked and the participants were categorized. The study population was split into 6 subgroups by age. Body mass index was categorized into normal (<25 Kg/m^2) and obese (≥25 Kg/m^2). Residency was categorized into urban and rural areas according to the official address of the subjects. Home income was calculated by equivalized gross household income per month in each year and grouped into 4 quartiles. In the analysis, the subjects were divided into low (1st and 2nd quartiles) and high (3rd and 4th quartiles) groups. Subjects who graduated from middle or elementary school were considered to have a low level of education, and those from high school or college were considered to have a high level of education. Occupation was categorized by following the Korean Standard Classification of Occupations. The occupational categories were as follows: white collar (included managers, professionals, clerks, service/sales workers, unemployed, retired, students, and housewives) and blue collar (agriculture, forestry, fishery workers, craft and related trade workers, plant and machine operators and assemblers, and simple labor). Smoking status was classified into two categories: non-smoker (a person who has smoked below 5 pack smoking) and smoker (a person who has smoked more than 5 pack smoking). Alcohol consumption was assessed by asking the participants about their average frequency of alcoholic beverage consumption during the month prior to the interview.

Blood and urine samples were also tested. All the collected blood and urine samples were transported to a single designated laboratory (Neodin Medical Institute, Seoul, Korea) and analyzed according to their qualified standard protocols and equipments.

Statistical Analyses

The data were analyzed with SPSS (version 18.0, SPSS Inc, Chicago, IL, USA), which is a software package that incorporates

Table 1. Prevalence of olfactory dysfunction according to the general characteristics of KNHNES participants according to the presence of olfactory dysfunction (weighted for the multistage sampling design of KNHANES 2009).

Characteristics	Unweighted total number	Olfactory dysfunction Weighted, % (SE)		P value
		No	Yes	
Overall	7306	95.5 (0.4)	4.5 (0.4)	
Age				<0.001
20–29	936	97.4 (0.5)	2.6 (0.5)	
30–39	1407	96.5 (0.6)	3.5 (0.6)	
40–49	1473	96.7 (0.6)	3.3 (0.6)	
50–59	1233	94.7 (0.8)	5.3 (0.8)	
60–69	1203	92.6 (1.0)	7.4 (1.0)	
≥70	1054	91.4 (1.0)	8.6 (1.0)	
Sex				0.302
Male	3149	95.3 (0.5)	4.7 (0.5)	
Female	4157	95.8 (0.4)	4.2 (0.4)	
Education				<0.001
≤Middle school	2867	93.7 (0.7)	6.3 (0.7)	
≥High school	4380	96.3 (0.4)	3.7 (0.4)	
Missing	59			
Income				<0.001
≤50 percent	3240	94.2 (0.6)	5.8 (0.6)	
>50 percent	3983	96.4 (0.4)	3.6 (0.4)	
Missing	83			
Occupation				0.022
White collar	5287	95.9 (0.4)	4.1 (0.4)	
Blue collar	1937	94.5 (0.6)	5.5 (0.6)	
Missing	82			
Residence				0.469
Rural	5408	95.4 (0.5)	4.6 (0.5)	
Urban	1898	96.0 (0.6)	4.0 (0.6)	
Body mass index				0.601
<25 kg/m^2	4926	95.4 (0.5)	4.6 (0.5)	
≥25 kg/m^2	2331	95.7 (0.5)	4.3(0.5)	
Air Pollutants				0.001
Exposure	399	92.0 (1.6)	8.0 (1.6)	
No exposure	6884	95.7 (0.4)	4.3 (0.4)	
Missing	23			
Chemical substance				0.399
Exposure	230	94.2 (1.9)	5.8 (1.9)	
No exposure	7053	95.6 (0.4)	4.4 (0.4)	
Missing	23			
Smoking				0.709
<5 packs	4382	95.6 (0.4)	4.4 (0.4)	
≥5packs	2864	95.4 (0.5)	4.6 (0.5)	
Missing	60			
Alcohol drinking				0.519
≤4times/month	5746	95.6 (0.4)	4.4 (0.4)	
≥2times/week	1573	95.2 (0.7)	4.8 (0.7)	
Missing	23			

SE, standard error; Air pollutants, habitual exposure to air pollutants in workplace; Chemical substance, habitual exposure to chemical substances in workplace.

Table 2. Prevalence of olfactory dysfunction according to the medical conditions of KNHANES participants.

Characteristic	Unweighted total number	Olfactory dysfunction Weighted, % (SE) No	Yes	P value
ENT diseases[†]				
Chronic rhinosinusitis	402	68.8 (3.3)	31.2 (3.3)	<0.001
Rhinitis	1696	91.6 (0.9)	8.4 (0.9)	<0.001
Chronic otitis media	261	90.5 (2.4)	9.5 (2.4)	0.001
Septal deviation	3090	95.2 (0.5)	4.8 (0.5)	0.405
Cardiovascular disease				
Hypertension	1513	93.8 (0.7)	6.2 (0.7)	0.001
Stroke	143	93.7 (1.8)	6.3 (1.8)	0.236
Angina	137	94.1 (2.0)	5.9 (2.0)	0.398
Arrhythmia	292	95.5 (1.1)	4.5 (1.1)	0.963
Metabolic disease				
Diabetes mellitus	573	95.7 (0.8)	4.3 (0.8)	0.850
Thyroid disease	254	89.9 (2.7)	10.1 (2.7)	0.001
Hyperlipidemia	574	94.4 (1.1)	5.6 (1.1)	0.205
Low-HDL cholesterolemia	1802	94.0 (0.9)	6.0 (0.9)	0.014
Pulmonary disease				
Tuberculosis	400	95.5 (0.4)	4.5 (0.4)	0.833
Asthma	224	87.9 (3.7)	12.1 (3.7)	0.001
COPD	43	94.7 (2.7)	5.3 (2.7)	0.749
Bronchiectasis	25	90.0 (7.5)	10.0 (7.5)	0.296
Gastro-intestinal disease				
Gastric ulcer	452	95.3 (1.1)	4.7 (1.1)	0.800
Hepatitis B	86	86.4 (4.6)	13.6 (4.6)	0.001
Neuromuscular disease				
Rheumatoid arthritis	148	93.6 (2.3)	6.4 (2.3)	0.336
Arthritis	967	92.4 (1.1)	7.6 (1.1)	<0.001
Osteoporosis	539	90.6 (1.5)	9.4 (1.5)	<0.001
Neoplastic disease				
Gastric cancer	49	86.3 (6.2)	13.7 (6.2)	0.016
Hepatic cancer	11	94.7 (10)	5.3 (5.3)	0.870
Colon cancer	17	89.4 (8.3)	10.6 (8.3)	0.260
Others				
Depression	277	93.8 (1.6)	6.2 (1.6)	0.195
Anemia	473	95.2 (1.0)	4.8 (1.0)	0.751
Atopic dermatitis	161	93.0 (2.1)	7.0 (2.1)	0.121
Cataract	660	92.5 (1.1)	7.5 (1.1)	<0.001

HDL, high density lipoprotein; COPD, chronic obstructive pulmonary disease;
[†]All ENT diseases were diagnosed by ENT residents.

sample weights and adjusts analysis for the complex sample design of the survey. We used the KNHANES sampling weight variables, along with masked variance primary sampling unit and stratum variables. This adjustment allowed the estimation of the entire non-institutionalized Korean population from the sample. Survey sample weights were used in all the analyses. The chi-square test was used to compare the characteristics between normosmic and hyposmic participants. The independent *t*-test was used to analyze blood and urine test results. After performing univariate analyses of variables, the multiple logistic regression analysis was performed including variables with values of $p<0.20$ in the univariate analyses to calculate adjusted odds ratios (OR) and their 95% confidence interval (CI). After conversion of continuous variables to dichotomous variables (normal *vs.* abnormal) according to their normal laboratory range, blood and urine test results were also included in the multivariate logistic model. A p-value <0.05 was considered significant. Missing data were considered to be missing completely at random.

Table 3. Comparison of blood/urine tests between normosmic and hyposmic participants.

	Olfactory dysfunction weighted (SE)		
Variable	**No**	**Yes**	**P value**
Blood test			
Fasting glucose (mg/dL)	96.90 (1.30)	97.39 (1.24)	0.705
$HbA1_C$ (%)	7.30 (0.23)	7.37 (0.21)	0.761
Insulin (μIU/mL)	9.89 (0.32)	9.56 (0.31)	0.300
Total cholesterol (mg/dL)	186.14 (2.20)	186.60 (2.17)	0.835
HDL cholesterol (mg/dL)	52.28 (0.88)	50.03 (0.88)	0.010
Triglyceride (mg/dL)	135.70 (7.80)	143.98 (7.76)	0.289
LDL cholesterol (mg/dL)	110.96 (4.79)	106.35 (4.71)	0.337
GOT (IU/L)	22.41 (0.88)	24.35 (0.85)	0.037
GPT (IU/L)	22.49 (1.73)	23.80 (1.71)	0.451
BUN (mg/dL)	14.17 (0.25)	14.81 (0.25)	0.013
Vitamin D (ng/mL)	17.74 (0.40)	17.76 (0.43)	0.953
Alkaline phosphatase (IU/L)	221.99 (3.43)	222.19 (3.32)	0.954
Lead (μg/dL)	2.49 (0.14)	2.89 (0.14)	0.006
Mercury (μg/L)	5.10 (0.41)	5.80 (0.41)	0.094
Cadmium (μg/L)	1.08 (0.07)	1.09 (0.07)	0.936
Urine test			
Creatinine (mg/L)	158.54 (6.03)	143.99 (5.86)	0.017
Sodium (g/day)	131.69 (4.00)	134.51 (3.94)	0.482
Cotinine (ng/mL)	445.86 (64.07)	393.55 (63.01)	0.415

HbA1c, hemoglobin A1c; HDL, high density lipoprotein; LDL, low density lipoprotein; GOT, glutamic oxaloacetic transaminase; GPT, glutamic pyruvic transaminase; BUN, blood urea nitrogen.

Results

Prevalence of Olfactory Dysfunction

Of 7,306 participants, 360 (4.9%) had olfactory dysfunction and its weighted prevalence was 4.5%. The prevalence of olfactory dysfunction according to the general characteristics of the participants is shown in **Table 1**. The chi-square analyses showed that sex, occupation, residence, body mass index, habitual exposure to chemical substances, smoking and drinking did not affect the prevalence of olfactory dysfunction. A logistic regression analysis using olfactory dysfunction as the outcome variable and their age as a continuous variable after adjustment for sex showed a significant increasing trend of the prevalence with age (adjusted OR, 1.026; CI, 1.019–1.033; P value, <0.001). There was no significant difference was found between two genders in each age subgroup, except 30–39 years when the participants were categorized by a decade of age (**Figure 1**). The increased prevalence was significantly associated with lower education level ($p<0.001$), lower income ($p<0.001$) and habitual exposure to air pollutants ($p=0.001$).

The chi-square analyses from the medical history data set demonstrated that several medical conditions were associated with olfactory dysfunction (**Table 2**). Among ENT diseases, chronic sinusitis, rhinitis and chronic otitis media were associated with olfactory dysfunction. In the univariate analyses, hypertension, thyroid diesease, low HDL (high density lipoprotein) cholesterolemia, asthma, hepatitis B, arthritis, osteoporosis, gastric cancer and cataract were also associated with olfactory dysfunction. In the independent t-test analyses of blood and urine test results, olfactory dysfunction was associated with decreased levels of blood HDL cholesterol and urine creatinine and increased levels of blood GOT (glutamic oxaloacetic transaminase), BUN (blood urea nitrogen) and lead (**Table 3**).

Multivariate Analyses of Risk Factors

In multivariate logistic regression analyses, the effect of each variable was adjusted for all the other variables (**Table 4**). In reference to the youngest subgroup aged 20–29 years, the odds of having olfactory dysfunction were significantly higher in the older subgroups aged 50–59 years (adjusted OR, 2.09; 95% CI, 1.16–3.77), 60–69 years (adjusted OR, 2.58; 95% CI, 1.31–5.08) and 70 years or older (adjusted OR, 3.17; 95% CI, 1.93–7.14). Lower income (adjusted OR, 1.43; 95% CI, 1.01–2.03), habitual exposure to air pollutants (adjusted OR, 2.18; 95% CI, 1.33–3.55), a history of hepatitis B (adjusted OR, 3.10; 95% CI, 1.25–7.68), rhinitis (adjusted OR, 1.78; 95% CI, 1.26–2.51) and chronic sinusitis (adjusted OR, 14.5; 95% CI, 10.06–21.05) were significantly associated with self-reported olfactory dysfunction. However, all of the blood and urine test results were not associated with the olfactory dysfunction.

Discussion

This is the first large population-based study reporting the prevalence of subjective olfactory dysfunction in Asia. The weighted prevalence of subjective olfactory dysfunction in the Korean population was 4.5%. In other words, it is equivalent approximately to 1.7 million (according to the current census population of South Korea) Korean adults. The prevalence was

Table 4. Adjusted odds ratio for the association between olfactory dysfunction and risk factors.

Exposure	Adjusted OR	CI	P value
Age group			
20–29	1.00		
30–39	1.08	0.60–1.95	0.792
40–49	1.31	0.73–2.37	0.360
50–59	2.09	1.16–3.77	0.014
60–69	2.58	1.31–5.08	0.006
≥70	3.71	1.93–7.14	<0.001
Income (≤50 percent)	1.43	1.01–2.03	0.042
Education (≤Middle school)	0.67	0.43–1.04	0.075
Occupation (Blue collar)	1.21	0.86–1.69	0.256
Air pollutants	2.18	1.33–3.55	0.002
Hypertension	0.94	0.63–1.41	0.796
Arthritis	0.99	0.62–1.58	0.968
Osteoporosis	1.32	0.81–2.13	0.259
Asthma	1.84	0.81–4.20	0.143
Depression	0.85	0.38–1.87	0.688
Atopic dermatitis	1.16	0.50–2.69	0.713
Thyroid disease	1.71	0.83–3.50	0.142
Cataract	1.06	0.69–1.63	0.784
Gastric cancer	1.44	0.34–6.08	0.612
Hepatitis B	3.10	1.25–7.68	0.014
Low HDL cholesterolemia	1.41	0.97–2.05	0.067
Rhinitis	1.78	1.26–2.51	0.001
Chronic otitis media	1.34	0.74–2.43	0.328
Chronic sinusitis	14.55	10.06–21.05	<0.001
HDL (<40 mg/dL)	0.82	0.62–1.10	0.192
GOT (>40 IU/L)	1.82	0.86–3.81	0.113
BUN (>24 mg/dL)	1.90	0.92–3.93	0.080
Lead (>10 µg/dL)	1.07	0.38–3.04	0.888
Mercury (>15 µg/dL)	1.20	0.42–3.34	0.730

OR, odds ratio; CI, confidence interval; HDL, high density lipoprotein; GOT, glutamic oxaloacetic transaminase; BUN, blood urea nitrogen; each exposure variable is adjusted for all other variables in the table; After conversion of continuous variables to dichotomous variables (normal *vs.* abnormal) according to their normal laboratory range, blood and urine test results were included in the multivariate logistic model.

obtained from a national survey, KNHANES 2009, which randomly recruited participants.

Prevalence studies based on psychophysical or semi-objective tests can provide more objective information. The prevalence evaluated by psychophysical tests in the Epidemiologic Hearing Loss Study was much higher than that obtained by questionnaires. The gap between the prevalence of self-reported olfactory dysfunction and that of semi-objective olfactory dysfunction increased with age. [4] This suggests that elderly persons would be likely to be unaware of their impaired sense of smell. The sensitivity of subjective olfactory dysfunction was low (20%) but the specificity was very high (94%). Furthermore, the sensitivity of subjective olfactory dysfunction decreased with age. [4] Another study also showed that self-reported olfactory dysfunction had low sensitivity (43.9%) and high specificity (85.4%). [5] These findings suggest that many people with an impaired sense of smell may be unaware of their olfactory problem. Given the low sensitivity of self-reported questionnaire, the present study shows that olfactory dysfunction is not a negligible disease because the self-reported prevalence was as high as 4.5% in our study. Therefore, it is important to identify the prevalence of self-reported olfactory dysfunction to make the patients seek medical attention, initiate medical care and prevent possible complications.

The present study also identified several risk factors associated with olfactory dysfunction. Age, socioeconomic status, environment and some medical conditions were likely to affect the impaired sense of smell. The odds of having olfactory dysfunction increased with age in both men and women in the present study. A previous large population study in the U.S. [1] showed similar results in that the prevalence of self-reported olfactory dysfunction was 1.99%, 2.56%, and 4.6% for subgroups aged 55–64 years, 65–74 years, and 75 years or older, respectively. Several mechanisms may be involved in age-related olfactory dysfunction. Long-lasting injuries to the olfactory mucosa such as various co-morbidities or exposure to environmental olfactory toxicants have been speculated to lead to the replacement of olfactory mucosa with respiratory epithelium. [7] Atrophy of the olfactory bulb and tract occurs in elderly people as the number of glomeuli and mitral cells decreases with age.[8–10] Structural magnetic resonance imaging showed an age-associated volume loss in the mesial temporal lobe which is important for olfactory processing. [11].

With regard to gender, some studies performed in the Western countries showed that olfactory dysfunction was more prevalent in men than in women. [12,13] However, our nationwide study demonstrated that its prevalence was not statistically different between men and women. The different result may be caused by differences in the study design and population. Given that our study samples were selected in a random fashion to represent the whole national population and asked by ENT residents who have been trained for olfactory disorders, there is unlikely to be sexual differences in olfactory dysfunction in Koreans.

As previously reported, [4] lower income was also found to be associated with olfactory dysfunction in our study. People with higher income may have better access to health care or have healthier lifestyles than people with lower incomes. [14] Environmental factors such as an exposure to chemicals or air pollution can also be factors associated with olfactory dysfunction. KNHANES showed that the habitual exposure to air pollutants in their workplaces was significantly associated with the higher prevalence of olfactory dysfunction. This finding suggests that we should also involve olfactory toxicity when the effects of air pollution in the workplaces on our health and quality of life are evaluated. On the other hand, a questionnaire about the habitual exposure to chemicals in their workplaces could not show its detrimental effects on olfaction in our study even if there were some reports for olfactory impairment caused by an occupational exposure to airborne industrial chemicals.[15–18] Tobacco smoking might also be a factor affecting the sense of smell. However, its harmful effects on olfaction are under controversy [5] and we could not observe its significant effects. Some studies have reported adverse effects of smoking on olfactory function, [19,20] whereas other studies reported no effect on olfaction. [21,22] Even if our cross-sectional study failed in identifying the olfactory toxicity of chemicals and smoking, further studies are required after several related factors such as exposure amount and duration are more strictly controlled.

Airway diseases can also affect the sense of smell. Although upper airway diseases such as chronic sinusitis and rhinitis were significantly associated with olfactory dysfunction, lower airway diseases such as bronchial asthma and chronic pulmonary airway disease showed no association. It is well known that hyposmia, dysosmia and dysgeusia are common symptoms in hepatitis. [23,24] Improvement in olfactory acuity was inversely related to the plasma-bilirubin level and directly related to the plasma-retinol-binding-protein level in acute hepatitis. [25] To our knowledge, this is the first large population-based study showing that a history of hepatitis B is associated with olfactory dysfunction although its mechanism is unclear.

The present study also has some limitations. First, olfactory function was evaluated by questionnaires instead of psychophysical tests. However, as ENT questionnaires were asked by trained ENT residents, the decision of olfactory dysfunction may be clinically quite relevant. Secondly, as this is neither a longitudinal study nor a strictly controlled experimental study, the causal relationship of the risk factors with olfactory dysfunction may be inconclusive. Nevertheless, the results may be reliable because this is a nationwide population-based study all around the country. Thirdly, the control of the diseases which may relate to olfactory dysfunction may pose influences on results. However, whether those diseases have been well controlled or not could not be evaluated in this study.

In conclusion, our population-based study showed that 4.5% of Korean population suffered from olfactory dysfunction. In addition to ENT diseases and medical conditions, old age, low income and habitual exposure to air pollutants were associated with an increased risk of impaired sense of smell. Given the quite high prevalence of olfactory dysfunction in Korea, further researches for its prevention and management are required.

Acknowledgments

We would like to express our sincerest gratitude and appreciation to 87 residents of the Department of Otorhinolaryngology-Head and Neck Surgery from 45 training hospitals in South Korea and the members of the Korea Centers for Disease Control & Prevention who participated in this survey.

Author Contributions

Study supervision: CSR CHL JWK. Critical revision of the manuscript for important intellectual content: CSR CHL JWK. Conceived and designed the experiments: WHL JWK. Performed the experiments: WHL JHW DKK JWK. Analyzed the data: WHL JHW SA JHL JWK. Contributed reagents/materials/analysis tools: YSC KHL KSK SWK CSR CHL AL. Wrote the paper: WHL JWK.

References

1. Hoffman HJ, Ishii EK, MacTurk RH (1998) Age-related changes in the prevalence of smell/taste problems among the United States adult population. Results of the 1994 disability supplement to the National Health Interview Survey (NHIS). Ann N Y Acad Sci 855: 716–722.
2. Denis L, Norman M (2009) Anosmia: Loss of smell in the elderly. Otolaryngol Clin N Am 42: 123–131.
3. Toller SV (1999) Assessing the impact of anosmia: review of a questionnaire's findings. Chem Senses 24: 705–712.
4. Murphy C, Schubert CR, Cruickshanks KJ, Klein BE, Klein R, et al. (2002) Prevalence of olfactory impairment in older adults. JAMA 288: 2307–2312.
5. Bramerson A, Johansson L, Ek L, Nordin S, Bende M (2004) Prevalence of olfactory dysfunction: the skovde population-based study. Laryngoscope 114: 733–737.
6. Nordin S, Bramerson A, Bende M (2004) Prevalence of self-reported poor odor detection sensitivity: the Skovde population-based study. Acta Otolaryngol 124: 1171–1173.
7. Nakashima T, Kimmelman CP, Snow JB Jr (1984) Structure of human fetal and adult olfactory neuroepithelium. Arch Otolaryngol 110: 641–646.
8. Bhatnagar KP, Kennedy RC, Baron G, Greenberg RA (1987) Number of mitral cells and the bulb volume in the aging human olfactory bulb: a quantitative morphological study. Anat Rec 218: 73–87.
9. Jones N, Rog D (1998) Olfaction: a review. J Laryngol Otol 112: 11–24.
10. Meisami E, Mikhail L, Baim D, Bhatnagar KP (1998) Human olfactory bulb: aging of glomeruli and mitral cells and a search for the accessory olfactory bulb. Ann N Y Acad Sci 855: 708–715.
11. Jernigan TL, Archibald SL, Fennema-Notestine C, Gamst AC, Stout JC, et al. (2001) Effects of age on tissues and regions of the cerebrum and cerebellum. Neurobiol Aging 22: 581–594.
12. Ship JA, Pearson JD, Cruise LJ, et al. (1996) Longitudinal changes in smell identification. The journals of gerontology Series A, Biological sciences and medical sciences; 51: M86–91.
13. Ship JA, Weiffenbach JM (1993) Age, gender, medical treatment, and medication effects on smell identification. Journal of gerontology 1993; 48: M26–32.
14. Schubert CR, Cruickshanks KJ, Fischer ME, Huang GH, Klein BE, et al. (2012) Olfactory impairment in an adult population: the Beaver Dam Offspring Study. Chem Senses 37: 325–334.
15. Gobba F (2006) Olfactory toxicity: long-term effects of occupational exposures. Int Arch Occup Environ Health 79: 322–331.
16. Amoore JE (1986) Effects of chemical exposure on olfaction in humans. In: Barrow CS (ed) Toxicology of the nasal passages. Hemisphere Publishing, Washington, DC, 155–190.
17. Doty RL (ed) (2003) Handbook of olfaction and gustation. Marcel Dekker, New York.
18. Doty RL, Mc Keown DA, Lee WW, Shaman P (1995) A study of the test-retest reliability of ten olfactory tests. Chem Senses 20: 645–656.
19. Weinstock RS, Wright HN, Smith DU (1993) Olfactory dysfunction in diabetes mellitus. Physiol Behav;53: 17–21.
20. Joyner RE (1964). Effect of cigarette smoking on olfactory acuity. Arch Otolaryngol;80: 576–579.
21. Hubert HB, Fabsitz RR, Feinleib M, Brown KS (1980) Olfactory sensitivity in human: genetic versus environmental control. Science;9: 607–609.
22. Venstrom D, Amoore JE (1968) Olfactory threshold in relation to age, sex or smoking. J Food Sci;33: 264–265.
23. Temmel AF, Pabinger S, Quint C, Munda P, Ferenci P, et al. (2005) Dysfunction of the liver affects the sense of smell. Wien Klin Wochenschr;117: 26–30.
24. Landis BN, Konnerth CG, Hummel T (2004) A study on the frequency of olfactory dysfunction. Laryngoscope;114: 1764–9.
25. Henkin RI, Smith FR (1971) Hyposmia in acute viral hepatitis. Lancet 1: 823–826.

12

Acute Effects of Particulate Matter and Black Carbon from Seasonal Fires on Peak Expiratory Flow of Schoolchildren in the Brazilian Amazon

Ludmilla da Silva Viana Jacobson[1]*, Sandra de Souza Hacon[2], Hermano Albuquerque de Castro[2], Eliane Ignotti[3], Paulo Artaxo[4], Paulo Hilário Nascimento Saldiva[5], Antonio Carlos Monteiro Ponce de Leon[6]

1 Department of Statistics, Fluminense Federal University, Niterói, Brazil, **2** National School of Public Health, Oswaldo Cruz Foundation, Rio de Janeiro, Brazil, **3** Institute of Natural and Technological Science, Mato Grosso State University, Cáceres, Brazil, **4** Physics Institute, University of São Paulo, São Paulo, Brazil, **5** Department of Pathology, University of São Paulo, São Paulo, Brazil, **6** Department of Epidemiology, Rio de Janeiro State University, Rio de Janeiro, Brazil

Abstract

Background: Panel studies have shown adverse effects of air pollution from biomass burning on children's health. This study estimated the effect of current levels of outdoor air pollution in the Amazonian dry season on peak expiratory flow (PEF).

Methods: A panel study with 234 schoolchildren from 6 to 15 years old living in the municipality of Tangará da Serra, Brazil was conducted. PEF was measured daily in the dry season in 2008. Mixed-effects models and unified modelling repeated for every child were applied. Time trends, temperature, humidity, and subject characteristics were regarded. Inhalable particulate matter (PM_{10}), fine particulate matter ($PM_{2.5}$), and black carbon (BC) effects were evaluated based on 24-hour exposure lagged by 1 to 5 days and the averages of 2 or 3 days. Polynomial distributed lag models (PDLM) were also applied.

Results: The analyses revealed reductions in PEF for PM_{10} and $PM_{2.5}$ increases of 10 μg/m^3 and 1 μg/m^3 for BC. For PM_{10}, the reductions varied from 0.15 (confidence interval (CI)95%: −0.29; −0.01) to 0.25 l/min (CI95%: −0.40; −0.10). For $PM_{2.5}$, they ranged from 0.46 (CI95%: −0.86 to −0.06) to 0.54 l/min (CI95%:−0.95; −0.14). As for BC, the reduction was approximately 1.40 l/min. In relation to PDLM, adverse effects were noticed in models based on the exposure on the current day through the previous 3 days (PDLM 0–3) and on the current day through the previous 5 days (PDLM 0–5), specially for PM_{10}. For all children, for PDLM 0–5 the global effect was important for PM_{10}, with PEF reduction of 0.31 l/min (CI95%: −0.56; −0.05). Also, reductions in lags 3 and 4 were observed. These associations were stronger for children between 6 and 8 years old.

Conclusion: Reductions in PEF were associated with air pollution, mainly for lagged exposures of 3 to 5 days and for younger children.

Editor: Qinghua Sun, The Ohio State University, United States of America

Funding: This work was supported by CNPq (http://www.cnpq.br)- research grant no. 555223/2006-0 (Edital 18/CNPq), CNPq (http://www.cnpq.br) - research grant no. 306620/2010-3, CNPq (http://www.cnpq.br) - research grant no. 573797/2008-0, FAPESP (http://www.fapesp.br) - research grant no. 2008/57719-9, and FAPEMAT (http://www.fapemat.mt.gov.br) - research grant no. 10037682/2006 (Edital PPSUS-MT 2006/FAPEMAT – N°. 010/2006). The funders had no role in study design, data collection and analysis, decision to publish, or preparation of the manuscript.

Competing Interests: The authors have declared that no competing interests exist.

* Email: ludmillaviana@yahoo.com.br

Introduction

Several studies have shown adverse health effects from air pollution in many parts of the world [1]–[3]. Panel studies with children and teenagers have revealed important associations between air pollution and episodes of respiratory symptoms or lung function [4]–[8]. Such effects were found in both asthmatic and healthy subjects aged between 6 and 18 years old. Because children are still developing physiologically, exposure to air pollutants is a major risk factor for health and may have consequences in adulthood [9]. However, air pollution studies in healthy children are scarce.

In Brazil, the health effects of air pollution from biomass burning have not been thoroughly studied, especially in the Amazon region. The Brazilian Amazon dry season is the most critical period for biomass burning. From June to October, levels of particulate matter are usually above the World Health Organisation guidelines [2].

Some studies undertaken in the Amazon region have shown adverse health effects from particulate matter due to biomass

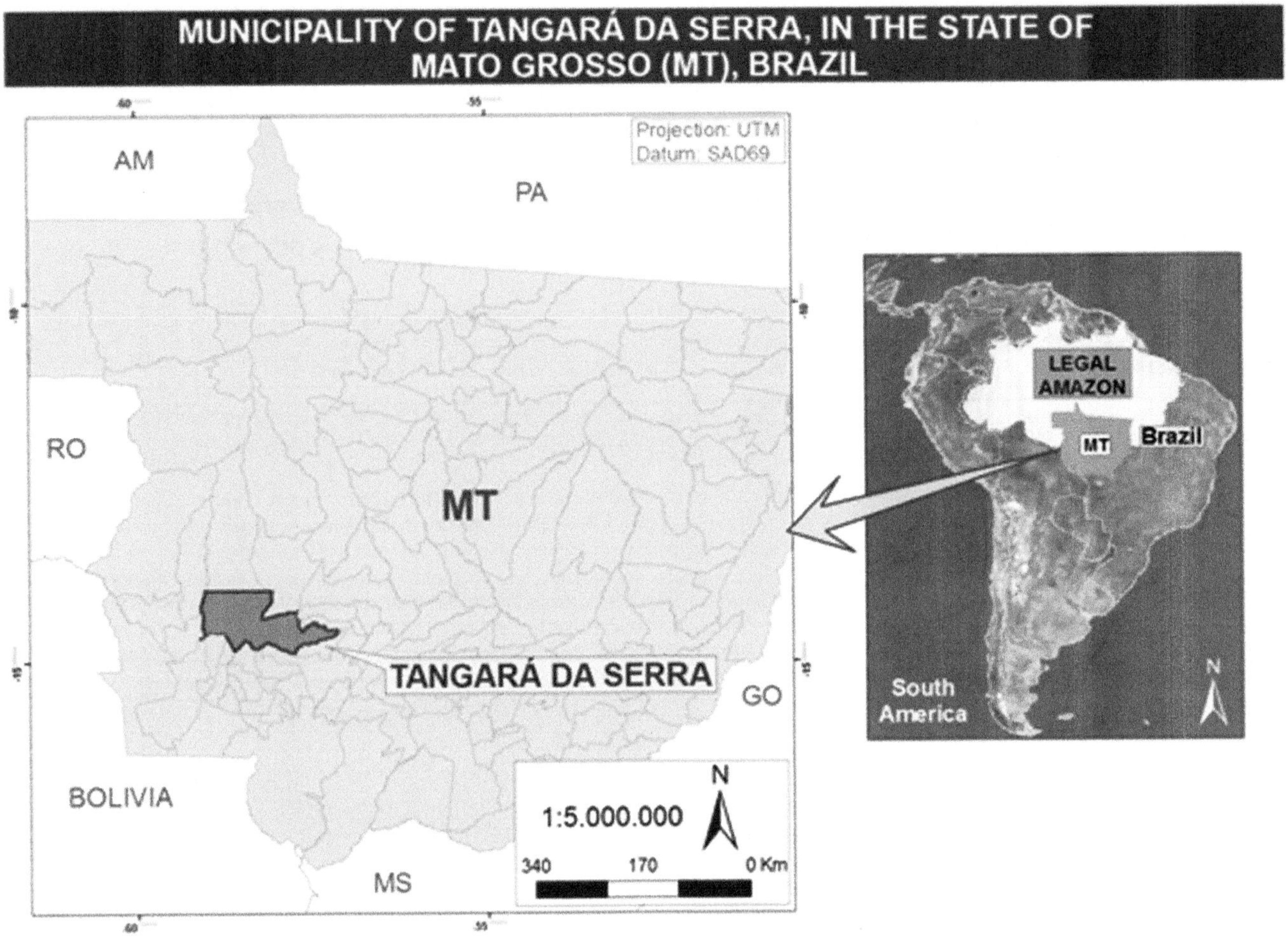

Figure 1. Geographic location of the municipality of Tangará da Serra, Mato Grosso State.

burning, for instance, increases in emergency room visits, outpatient visits, and hospital admissions as well as decreases in peak expiratory flow (PEF) and increases in the frequency of micronuclei in oral epithelial cells [10]–[15]. These effects were found in children and the elderly in municipalities around the Arc of Deforestation region.

The Brazilian Amazon covers an area of approximately 5,000,000 km^2 and 9 Brazilian states. Over 25 million inhabitants are changing the land use [16]. The majority of forest fire hotspots takes place in the Arc of Deforestation [17], whilst the state of Mato Grosso usually shows the worst records. Ignotti et al. [17] compared the indicators of morbidity and mortality caused by respiratory diseases in municipalities of Mato Grosso. Alta Floresta and Tangará da Serra were identified as priority areas for assessing the air pollution health effects of biomass burning in the Brazilian Amazon.

A panel study of schoolchildren carried out in the municipality of Alta Floresta found adverse effects of particulate matter with an aerodynamic diameter less than 2.5 μm ($PM_{2.5}$) on PEF [10]. The results were stratified by the time of day the children attended school, morning or afternoon. Stronger associations were detected in the afternoon group. However, only single-lag exposures were estimated, namely, levels of $PM_{2.5}$ on the current day or lagged by 1 or 2 days and the averages of 0- to 1-day lags, 1- to 2-day lags, and 0- to 2-day lags.

A similar panel study was carried out in Tangará da Serra in 2008. Its results are reported in this article. The primary aim of this study was to estimate the effect of current air pollution levels in the Amazonian dry season on schoolchildren's PEF. Three separate secondary aims were investigated, namely:

(i) To compare two methodological approaches in the analysis of repeated measures data: mixed-effects models (MEM) and a unified model to be applied for every child;

(ii) To assess the role played by age in explaining the difference between the morning and afternoon groups of children that Jacobson et al. [10] uncovered. In Brazil, most primary schools adopt the policy of two periods of study classes, morning and afternoon. Further, younger children attend school in the afternoon and older children go to school in the morning. Therefore, age could explain the difference observed in the previous study;

(iii) To extend the lag structures that Jacobson et al. [10] explored but to also use a polynomial distributed-lag model (PDLM). In the Amazon, particles emitted by biomass burning remain in the atmosphere for one week or so [18]. Therefore, to evaluate the cumulative effects based on polynomial distributed-lag models seems to be more appropriate. Moreover, few panel studies have used PDLM to estimate air pollution effects, especially those related to children's PEF [19]–[21].

Material and Methods

Study Design and Variables

The study was conducted in the Tangará da Serra municipality, in the southwest of Mato Grosso State, in a transition area of the Amazon and Cerrado biomes (Figure 1). Its total population was 83,431 inhabitants (population density was 7.32 inhab./km^2), and the population from 6 to 15 years old was 14,398 inhabitants (17.3%) [16]. Its main economic activities are agriculture and livestock. Despite its increasing production of sugarcane, the harvest is still manual and occurs during the dry season [22]. The pre-harvest burning of sugarcane and the plumes of biomass burnings from neighbouring regions are the main sources of air pollution in the city.

In 2008, a panel study of schoolchildren from a local public school was carried out (Lat 14°37′59.05″ S; Long 57°31′25.34″ W). The school had 875 students from 6 to 15 years old attending school in the morning or in the afternoon. The study sample comprised 234 schoolchildren who were followed up daily, except weekends and holidays, from August to November. Children were randomly selected according by age group (between 6 and 15) and school schedule.

Ethical approval was obtained from the UNEMAT (Mato Grosso State University) Ethics Committee, and a consent form to participate in the study was signed by the children's parents or legal guardians. The protocol described by the American Thoracic Society [23] for PEF exams was adopted throughout the study. After careful training, nursing undergraduate students collected PEF data under the supervision of a senior pulmonologist and a senior nurse.

Three daily PEF measures were sequentially taken. To avoid disruption of the school routine, groups of 4 or 5 children performed the tests and returned to class afterwards. The device used was the Mini-Wright Peak Flow Meter (Clement Clarke International Ltd, United Kingdom). The study outcome was the maximum of the three sequential PEF measures.

Weight and height were measured at the study's onset, using a Welmy anthropometric mechanical scale with 150 kg maximum capacity. Face-to-face questionnaires comprising health conditions and exposure to tobacco smoking at home were answered by the children's guardians/parents. To assess asthma symptoms, the questionnaire included phase I of the *International Study of Asthma and Allergies in Childhood* [24].

Exposure to air pollution was measured during the entire study period. The devices were placed at the UNEMAT campus in Tangará da Serra. The selected pupils lived within a 5 km radius of the campus and a maximum of 1 km radius of the study school. PM_{10} concentrations (particulate matter with an aerodynamic diameter less than 10 μm) were measured continuously using the Thermo Electron Corporation (DATARAM) device. Moreover, the Aethalometer was used to measure black carbon (BC) levels, and stacked filter units (SFU) were also used to measure $PM_{2.5}$. The following were regarded as exposures to PM_{10} and BC: (i) the 24-hour mean of the previous day (lag 1) until 5 days before (lag 5); and cumulative exposures, such as the average of the daily means of the same day and the previous one (lags 0–1), of the previous two days (lags 1–2), and of the same day and the two previous days (lags 0–2). For $PM_{2.5}$, there were a few missing data; therefore, to minimise loss of information, cumulative exposures were not investigated.

Measures of relative humidity and temperature were provided by INMET (National Institute of Meteorology, Brazil). Levels of relative humidity and mean temperature were regarded in all statistical models.

Statistical Analysis

The study database included variables that vary over time, for example, the outcome, the exposure, the weather, and outdoor leisure time, but also time-invariant covariates, such as age, sex, passive smoking, asthma, weight, height, and body mass index (BMI). Despite the data collection from August to November, this study presents the results for the dry season period only, from August to October.

Five types of statistical analysis were implemented in order of complexity, namely: (i) separate data analysis for each child; (ii) a mixed-effects core model regarding climate and individual covariates; (iii) air pollution effect estimation for single lags; (iv) air pollution effect estimation for PDLM; (v) and a sensitivity analysis.

For item (i), the same modelling steps were applied to the PEF time series for each child. Several strategies were examined for the adjustment of time trends, temperature and humidity, such as linear, quadratic or parametric splines with one knot or two knots, as well as linear or quadratic polynomials. The modelling proceeded including a 1st order autoregressive term, later estimated for each child. To summarise the results, averages were calculated for the Akaike Information Criterion (AIC), the estimate of the 1st order autoregressive term ($\hat{\varphi}$), and the set of PM_{10} lag 1 effects. Positive ($p<0.05$), negative ($p<0.05$), or null effects ($p>0.05$) of PM_{10} lag 1 were classified into three groups. The unified model that presented the smallest AIC average was selected. Finally, tables of counts (proportions) or averages by sex, age, passive smoking, presence of asthma, BMI, weight, and height allowed comparison of the three groups.

Because this was a study of repeated measures, MEMs were also appropriate for the statistical analysis. Regarding this approach, the model adjustment considered a common autoregressive lag 1 (AR(1)) correlation structure whereas the variance function comprised three time-invariant covariates, namely, BMI in four categories, age in three, and asthma status. Four adjustments based on polynomials of time were evaluated for the outcome long-term trend: a cubic parametric spline with one knot; a quadratic parametric spline with two equally spaced knots; a quadratic parametric spline with one knot; and a quadratic polynomial of time [25]. The above considered time to be centred in the middle of the study period as well as random coefficients. The same approach was applied for temperature and relative humidity, but also linear effects were explored. Furthermore, exposures on the same day, the previous day, and the day before that were examined. Finally, height, weight, BMI, age, sex, ISAAC asthma status, passive smoking, and outdoor leisure time, with fixed effects, were regarded in the models. Because of the number of model comparisons, the AIC and the deviance test were used to select the most parsimonious model, which then was scrutinised with regard to residual diagnostics to confirm the normality assumptions of the random effects. Jacobson et al.[10] describe the details of this approach.

Regarding the choice of core model based on the modelling steps mentioned above, single-lag and PDLM were applied to estimate the effects. Single-lag models were fitted for all exposures mentioned previously. PDLM estimates a fixed number of lagged effects constrained to a given choice of polynomial as well as the overall sum of effects [26]. In this study, quadratic polynomials of the effects lagged up to 3 (0–3) or 5 (0–5) days were calculated.

Sensitivity analyses were performed to assess the robustness of the results. For instance, measures of $PM_{2.5}$ were missing for 25% of the study days. Therefore, missing values of $PM_{2.5}$ were replaced by the $PM_{2.5}$ average on the previous day and the day after. The effect of this exposure on PEF was estimated using single lags and PDLM. Because of the methods used in PDLM, we evaluated the effects of $PM_{2.5}$ only considering levels after imputation. Furthermore, all of the analyses were repeated excluding outliers, the definition of which was standardised residuals greater than four in absolute value. We also explored the presence of a threshold effect of PM_{10} using a linear parametric spline. Each exposure (lag 1, lag 2, lag 3, lag 4, lag 5, and cumulative lag 1–5) was centred at different thresholds, namely, 20 μg/m^3, 30 μg/m^3, 40 μg/m^3, 50 μg/m^3, 60 μg/m^3, 70 μg/m^3, and 80 μg/m^3.

All analyses were performed using R 2.15.1 [27] and the library *nlme* [28].

Results

Of the 234 students selected for the sample, 7 left school during the study and were considered lost to follow-up. A group of 7 students were excluded from the regression analysis because their guardians/parents did not answer the questionnaire. The total number of study days was 53.

Table 1 displays the summary statistics of individual characteristics (sex, age, weight, height, asthma, and passive smoking), PEF measures, and weather and air pollution variables. Forty-seven percent of the 227 students were boys, the average age was 10.3 years old, the average weight was 36.2 kg (17 to 82.2 kg), and the average height was 1.40 m (1.11 to 1.73 m). According to the ISAAC criteria, 18% of the students (n = 39) reported asthma symptoms (21 girls and 18 boys). Passive smoking at home was reported by 33% of the students' parents (n = 72). Overall mean PEF was 289.9 l/m and it ranged from 70 to 780 l/min.

The average relative humidity and the daily mean temperature were 57% (29.4 to 90.5%) and 26°C (16.5 to 29.9°C), respectively. The averages of the daily mean PM_{10}, $PM_{2.5}$, and BC were 62.7 μg/m^3 (12.4 to 164.5 μg/m^3), 19.6 μg/m^3 (4.4 to 55.9 μg/m^3), 0.998 μg/m^3 (0.271 to 2.146 μg/m^3), respectively (Table 1). Typically in this part of Brazil in October and November, humidity levels increase and air pollution levels decrease because of the beginning of the rainy season, whereas temperature levels are somewhat constant. Figure 2 shows the time series for temperature, humidity, PM_{10}, $PM_{2.5}$, and BC.

In this section, the effects reported are in litres per minute and are for increases of 10 μg/m^3 for PM_{10} and $PM_{2.5}$, and 1 μg/m^3 for BC.

Results for the Unified Model Approach for Each Child

Table 2 shows the summary statistics of the unified model approach. Regarding AIC, 1st order autocorrelation ($\hat{\varphi}$), PM_{10} lag 1 effect averages, and counts of individuals with positive or negative effects of PM_{10} lag 1 are displayed. The most parsimonious 1st order autoregressive model consisted of a quadratic parametric spline with two equally spaced knots for the long-term trend, a quadratic polynomial for humidity lagged by two days, and temperature lagged by two days. The $\hat{\phi}$ average was 0.0232 (sd 0.246; −0.812 to 0.549), and for 34 children the fitted models presented $\hat{\varphi}$ larger than zero and $p<0.05$. With respect to PM_{10} lag 1, 12 students presented positive effects ($p<0.05$), 13 students presented adverse effects ($p<0.05$), and 202 presented no effects ($p \geq 0.05$).

Table 3 summarizes the results of the AR(1) estimated model according to individuals characteristics. Averages of weight, height, BMI, and age were calculated according to these groups (positive, negative, and non-significant effect) and compared using the Kruskal-Wallis test. Of the four individual covariates, only age showed differences ($p<0.05$). Similarly, proportions of sex, asthma, and passive smoking were compared across groups using Fisher's exact test. There were no proportion differences. Of the individual covariates investigated, only age affected the magnitude of PM_{10} effects (Table 3).

Moreover, we computed combined-effect estimates of the exposure to PM_{10} levels lagged by between 1 and 5 days, using averages weighed by the inverse of their variances. The most adverse effect found was for lag 4. For PM_{10} lag 4, there was a reduction in PEF equal to 0.16 (95% confidence interval(CI): −0.31; −0.02). The other results are not shown.

Results for the Mixed-Effects Models

MEMs were applied in the next step of the data analysis. The choice of core model consisted of long-term trend adjusted by a quadratic parametric spline with two equally spaced knots and random coefficients, a quadratic polynomial curve for humidity lagged by two days with random coefficients, temperature lagged by two days and a random coefficient, and fixed-effects covariates such as BMI, age, sex, asthma status, and exposure to air pollution; it also regarded an AR(1) and sixteen different subject-error variances depending on the subject status with respect to asthma, age, and BMI. Table 4 summarises the adjusted mixed models.

Single-Lag Effects

Table 5 displays the results of the models for all students and stratified by three age groups and considering single-lag exposures. This approach revealed some adverse effects of air pollution on PEF. Taking into account all children: (i) for PM_{10}, the results showed reductions of 0.25 (95%CI: −0.40; −0.10), 0.20 (95%CI: −0.32; −0.07), and 0.15 (95%CI: −0.29; −0.01) in lags 3, 4, and 5, respectively; (ii) for $PM_{2.5}$ lag 4, there was a reduction of 0.54 (95%CI: −0.95; −0.14). However, after $PM_{2.5}$ imputation, the models revealed associations not only with lag 4 (−0.50; 95%CI: −0.89; −0.12) but also with lag 3 (−0.46; 95%CI: −0.86; −0.06); (iii) for BC, the reduction was roughly 1.40 (95%CI: −2.50; −0.29) in both lags 4 and 5.

The results were stratified by age: 6 to 8 years old, 9 to 11, and 12 to 15. Children aged 6–8 were the most susceptible group. The results regarding this group were: (i) for PM_{10}, the reduction varied from 0.28 (95%CI: −0.50; −0.06) to 0.43 (95%CI: −0.65; −0.20) in lags 2, 3, 4, and 5, and there was also a reduction of 0.30 (95%CI: −0.54; −0.06) for the cumulative exposure in lag 1–2; (ii) for $PM_{2.5}$, important reductions were observed for lag 2 (−0.91; 95%CI: −1.65; −0.16) and lag 4 (−0.93; 95%CI: −1.55; −0.32). After imputation, the adverse effect of $PM_{2.5}$ was noticed only for lag 4 (−0.87; 95%CI: −1.47; −0.28); (iii) for BC, lagged exposures of 4 and 5 days reduced PEF in this age group. Further, for children aged 9–11, the adverse effect was observed only for BC. For lag 5, there was a reduction of 1.88 (95%CI: −3.64; −0.12). No associations were found for children aged 12–15.

Polynomial Distributed-lag Effects

Table 6 presents the results of the PDLM 0–3 for all children and stratified by age. Some adverse effects were revealed under this approach. For PM_{10}, the overall effect was not significant for all children; however, lag 3 showed a reduction of 0.25 (95%CI: −0.43; −0.08). There was also an overall adverse effect of PM_{10} for children 6–8 (−0.34; 95%CI: −0.66; −0.01), particularly for lag 3 (−0.35; 95%CI: −0.62; −0.08). For $PM_{2.5}$ lag 3, the group of all children suffered a reduction of 0.41 (95%CI: −0.81; −0.01). There were no adverse effects for BC. Likewise, no evidence of adverse effects was found for children aged 9–15.

The PDLM 0–5 effect estimates revealed stronger evidence of adverse effects (Table 7). Regarding PM_{10} for all children, the

Table 1. Summary statistics of weather variables, PM_{10}, $PM_{2.5}$, BC, peak expiratory flow measurements, and children's individual characteristics.

Variables	n	%missing	average (sd)	min	Percentile 10	25	50	75	90	max
Average daily temperature (°C)	82	0.0	26.0 (2.8)	16.5	22.2	24.2	26.8	28.0	29.1	29.9
Average daily humidity (%)	82	0.0	56.8 (17.1)	29.4	34.2	42.0	54.9	70.0	81.6	90.5
Average daily PM_{10} (μg/m^3)	75	8.5	62.7 (40.7)	12.4	20.5	29.7	52.7	89.7	127.3	164.5
Average daily $PM_{2.5}$ (μg/m^3)	61	25.6	19.6 (11.9)	4.4	6.6	9.6	16.1	28.3	35.8	55.9
Average daily black carbon (μg/m^3)	72	12.2	0.998 (0.481)	0.271	0.403	0.597	0.949	1.254	1.834	2.146
Peak flow (l/m)	11198	41.14*	289.9	70.0	180.0	220.0	280.0	350.0	410.0	780.0
Age	227	0.0	10.3 (2.5)	6.0	7	8	10.0	13.0	14.0	15.0
Weight (Kg)	227	0.0	36.2 (21.4)	17.0	21.4	26.1	32.8	43.5	55.4	82.2
Height (cm)	227	0.0	140.2 (120.0)	111.0	120.0	129.0	138.0	152.0	162.0	173.0
BMI	227	0.0	17.9 (3.5)	11.9	14.5	15.2	16.9	19.7	22.7	32.1
Asthmatic (%)	220	3.08	18.0							
Passive smoke (%)	218	4.0	33.0							
Boys (%)	227	0.0	46.7							

Tangará da Serra-MT, Brazil, 2008.
n = number of observations; sd = standard deviation; min = minimum; max = maximum; * includes holidays, weekends and actual missing values.

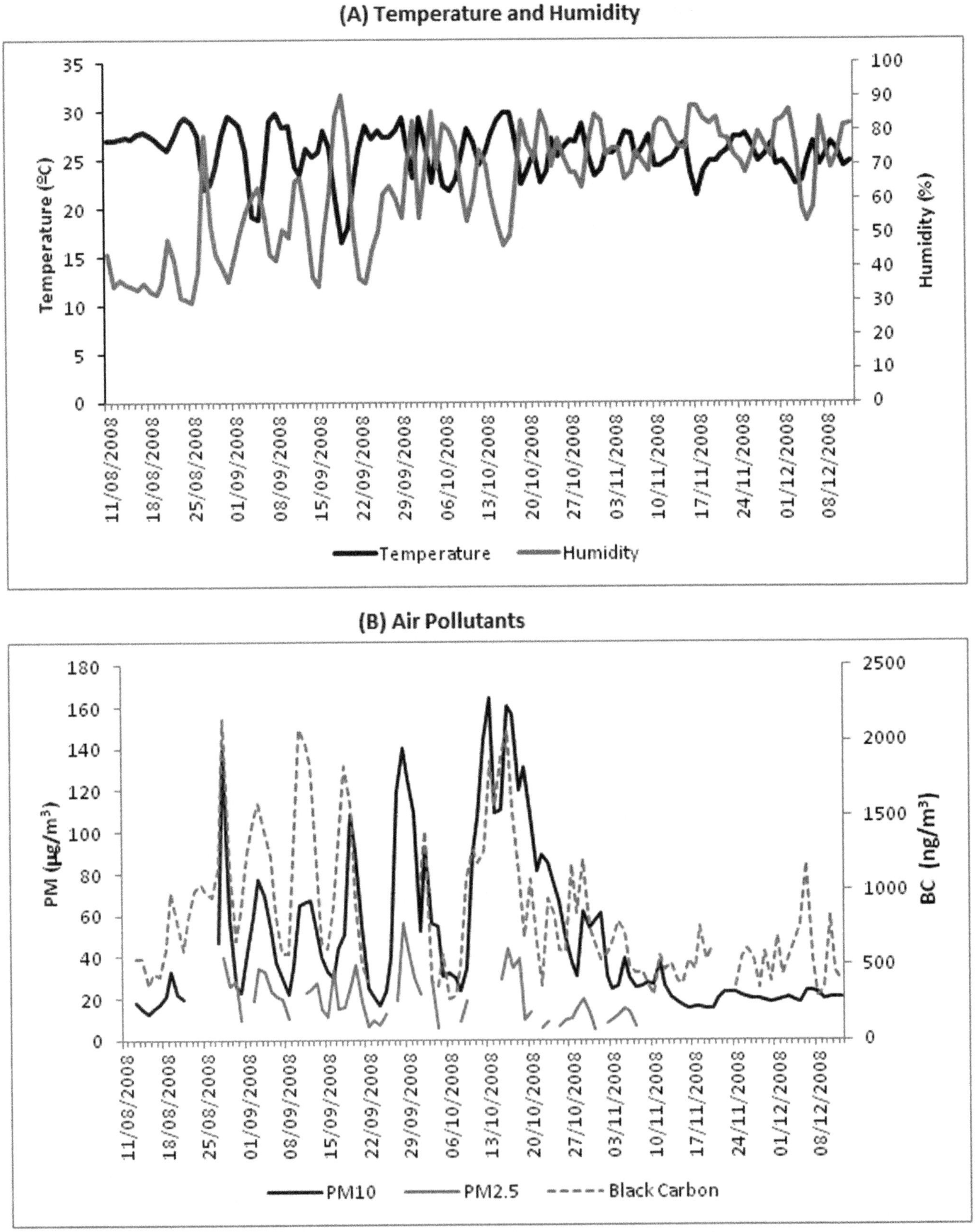

Figure 2. Daily temperature, humidity and air pollutant readings. Tangará da Serra, MT, Brazil-2008.

Table 2. Summary adjustment strategies of the unified model approach.

Model Adjustment – AR(1) (n = 227)			AIC	$\hat{\varphi}$		Change in PEF (PM10 lag 1)***		
Time trend	**Humidity (lag 2)**	**Temperature (lag 2)**	**Average***	**Average***	**p<0.05 and $\hat{\varphi}$ higher than zero (n)**	**Average ****	**Positive and p<0.05 (n)**	**Negative and p<0.05 (n)**
Linear Spline	Linear	Linear	401.74	0.1468	61	0.038	11	10
Linear Spline	Linear Spline	Linear Spline	402.91	0.1453	57	0.047	15	12
Quadratic Polynomial	Quadratic Polynomial	Quadratic Polynomial	403.17	0.1483	65	0.027	15	12
Quadratic Spline$^{\perp}$ (1 Knot)	Quadratic Polynomial	Quadratic Polynomial	402.49	0.0854	46	0.039	17	13
Quadratic Spline$^{\perp}$ (2 Knot)	Quadratic Polynomial	Quadratic Polynomial	401.41	0.0173	35	−0.017	15	12
Quadratic Spline$^{\perp}$ (2 Knot)	*Quadratic Polynomial*	*Linear*	*400.83*	*0.0232*	*34*	*−0.033*	*12*	*13*
Cubic Spline$^{\perp}$ (1 Knot)	Quadratic Polynomial	Quadratic Polynomial	401.78	0.0270	35	−0.035	14	13

Tangará da Serra-MT, Brazil, 2008.
*Simple arithmetic average; **Averages weighed by the inverse of their variances; *** Change in PEF for an increase of 10 μg/m^3 in PM_{10}lag1; n – Number of cases; $^{\perp}$Parametric Splines.

Table 3. Characteristics of the schoolchildren according to the results of the AR(1) estimated model.

Individual Characteristics	Change in PEF - PM_{10} Lag1		
	Positive and Significant (n = 12)	Negative and Significant (n = 13)	Non-Significant (n = 202)
Sex (%)			
Girls	66.7 (n = 8)	46.2 (n = 6)	53.0 (n = 107)
Boys	33.3 (n = 4)	53.8 (n = 7)	47.0 (n = 95)
Asthma Status (%)			
Yes	8.3 (n = 1)	23.1 (n = 3)	18.1 (n = 35)
No	91.7 (n = 11)	76.9 (n = 10)	81.9 (n = 158)
Passive Smoke (%)			
Yes	33.3 (n = 4)	30.8 (n = 4)	33.2 (n = 64)
No	66.7 (n = 8)	69.2 (n = 9)	66.8 (n = 129)
Age			
Average (SD)	12 (2.17)	9.62 (2.43)	10.23 (2.55)
Weight			
Average (SD)	43.57 (15.01)	35.26 (11.42)	35.92 (13.40)
Height			
Average (SD)	149.3 (14.16)	138.69 (11.93)	139.71 (15.28)
BMI			
Average (SD)	19.05 (3.87)	17.97 (4.03)	17.79 (3.50)

Tangará da Serra-MT, Brazil, 2008.

overall effect was −0.31 (95%CI: −0.56; −0.05). Additionally, reductions of 0.09 (95%CI: −0.16; −0.01) and 0.10 (95%CI: −0.15; −0.05) in lags 3 and 4, respectively, were estimated. For children aged 6–8, there were strong effects for lags 2, 3, and 4. The overall effect for this age group indicated a reduction of 0.52 (95%CI: −0.92; −0.12) caused by PM_{10}.

Moreover, for all children, the PDLM 0–5 effect estimates of $PM_{2.5}$ showed reductions for lag 2 (−0.21; 95%CI: −0.42; −0.003), lag 3 (−0.27; 95%CI: −0.48; −0.06), and lag 4 (−0.19; 95%CI: −0.36; −0.02), but there was no overall effect. Additionally, for children aged 6–8, the results revealed reductions for lags 3 and 4.

BC presented associations with PEF. For all children, PDLM 0–5 revealed a reduction only for lag 4 (−0.73; 95%CI: −1.22; −0.24). Furthermore, for children aged 6–8, there were reductions of 0.97 (95%CI: −1.74; −0.20) and 1.77 (95%CI: −3.26; −0.27) in lags 4 and 5, respectively.

Sensitivity Analysis

Exclusion of outliers hardly affected the results. However, other exposures showed adverse effects, for instance: (i) For single-lag models, PM_{10} lag 3 presented a negative association with PEF for children aged 6–8, and $PM_{2.5}$ lagged by 3 days for all children; (ii) for PDLM, a reduction was observed with $PM_{2.5}$ lag 2 in children aged 6–8.

Table 4. Summary adjustment strategies of the mixed-effect models.

Model Adjustment* (n = 220)			AIC	PHI (AR(1))	Change in PEF **(IC95%)
Time trend	Humidity (lag 2)	Temperature (lag 2)			
Linear	Linear	Linear	90301.19	0.332	−0.020 (−0.165; 0.125)
Linear	Quadratic Polynomial	Linear	90292.05	0.330	−0.021 (−0.165; 0.123)
Linear Spline⊥	Linear	Linear	90207.98	0.291	−0.016 (−0.157; 0.125)
Quadratic Polynomial	Quadratic Polynomial	Linear	90217.62	0.295	−0.026 (−0.168; 0.115)
Quadratic Spline⊥ (1 Knot)	Quadratic Polynomial	Linear	90147.36	0.261	−0.048 (−0.188; 0.092)
Quadratic Spline⊥ (2 Knot)	*Quadratic Polynomial*	*Linear*	*90104.27*	*0.238*	*−0.108 (−0.251; 0.035)*
Cubic Spline⊥ (1 Knot)	Quadratic Polynomial	Linear	90120.02	0.242	−0.114 (−0.259; 0.032)

Tangará da Serra-MT, Brazil, 2008.
*Model adjusted for age, BMI, gender, and asthma status; random coefficients for the intercept, time trend, humidity, and temperature; variance function of the random error included age, BMI, and asthma status.
** Change in PEF for an increase of 10 $\mu g/m^3$ in PM_{10}lag1.
⊥Parametric Splines.

Table 5. Estimated changes in peak expiratory flow (in l/min) for an increase of 10 μg/m³ in PM_{10} and $PM_{2.5}$ and an increase of 1 μg/m³ in black carbon for all children and stratified by age groups.

Exposure	All Children (n = 220)		6 to 8 years (n = 69)		9 to 11 years (n = 69)		12 to 15 years (n = 82)	
	Change in PEF (95%CI)		Change in PEF (95%CI)		Change in PEF (95%CI)		Change in PEF (95%CI)	
PM_{10}								
Lag 1	−0.108	(−0.251; 0.035)	−0.215	(−0.436; 0.006)	−0.102	(−0.375; 0.172)	0.082	(−0.175; 0.338)
Lag 2	−0.104	(−0.251; 0.042)	−0.280	(−0.504; −0.055)	−0.015	(−0.301; 0.271)	0.100	(−0.162; 0.362)
Lag 3	−0.252	(−0.399; −0.104)	−0.427	(−0.654; −0.199)	−0.086	(−0.376; 0.205)	−0.112	(−0.372; 0.148)
Lag 4	−0.196	(−0.322; −0.070)	−0.283	(−0.477; −0.089)	−0.139	(−0.385; 0.108)	−0.125	(−0.347; 0.097)
Lag 5	−0.151	(−0.293; −0.010)	−0.296	(−0.515; −0.077)	−0.115	(−0.393; 0.162)	−0.055	(−0.304; 0.195)
Lag 0–1	−0.071	(−0.245; 0.103)	−0.171	(−0.441; 0.099)	−0.067	(−0.40; 0.266)	0.117	(−0.191; 0.426)
Lag 1–2	−0.126	(−0.284; 0.032)	−0.298	(−0.540; −0.055)	−0.063	(−0.370; 0.245)	0.107	(−0.174; 0.388)
Lag 0–2	−0.104	(−0.287; 0.078)	−0.262	(−0.545; 0.021)	−0.049	(−0.406; 0.309)	0.123	(−0.201; 0.447)
$PM_{2.5}$								
Lag 1	0.129	(−0.316; 0.573)	0.108	(−0.576; 0.791)	0.084	(−0.748; 0.915)	0.175	(−0.590; 0.941)
Lag 2	−0.286	(−0.773; 0.201)	−0.905	(−1.646; −0.163)	0.202	(−0.700; 1.103)	0.132	(−0.702; 0.966)
Lag 3	−0.377	(−0.792; 0.037)	−0.322	(−0.954; 0.311)	−0.558	(−1.331; 0.216)	−0.191	(−0.926; 0.544)
Lag 4	−0.541	(−0.946; −0.137)	−0.931	(−1.545; −0.317)	−0.560	(−1.296; 0.175)	−0.159	(−0.866; 0.549)
Lag 5	−0.029	(−0.437; 0.429)	−0.137	(−0.835; 0.560)	0.297	(−0.556; 1.150)	−0.403	(−1.212; 0.406)
$PM_{2.5}$ Imputation								
Lag 1	0.134	(−0.233; 0.550)	0.042	(−0.603; 0.688)	0.269	(−0.531; 1.070)	0.176	(−0.566; 0.918)
Lag 2	−0.181	(−0.601; 0.239)	−0.581	(−1.227; 0.065)	0.443	(−0.366; 1.252)	−0.016	(−0.770; 0.738)
Lag 3	−0.462	(−0.862; −0.062)	−0.433	(−1.043; 0.178)	−0.515	(−1.294; 0.263)	−0.253	(−0.965; 0.459)
Lag 4	−0.504	(−0.839; −0.118)	−0.374	(−1.466; −0.283)	−0.575	(−1.326; 0.175)	−0.087	(−0.775; 0.600)
Lag 5	−0.051	(−0.478; 0.377)	−0.474	(−1.135; 0.186)	0.201	(−0.620; 1.022)	−0.017	(−0.776; 0.743)
Lag 0–1	0.261	(−0.231; 0.753)	0.298	(−0.463; 1.058)	0.301	(−0.644; 1.246)	0.242	(−0.636; 1.122)
Lag 1–2	−0.094	(−0.616; 0.429)	−0.480	(−1.127; 0.749)	0.509	(−0.496; 1.514)	0.050	(−0.876; 0.975)
Lag 0–2	0.048	(−0.556; 0.651)	−0.189	(−1.127; 0.749)	0.535	(−0.623; 1.693)	0.115	(−0.952; 1.181)
BC								
Lag 1	0.610	(−0.435; 1.655)	1.950	(0.323; 3.577)	0.120	(−1.762; 2.002)	−0.730	(−2.592; 1.132)
Lag 2	0.074	(−1.284; 1.432)	0.630	(−1.467; 2.727)	0.470	(−1.980; 2.920)	−1.330	(−3.760; 1.100)
Lag 3	−0.277	(−1.635; 1.081)	−0.900	(−2.978; 1.178)	0.640	(−1.810; 3.090)	−0.430	(−2.880; 2.020)
Lag 4	−1.396	(−2.499; −0.293)	−2.140	(−3.826; −0.454)	−1.200	(−3.180; 0.780)	−0.060	(−2.059; 1.939)
Lag 5	−1.539	(−2.525; −0.553)	−2.510	(−4.019; −1.001)	−1.880	(−3.644; −0.116)	0.200	(−1.584; 1.984)
Lag 0–1	1.428	(0.230; 2.626)	2.630	(0.768; 4.492)	0.940	(−1.216; 3.096)	0.080	(−2.037; 2.197)
Lag 1–2	0.620	(−0.705; 1.945)	1.920	(−0.158; 3.998)	0.370	(−2.002; 2.742)	−1.070	(−3.442; 1.302)
Lag 0–2	1.410	(−0.005; 2.825)	2.854	(0.639; 5.069)	0.950	(−1.578; 3.478)	−0.310	(−2.819; 2.199)

Tangará da Serra-MT, Brazil – 2008.

Table 6. Estimated changes in peak expiratory flow (in l/min) for an increase of 10 μg/m³ in PM_{10} and $PM_{2.5}$ and an increase of 1 μg/m³ in black carbon for all children and stratified by age groups, according to PDLM based on the exposures of the current day to the previous 3 days.

Exposure	All Children (n = 220)		6 to 8 years (n = 69)		9 to 11 years (n = 69)		12 to 15 years (n = 82)	
	Change in PEF (95%CI)		Change in PEF (95%CI)		Change in PEF (95%CI)		Change in PEF (95%CI)	
PM_{10}								
Lag 0	0.037	(−0.161; 0.236)	0.125	(−0.174; 0.425)	0.047	(−0.321; 0.414)	−0.107	(−0.465; 0.252)
Lag 1	0.066	(−0.054; 0.186)	0.023	(−0.161; 0.207)	−0.034	(−0.258; 0.190)	0.206	(−0.006; 0.419)
Lag 2	−0.030	(−0.150; 0.089)	−0.135	(−0.316; 0.046)	−0.066	(−0.288; 0.156)	0.148	(−0.068; 0.364)
Lag 3	−0.252	(−0.429; −0.075)	−0.349	(−0.623; −0.076)	−0.050	(−0.381; 0.281)	−0.281	(−0.591; 0.028)
Overall	−0.179	(−0.390; 0.031)	−0.336	(−0.661; −0.010)	−0.104	(−0.496; 0.288)	−0.034	(−0.402; 0.335)
$PM_{2.5}$								
Lag 0	0.199	(−0.270; 0.669)	0.561	(−0.147; 1.269)	−0.072	(−0.926; 0.781)	0.018	(−0.834; 0.870)
Lag 1	0.084	(−0.210; 0.378)	−0.163	(−0.613; 0.288)	0.368	(−0.170; 0.906)	0.121	(−0.402; 0.645)
Lag 2	−0.119	(−0.432; 0.195)	−0.443	(−0.921; 0.034)	0.216	(−0.357; 0.789)	0.012	(−0.549; 0.574)
Lag 3	−0.408	(−0.809; −0.007)	−0.281	(−0.893; 0.331)	−0.528	(−1.266; 0.209)	−0.308	(−1.022; 0.406)
Overall	−0.243	(−0.922; 0.435)	−0.326	(−1.373; 0.720)	−0.016	(−1.267; 1.234)	−0.156	(−1.352; 1.040)
BC								
Lag 0	1.797	(0.504; 3.090)	1.760	(−0.226; 3.746)	1.774	(−0.511; 4.059)	1.753	(−0.574; 4.082)
Lag 1	0.111	(−0.719; 0.942)	0.963	(−0.309; 2.235)	−0.286	(−1.751; 1.179)	−0.771	(−2.278; 0.737)
Lag 2	−0.566	(−1.484; 0.352)	0.069	(−1.355; 1.494)	−0.733	(−2.353; 0.888)	−1.304	(−2.945; 0.336)
Lag 3	−0.235	(−1.649; 1.179)	−0.920	(−3.072; 1.232)	0.434	(−2.066; 2.935)	0.152	(−2.433; 2.737)
Overall	1.108	(−0.739; 2.955)	1.873	(−1.031; 4.777)	1.189	(−2.091; 4.470)	−0.169	(−3.427; 3.089)

Tangará da Serra-MT, Brazil – 2008.

Table 7. Estimated changes in peak expiratory flow (in l/min) for an increase of 10 μg/m^3 in PM_{10} and $PM_{2.5}$ and an increase of 1 μg/m^3 in black carbon for all children and stratified by age groups, according to PDLM based on the exposures of the current day to the previous 5 days.

Exposure	All Children (n=220)		6 to 8 years (n=69)		9 to 11 years (n=69)		12 to 15 years (n=82)	
	Change in PEF (95%CI)		Change in PEF (95%CI)		Change in PEF (95%CI)		Change in PEF (95%CI)	
PM_{10}								
Lag 0	0.046	(−0.111; 0.202)	0.076	(−0.162; 0.313)	−0.039	(−0.330; 0.252)	0.040	(−0.241; 0.321)
Lag 1	0.012	(−0.046; 0.070)	−0.045	(−0.134; 0.045)	−0.022	(−0.130; 0.087)	0.038	(−0.064; 0.140)
Lag 2	−0.056	(−0.125; 0.013)	−0.124	(−0.229; −0.019)	−0.020	(−0.149; 0.108)	0.022	(−0.102; 0.145)
Lag 3	−0.085	(−0.158; −0.012)	−0.161	(−0.272; −0.051)	−0.035	(−0.171; 0.100)	−0.008	(−0.139; 0.122)
Lag 4	−0.099	(−0.153; −0.046)	−0.157	(−0.240; −0.074)	−0.066	(−0.166; 0.033)	−0.052	(−0.145; 0.041)
Lag 5	−0.100	(−0.228; 0.029)	−0.112	(−0.308; 0.084)	−0.113	(−0.352; 0.125)	−0.109	(−0.339; 0.120)
Overall	−0.306	(−0.564; −0.048)	−0.523	(−0.922; −0.124)	−0.296	(−0.775; 0.184)	−0.070	(−0.522; 0.382)
$PM_{2.5}$								
Lag 0	0.321	(−0.070; 0.711)	0.288	(−0.306; 0.882)	0.435	(−0.277; 1.146)	0.164	(−0.538; 0.867)
Lag 1	−0.014	(−0.197; 0.169)	−0.053	(−0.335; 0.229)	0.057	(−0.281; 0.395)	0.008	(−0.314; 0.331)
Lag 2	−0.211	(−0.418; −0.003)	−0.280	(−0.599; 0.039)	−0.156	(−0.541; 0.229)	−0.083	(−0.449; 0.283)
Lag 3	−0.270	(−0.476; −0.064)	−0.393	(−0.708; −0.078)	−0.203	(−0.584; 0.178)	−0.109	(−0.473; 0.255)
Lag 4	−0.191	(−0.363; −0.020)	−0.392	(−0.659; −0.126)	−0.085	(−0.401; 0.232)	−0.070	(−0.370; 0.231)
Lag 5	0.024	(−0.350; 0.399)	−0.278	(−0.855; 0.299)	0.193	(−0.491; 0.876)	0.034	(−0.629; 0.697)
Overall	−0.340	(−1.206; 0.526)	−1.109	(−2.448; 0.231)	0.247	(−1.339; 1.833)	−0.055	(−1.586; 1.476)
BC								
Lag 0	0.831	(−0.174; 1.836)	0.968	(−0.592; 2.529)	0.441	(−1.345; 2.227)	0.910	(−0.885; 2.705)
Lag 1	0.339	(−0.103; 0.782)	0.668	(−0.033; 1.370)	0.238	(−0.551; 1.027)	−0.087	(−0.862; 0.688)
Lag 2	−0.085	(−0.710; 0.540)	0.245	(−0.729; 1.219)	−0.733	(−1.842; 0.377)	−0.633	(−1.747; 0.482)
Lag 3	−0.442	(−1.092; 0.208)	−0.302	(−1.311; 0.708)	−0.288	(−1.442; 0.865)	−0.728	(−1.892; 0.436)
Lag 4	−0.731	(−1.218; −0.245)	−0.972	(−1.739; −0.204)	−0.611	(−1.477; 0.256)	−0.373	(−1.234; 0.489)
Lag 5	−0.953	(−1.902; −0.004)	−1.765	(−3.260; −0.271)	−0.973	(−2.657; 0.710)	0.434	(−1.239; 2.106)
Overall	−1.040	(−3.301; 1.220)	−1.157	(−4.795; 2.482)	−1.198	(−5.244; 2.848)	−0.476	(−4.380; 3.428)

Tangará da Serra-MT, Brazil – 2008.

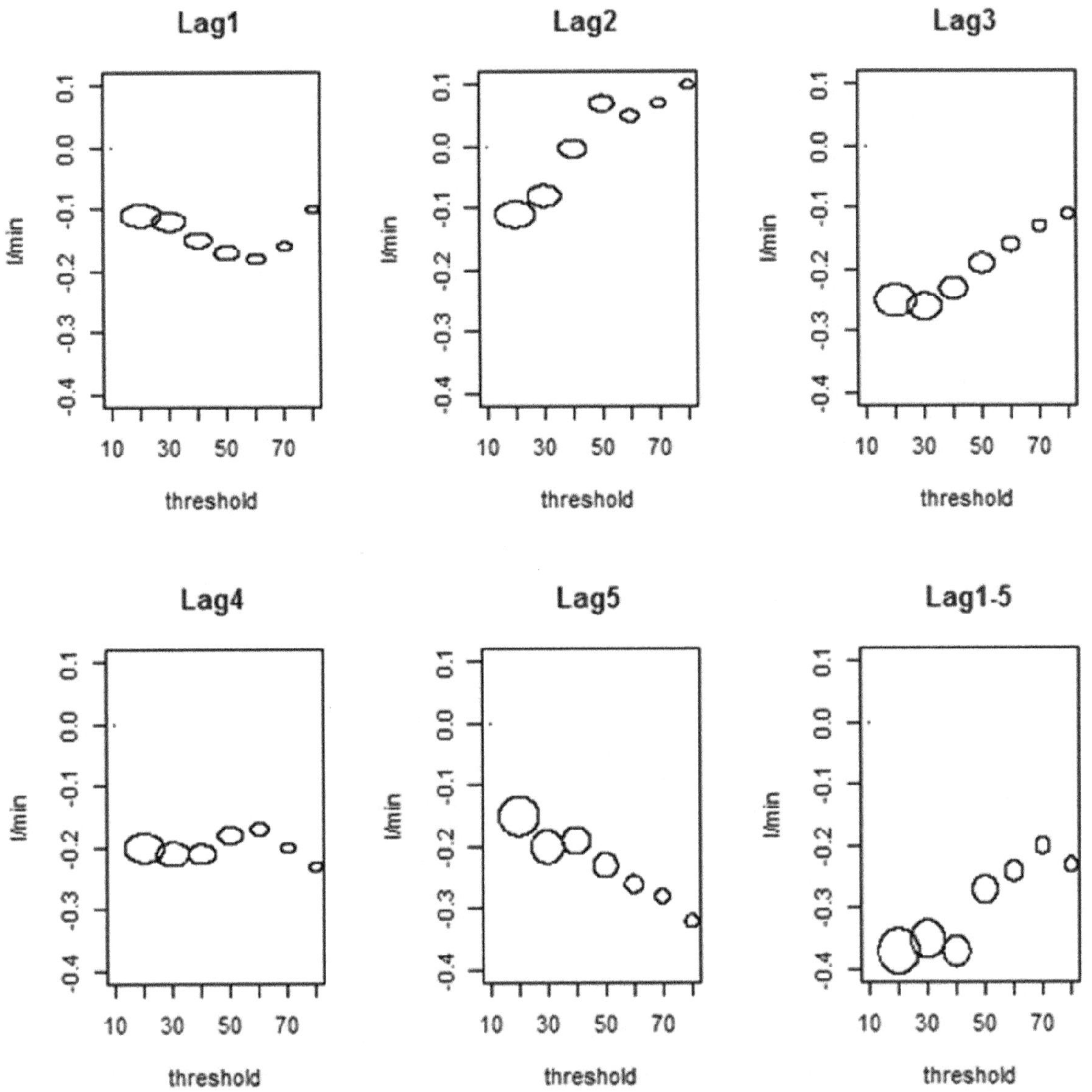

Figure 3. Estimated changes in peak expiratory flow (in l/min) for an increase of 10 μg/m³ in PM_{10}, above the thresholds for all children. Tangará da Serra-MT, Brazil – 2008.

We found evidence for PM_{10} thresholds. Under the linear parametric spline approach, there were two slopes, one for the effect estimate below the threshold and one for the effect above it. Concerning the former, there were no important findings, except for lags of PM_{10} such as: (i) lag 2: there was a reduction of 0.70 (95%CI: −1.30; −0.99) for exposures below 50 μg/m³; (ii) lag 3: the effects of exposures below 60 μg/m³, 70 μg/m³, and 80 μg/m³ were reductions of 0.49 (95%CI: −0.93; −0.05), 0.45 (95%CI: −0.83; −0.07), and 0.40 (95%CI: −0.72; −0.08), respectively; (iii) lag 1–5: reductions for the exposures below 60 μg/m³ (−0.63; 95%CI: −1.18; −0.08), 70 μg/m³ (−0.55; 95%CI: −0.99; −0.11), and 80 μg/m³ (−0.46; 95%CI: −0.84; −0.09).

Figure 3 presents a bubble plot for the estimated effects above the thresholds for all exposures. The size of the bubbles is proportional to the number of days above the thresholds. Negative associations between PEF and PM_{10} were noticed for lags 3, 4, 5, and lag 1–5. More specifically, for lag 3 the associations were important for the following thresholds: 20 μg/m³ (−0.25; 95%CI: −0.40; −0.10); 30 μg/m³ (−0.26; 95%CI: −0.41; −0.11); and 40 μg/m³ (−0.23; 95%CI: −0.40; −0.06). For lag 4, the effects were considerable for the same thresholds, varying from −0.20 μg/m³ to −0.21 μg/m³ (95%CI: −0.35, −0.08), as well as for 50 μg/m³ (−0.18; 95%CI: −0.34; −0.02). Concerning lag 5, there were reductions for all thresholds: (i) 20 μg/m³, −0.15 (95%CI: −0.30; −0.01); (ii) 30 μg/m³, −0.20 (95%CI: −0.35; −0.04); (iii) 40 μg/m³, −0.19 (95%CI: −0.36; −0.03); (iv) 50 μg/m³, −0.23 (95%CI: −0.42; −0.05); (v) 60 μg/m³, −0.26 (95%CI: −0.47; −0.05); (vi) 70 μg/m³, −0.28 (95%CI: −0.53; −0.03); (vii) 80 μg/m³, −0.32 (95%CI: −0.62; −0.03). For the cumulative exposure (lag1–5), the reductions were associated with the following thresholds: 20 μg/m³ (−0.37; 95%CI: −0.57; −0.16); 30 μg/m³ (−0.35; 95%CI: −0.55; −0.14); and 40 μg/m³ (−0.37; 95%CI: −0.61; −0.13).

Discussion

This study's results suggest that air pollution from biomass burning may be a respiratory health risk factor for schoolchildren aged 6 to 15 years old living in the Brazilian Amazon. This association is consistent with the previous panel study undertaken in the municipality of Alta Floresta [10]. The current study extended previous analyses by including PDLM as well as adding the effects of BC and PM_{10} exposures. Further, we explored two methods for analysing panel studies: MEM and univariate time series modelling, which was applied to every child. Although the combined-effect estimates obtained using the univariate approach and those from MEM were similar, their precision was different. However, it was particularly important to show that some students were more vulnerable to air pollution than were others.

Our findings corroborate the results of panel studies describing a linear effect of air pollution on PEF [5], [8], [29]. No association was found with $PM_{2.5}$ or BC in a rural panel study undertaken in the southeast of Brazil to investigate the effects of pre-harvest sugarcane burning on PEF [30]. Systematic reviews of panel studies demonstrated negative pooled effects of PM_{10} and $PM_{2.5}$ on PEF [31], [32], [33].

Although we found associations for all children, the effects were stronger after stratification by age, with PEF decrements for the youngest group, confirming that younger children are more susceptible to air pollution effects [1]. In the Alta Floresta study [10], a negative association of $PM_{2.5}$ with PEF was found for children who studied in the afternoon. However, the authors noted that the majority of children in this group were from 6 to 9 years old. Our results confirm the hypothesis that age is most likely to explain this finding.

A multitude of factors are related to children's PEF. In the present study, PEF measurements were associated with temperature, humidity, BMI, age, sex, and asthma status. Weather conditions can influence children's health [34] but also interfere with pollutants in the atmosphere [35]. Other environmental factors could affect PEF, for instance, time spent outdoors and passive smoking. However, in this study these variables were not important in the modelling, most likely because in the dry season children spend most of their time outdoors when they are not at school because of lack of leisure indoors as well as the high temperatures.

Most adverse effects found in this study were lagged by 3, 4, and 5 days. Lagged effects as well as cumulative effects are expected in this region not only because of the characteristics of Amazon biomass burning (every day during the dry season) but also because of meteorological factors, which allow for pollutants to remain in the atmosphere for long periods of time. The PDLM approach allowed exploring the effect of air pollutants on PEF distributed over time and also the overall effect up to 3 or 5 lagged days. The results revealed negative associations between air pollution and PEF, mainly for PDLM 0–5, which is consistent with the Mexico City panel study that found reductions in children's PEF associated with O_3, $PM_{2.5}$, and PM_{10} [21]. In the literature, this methodology had been broadly applied in time series studies [26] but not as much in children's panel studies.

The PM_{10} effect was scrutinised using different thresholds. Our results suggest hazardous effects below 50 $\mu g/m^3$. For lag 5, there was a clear negative gradient. The World Health Organisation guidelines for PM_{10} are 20 $\mu g/m^3$ for the annual average and 50 $\mu g/m^3$ for the daily average in urban areas [2]. During the 2008 dry season in Tangará da Serra, approximately 53% of the measurements were above 50 $\mu g/m^3$.

This study could be taken into consideration by the Brazilian Ministry of Health and other health agencies to establish guidelines for health protection in regions where biomass burning takes place. Reducing air pollutant levels is a challenge for local authorities; however, more effective actions should be taken to minimise fires, such as intensified patrolling in the region, heavy fines, policy reforms on taxes and credits, and tougher legislation concerning land occupation conflicts.

Air pollution was measured at the local university campus that is near the study school. In fact, all study children lived within a 5 km distance from the university campus. We believe that the device location did not sub or super estimated the observed adverse effects because particles emitted through biomass burning have relatively long resident time in the atmosphere and can be transported over long distances, crossing international boundaries [18]. This study would have benefited from a more specific questionnaire, for instance to evaluate patterns of physical activity and time spent outdoors.

In conclusion, this study showed a negative association between exposure to air pollution and PEF in schoolchildren living in Tangará da Serra. The analysis per child indicated that age was an effect modifier and that air pollution mostly affects younger children.

Acknowledgments

This paper is a contribution of the Brazilian National Institute of Science and Technology (INCT) for Climate Change. The authors would like to thank Beatriz Fátima Alves de Oliveira for the logistical support and Jose Eduardo Ernesto Pinheiro from Universidade Federal do Rio de Janeiro for the spirometry exams.

Author Contributions

Conceived and designed the experiments: LSVJ SSH EI ACMPL. Performed the experiments: EI HAC PA PHNS. Analyzed the data: LSVJ ACMPL. Contributed to the writing of the manuscript: LSVJ SSH HAC EI PA PHNS ACMPL.

References

1. Pope CA, Dockery DW (2006) Health Effects of Fine Particulate Air Pollution: Lines that Connect. J Air & Waste Manage Assoc 56: 709–742.
2. WHO (World Health Organization) (2005) Air quality guidelines for particulate matter, ozone, nitrogen dioxide and sulfur dioxide. Global update 2005. Summary of risk assessment. Geneva.
3. Cohen AJ, Anderson HR, Ostro B, Pandey KD, Krzyzanowski M, et al. (2004) Urban Air Pollution. In: Ezzati M, Lopez AD, Rodgers A, Murray CJL. Comparative Quantificatio Of Health Risks. Global and Regional Burden of Disease Attributable to Selected Major Risk Factors. Vol. 2, Chapter: 17.
4. Jedrychowski WA, Perera FP, Maugeri U, Mroz E, Klimaszewska-Rembiasz M, et al. (2010) Effect of prenatal exposure to fine particulate matter on ventilatory lung function of preschool children of non-smoking mothers. Paediatr Perinat Epidemiol 24(5):492–501.
5. Castro HA, Cunha MF, Mendonça GAS, Junger WL, Cunha-Cruz J, et al. (2009) Effect of air pollution on lung function in schoolchildren in Rio de Janeiro, Brazil. Rev Saúde Públ 43: 26–34.
6. Gotschi T, Heinrich J, Sunyer J, Kunzli N (2008) Long-Term Effects of Ambient Air Pollution on Lung Function: A Review. Epidemiol 19: 690–701.
7. Escamilla-Nuñez MC, Barraza-Villarreal A, Hernandez-Cadena L, Moreno-Macias H, Ramirez-Aguilar M, et al. (2008) Traffic-related air pollution and respiratory symptoms among asthmatic children, resident in Mexico City: the EVA cohort study. Respir Res 9: 74.
8. Epton MJ, Dawson RD, Brooks WM, Kingham S, Aberkane T, et al. (2008) The effect of ambient air pollution on respiratory health of school children: a panel study. Environ Health 7: 16.
9. PAHO (Pan American Health Organization) (2005) An Assessment of Health Effects of Ambient Air Pollution in Latin America and the Caribbean. Washington, DC.

10. Jacobson LSV, Hacon SS, Castro HA, Ignotti E, Artaxo P, et al. (2012) Association between fine particulate matter and the peak expiratory flow of schoolchildren in the Brazilian subequatorial Amazon: a panel study. Environ Res 117: 27–35.
11. Sisenando HA, Batistuzzo de Medeiros SR, Artaxo P, Saldiva PH, Hacon SS (2012) Micronucleus frequency in children exposed to biomass burning in the Brazilian Legal Amazon region: a control case study. BMC Oral Health 12: 6.
12. Ignotti E, Hacon SS, Junger WL, Mourão D, Longo K, et al. (2010) Air pollution and hospital admissions for respiratory diseases in the subequatorial Amazon: a time series approach. Cad Saude Publica 26(4):747–61.
13. Carmo CN, Hacon S, Longo KM, Freitas S, Ignotti E, et al. (2010) Association between particulate matter from biomass burning and respiratory diseases in the southern region of the Brazilian Amazon. Rev Panam Salud Publica 27(1):10–6.
14. Silva AMC, Mattos IE, Freitas SR, Longo KM, Hacon SS (2010) Material particulado ($PM_{2.5}$) de queima de biomassa e doenças respiratórias no sul da Amazônia brasileira. Rev Bras Epidemiol 13(2):337–51.
15. Mascarenhas MDM, Vieira LC, Lanzieri TM, Leal APPR, Duarte AF, et al. (2008) Poluição atmosférica devido à queima de biomassa florestal e atendimentos de emergência por doença respiratória em Rio Branco, Brasil – setembro, 2005. J Bras Pneumol 34: 42–46.
16. Fundação Instituto Brasileiro de Geografia e Estatística – IBGE (2010) Censo Demográfico, 2010. Available: http://www.censo2010.ibge.gov.br/. Accessed 2012 Dec.
17. Ignotti E, Hacon SS, Silva AMC, Junger WL, Castro HÁ (2007) Efeitos das queimadas na Amazônia: método de seleção dos municípios segundo indicadores de saúde. Rev Bras Epidemiol 10: 453–464.
18. Freitas SR, Longo KM, Silva Dias MAF, Silva Dias PL (2005) Emissões de queimadas em ecossistemas da América do Sul. Estudos Avançados 19: 167–185.
19. Lin W, Huang W, Zhu T, Hu M, Brunekreef B, et al. (2011) Acute respiratory inflammation in children and black carbon in ambient air before and during the 2008 Beijing Olympics. Environ Health Perspect 119(10):1507–12.
20. Malloy EJ, Morris JS, Adar SD, Suh H, Gold DR, et al. (2010) Wavelet-based functional linear mixed models: an application to measurement error-corrected distributed lag models. Biostatistics 11(3):432–452.
21. Gold DR, Damokosh AI, Pope CA 3rd, Dockery DW, McDonnell WF, et al. (1999) Particulate and ozone pollutant effects on the respiratory function of children in southwest Mexico City. Epidemiol 10(1):8–16.
22. Oliveira BF, Ignotti E, Artaxo P, Saldiva PH, Junger WL, et al. (2012) Risk assessment of PM2.5 to child residents in Brazilian Amazon region with biofuel production. Environ Health 11(1):64.
23. American Thoracic Society (1995) Standardization of Spirometry. Am J Respir Crit Care Med 152: 1107–1136.
24. ISAAC Steering Committee (2007) [Homepage in internet]. New Zealand. Available: http://isaac.auckland.ac.nz. Accessed 2012 Nov.
25. Snijders TAB, Bosker RJ (1999) Multilevel Analysis: An Introduction to basic and advanced multilevel modeling. SAGE Publications, pgs 261.
26. Schwartz J (2000) The Distributed Lag between Air Pollution and Daily Deaths. Epidemiol 11(3):320–326.
27. R Development Core Team (2012) R: A language and environment for statistical computing. Vienna: R Foundation for Statistical Computing.
28. Pinheiro J, Bates D, DebRoy S, Sarkar D, R Development Core Team (2012) nlme: Linear and Nonlinear Mixed Effects Models. R package version 3.1–104. Vienna: R Foundation for Statistical Computing.
29. Roemer W, Hoek G, Brunekreef B, Haluszka J, Kalandidi A, et al. (1998) Daily variations in air pollution and respiratory health in a multicentre study: the PEACE project. Pollution Effects on Asthmatic Children in Europe. Eur Respir J 12(6):1354–61.
30. Riguera D, Andre PA, Zanetta DMT (2011) Poluição da queima de cana e sintomas respiratórios em escolares de Monte Aprazível, SP. Rev Saúde Pública 45(5): 878–886.
31. Ward DJ, Ayres JG (2004) Particulate air pollution and panel studies in children: a systematic review. Occup Environ Med 61(4): e13.
32. Weinmayr G, Romeo E, De Sario M, Weiland SK, Forastiere F (2010) Short-Term Effects of PM_{10} and NO_2 on Respiratory Health among Children with Asthma or Asthma-like Symptoms: A Systematic Review and Meta-Analysis. Environ Health Perspect 118: 449–457.
33. Li S, Williams G, Jalaludin B, Baker P (2012) Panel Studies of Air Pollution on Children's Lung Function and Respiratory Symptoms: A Literature Review. J Asthma 49(9):895–910.
34. Xu Z, Etzel RA, Su H, Huang C, Guo Y, et al. (2012) Impact of ambient temperature on children's health: a systematic review. Environ Res 117: 120–31.
35. Davidson EA, de Araújo AC, Artaxo P, Balch JK, Brown IF, et al. (2012) The Amazon basin in transition. Nature 481(7381):321–8.

13

The Health Effects of a Forest Environment on Subclinical Cardiovascular Disease and Heath-Related Quality of Life

Tsung-Ming Tsao[1], Ming-Jer Tsai[1,2*]9, Ya-Nan Wang[1,2], Heng-Lun Lin[3], Chang-Fu Wu[4], Jing-Shiang Hwang[5], Sandy-H.J. Hsu[6], Hsing Chao[7], Kai-Jen Chuang[7], Charles- C.K. Chou[8], Ta-Chen Su[3*]9

1 The Experimental Forest, National Taiwan University, Nantou, Taiwan, **2** The School of Forestry and Resource Conservation, National Taiwan University, Taipei, Taiwan, **3** Department of Internal Medicine and Cardiovascular Center, National Taiwan University Hospital, Taipei, Taiwan, **4** Department of Public Health, National Taiwan University, Taipei, Taiwan, **5** Institute of Statistical Science, Academia Sinica, Taipei, Taiwan, **6** Laboratory Medicine, National Taiwan University Hospital, Taipei, Taiwan, **7** School of Public Health, Taipei Medical University, Taipei, Taiwan, **8** Research Center for Environmental Changes, Academia Sinica, Taipei, Taiwan

Abstract

Background: Assessment of health effects of a forest environment is an important emerging area of public health and environmental sciences.

Purpose: To demonstrate the long-term health effects of living in a forest environment on subclinical cardiovascular diseases (CVDs) and health-related quality of life (HRQOL) compared with that in an urban environment.

Materials and Methods: This study included the detailed health examination and questionnaire assessment of 107 forest staff members (FSM) and 114 urban staff members (USM) to investigate the long-term health effects of a forest environment. Air quality monitoring between the forest and urban environments was compared. In addition, work-related factors and HRQOL were evaluated.

Results: Levels of total cholesterol, low-density lipoprotein cholesterol, and fasting glucose in the USM group were significantly higher than those in the FSM group. Furthermore, a significantly higher intima-media thickness of the internal carotid artery was found in the USM group compared with that in the FSM group. Concentrations of air pollutants, such as NO, NO_2, NO_x, SO_2, CO, $PM_{2.5}$, and PM_{10} in the forest environment were significantly lower compared with those in the outdoor urban environment. Working hours were longer in the FSM group; however, the work stress evaluation as assessed by the job content questionnaire revealed no significant differences between FSM and USM. HRQOL evaluated by the World Health Organization Quality of Life-BREF questionnaire showed FSM had better HRQOL scores in the physical health domain.

Conclusions: This study provides evidence of the potential beneficial effects of forest environments on CVDs and HRQOL.

Editor: Una Macleod, Supportive care, Early Diagnosis and Advanced disease (SEDA) research group, United Kingdom

Funding: This study was financially supported by the Experimental Forest, National Taiwan University under grant no. A14 and no. 3 projects. National Science Council, Taiwan (NSC 101-2314-B-002-184-NY3) and National Taiwan University (NTU-CESRP-102R7620) also partially supported this study. The funders had no role in study design, data collection and analysis, decision to publish, or preparation of the manuscript.

Competing Interests: The authors have declared that no competing interests exist.

* Email: tachensu@ntu.edu.tw (TCS); tmj@ntu.edu.tw (MJT)

9 These authors contributed equally to this work.

Introduction

Cardiovascular diseases (CVDs), including coronary heart disease, strokes, heart failure, hypertension, rheumatic heart disease, myocardial infarction, cardiomyopathy, and other heart diseases, are the leading causes of morbidity and mortality worldwide [1]. In particular, the increasing rate of CVDs morbidity and mortality has globally become a major focus of public health policies and epidemiological studies. The causes of CVDs are very complex; consistent evidence from both epidemiological and experimental studies have demonstrated that environmental pollution, genetics, dietary habits, diabetes, hypertension, smoking habits, and psychosocial factors increase the risk of developing CVDs [2–8]. In particular, environmental pollution is a major factor affecting CVDs; a large number of studies have confirmed the association of air pollution with human diseases, which is not confined to illness and also involves a higher impact on CVDs morbidity and mortality [4]. We have provided evidence that exposure to air pollution may decrease pulse pressure in the general population [9]. Moreover, in middle-aged adults, the maximum intima-media thickness (IMT) of the carotid artery is associated with the individual's exposure to air pollution of PM_{10}, $PM_{2.5}$ absorbent, NO_2, and NO_x as derived from long-term air pollution exposure estimated by land-use regression models of the European Study of Cohorts for Air Pollution Effects [10]. Ambient PM air pollution from combustion sources including elemental carbon, organic carbon, and several metals are significantly associated with cardiovascular mortality [4,11–13]. Furthermore, significantly positive associations have been observed in cardiovascular hospitalizations and emergency department visits [10,13].

Silva et al. estimated that approximately 2,100,000 premature deaths each year occur due to air pollution [14].

Forest environments are associated with positive health effects compared to urban environments. Specifically, living in a forest environment lowers cholesterol concentrations, blood pressure (BP), and pulse rates, increases levels of natural killer cell activity, has beneficial effects on cardiovascular and metabolic parameters, improves psychological wellbeing, enhances human immune function, decreases sympathetic nerve activity, and enhances parasympathetic nerve activity in the human body [15–19]. Experimental studies in different countries have demonstrated that forest environments can better enhance mood and work performance compared with urban environments [20,21]. Thus, exposure to a forest environment may be therapeutic and may provide potential health benefits. Nevertheless, the studies above have investigated the effects of a forest environment on human physiological and psychological activities compared with an urban environment in experiments of short-term forest bathing trips. Therefore, studies regarding the long-term health effects for workers living in a forest environment compared with those living in an urban environment are still limited.

The objectives of this study were (1) to compare the measurements of cardiovascular parameters and health-related quality of life (HRQOL) in people working in forest and urban environments and (2) to demonstrate the forest environment has superior air quality compared with that in an urban environment.

Materials and Methods

Study design and population

This research has been approved by the 37th meeting (Jan. 30, 2013) of Research Ethics Committee of the National Taiwan University Hospital. The committee is organized under, and operates in accordance with, the Good Clinical Practice guidelines and governmental laws and regulations. The study participants have provided their written informed consent before received a series of detailed examinations and questionnaires.

Because the staff members in forest environment (The Experimental Forest, College of Bio-Resources and Agriculture, National Taiwan University) were 160 in total, excluding those of clinical diabetes and documented cardiovascular diseases, we determined sample size (n = 100) for healthy volunteers in forest and urban groups and sent 120 invitations to staff members for each environment for this study. All participants provided informed consent for our study before examination. Finally, the study participants included 114 urban staff members (USM) working in an urban environment in Taipei city and 107 forest staff members (FSM) working in an experimental forest of National Taiwan University, Nantou County, Taiwan. The FSM group or USM group included participants who had worked in the forest or urban environment for more than 1 year.

All participants received a series of detailed examinations and questionnaires, including baseline examination, oral glucose tolerance test (OGTT), HRQOL questionnaires, job stress assessment, cardio-ankle vascular index (CAVI), and carotid artery IMT assessment. The health examination and questionnaires (including QOL questionnaire and job stress assessments) were conducted during the initial spring season from January 31, 2013 to February 25, 2013 in Taipei and Nantou County, Taiwan.

Site descriptions of forest and urban environments

Forest and urban environmental samples were collected from the Xitou experimental forest of National Taiwan University, and an interior office of a commercial building in Taipei, respectively. The Xitou experimental forest, located in central Taiwan, covers 2,349 ha and ranges in elevation from 500 to 2,025 m. The mean relative humidity and temperature data from 2010 to 2012 were 89% and 17°C, respectively, as indicated by the Xitou monitoring station of the Central Weather Bureau [22]. The mean annual precipitation is 2,317 mm with a distinct dry season from October to April. Our sampling site is located at an elevation of 1,150 m near the meteorological station of Xitou experimental forest (23°40′ N 120°47′ E). The commercial building where urban measurements were taken is located near the Taipei main station in Taipei city within the north of Taiwan. The mean relative humidity and temperature data were 73% and 23°C, respectively, as provided by the Taipei monitoring station of the Central Weather Bureau [22]. Interior office samples were collected in the commercial building's 7th, 19th, and 21st floors where study subjects worked on.

Baseline examination and OGTT

BP was measured twice after at least 5 min of rest in a sitting position. The systolic BP used in the analyses was the average of two measurements. In both groups, a standard 75-g OGTT was performed after an overnight fast of at least 10 h, with measurements from blood samples at fasting via an ante-cubital vein, and at 30 min, 1 h, 90 min, and 2 h post-challenge plasma glucose *by using Finger Stick Capillary Dried* Blood Spots. Plasma glucose and serum levels of cholesterol, triglycerides, and low- and high-density lipoprotein cholesterol (LDL-C and HDL-C) were measured using an auto-analyzer (Toshiba, TBA-200FR; Toshiba, Tokyo, Japan). Small dense LDL-C also was measured by commercial kit (Denka Seken, Tokyo, Japan).

Exposure assessments

Instruments for forest and urban environmental monitoring were set in the Xitou experimental forest and in an interior office of a commercial building (indoor environment) in Taipei. Each monitoring system comprised a carbon monoxide analyzer (CO, Model 9841, Ecotech Inc., USA), an ozone analyzer (O_3, TECO 49, API 400A: based on the principle of UV absorption), a nitrogen oxide analyzer (NO, NO_2, NO_x, Model 9841, Ecotech Inc., USA), a sulfur dioxide analyzer (SO_2, Model 9850, Ecotech Inc., USA), a PM_{10} monitor (BAM1020, Met One Inc., Washington, USA), temperature, and relative humidity probe (Metone 083C, Met One Inc., Oregon, USA). CO, O_3, NO, NO_2, NO_x, SO_2, PM_{10}, temperature, and relative humidity measurements were recorded daily every minute during the examination period. $PM_{2.5}$ was collected by a Harvard impactor (Air Diagnostics and Engineering Inc., Maine, USA) [23]. Monitoring data of forest environment were collected by the environmental monitoring system from February 1 to 6, 2013. Indoor monitoring data of urban environment were collected by the same environmental monitoring system from February 20 to 25, 2013. Outdoor monitoring data of urban environment were provided by the Taipei air quality monitoring stations of the Taiwan Environmental Protection Administration.

CAVI and Ankle Brachial Index (ABI)

Arterial stiffness was measured using the CAVI and ABI. CAVI is measured from the electrocardiogram, phonocardiogram, brachial artery waveform, and ankle artery waveform and was calculated by CAVI-VaSera VS-1500N (Fukuda Denshi, Tokyo, Japan). ABI and 4-limb blood pressure were also measured. The vascular screening system was placed in a quiet and independent room in the Xitou natural education area. CAVI and ABI for each participant were conducted at least twice. The test-retest reliability

Table 1. General characteristics of urban and forest staff member groups.

Variables	Urban N=114	Forest N=107	*p* value
Age (years)	43.2±7.2	44.3±10.5	0.369
Male sex (%)	59.7	67.3	0.239
BMI (kg/m^2)	24.4±3.8	24.9±3.7	0.418
Waist circumference (cm)	83.3±10.1	84.0±10.6	0.662
Systolic BP (mmHg)	115.1±15.0	119.2±16.5	0.103
Diastolic BP (mmHg)	73.1±10.7	72.9±10.9	0.659
Hypertension (%)	20.2	23.3	0.566
Hypertension with medication (%)	9.7	4.7	0.154
Cholesterol (mg/dL)	225.3±33.3	207.1±30.1	<0.001
Cholesterol ≥200 mg/dL (%)	74.6	59.8	0.019
Hyperlipidemia with medication (%)	1.8	1.9	0.949
Triglycerides (mg/dL)	135.7±111.5	140.9±110.0	0.730
HDL-C (mg/dL)	60.7±15.5	57.6±14.4	0.123
LDL-C (mg/dL)	142.2±31.9	129.7±29.0	0.003
LDL-C ≥130 mg/dL (%)	60.5	46.7	0.040
Small dense LDL-C (mg/dL)	34.6±15.6	33.8±17.3	0.724
Small dense LDL-C/LDL-C	0.24±0.1	0.26±0.1	0.279
Coffee (%)	71.7	53.9	0.007
<360 ml/day	40.7	46.2	
≥360 ml/day	31.0	7.7	
Tea (%)	57.5	92.5	<0.001
<500 ml/day	46.0	77.4	
≥500 ml/day	11.5	15.1	
Alcohol drinking (%)	20.2	41.1	<0.001
1–2 times/week	11.4	16.8	
≥3 times/week	8.8	24.3	
Smoking (%)	27.2	39.3	0.057
ex-smoker	12.3	15.9	
current	14.9	23.4	
Exercise (%)	45.9	54.4	0.216

Continuous variables were expressed as mean ± SD and *t*-test were used to make comparisons. For categorical data, χ^2 test was used.

of CAVI and ABI in the Taiwanese population has been demonstrated as excellent [24].

Carotid arteries IMT assessments

Carotid atherosclerosis in the common carotid artery was assessed by measuring carotid IMT, using a high-resolution B-mode, GE Vivid i ultrasound system (Horten, Norway), equipped with a 3.5–10 MHz real-time B-mode scanner. Details concerning the methods of carotid IMT measurements have been reported previously [10,25]. In addition, a software package for vascular ultrasound was used. In general, duplex scanning refers to an ultrasound scanning procedure, recording both B-mode images of gray scale from the arteries of interest, and Doppler information about velocity and resistance in the relevant segments. The maximum and mean carotid IMT proximal to the carotid bifurcation, bulb, and internal carotid artery were measured bilaterally. CCA1 and CCA2 are points located at 0–1 cm and 1–2 cm, respectively, on CCA, distal to the carotid bifurcation. All scans were recorded on a digitalized memory system in DICOM format for subsequent off-line analysis. The carotid IMT measurement had excellent intraobserver coefficients of correlation reliability for maximum and mean carotid IMT with 0.976 and 0.988 at LCCA, and 0.970 and 0.973 at RCCA, respectively.

Quality of life assessment questionnaire

The brief Taiwanese version of the World Health Organization Quality of Life (WHOQOL-BREF) [26] questionnaire was completed when subjects were receiving a health examination. Moreover, self-assessment of perceived health status by visual analogue scale was collected. The Taiwanese version of the WHOQOL-BREF has been demonstrated to effectively show a significant difference between normal population controls and patients in Taiwan [26,27].

Statistical analyses

The general characteristics, OGTT, environmental factors, working hours, and job stress scores were compared between the groups FSM and USM. Continuous variables were expressed as the mean ± standard deviation and binary variables were expressed in percentage. We used t-test to test the group mean difference of the characteristic variable if the variable is continuous and normally distributed. The Chi-square test was applied for categorical data.

Data regarding the difference between FSM and USM including CAVI, ABI, and maximum and mean values of IMT at CCA, ICA, BULB, and mean IMT (mean values of 3 sites of IMT measurements) were compared using fitted regression models with the adjustment of age, gender, fasting sugar, SBP, BMI, LDL-C, and habits of smoking and alcohol. Work stress scores and HRQOL scores also were compared between two groups in regression models with the adjustment of age, gender, fasting sugar, SBP, BMI, LDL-C, smoking, and alcohol in multiple regression models. Data analysis was performed with SAS statistical software (version 11.1, SAS Institute Inc., Cary, NC, USA).

Results

General characteristics among participants in the FSM and USM are summarized in Table 1. The mean age of the FSM group was 44.3±10.5 years and was greater than the USM group. The FSM group had more male participants than the USM group (67.3% vs. 59.7%). The USM group had higher cholesterol and LDL-C levels than the FSM group. Coffee, tea, and drinking habits in the two groups had a significant difference. In particular, tea drinking in the FSM group (92.5%) was higher than that in the USM group (57.5%).

OGTT results are presented in Table 2. Mean fasting plasma glucose in the USM group was 104.1±20.1 mg/dL and was higher than that in the FSM group (99.1±14.2 mg/dL). After OGTT, the 2-h plasma glucose concentration of the USM and FSM groups were 133.8±41.5, and 129.0±40.4 mg/dL, respectively. There was a significantly higher prevalence of impaired glucose tolerance and pre-diabetes mellitus in the USM group.

The working hours, job stress scores, and WHOQOL-BREF domain scores are presented in Table 3. The mean of working hours per week was significantly higher in the FSM group (51.4±11.5 h) compared to the USM group (45.5±11.1 h). No significant difference was observed between scores of job stress and work support between the two groups. Furthermore, according to mean WHOQOL-BREF domain scores, the FSM group scored higher than the USM group in all four domains. In particular, after controlling associated covariates in a regression model, the score of physical health was significantly higher in the FSM group than the USM group (29.6±3.6 vs. 29.0±4.2).

The mean concentration of air pollutants (SO_2, NO, NO_2, NO_x, CO, O_3, and PM_{10}), temperature, and relative humidity are presented in Table 4. Considering the results of the monitoring analysis, there were significant differences in SO_2, NO, NO_2, NO_x, CO, PM_{10}, temperature, and relative humidity factors between the urban (indoors and outdoors) and forest environments. The forest environment had higher O_3 concentration than the indoor urban environment, however there were no significant differences between of forest environment (23.1 ppb) and outdoor urban environments (25.4 ppb). The mean concentration of $PM_{2.5}$ in the forest environment was lower than in the urban environment (indoor and outdoor). Despite a non-significant trend for $PM_{2.5}$ between the forest and indoor urban environments, the mean concentration of $PM_{2.5}$ in the outdoor urban environment was 37.2±24.0 μg/m^3, which was higher than in the forest environment (7.2±3.9 μg/m^3), with a significant difference ($p <$ 0.028).

The CAVI and ABI among participants of the FSM and USM groups are presented in Table 5. Although the mean CAVI values of the USM group (7.48±0.81) were slightly higher than those of the FSM group (7.41±1.01), there was no significant difference from the fitted regression models with the adjustment of age, gender, fasting sugar, systolic BP, BMI, LDL-C, smoking, and alcohol consumption. However, the mean ABI was 1.13±0.07 in the USM group, which was significantly higher than that in the FSM group (1.06±0.08). Table 6 shows the carotid IMT measurements among participants of both groups. A significantly lower mean carotid IMT at ICA was observed in the FSM group than that in the USM group. The results of maximum and mean of all carotid IMT measurements in the USM group were higher than in the FSM group.

Discussion

This is the first study to demonstrate the health effects of forest environment by comparing traditional cardiovascular risk factors, noninvasive cardiovascular assessments, and detailed environmental monitoring simultaneously in middle-aged workers living in forest and urban environments. Studies have demonstrated that psychosocial and environmental factors play an important role in

Table 2. Oral glucose tolerance test of urban and forest staff member groups.

	Urban	Forest	*p* value[a]
	N = 114	N = 107	
Fasting plasma glucose (mg/dL)	104.1±20.1	99.1±14.2	0.031
30 min (mg/dL)	168.4±39.1	161.7±33.9	0.177
60 min (mg/dL)	165.3±49.4	162.2±47.8	0.641
90 min (mg/dL)	149.7±45.3	142.2±46.6	0.225
120 min (mg/dL)	133.8±41.5	129.0±40.4	0.380
HbA1C (%)	5.6±0.7	5.6±0.6	0.710
OGTT Diabetes mellitus (%)	12.3	14.0	0.702
Impaired glucose tolerance (%)	26.3	11.2	0.004
Pre-diabetes mellitus (%)	19.3	6.5	0.005

[a]Comparisons of means between the two groups were based on t-statistic for continuous variables and Chi-square test for the categorical variables.

Table 3. Work hours, job stress scores (Job Content Questionnaires), and WHOQOL-BREF domain scores of urban and forest staff member groups.

	Urban N=114	Forest N=107	*p* value[a]
Work hours per week	45.5±11.1	51.4±11.5	<.0001
Job stress			
Control	25.9±3.2	25.8±3.3	0.622
Demand	15.8±2.5	15.3±2.2	0.131
Boss support	12.0±1.6	11.8±1.8	0.479
Colleague Support	12.3±1.2	12.1±1.2	0.4255
Work insecurity	14.9±2.5	14.2±3.0	0.084
Work place justice	24.1±3.1	24.6±3.5	0.2981
WHOQOL-BREF, domains			
Physical health	29.0±4.2	29.6±3.6	0.0069
Psychological	20.5±3.4	21.7±2.8	0.1656
Social relationship	13.5±2.1	14.2±2.1	0.0974
Environment	34.5±5.2	34.5±4.5	0.7888

[a]The test was based on regression models adjusted for age, gender, fasting sugar, systolic BP, BMI, LDL-C, smoking, and alcohol.

predicting cardiovascular health [4,20]. In this study, in addition to the higher levels of cholesterol and glucose and a prevalence of pre-diabetes, environmental factors may be accountable for the significant differences observed between the two groups. The statistically significant differences observed in this study imply that subclinical CVDs markers of carotid IMT are better in subjects working and living in a forest environment compared to those working and living in urban environments.

In middle-aged adults, the levels of blood pressure, cholesterol, fasting glucose, and carotid IMT increased with age and were higher in males [28]. The carotid IMT of the FSM group was better than those of the USM group, despite having a greater mean age and a higher percentage of males. These results provide evidence of the beneficial health effects of living in a forest environment compared to living in an urban environment. The prevalence of hypercholesterolemia and pre-diabetes in the USM group was higher than in the FSM group. Higher glucose and LDL-C levels in the blood may lead to endothelial cell lesion and damage by glycation, allowing easy access of LDL-C through arterial endothelial cells, resulting in developing CVDs with vascular stenosis and obstruction.

Urban air pollution associated with CVDs morbidity and mortality has been well documented [4,29,30], and the scientific statement by the American Heart Association suggested that the inhalation of air pollution may stimulate and trigger the development of atherosclerosis [4]. Our study showed a significantly lower concentration of gaseous air pollutants (i.e., NO, NO_2, NO_x, SO_2, and CO) in forest environments compared to urban (indoor and outdoor) environments. In a previous study, we have proved that short-term O_3 exposure had a significant adverse

Table 4. Environmental monitoring in forest and urban environments.

	Forest environment	Urban environment		P_1 value[a]	P_2 value[b]
		Indoor	Outdoor		
	N[c]=144	N=144	N=144		
SO_2 (ppb)	2.0±0.1	3.8±1.4	2.7±1.4	<.001	<.001
NO (ppb)	2.5±1.0	20.5±20.4	6.3±6.2	<.001	<.001
NO_2 (ppb)	3.1±1.6	10.7±2.8	25.4±11.8	<.001	<.0001
NO_x (ppb)	5.6±2.0	31.3±20.8	31.7±16.5	<.001	<.001
CO (ppm)	0.3±0.1	1.3±0.8	0.6±0.3	<.001	<.001
O_3 (ppb)	23.1±13.4	1.9±0.8	25.4±16.1	<.001	0.137
Temperature (°C)	14.5±3.8	22.0±1.3	18.0±2.6	<.001	<.001
Relative humidity	87.3±12.5	55.2±3.7	78.3±9.1	<.001	<.001
PM_{10} (μg/m^3)	20.3±9.2	15.9±4.3	48.0±29.4	<.001	<.001

[a]P_1 value corresponds to t-test on difference between the XiTou and Urban site (Indoor).
[b]P_2 value corresponds to t-test on difference between the XiTou and Wanhua site (Outdoor) of EPA, Taipei, Taiwan.
[c]N corresponds to the sample size of hourly average data.

Table 5. Cardio-ankle vascular index (CAVI) and ankle brachia index (ABI) of urban and forest staff member groups.

	Urban	Forest	*p* value[a]
	N = 114	N = 107	
CAVI			
Right	7.48±0.89	7.44±1.04	0.342
Left	7.47±0.76	7.37±1.00	0.177
Mean	7.48±0.81	7.41±1.01	0.242
ABI			
Right	1.129±0.08	1.061±0.08	<0.001
Left	1.131±0.08	1.061±0.08	<0.001
Mean	1.130±0.07	1.060±0.08	<0.001

[a]Tests of difference in each mean index between urban and forest groups using regression models with adjustment of age, gender, fasting sugar, systolic BP, BMI, LDL-C, smoking and alcohol drinking habit.

effect on aortic stiffness by CAVI measurement in young mail-carriers [31]. Since the O_3 level in forest environment were not lower than that in urban environment, the potential hazardous effect arose from O_3 exposure should be taken into account. This partially explains why there are no significant differences in CAVI measurement between the two groups in our study, even though all air quality indicators other than O_3 were better in forest environment. Because CAVI measurements can be affected by short-term air pollution exposure [31,32], the beneficial health effects of lower levels of PM and other gaseous components might be attenuated by the relative higher O_3 concentration in forest environment. There is a need to continuously monitor the concentration of O_3 and the changes in different seasons. As for PM_{10} and $PM_{2.5}$, the mean concentration in the outdoor urban environment was also higher than those in the forest environment.

It is commonly recognized that PM consists of soil dust, nanoparticles, industrial emissions, sea salts, smog particles, and combustion particles from vehicle sources [20]. Within the past 10 years we have proved that air pollution in the Taipei metropolis conferred short-term adverse cardiovascular effects, such as impaired heart rate variability, impaired coagulation markers, increased inflammation indices, oxidation stress, and increased aortic stiffness, in susceptible patients as well as in healthy subjects [5,9,31–35]. Our recent study, echoing findings of studies in the US [36] and Europe [37] found that long-term residential air pollution may increase subclinical atherosclerosis indexed by carotid IMT. Furthermore, the current study confirms that a forest environment with less air pollution may benefit cardiovascular health in subclinical atherosclerosis.

The concept of QOL complements the WHO definition of health as "not only the absence of disease and infirmity but also the presence of physical, mental, and social well-being" [38]. HRQOL is the primary concern of healthcare professionals and is becoming an important health outcome indicator. Therefore, patient's self-reported outcomes are being increasingly emphasized in recent years [39–45] and have become an integral component of several ongoing clinical trials [42,43]. The Taiwanese version of WHOQOL-BREF has been demonstrated with good reliability

Table 6. Carotid intima-media thickness (IMT) of the urban and forest staff member groups.

	Urban	Forest	*p* value[a]
	N = 114	N = 107	
Carotid IMT, mm			
Common carotid artery			
Max	0.674±0.115	0.682±0.132	0.280
Mean	0.552±0.093	0.562±0.114	0.432
Internal carotid artery			
Maximum	0.597±0.095	0.582±0.104	0.066
Mean	0.498±0.076	0.481±0.083	0.033
Bulb			
Maximum	0.702±0.137	0.686±0.159	0.169
Mean	0.582±0.103	0.567±0.129	0.150
IMT maximum[b]	0.658±0.092	0.650±0.110	0.047
IMT mean[b]	0.544±0.072	0.536±0.090	0.046

[a]Based on regression models with the adjustment of age, gender, fasting sugar, systolic BP, BMI, LDL-C, smoking and alcohol drinking habit.
[b]IMT mean and maximum are the mean and maximum values of combining three sites of carotid arteries.

and validity in Taiwan [26,27,41,45]. In the present study, the FSM group was higher than the USM group in all four domain scores. In particular, the score of physical health (including pain, energy, sleep, mobility, activity, medication, and work) of the FSM group was significantly higher than those of the USM group. The results indicated the long-term health effects of a forest environment on CVDs as well as on HRQOL. HRQOL measurement in the USM group provides a clear understanding of the well-being of participants. In addition, viewing aggregated health and environment outcomes data helps professionals address environmental risk factors and provide beneficial health information on CVDs and HRQOL for most of the working population living in urban environments.

This study has several strengths and limitations, as follows: this is the first study to demonstrate the long-term health effects of a forest environment on subclinical CVDs and HRQOL compared with those of an urban environment. Even though the study indicated the potential beneficial health effects of living in a forest environment on CVDs and HRQOL, we indeed cannot infer the better HRQOL and subclinical CVDs in workers living in forest environment because of their better air quality by real-time monitoring of air pollutants in forest compared with those of an urban environment.

First, the CV effects of changes in seasons have not been considered in this paper. Seasonal variation of cardiovascular events may be linked to the changes of cardiovascular and endocrine/metabolic markers in different seasons [34]. However, the carotid IMT demonstrated a long-term surrogate outcome of subclinical atherosclerosis, which provides clear evidence of potential cardiovascular beneficial effects of forest environments. Second, this study included multi-discipline professionals such as healthcare providers (cardiologist, medical laboratory, and case managers), public health professionals, forest environment specialists, statisticians, and atmospheric science specialists. Because of the different occupational characteristics between the FSM and USM groups, the job content questionnaire and working conditions were compared to clarify the possible work-related factors that may bias the outcomes. Third, the beneficial health factors of a forest environment, induced by agents such as phytoncide and negative ion, have not been monitored and assessed in this paper. Information regarding the specific health benefits of organic compounds found in forest environments and on the types of mechanisms mediating the effects of phytoncide on the cardiovascular system are an important emerging area of public health and environmental sciences.

Fourth, the sample size is small, and we may not have had the statistical power to detect a significant effect on the CAVI and OGTT after 2 h post-challenge plasma glucose, and job stress with different variables. Therefore, increasing the number of samples for detailed studies of their significant differences is very important in the future.

Even though the preliminary results report the first detailed survey and environmental monitoring during early spring. The ongoing study of this project will explore the seasonal changes in the health effects of both FSM and USM groups by 4-seasons' field environmental monitoring, and follow-up health examinations to corroborate and provide important evidence on the health effects of a natural environment as an alternative therapeutic option for CVDs.

In conclusion, this study indicated the potential health effects on subclinical marker of cardiovascular disease, in terms of CIMT and subjective HRQOL in workers living in forest environment. A large-scale and cohort study in peoples living in forest comparing to living in urban environments should be warranted.

Acknowledgments

We also express our deepest gratitude to the Xitou and Fenghuang natural education area for supporting the experimental investigation of the Experimental Forest, National Taiwan University.

Author Contributions

Conceived and designed the experiments: MJT CFW JSH CCKC TCS. Performed the experiments: TMT CFW YNW HLL TCS. Analyzed the data: JSH TCS. Contributed reagents/materials/analysis tools: HC KJC TCS SHJH. Wrote the paper: TMT CFW MJT.

References

1. Mathers CD, Saloman SJ, Ezati M (2006) Global burden of disease and risk factors. New York, NY: Oxford University Press.
2. Kim C, Beckles GL, MSc MD (2004) Cardiovascular disease risk reduction in the behavioral risk factor surveillance system. Am J Prev Med 27: 1–7.
3. Lavie CJ, Milani RV, Ventura HO (2009) Obesity and cardiovascular disease: Risk factor, paradox, and impact of weight loss. J Am College Cardiovasc 53: 1925–1932.
4. Brook RD, Rajagopalan S, Pope III CA, Brook JR, Bhatnagar A, et al. (2010) American Heart Association council on epidemiology and prevention, council on the kidney in cardiovascular disease, and council on nutrition, physical activity and metabolism. Particulate matter air pollution and cardiovascular disease: an update to the scientific statement from the American Heart Association. Circulation 121: 2331–2378.
5. Su TC, Chan CC, Liau CS, Lin LY, Kao HL, et al. (2006) Urban air pollution increases plasma activator inhibitor-1 and fibrinogen levels in susceptible patients. Eur J Cardiovasc Prev Rehabil 13: 849–852.
6. Vaccarino V, Bremner JD (2012) Psychiatric and behavioral aspects of cardiovascular disease RO Bonow, DL Mann, DP Zipes, P Libby (Eds.), Braunwald's heart disease - a textbook of cardiovascular medicine (9th ed.), Elsevier-Saunders, Philadelphia, PA.
7. Mulle JG, Vaccarino V (2013) Cardiovascular disease, psychosocial factors, and genetics: The case of depression. Pro Cardiovasc Diseases 55: 557–5562.
8. Elderon L, Whooley MA (2013) Depression and cardiovascular disease. Pro Cardiovasc diseases 55: 511–523.
9. Chen SY, Su TC, Lin YL, Chan CC (2012) Short-term effects of air pollution on pulse pressure among nonsmoking adults. Epidemiology 23: 341–348.
10. Su TC, Hwang JJ, Shen YC, Chan CC (2014) Carotid intima-media thickness is associated with long-term exposure to traffic-related air pollution in middle-aged adults. Environ Health Perspect (In press).
11. Ito K, Mathes R, Ross Z, Nádas A, Thurston G, et al. (2011) Fine particulate matter constituents associated with cardiovascular hospitalizations and mortality in New York City. Environ Health Perspect 119: 467–473.
12. Ostro B, Lipsett M, Reynolds P, Goldberg D, Hertz A, et al. (2010) Long-term exposure to constituents of fine particulate air pollution and mortality: results from the California Teachers Study. Environ Health Perspect 118: 363–369.
13. Brunekreef B, Forsberg B (2005) Epidemiological evidence of effects of coarse airborne particles on health. Eur Respir J 26: 309–18.
14. Silva R, West JJ, Zhang Y, Anenberg SC, Lamarque JF, et al. (2013) Global premature mortality due to anthropogenic outdoor air pollution and the contribution of past climate change. Environ Res Letters 8: doi:10.1088/1748-9326/8/3/034005.
15. Li Q, Morimoto K, Nakadai A, Inagaki H, Katsumata M, et al. (2007) Forest bathing enhances human natural killer activity and expression of anti-cancer proteins. Int J Immunopathol Pharmacol 20: 3–8.
16. Li Q, Morimoto K, Kobayashi M, Inagaki H, Katsumata M, et al. (2008) Visiting a forest, but not a city, increases human natural killer activity and expression of anti-cancer proteins. Int J Immunopathol Pharmacol 21: 117–127.
17. Li Q, Morimoto K, Kobayashi M, Inagaki H, Katsumata M, et al. (2008) A forest bathing trip increases human natural killer activity and expression of anti-cancer proteins in female subjects. J Biol Regul Homeost Agents 22: 45–55.
18. Lee J, Park BJ, Tsunetsugu Y, Kagawa T, Miyazaki Y (2009) Restorative effects of viewing real forest landscapes, based on a comparison with urban landscapes. Scandinavian J Forest Res 24: 227–234.
19. Lee J, Park BJ, Tsunetsugu Y, Ohira T, Kagawa T, et al. (2011) Effect of forest bathing on physiological and psychological responses in young Japanese male subjects? Public Health 125: 93–100.
20. Morita E, Fukuda S, Nagano J, Hamajima N, Yamamoto H, et al. (2006) Psychological effects of forest environments on healthy adults: Shinrin-yoku (forest-air bathing, walking) as a possible method of stress reduction. Public Health 121: 54–63.

21. Shin WS (2007) The influence of forest view through a window on job satisfaction and job stress. Scand J Forest Res 22: 248–53.
22. Central Weather Bureau (CWB). 2010–2012. Monthly climate data Available: http://www.cwb.gov.tw/V7/climate/monthlyData/mD.htm.
23. Chow J (1995) Measurement methods to determine compliance with ambient air quality standards for suspended particles. J Air Waste Manage 45: 320–382.
24. Huang CL, Chen MF, Jeng JS, Lin LY, Wang WL, et al. (2007) Postchallenge hyperglycaemic spike associate with arterial stiffness. Int J Clin Pract 61: 397–402.
25. Lin CY, Wen LL, Lin LY, Liend GW, Chien KL, et al. (2013) Association between levels of serum perfluorinated chemicals and carotid artery intima-media thickness in adolescents and young adults. Int J Cardiol 168(4): 3309–3316.
26. Yao G, Chung CW, Yu CF, Wang JD (2002) Development and verification of validity and reliability of the WHOQOL-BREF Taiwan version. J Formos Med Assoc 101: 342–351.
27. Horng YS, Hwang YH, Wu HC, Liang HW, Jang Y, et al. (2005) Predicting health-related quality of life in patients with low back pain. Spine 30: 551–555.
28. Su TC, Chien KL, Jeng JS, Chen MF, Hsu HC, et al. (2012) Age- and gender-associated determinants of carotid intima media-media thickness: A community-based study. J Atheroscler Thromb 19: 872–880.
29. Pope III CA, Burnett RT, Thurston GD, Thun MJ, Calle EE, et al. (2004) Cardiovascular mortality and long-term exposure to particulate air pollution: epidemiological evidence of general pathophysiological pathways of disease. Circulation 109: 71–77.
30. Pope III CA, Muhlestein JB, May HT, Thun MJ, Calle EE, et al. (2006) Ischemic heart disease events triggered by short-term exposure to fine particulate air pollution. Circulation 114: 2443–2448.
31. Wu CF, Kuo IC, Su TC, Li YR, Lin LY, et al. (2010) Effects of personal exposure to particulate matter and ozone on arterial stiffness and heart rate variability in healthy adults. Am J Epidemiol 12: 1299–1309.
32. Wu CF, Li YR, Kuo IC, Hsu SC, Lin LY, et al. (2012) Investigating the association of cardiovascular effects with personal exposure to particle components and sources. Sci Total Environ 431: 176–182.
33. Chen SY, Chan CC, Lin YL, Hwang JS, Su TC (2014) Fine particulate matter results in hemodynamic changes in subjects with blunted nocturnal blood pressure dipping. Environ Res. 131: 1–5.
34. Su TC, Wu CF, Chan CC, Chen MF (2011) C-reactive protein and particulate matter predict plasma fibrinogen levels. Int J Cardiol 153: 100–101.
35. Chuang KJ, Chan CC, Su TC, Lee CT, Tang CS (2007) Urban air pollution, inflammation, oxidative stress, coagulation and autonomic dysfunction in young adults. Am J Respir Crit Care Med 176: 370–376.
36. Künzli N, Jerrett M, Garcia-Esteban R, Basagaña X, Beckermann B, et al. (2010) Ambient air pollution and the progression of atherosclerosis in adults. PLoS One 5: e9096.
37. Bauer M, Moebus S, Möhlenkamp S, Dragano N, Nonnemacher M, et al. (2010) HNR Study Investigative Group. Urban particulate matter air pollution is associated with subclinical atherosclerosis: results from the HNR (Heinz Nixdorf Recall) study. J Am Coll Cardiol 56: 1803–1808.
38. The World Health Organization (1955) In: World Health Organization. Handbook of Basic Documents. 5th ed. Geneva: Palais des Nations, 3–52 p.
39. Strine TW, Okoro CA, Chapman DP, Balluz LS, Ford ES, et al. (2005) Health-related quality of life and health risk behaviors among smokers. Am J Prev Med 58: 182–187.
40. Groessl EJ, Kaplan RM, Jack RW, Katula JA, King AC, et al. (2007) Health-related quality of life in older adults at risk for disability. Am J Prev Med 33: 214–218.
41. Chiu MH, Hwang HF, Lee HD, Chien DK, Chen CY, et al. (2012) Effect of fracture type on health-related quality of life among older women in Taiwan. Arch Phys Med Rehabil 93: 512–519.
42. Atherton PJ, Sloan JA, Atherton PJ, Sloan JA (2006) Rising importance of patient-reported outcomes. Lancet Oncol 7: 883–884.
43. Leidy NK, Beusterien K, Sullivan E, Richner R, Muni NI (2006) Integrating the patient's perspective into device evaluation trials. Value Health 9: 394–401.
44. Jia H, Lubetkin EI (2010) Obesity-related quality-adjusted life years lost in the U.S. from 1993 to 2008. Am J Prev Med 39: 220–227.
45. Hwang HF, Liang WM, Chiu YN, Lin MR (2003) Suitability of the WHOQOL-BREF for community-dwelling older people in Taiwan. Age Ageing 32: 593–600.

14

Increased Risk of Dementia in Patients Exposed to Nitrogen Dioxide and Carbon Monoxide: A Population-Based Retrospective Cohort Study

Kuang-Hsi Chang[1,2], Mei-Yin Chang[3], Chih-Hsin Muo[4], Trong-Neng Wu[1], Chiu-Ying Chen[1], Chia-Hung Kao[5,6]*

1 Department of Public Health, China Medical University, Taichung, Taiwan, **2** Department of Medical Research, Taichung Veterans General Hospital, Taichung, Taiwan, **3** Department of Medical Laboratory Science and Biotechnology, School of Medical and Health Sciences, Fooyin University, Kaohsiung, Taiwan, **4** Management Office for Health Data, China Medical University Hospital, Taichung, Taiwan, **5** Graduate Institute of Clinical Medical Science, College of Medicine, China Medical University, Taiwan, **6** Department of Nuclear Medicine and PET Center, China Medical University Hospital, Taichung, Taiwan

Abstract

Background: The air pollution caused by vehicular emissions is associated with cognitive decline. However, the associations between the levels of nitrogen dioxide (NO_2) and carbon monoxide (CO) exposure and dementia remain poorly defined and have been addressed in only a few previous studies.

Materials and Methods: In this study, we obtained data on 29547 people from the National Health Insurance Research Database (NHIRD) of Taiwan, including data on 1720 patients diagnosed with dementia between 2000 and 2010, and we evaluated the risk of dementia among four levels of air pollutant. Detailed data on daily air pollution were available from January 1, 1998 to December 31, 2010. Yearly average concentrations of pollutants were calculated from the baseline to the date of dementia occurrence, withdrawal of patients, or the end of the study, and these data were categorized into quartiles, with Q1 being the lowest level and Q4 being the highest.

Results: In the case of NO_2, the adjusted hazard ratios (HRs) of dementia for all participants in Q2, Q3, and Q4 compared to Q1 were 1.10 (95% confidence interval (CI), 0.96–1.26), 1.01 (95% CI, 0.87–1.17), and 1.54 (95% CI, 1.34–1.77), and in the case of CO, the adjusted HRs were 1.07 (95% CI, 0.92–1.25), 1.37 (95% CI, 1.19–1.58), and 1.61 (95% CI, 1.39–1.85).

Conclusion: The results of this large retrospective, population-based study indicate that exposure to NO_2 and CO is associated with an increased risk of dementia in the Taiwanese population.

Editor: Gianluigi Forloni, "Mario Negri" Institute for Pharmacological Research, Italy

Funding: The study was supported in part by China Medical University (CMU102-BC-2), Taiwan Ministry of Health and Welfare Clinical Trial and Research Center of Excellence (MOHW103-TDU-B-212-113002), Taiwan Ministry of Health and Welfare Cancer Research Center for Excellence (MOHW103-TD-B-111-03), and International Research-Intensive Centers of Excellence in Taiwan (I-RiCE) (NSC101-2911-I-002-303). The funders had no role in study design, data collection and analysis, decision to publish, or preparation of the manuscript.

Competing Interests: The authors have declared that no competing interests exist.

* Email: d10040@mail.cmuh.org.tw

Introduction

Ambient air pollution includes solid and gaseous pollutants [1,2]. Most of the studies that have investigated the effects of pollutants on cognitive functions have examined the influence of solid pollutants [3–8]. However, exposure to ambient gaseous pollutants such as nitrogen dioxide (NO_2) is known to increase the risk of cerebrovascular and neurodegenerative diseases and ischemic stroke [9–12]. Cerebrovascular disease is the principal contributor to dementia [13,14], and Alzheimer's disease (AD) is the most common neurodegenerative disease. Moreover, a population-base study reported that dementia often developed after the occurrence of an ischemic stroke [15]. Several previous studies have suggested negative associations between NO_2 exposure and cognitive development in children, including preschool children [16–18], and animal studies have indicated that NO_2 exposure inhibits the recovery of nerve function after a stroke [19,20]. In addition, one animal study reported that nitration can induce beta-amyloid aggregation and plaque formation [21]; beta-amyloid aggregation is a pathologic hallmark of AD. However, a literature search indicated that only a few studies have been conducted to address the link between NO_2 exposure and cognitive function in adults. In a recent study conducted on 1496 middle-aged people living in Los Angeles, no statistically significant correlation was detected between the level of NO_2 exposure and cognitive functions [22]. Therefore, we conducted a retrospective cohort study to determine the association between NO_2 and dementia risk. Furthermore, in this study, we evaluated the influence of carbon monoxide (CO), because acute CO poisoning may cause headache, nausea, malaise, and fatigue [23], and chronic CO exposure has been linked to depression, confusion, memory loss, and cognitive decline [24,25]. Comparison between this study with other environmental study of Taiwan NHRID, the main difference is the residential area definition. In previous studies, the residential area is as the insurance area [26]. In present study, we defined the residential areas as the location of clinics which subjects sought treatment for acute upper respiratory infections.

Materials and Methods

Data sources and study population

In March 1995, the Taiwan National Health Insurance (NHI) program, which is a single-payer, compulsory social insurance system that has provided insurance coverage to almost every citizen in Taiwan, was established. The NHI covered approximately 99% of the 22.96 million citizens in Taiwan at the end of 2007 [27]. To protect patient privacy, the data on patient identities are encrypted in the National Health Insurance Research Database (NHIRD), and the database is accessible to researchers and the public in Taiwan. In this study, we used a subset of the NHIRD data containing comprehensive health-care data, including files on ambulatory care claims, inpatient claims, and prescriptions received by 1000000 people who were randomly selected from all insured beneficiaries. These data files can be linked through an encrypted but unique personal identification number and, thus, provide a longitudinal medical history of each patient. The health status of each person was identified according to the International Classification of Disease, Ninth Revision, Clinical Modification (ICD-9-CM).

Exposure assessment

Across Taiwan, 74 ambient air quality monitoring stations are located based on population density. Air quality data are maintained by Taiwan Environmental Protection Administration. [28]. A database containing daily NO_2 and CO concentrations measured at the monitoring stations was available for the period from January 1, 1998 to December 31, 2010. The people included in this study were assigned pollutant-exposure values based on the data obtained from the monitoring station present in the residential district in which the clinic where the people most frequently sought treatment for acute upper respiratory infection was located (ICD-9-CM Code 460). Yearly average concentrations of pollutants were calculated from the baseline to the date of dementia occurrence, the withdrawal of patients, or the end of the study period, and the data were categorized into quartiles.

Study patients

We identified 29547 people who were aged 50 years or older and for whom estimable air pollution data were available, but who did not present a history of head injury (ICD-9-CM Codes 800.804, 850.854.1, 310.2, and 959.01), stroke (ICD-9-CM Codes 430–438), or dementia (ICD-9-CM Codes 290.0–290.4, 294.1, and 331.0) before 2000.

Ethics statement

Because identification numbers of patients had been encrypted, patient consent was not required for this study. This study was approved by the Research Ethic Committee at China Medical University (CMU-REC-101-012). The committee waived the requirement for consent.

Statistical analysis

We used x^2 tests to examine the distributions of sex, monthly income (New Taiwan Dollar<14 400, 14 400–18 300, 18 301–21 000, and >21 000), diabetes (DM, ICD-9-CM Code 250), ischemic heart disease (IHD, ICD-9-CM Codes 410–414), hypertension (HT, ICD-9-CM Codes 401–405),chronic obstructive pulmonary disease(COPD, ICD-9-CM Codes 490–496), alcoholism (ICD-9-CM Codes 303.305.0andV113),and the quartiles of NO_2 concentration (ppb; <6652.3, 6652.3–8349.0, 8349.1–9825.5,>9825.5) and CO concentration (ppm; <196.2, 196.2–241.6, 241.7–296.9, >296.9). A one-way analysis of variance (ANOVA) was performed to compare the age among the quartiles of NO_2 and CO concentrations. We calculated the incidence density rates of dementia in person-years in each quarter stratified according to sex. The incidence rate ratio (IRR) was estimated using a Poisson regression. Univariate and multivariate Cox proportional hazard regression analyses were performed to calculate the hazard ratios (HRs) and 95% confidence intervals (CIs) of the risk of dementia in association with pollutant levels. Multiple models were tested by controlling for age, sex, monthly income, DM, HT, IHD, COPD, alcoholism, and urbanization. Plots of the Kaplan-Meier analysis were used to determine the probability of people remaining without dementia, and the log-rank test was used to evaluate the differences among quartiles of pollutant concentrations. All analyses were performed using SAS 9.2 software (SAS Institute Inc., Cary, NC, USA), and the Kaplan-Meier survival curve was plotted using the Statistical Package for the Social Sciences (Version 15.1; SPSS Inc, Chicago, IL, USA). All tests were considered statistically significant when two-tailed P values were <.05.

Results

We obtained a total of 29547 and 29537 data on daily NO_2 and CO exposure, respectively. Dementia was not present at the baseline (2000), and 1720 people developed dementia after follow-up (yearly CO data were available for 1718 people). We categorized the NO_2 and CO levels into quartiles, with Q1 being the lowest level and Q4 being the highest. The people included in this study had a mean age of 61.4 years (SD 8.5 y). In both the NO_2 and CO groups, the highest level of the quartiles was associated with the people being slightly younger, more frequently earning a high monthly income, and living in a highly urbanized residential area, but less frequently exhibiting IHD and COPD compared with other quartiles (Tables 1 and 2). Table 3 shows the associations between the gaseous pollutant levels and the risk of dementia. Among the quartiles Q1, Q2, Q3, and Q4 of NO_2 in all patients, the IRRs in Q2, Q3, and Q4 compared with that in Q1 were 1.05, 0.90, and 1.35, and the adjusted HRs of dementia were 1.10 (95% CI, 0.96–1.26), 1.01 (95% CI, 0.87–1.17), and 1.54 (95% CI, 1.34–1.77), respectively. Among men, we determined that the IRRs in Q2, Q3, and Q4 compared with that in Q1 were 1.08, 0.79, and 1.28, and the adjusted HRs were 1.16 (95% CI, 0.95–1.43), 0.89 (95% CI, 0.71–1.11), and 1.52 (95% CI, 1.23–1.88), respectively. Among women, the IRRs in Q2, Q3, and Q4 compared with that in Q1 were 1.05, 1.11, and 1.56, and the adjusted HRs were 1.05 (95% CI, 0.87–1.27), 1.11 (95% CI, 0.92–1.35), and 1.56 (95% CI, 1.29–1.87), respectively. When the data on sex were stratified or merged for analysis, statistically significant correlations of IRRs and adjusted HRs were measured in Q4 compared with those in Q1.

Among the quartiles of CO concentration, the IRRs in Q2, Q3, and Q4 compared with that in Q1 were 0.96, 1.23, and 1.36, and the adjusted HRs were 1.07 (95% CI, 0.92–1.25), 1.37 (95% CI,1.19–1.58),and 1.61 (95% CI, 1.39–1.85), respectively, in all people included in the study. Among men, the IRRs in Q2, Q3, and Q4 compared with that in Q1 were 0.97, 1.18, and 1.28, and the adjusted HRs were 1.16 (95% CI, 0.93–1.45), 1.28 (95% CI, 1.04–1.58), and 1.57 (95% CI, 1.26–1.94), respectively. Among women, the IRRs in Q2, Q3, and Q4 compared with that in Q1 were 0.95, 1.28, and 1.43, and the adjusted HRs were 1.01 (95% CI, 0.82–1.24), 1.46 (95% CI, 1.21–1.77), and 1.64 (95% CI, 1.36–1.98), respectively. A clear trend that was detected was an increase in the risk of dementia as CO exposure increased. Figures 1 and 2 show the Kaplan-Meier curves of freedom that were calculated for dementia and are separated according to pollutant level. Statistically significant differences in the occurrence of dementia were observed among the quartiles of NO_2 and CO concentrations (log-rank test, $P<.001$).

Discussion

The major finding of previous animal study was that nitration was highly correlated with beta-amyloid aggregation and plaque formation, and beta-amyloid aggregation is a pathologic hallmark of AD [21]. Another animal study indicated that NO_2 expose can exacerbate the ultra structural impairment of synapses in stroke rats, and induce neuronal damage in healthy rats [29]. The apolipoprotein E (APOE) e4 allele was a well know genetic risk factor or AD, and a randomized clinical trial has found CO poisoning can induce APOE e4 carriers suffer greater morbidity [30].

The major finding of our study was that increased exposure to NO_2 (Q4) is associated with an enhanced risk of dementia in men and women. The probability of dementia occurrence was increased by 52%–56% in Q4 compared with Q1. A similar trend was observed in the CO group, and the results collectively showed that increasing levels of the 2 pollutants increased the risk of dementia in a dose-dependent manner.

This study was a national population-based investigation on ambient air pollution and dementia. Therefore, collecting individual exposure data was not feasible. To obtain exposure data associated with study patients, previous studies have identified the residential areas of patients by employing a GIS-based system. To protect the privacy of patients, the NHIRD does not provide patients' addresses. Therefore, we identified the residential areas of the patients based on the location of the clinic at which the patients most frequently sought treatment for acute upper respiratory tract infection. In the United States, upper respiratory tract infections are the most common type of infectious disease, and each adult experiences approximately 3 respiratory infections annually [31]. Identifying residential areas in the accessible medical resources, as we did in this study, is more accurate than listing patients according to insurance area [32,33].

Previous studies have suggested that smoking and drinking alcohol are highly correlated with the risk of AD [34–40]. Because of the limitations of the NHIRD, we could not obtain data on the smoking or drinking status of the patients. Therefore, we performed multivariate analysis with COPD and alcoholism adjusted in accordance with previous studies that indicated that smoking is a major causative factor in the development of COPD, and in which alcoholism was diagnosed based on drinking patterns and the attitudes of patients [41–43]. In Taiwan, women are not encouraged to smoke or drink alcohol, as reflected in the low prevalence of these behaviors among women (3% and 1%, respectively) [44,45]. We were able to overcome this limitation by stratifying and adjusting the data according to sex [46].

We adjusted for urbanization in the multivariate analysis. The level of urbanization was determined according to population density (number of people/km^2), the population ratio of people with a college-level education or higher, the population ratio of people aged over 65 years, the population ratio of agricultural workers, and the number of physicians per 100000 people [47]. The 359 communities in Taiwan were classified into 7groups: highly urbanized area, moderately urbanized area, boomtown, general town, aging town, agricultural town, and remote town. This classification method has been used in several studies [48–50].

In addition, we obtained results contrasting those related to dementia, as shown in Tables 1 and 2: the frequency of IHD and COPD were low at the highest level of the pollutants. These results agree with the explanation provided by previous studies suggesting that patients who are highly educated and earn a high monthly income live in areas where the level of air pollutants is high [6,22].

The strengths of this study are the following. First, this study was based on a long follow-up period, which allowed the possible occurrence of dementia to be assessed. Second, Taiwan launched a national health insurance (NHI) in 1995, operated by a single-buyer, the government. All insurance claims should be scrutinized by medical reimbursement specialists and peer review. The diagnoses of dementia were based on the ICD-9 code determined by qualified clinical neurology physicians under strict audit in the reimbursement process. Therefore the diagnoses and codes for dementia should be accurate and reliable. Third, this study was conducted using a large population derived from the NHIRD. In Taiwan, the government is the only compulsory social insurance provider; approximately 99% of the 23.74 million citizens of Taiwan are enrolled in the NHI program. Because this was a nationwide study, we considered urbanized towns throughout Taiwan. Lastly, in this study, cerebrovascular and cardiovascular diseases were considered and the association between pollutants and dementia was evaluated. We excluded subjects with cardiovascular before the index date in this study because cardiovascular was a widely known predictor for dementia. IHD increased 27% risk for dementia in both model 1 and model 2. (Table S1).

Certain limitations of this study should be considered. First, the evidence derived from a retrospective cohort study is generally lower in statistical quality than that obtained from randomized trials because, in such retrospective studies, potential biases exist that are related to the adjustment of confounding variables. Despite our meticulous study design and the measures adopted to control for confounding factors, bias resulting from unknown confounders may have affected our results. Second, all data in the

Table 1. Comparison of Baseline Characteristics among quartiles of NO_2 yearly average.

	Quartiles of NO_2 yearly average								p	Total (n = 29547)	
	Q1 (n = 7349)		Q2 (n = 7425)		Q3 (n = 7572)		Q4 (n = 7201)				
Dementia	406	5.5	425	5.7	374	4.9	515	7.2	<0.001	1720	5.8
Age (mean, SD)	61.8	8.4	61.4	8.5	61.0	8.4	61.4	8.8	<0.001†	61.4	8.5
Male	3365	45.8	3469	46.7	3474	45.9	3298	45.8	0.611	13606	46.0
Monthly income											
<14400	1481	20.2	1814	24.4	2004	26.5	1991	27.7	<0.001	7290	24.7
14400–18300	1054	14.3	1324	17.8	1511	20.0	1480	20.6		5369	18.2
18301–21000	3255	44.3	2399	32.3	1992	26.3	1785	24.8		9431	31.9
>21000	1559	21.2	1887	25.4	2062	27.2	1944	27.0		7452	25.2
DM	845	11.5	837	11.3	916	12.1	850	11.8	0.421	3448	11.7
IHD	1347	18.3	1354	18.2	1295	17.1	1222	17.0	0.047	5218	17.7
HT	2899	39.4	2906	39.1	2889	38.2	2785	38.7	0.391	11479	38.8
COPD	2612	35.5	2608	35.1	2579	34.1	2376	33.0	0.005	10175	34.4
Alcoholism	19	0.3	19	0.3	22	0.3	10	0.1	0.250	70	0.2
Urbanization											
Highly urbanization	1330	18.1	1668	22.5	2503	33.1	3720	51.7	<0.001	9221	31.2
Moderate urbanization	2157	29.4	2782	37.5	2908	38.4	1828	25.4		9675	32.7
Boomtown	907	12.3	986	13.3	1485	19.6	1126	15.6		4504	15.2
General town	1692	23.0	1160	15.6	412	5.4	298	4.1		3562	12.1
Aging town	304	4.1	56	0.8	68	0.9	72	1.0		500	1.7
Agricultural town	658	9.0	321	4.3	111	1.5	88	1.2		1178	4.0
Remote town	301	4.1	452	6.1	85	1.1	69	1.0		907	3.1

Chi-square test;
†T-test;

Table 2. Comparison of Baseline Characteristics among quartiles of CO yearly average.

	Quartiles of CO yearly average								p	Total (n = 29537)	
	Q1 (n = 7565)		Q2 (n = 6428)		Q3 (n = 7681)		Q4 (n = 7863)				
Dementia	391	5.2	321	5.0	476	6.2	530	6.7	<0.001	1718	5.8
Age (mean, SD)	61.8	8.3	51.1	8.3	61.4	8.6	61.3	8.8	<0.001†	61.4	8.5
Male	3532	46.7	2882	44.8	3587	46.7	3597	45.7	0.084	13598	46.0
Monthly income											
<14400	1477	19.5	1477	23.0	2190	28.5	2144	27.3	<0.001	7288	24.7
14400–18300	1074	14.2	1189	18.5	1513	19.7	1591	20.2		5367	18.2
18301–21000	3401	45.0	2095	32.6	1887	24.6	2046	26.0		9429	31.9
>21000	1613	21.3	1667	25.9	2088	27.2	2080	26.5		7448	25.2
DM	862	11.4	712	11.1	918	12.0	954	12.1	0.173	3446	11.7
IHD	1430	18.9	1054	16.4	1394	18.1	1339	17.0	<0.001	5217	17.7
HT	2980	39.4	2455	38.2	3021	39.3	3017	38.4	0.306	11473	38.8
COPD	2785	36.8	2189	34.1	2607	33.9	2587	32.9	<0.001	10168	34.4
Alcoholism	19	0.3	15	0.2	24	0.3	12	0.2	0.232	70	0.2
Urbanization											
Highly urbanization	912	12.1	1697	26.4	2694	35.1	3918	49.8	<0.001	9221	31.2
Moderate urbanization	2388	31.6	2615	40.7	2323	30.2	2346	29.8		9672	32.7
Boomtown	1084	14.3	819	12.7	1576	20.5	1024	13.0		4503	15.2
General town	1684	22.3	772	12.0	781	10.2	322	4.1		3559	12.0
Aging town	336	4.4	22	0.3	65	0.8	74	0.9		497	1.7
Agricultural town	699	9.2	253	3.9	120	1.6	106	1.3		1178	4.0
Remote town	462	6.1	250	3.9	122	1.6	73	0.9		907	3.1

Chi-square test;
†T-test;

Table 3. Comparisons of difference dementia incidences and associated hazard ratios among four levels of air pollutants by gender stratification.

			Dementia	PY	Incidence rate#	IRR*	95%CI	aHR†	95%CI
NO_2	Total	Q1	406	75461.4	5.38	1.00		1.00	
		Q2	425	75246.1	5.65	1.05	0.92, 1.20	1.10	0.96, 1.26
		Q3	374	77576.5	4.82	0.90	0.78, 1.03	1.01	0.87, 1.17
		Q4	515	71461.0	7.21	1.35	1.18, 1.54	1.54	1.34, 1.77
	Male	Q1	186	33853.8	5.49	1.00		1.00	
		Q2	206	34587.2	5.96	1.08	0.89, 1.32	1.16	0.95, 1.43
		Q3	152	34973.3	4.35	0.79	0.64, 0.98	0.89	0.71, 1.11
		Q4	224	31976.0	7.01	1.28	1.05, 1.56	1.52	1.23, 1.88
	Female	Q1	220	41607.6	5.29	1.00		1.00	
		Q2	219	40658.9	5.39	1.02	0.85, 1.23	1.05	0.87, 1.27
		Q3	222	42603.2	5.21	0.99	0.82, 1.19	1.11	0.92, 1.35
		Q4	291	39485.0	7.37	1.41	1.18, 1.67	1.56	1.29, 1.87
CO	Total	Q1	391	77816.4	5.02	1.00		1.00	
		Q2	321	66509.7	4.83	0.96	0.83, 1.11	1.07	0.92, 1.25
		Q3	476	77215.4	6.16	1.23	1.08, 1.41	1.37	1.19, 1.58
		Q4	530	78172.7	6.78	1.36	1.19, 1.55	1.61	1.39, 1.85
	Male	Q1	182	35681.8	5.10	1.00		1.00	
		Q2	145	29334.5	4.94	0.97	0.78, 1.20	1.16	0.93, 1.45
		Q3	212	35371.7	5.99	1.18	0.97, 1.44	1.28	1.04, 1.58
		Q4	227	34977.3	6.49	1.28	1.05, 1.55	1.57	1.26, 1.94
	Female	Q1	209	42134.6	4.96	1.00		1.00	
		Q2	176	37175.2	4.73	0.95	0.78, 1.16	1.01	0.82, 1.24
		Q3	264	41843.8	6.31	1.28	1.07, 1.54	1.46	1.21, 1.77
		Q4	303	43195.4	7.01	1.43	1.20, 1.70	1.64	1.36, 1.98

Incidence rate#, per 1,000 person-years;
IRR*, incidence rate ratio;
Adjusted HR† : multiple analysis including age, sex, monthly income, DM, IHD, HT, COPD, alcoholism and urbanization.

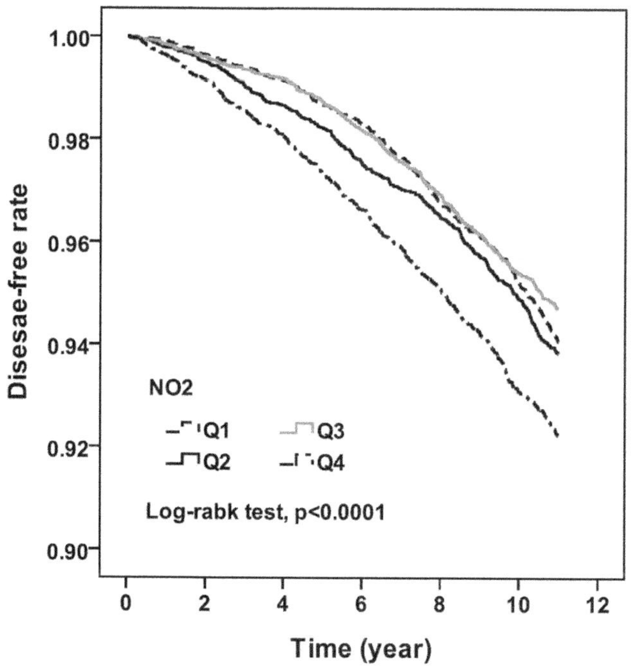

Figure 1. Probability free of dementia among quartiles of yearly average concentration in NO_2.

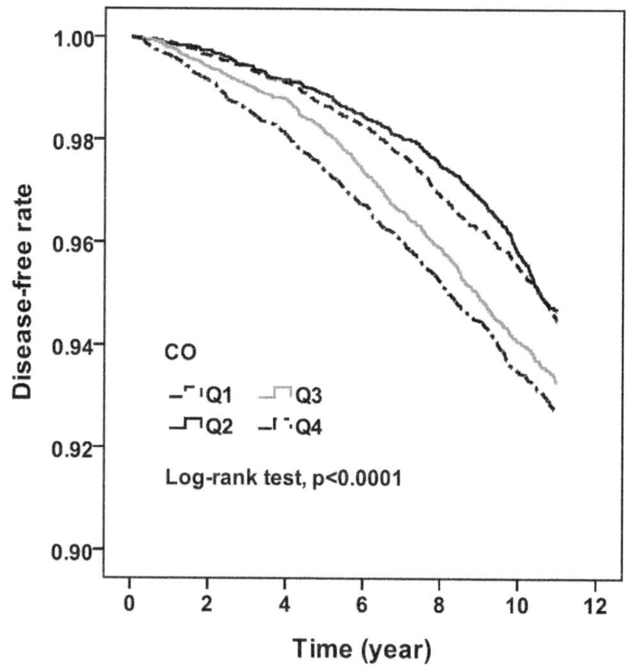

Figure 2. Probability free of dementia among quartiles of yearly average concentration in CO.

NHIRD are anonymous. Thus, relevant clinical variables, such as imaging results and pathology findings, were unavailable for the patient cases included in this study. Third, the participants were assigned to residential districts based on the clinic where they most frequently sought treatment for acute upper respiratory infection. Therefore, the resident who has no acute upper respiratory infection during study period had being excluded in this study. In our opinion, the resident without respiratory infection related medical record exposed to low level air pollutants. It might under the estimated risk of dementia. Nevertheless, the data on air pollutants and dementia diagnoses were reliable.

Conclusions

Understanding the regional distribution of human health statuses can facilitate the investigation of the spread of diseases and the related risk factors as well as the assessment of medical resources and the planning of the use of these resources. In future research, animal studies can be conducted to further examine the association between air pollutants and neurological disorders.

Author Contributions

Study concept and design: KHC CHK. Acquisition of data: KHC MYC CHM TNW CYC CHK. Analysis and interpretation of data: KHC CHM CHK. Drafting of the manuscript: KHC MYC CHM TNW CYC CHK. Critical revision of the manuscript for important intellectual content: KHC CHM CHK. Statistical analysis: CHM. Obtained funding: CHK. Administrative, technical, or material support: KHC MYC CHM TNW CYC CHK. Study supervision: CHK.

References

1. Dickey JH, Part VII (2000) Air pollution: overview of sources and health effects. Dis Mon 46:566–89.
2. Lewtas J (2007) Air pollution combustion emissions: characterization of causative agents and mechanisms associated with cancer, reproductive, and cardiovascular effects. Mutat Res 636:95–133.
3. Weuve J, Puett RC, Schwartz J, Yanosky JD, Laden F, et al. (2012) Exposure to particulate air pollution and cognitive decline in older women. Arch Intern Med 172:219–27.
4. Srám RJ, Benes I, Binková B, Dejmek J, Horstman D, et al. (1996) Teplice program–the impact of air pollution on human health. Environ Health Perspect 104 Suppl 4:699–714.
5. Suglia SF, Gryparis A, Wright RO, Schwartz J, Wright RJ (2008) Association of black carbon with cognition among children in a prospective birth cohort study. Am J Epidemiol 167:280–6.
6. Chen JC, Schwartz J (2009) Neurobehavioral effects of ambient air pollution on cognitive performance in US adults. Neurotoxicology 30:231–9.
7. Ranft U, Schikowski T, Sugiri D, Krutmann J, Krämer U (2009) Long-term exposure to traffic-related particulate matter impairs cognitive function in the elderly. Environ Res 109:1004–11.
8. Power MC, Weisskopf MG, Alexeeff SE, Coull BA, Spiro A 3rd, et al. (2011) Traffic-related air pollution and cognitive function in a cohort of older men. Environ Health Perspect 119:682–7.
9. Lisabeth LD, Escobar JD, Dvonch JT, Sánchez BN, Majersik JJ, et al. (2008) Ambient air pollution and risk for ischemic stroke and transient ischemic attack. Ann Neurol 2008;64:53–9.
10. Migliore L, Coppedè F (2009) Environmental-induced oxidative stress in neurodegenerative disorders and aging. Mutat Res 674:73–84.
11. Turin TC, Kita Y, Rumana N, Nakamura Y, Ueda K, et al (2012) Ambient air pollutants and acute case-fatality of cerebro-cardiovascular events: Takashima Stroke and AMI Registry, Japan (1988–2004). Cerebrovasc Dis 34(2):130–9.
12. Andersen ZJ, Kristiansen LC, Andersen KK, Olsen TS, Hvidberg M, et al. (2012) Stroke and long-term exposure to outdoor air pollution from nitrogen dioxide: a cohort study. Stroke 43:320–5.
13. Knopman DS (2007) Cerebrovascular disease and dementia. Br J Radiol 80 :S121–7.
14. O'Brien JT (2006) Vascular cognitive impairment. Am J Geriatr Psychiatry 14:724–33.

15. Kokmen E, Whisnant JP, O'Fallon WM, Chu CP, Beard CM (1996) Dementia after ischemic stroke: a population-based study in Rochester, Minnesota (1960–1984). Neurology 46:154–9.
16. Morales E, Julvez J, Torrent M, de Cid R, Guxens M, et al. (2009) Association of early-life exposure to household gas appliances and indoor nitrogen dioxide with cognition and attention behavior in preschoolers. Am J Epidemiol 169:1327–36.
17. Freire C, Ramos R, Puertas R, Lopez-Espinosa MJ, Julvez J, et al. (2010) Association of traffic-related air pollution with cognitive development in children. J Epidemiol Community Health 64:223–8.
18. Clark C, Crombie R, Head J, van Kamp I, van Kempen E, et al. (2012) Does traffic-related air pollution explain associations of aircraft and road traffic noise exposure on children's health and cognition? A secondary analysis of the United Kingdom sample from the RANCH project. Am J Epidemiol 176:327–37.
19. Zhu N, Li H, Han M, Guo L, Chen L, et al. (2012) Environmental nitrogen dioxide (NO2) exposure influences development and progression of ischemic stroke. Toxicol Lett 214:120–30.
20. Li H, Xin X (2013) Nitrogen dioxide (NO(2)) pollution as a potential risk factor for developing vascular dementia and its synaptic mechanisms. Chemosphere 92:52–8.
21. Kummer MP, Hermes M, Delekarte A, Hammerschmidt T, Kumar S, et al. (2011) Nitration of tyrosine 10 critically enhances amyloid β aggregation and plaque formation. Neuron 71:833–44.
22. Gatto NM, Henderson VW, Hodis HN, St John JA, Lurmann F, et al. (2013) Components of air pollution and cognitive function in middle-aged and older adults in Los Angeles. Neurotoxicology 40C:1–7.
23. Blanco F, Alkorta I, Solimannejad M, Elguero J (2009) Theoretical study of the 1:1 complexes between carbon monoxide and hypohalous acids. J Phys Chem A 113:3237–44.
24. Roberts GP, Youn H, Kerby RL (2004) CO-sensing mechanisms. Microbiol Mol Biol Rev 68:453–73, table of contents.
25. Chen HL, Chen PC, Lu CH, Hsu NW, Chou KH, et al. (2013) Structural and cognitive deficits in chronic carbon monoxide intoxication: a voxel-based morphometry study. BMC Neurol 13:129.
26. Jung CR, Lin YT, Hwang BF (2013) Air pollution and newly diagnostic autism spectrum disorders: a population-based cohort study in Taiwan. PLoS One 8:e75510.
27. National Health Insurance Research Database (NHIRD): Introduction to the National Health Insurance Research Database (NHIRD), Taiwan (2010) Available: http://w3.nhri.org.tw/nhird//date_01.html
28. Taiwan Air Quality Monitoring Network in Taiwan Environmental Protection Administration. Available: http://taqm.epa.gov.tw/taqm/en/PsiMap.aspx
29. Li H, Xin X (2013) Nitrogen dioxide (NO (2)) pollution as a potential risk factor for developing vascular dementia and its synaptic mechanisms. Chemosphere 92:52–8
30. Hopkins RO, Weaver LK, Valentine KJ, Mower C, Churchill S, et al. (2007) Apolipoprotein E genotype and response of carbon monoxide poisoning to hyperbaric oxygen treatment. Am J Respir Crit Care Med 176: 1001–6.
31. Garibaldi RA (1985) Epidemiology of community-acquired respiratory tract infections in adults. Incidence, etiology, and impact. Am J Med 78:32–7.
32. Kuo SS, Chang RE (2010) Geographical analysis of ESRD incidence and environment [Dissertation]. Taipei: Graduate Institute of Health Care Organization Administration, National Taiwan University. [In Chinese: English abstract]
33. Ministry of the Interior, R.O.C. (Taiwan). Monthly bulletin of interior statistics. Available at: http://sowf.moi.gov.tw/stat/month/list.htm. Accessed 2011 March 3. [In Chinese: English abstract]
34. Cataldo JK, Prochaska JJ, Glantz SA (2010) Cigarette smoking is a risk factor for Alzheimer's Disease: an analysis controlling for tobacco industry affiliation. J Alzheimers Dis 19:465–80.
35. Deng J, Shen C, Wang YJ, Zhang M, Li J, et al. (2010) Nicotine exacerbates tau phosphorylation and cognitive impairment induced by amyloid-beta 25–35 in rats. Eur J Pharmacol 637:83–8.
36. Oddo S, Caccamo A, Green KN, Liang K, Tran L, et al. (2005) Chronic nicotine administration exacerbates tau pathology in a transgenic model of Alzheimer's disease. Proc Natl Acad Sci U S A 102:3046–51.
37. Juan D, Zhou DH, Li J, Wang JY, Gao C, et al. (2004) A 2-year follow-up study of cigarette smoking and risk of dementia. Eur J Neurol 11:277–82.
38. Peters R, Peters J, Warner J, Beckett N, Bulpitt C (2008) Alcohol, dementia and cognitive decline in the elderly: a systematic review. Age Ageing 37:505–12.
39. Deng J, Zhou DH, Li J, Wang YJ, Gao C, et al. (2006) A 2-year follow-up study of alcohol consumption and risk of dementia. Clin Neurol Neurosurg 108:378–83.
40. Anstey KJ, Mack HA, Cherbuin N (2009) Alcohol consumption as a risk factor for dementia and cognitive decline: meta-analysis of prospective studies. Am J Geriatr Psychiatry 17:542–55.
41. Pauwels RA, Rabe KF (2004) Burden and clinical features of chronic obstructive pulmonary disease (COPD). Lancet 364:613–20.
42. Patel BD, Loo WJ, Tasker AD, Screaton NJ, Burrows NP, et al. (2006) Smoking related COPD and facial wrinkling: is there a common susceptibility? Thorax 61:568–571.
43. Enoch MA, Goldman D (2002) Problem drinking and alcoholism: diagnosis and treatment. Am Fam Physician 65:441–8.
44. Liang CY, Chou TM, Ho PS, Shieh TY, Yang YH (2004) Prevalence Rates of Alcohol Drinking in Taiwan. Taiwan Journal of Oral Medicine & Health Sciences 20:91–104
45. Chuang YC, Chuang KY (2008) Gender differences in relationships between social capital and individual smoking and drinking behavior in Taiwan. Soc Sci Med 67:1321–30.
46. Chang KH, Chung CJ, Lin CL, Sung FC, Wu TN, et al. (2014) Increased risk of dementia in patients with osteoporosis: a population-based retrospective cohort analysis. Age (Dordr) 36:967–75.
47. Liu C, Hung Y, Chuang Y, Chen Y, Weng W, et al. (2006) Incorporating development stratification of Taiwan townships into sampling design of large scale health interview survey. Journal of Health Management 4:1–22 [in Chinese].
48. Chiang PH, Chang YC, Lin JD, Tung HJ, Lin LP, et al. (2013) Healthcare utilization and expenditure analysis between individuals with intellectual disabilities and the general population in Taiwan: a population-based nationwide child and adolescent study. Res Dev Disabil 34:2485–92.
49. Lin YJ, Tian WH, Chen CC (2011) Urbanization and the utilization of outpatient services under National Health Insurance in Taiwan. Health Policy 103:236–43.
50. Lin HC, Lin YJ, Liu TC, Chen CS, Lin CC (2007) Urbanization and place of death for the elderly: a 10-year population-based study. Palliat Med 21:705–11.

15

Relationship between Air Pollutants and Economic Development of the Provincial Capital Cities in China during the Past Decade

Yunpeng Luo[1], Huai Chen[1,2]*, Qiu'an Zhu[1], Changhui Peng[1,3]*, Gang Yang[1], Yanzheng Yang[1], Yao Zhang[1]

1 State Key Laboratory of Soil Erosion and Dryland Farming on the Loess Plateau, College of Forestry, Northwest A&F University, Yangling, Shaanxi, China, **2** Chengdu Institute of Biology, Chinese Academy of Sciences, Chengdu, China, **3** Center of CEF/ESCER, Department of Biology Science, University of Quebec at Montreal, Montreal, Canada

Abstract

With the economic development of China, air pollutants are also growing rapidly in recent decades, especially in big cities of the country. To understand the relationship between economic condition and air pollutants in big cities, we analysed the socioeconomic indictorssuch as Gross Regional Product per capita (GRP per capita), the concentration of air pollutants (PM_{10}, SO_2, NO_2) and the air pollution index (API) from 2003 to 2012 in 31 provincial capitals of mainland China. The three main industries had a quadratic correlation with NO_2, but a negative relationship with PM_{10} and SO_2. The concentration of air pollutants per ten thousand yuan decreased with the multiplying of GRP in the provinical cities. The concentration of air pollutants and API in the provincial capital cities showed a declining trend or inverted-U trend with the rise of GRP per capita, which provided a strong evidence for the Environmental Kuznets Curve (EKC), that the environmental quality first declines, then improves, with the income growth. The results of this research improved our understanding of the alteration of atmospheric quality with the increase of social economy and demonstrated the feasibility of sustainable development for China.

Editor: Xiaoyan Yang, Chinese Academy of Sciences, China

Funding: This study was supported by 100 Talents Program of The Chinese Academy of Sciences, by Program for New Century Excellent Talents in University (NCET-12-0477), by the National Natural Science Foundation of China (No. 31100348), by a Natural Sciences and Engineering Research Council of Canada (NSERC) discovery grant, and by China's QianRen program. The funders had no role in study design, data collection and analysis, decision to publish, or preparation of the manuscript.

Competing Interests: The authors have declared that no competing interests exist.

* Email: chenhuai81@gmail.com (HC); cpeng86@gmail.com (CP)

Introduction

China has seen economic soaring in the past three decades, with its gross domestic product (GDP) expanding 140 times during 1978–2012 (National Bureau of Statistics, 2013). However, such economic soaring is accompanied with deterioration of the atmospheric quality. In the first three months of 2013, just like what happened in London in 1952 [1], long-time haze influenced large area of China (Fig S1), which further stimulated the strong demand for improvement of air quality.

Air pollution has significant influence on both climate and human health [2,3]. Oxidising air pollutants like ozone stimulate reactions to produce more greenhouse gases which exacerbate global warming [4]. Besides, decreasing precipitation and increasing dimness [5,6], widening of the tropics [7], weakening of summer moonsoon in South Asian [4,8], as well as large-scale ocean circulation and some extreme weather like hurricane [9,10], are all linkd to air pollution. Moreover, anthropogenic air pollutants, especially particulate matter is extremely harmful to human health. According to Silva *et al.* (2013), more than 2 million premature deaths are associated with $PM_{2.5}$-related diseases [11]. Research results from Spain and England reported that long-term exposure to air pollution mainly explained heart disease morbidity and mortality [12,13]. Similarly, the heating policy in Northern China was found to cause reduction in life expectancies of Northern residents by about 5.52 years [14]. Given the great influece of air pollution on natural environment and human life, researches are attaching ever greater importance to the causes and effects of air pollution [4,15–17].

Environmental problems result from economic expansion which increases extraction of natural resources and accumulation of waste, in the end exceeding the carrying capacity of the biosphere to the pollutants [18]. From the perspective of the history of human society economy development, the environmental quality is not fixed along a country's development path [19–21]. In the 1990s, scientists found an inverted U-shaped relationship between environmental quality and social income [22–25]. Such relationship was defined as Environmental Kuznets Curve (EKC), showing that the environmental quality would first deteriorate with the increase in revenue, and then it would improve when incomes rise to a certain level [25]. Numerous research results related to developed countries have identified EKC curve between income and air pollutants, especially in the Organization for Economic Co-operation and Development (OECD) countries

[18,23,26]. However, the relationship between income and air pollutants varies considerably among developing countries. For example, the same air pollutant sulfur dioxide (SO_2) showed an inverted U-shape relationship with income for Tunisia, but an N-shape relationship for Turkey [27,28].

China is the biggest developing country in the world, whose high-speed economic development as well as environmental changes and protection may provide experiences and lessons for other developing countries in this respect. There are also some studies regarding EKC in China [29,30]. However, most of them focused on econometrics, without relating air pollutants to specific levels of economic development. Considering the disparate economy development paces of different provinces in this big country, analysis about how particular air pollution is related to economic development of each region is needed [31]. Moreover, EKC researches in China concentrated on comprehensive indictors like the total amount of atmospheric emission [32–35]. Although some researches studied specific pollutants like SO_2 or PM [36–37], the relationship between the most important three categories of air pollutants and socioeconomic indicators is not adequately reported.

In this study, we aimed to establish regression models to fit the relationship between air pollutants and the three major industries (the primary, secondary and tertiary industries), so as to reveal the relationship between industry and air quality deterioration. We also caculated the ratio of air pollutant concentration to Gross Regional Product (GRP) per capita in order to know the contribution of economic development to air pollution over time. Finally, regression analysis was conducted to verify the existence of EKC in Chinese cities, or to define the otherwise relationship between air pollutants and revenue of Chinese citizens.

Datasets and Methods

1. Study area and data source

Data were collected for 31 provincial capitals in mainland China, which are representative of the general condition of each province (Fig. 2). In order to investigate the relationship between social economy and concentration of environmental pollutants in China, we downloaded data about these two aspects in the database of the National Bureau of Statistics of China (Table S1). The economic data included GRP, population, primary industry output, secondary industry output and tertiary industry output. The pollutant data collected included the concentration of PM_{10}, SO_2 and NO_2 (the three most important air pollutants in China [38]). The air pollution index (API) was calculated with the following formula:

$$I_j = \frac{I_{high} - I_{low}}{C_{high} - C_{low}}(C - C_{low}) + I_{low}$$

Where:

I = (Air pollution) index of one specific pollutant,
C = pollutant concentration,
C_{low} = the concentration breakpoint ≤C,
C_{high} = the concentration breakpoint ≥C,
I_{low} = the index breakpoint corresponding to C_{low},
I_{high} = the index breakpoint corresponding to C_{high},
j = Air pollutants indicators (PM_{10}, SO_2, NO_2).

$$API = Max(I_j)$$

The criteria of breakpoints for air pollutants were taken from the website of Ministry of Environment Protection of the People's Republic of China [39].All data from 2003 to 2012 used for statistical analyses were retrieved from National Bureau of Statistics of China. As the demographic data for Lhasa during 2003–2006 and 2010 was missing, we did not do analyze the city for these years.

2. Relationship between air pollutants and the three main industries

Linear, quadratic and cubic regression analysis was conducted to examine whether there existed simple positive or negative relationship between the concentration of air pollutants (dependent variables) and the output per capita of the three industries (independent variables) in the provincial capital cities. We chose the best appropriate regression model for each air pollutant and industry and plotted the regression line for those which were significantly correlated.

3. Trend analysis for the ratio of pollutant concentration to industry output

For comparing socioeconomic development level in different regions in China, we classified all the 31 provincial capital cities of mainland China into four economic regions including East Coast (East), Central China (Central), Northeastern China (Northeast), and Western China (West), according to strategies promulgated by the Central People's Government [40]. The ratio of annual air pollutant concentration to GRP per capita (c PM_{10}/GRP per capita, c SO_2/GRP per capita, and c NO_2/GRP per capita) was calculated year by year for each region. Hereafter, the line trend plots of the ratios were constructed to illustrate the variation of energy efficiency during 2003 to 2012.

4. Analysis associated with EKC

In order to investigate whether EKC exists in China, regression methods were applied to the panel data of GRP per capita and pollutants' indicators in all provincial capital cities. We also conducted regression fitting for the four economic regions for further information. The relationships between air pollutants and GRP per capita were estimated by the simplified EKC model provided below, which was also described by Agras et al. (41) and Li et al. (42)[41–42]:

$$E_{ij} = \alpha_{ij} + \beta_1 X_{ij} + \beta_2 X_{ij}^2 + \beta_3 X_{ij}^3$$

Where E is the concentration of air pollutant; X is GRP per capita; α_{ij} is a fixed effect; ε_{ij} is a stochastic error term; i is a region index (region values are "East, Central, Northeast, West and All provincial capital cities"); j is an air pollutant indicator (PM_{10}, SO_2, NO_2 or API); β_1, β_2, β_3 are the coefficient for the income variable, for the income squared variable and for the income cubic term, respectively.

5. Statistical analysis

All the regression analysis related to air pollutants and socioeconomic indicators was performed with SPSS for Windows (IBM SPSS statistics; Version 20). The effect of a certain variable was considered statistically significant for $P<0.05$. Annual mean values of data used for trend analysis of energy effiency between 2003 to 2012 were caculated by Excel 2010.

Table 1. Regression for concentration of PM_{10}, SO_2, NO_2 and the three main industries.

	Model summary						
	Regression					Coefficients T test	
Described Relationship	Model	n	R^2	SE	Sig.	Independent variable	Constant
PM_{10}	Linear		0.147	0.026	0.000**	0.000**	0.000**
&	Quadratic		0.178	0.025	0.000**	0.000**	0.000**
Primary industry	Cubic		0.203	0.025	0.000**	0.000**	0.000**
SO_2	Linear		0.141	0.020	0.000**	0.000**	0.000**
&	Quadratic	305	0.163	0.020	0.000**	0.000**	0.000**
Primary industry	Cubic		0.172	0.020	0.000**	0.001**	0.000**
NO_2	Linear		0.004	0.013	0.280		
&	Quadratic		0.007	0.013	0.348		
Primary industry	Cubic		0.016	0.013	0.172		
PM_{10}	Linear		0.025	0.027	0.005**	0.005**	0.000**
&	Quadratic		0.034	0.027	0.006**		
Secondary industry	Cubic		0.046	0.027	0.003**	0.035*	0.000**
SO_2	Linear		0.018	0.022	0.018*	0.018*	0.000**
&	Quadratic	305	0.022	0.022	0.041*	0.873	0.000
Secondary industry	Cubic		0.021	0.022	0.092		
NO_2	Linear		0.097	0.012	0.000**	0.000**	0.000**
&	Quadratic		0.118	0.012	0.000**	0.000**	0.000**
Secondary industry	Cubic		0.118	0.012	0.000**	0.100**	0.000**
PM_{10}	Linear		0.029	0.027	0.003**	0.003**	0.000**
&	Quadratic		0.037	0.027	0.003**	0.012*	0.000**
Tertiary industry	Cubic		0.044	0.027	0.003**	0.015*	0.000**
SO_2	Linear		0.039	0.021	0.001**	0.001**	0.000**
&	Quadratic	305	0.039	0.022	0.002**	0.235	0.000**
Tertiary industry	Cubic		0.046	0.021	0.003**	0.064	0.000**
NO_2	Linear		0.120	0.012	0.000**	0.000**	0.000**
&	Quadratic		0.135	0.012	0.000**	0.000**	0.000**
Tertiary industry	Cubic		0.136	0.012	0.000**	0.225	0.000**

* $P<0.05$; ** $P<0.01$.

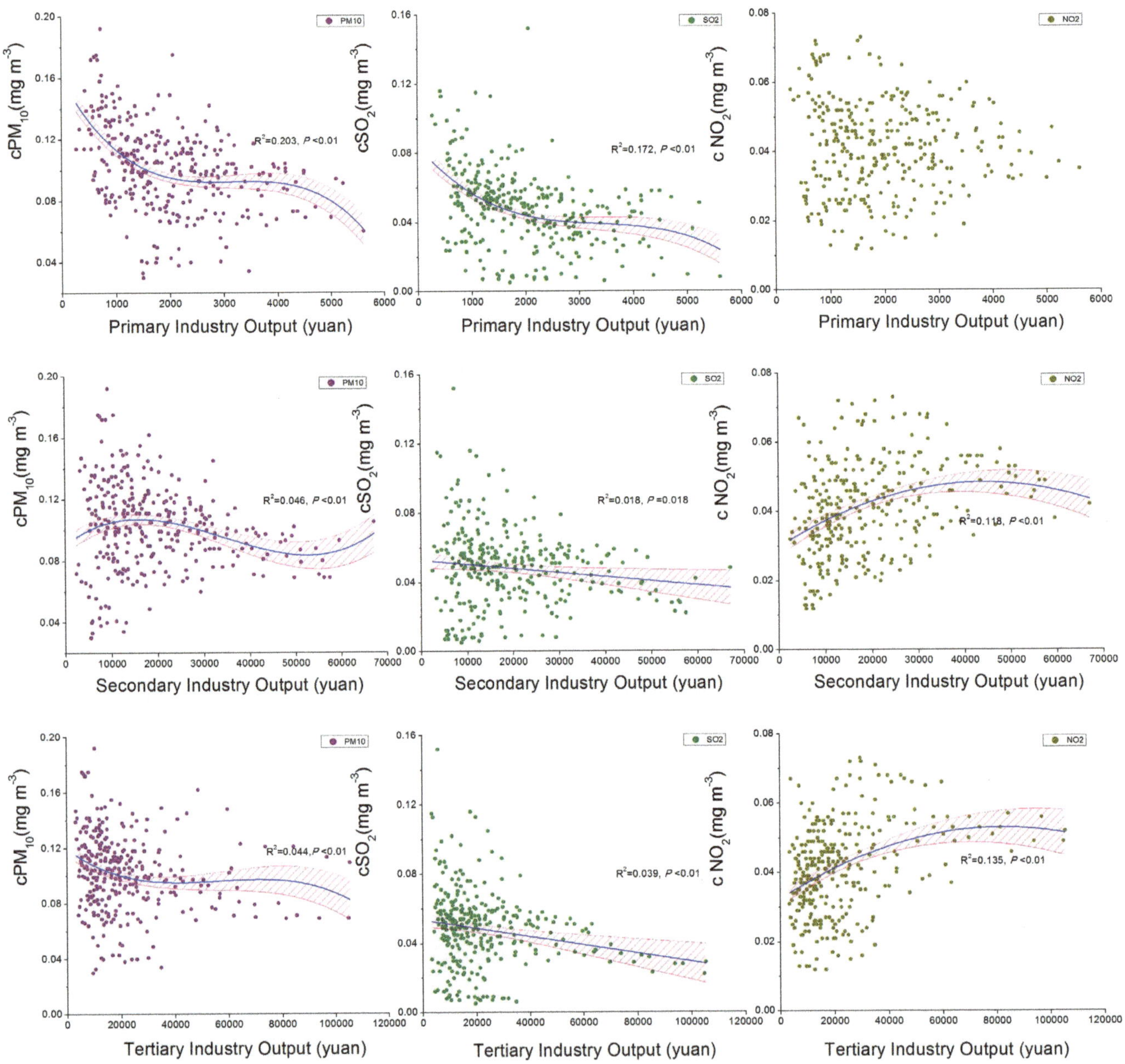

Figure 1. Air pollutant concentrations as related to the output per capita of three industries in the provincial capitals of China. (a) The output per capita of the primary industry; (b) The output per capita of the secondary industry; (c) The output per capita of the tertiary industry.

Results

1. Relationship between air pollutants and the three main industries in Chinese cities

Analysis on the provincial capital cities illustrated quadratic relationships between concentration of NO_2 and the output per capita of secondary and tertiary industries (Table 1, Fig. 1). The NO_2 concentration rose with the increase of the output per capita of the secondary and tertiary industries at the first stage, then began to decrease when the output reached around 45,000 and 70,000 yuan respectively. However, there was no remarkable relationship between that of the primary industry and the NO_2 concentration. The results also indicated that all the three industries had significantly negative relationship with the concentration of PM_{10} and SO_2.

2. Variation of efficiency ratio in recent years

The ratio of the air pollutant concentration to GRP per capita (cPM_{10}/GRP per capita, cSO_2/GRP per capita and cNO_2/GRP per capita) had a steady declining trend in the four economic regions, especially in the western mainland China, showing a notable enhancement of energy efficiency (Fig. 2). The cPM_{10}/GRP per capita ratio fell from 0.103 mg m^{-3} (ten thousand yuan)$^{-1}$ in 2003 to 0.018 in the year of 2012 by 470% in the West, from 0.064 to 0.015 in the Northeast by 320%, from 0.081 to 0.014 in the Central by 470%, and from 0.043 to 0.009 mg m^{-3} (ten thousand yuan)$^{-1}$ in the East by 370%. The ratios of cSO_2/GRP per capita and cNO_2/GRP per capita also showed analogous disparity among the four regions, with the variation range of efficiency ratio the smallest in the East (Fig. 2).

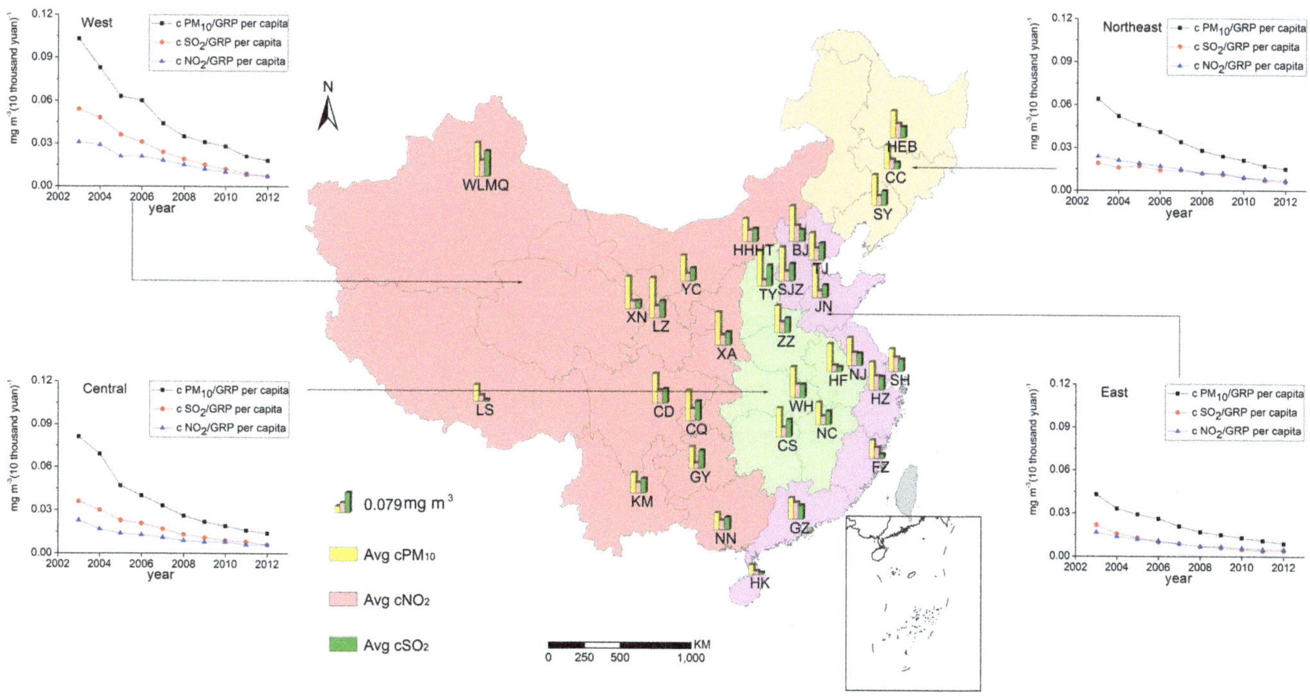

Figure 2. Annual mean concentration of PM_{10}, SO_2 and NO_2 from 2003 to 2012 in different province capitals of mainland China (bar charts on the Chinese map). Four line charts represent the relationships between annual mean air pollutant and GRP per capita of the East, Central, Northeast and West China respectively from 2003 to 2012.

3. EKC analysis in all the provincial capital cities and the four economic regions

The relationship between air pollutants and GRP per capita in all the provincial capitals is presented in Fig. 3. The concentration of PM_{10} and SO_2 or API had a significantly negative linear relationship with GRP per capita; meanwhile the concentration of NO_2 had a quadratic relationship with GRP per capita. However, the relationship between air pollutants and GRP per capita was not the same for the four economic regions (Table 2, 3). The PM_{10} concentration was significantly related to GRP per capita only in the Central. Similarly, API was also significantly related to GRP per capita only in the Central. The SO_2 concentration had significant negative linear relationship with GRP per capita in the Central but positive linear relationship in the Northeast. The NO_2 concentration was positively related to the GRP per capita in the Central, West, and quadratic for the East region, but not significantly related to that in the Northeast.

Discussion

1. The relationship between the three main industries and air pollutants

Our results (Fig. 1) showed the quadratic relationship between the secondary and tertiary industries and NO_2 in the provincial capital cities of China. The increase of NO_2 concentration was probably caused by the continuous increase of civil vehicles (The civil vehicles number increased from 1.36 million to 78.0 million according to the National Bureau of Statistics) and the widespread use of transportation in many fields such as tourism. This was in agreement with the point of view ascribing anthropogenic pollution to combustion of fossil fuel [43–44]. Fortunately, the concentration of NO_2 began to descend as the output of secondary and tertiary industries came to a certain level, probably due to the increasing energy efficiency (Fig. 2) and environmental-friendly measurements such as transportation control during traffic peaks [45]. Different from NO_2, the concentration of PM_{10} and SO_2 decreased with the increase of the industry output per capita, which could also be explained by the improved energy efficiency. The negative relationship between the output per capita of tertiary industry and PM_{10} and SO_2 might, to a large part, attributable to the rapid development of low energy-consumption industries such as high-tech industry, though this explanation needed further confirmation. Besides, unadvanced managements such as straw burning were restricted in suburbs with the improved living standard in cities [46], which also helped to decrease the concentration of PM_{10} and SO_2.

2. Variation of energy efficiency in Chinese cities

The results showed that pollutant emissions at every ten thousand yuan fell with sustainable growth of GRP in the provincial capital cities from 2003 to 2012 (Fig. 2), which was coincident with the improvement in energy and technology in Chinese industries [47–48]. Besides, there was a distinct difference between the high income regions and less developed ones; the more developed cities had lower concentration of air pollutants with smaller variation ranges than the less developed cities. This was probably because of the lower energy intensity and more advanced technology [49–50] as well as better-implemented environment-friendly policies [51] in developed cities like Beijing.

3. The EKC in China

The relationship curves of social economy and some air pollution indicators in a period of time present different shape at different stages of development level of the country or state [52]. SO_2 per capita, for instance, seemed to tail off in 12 selected European countries when GRP per capita reached around 10000

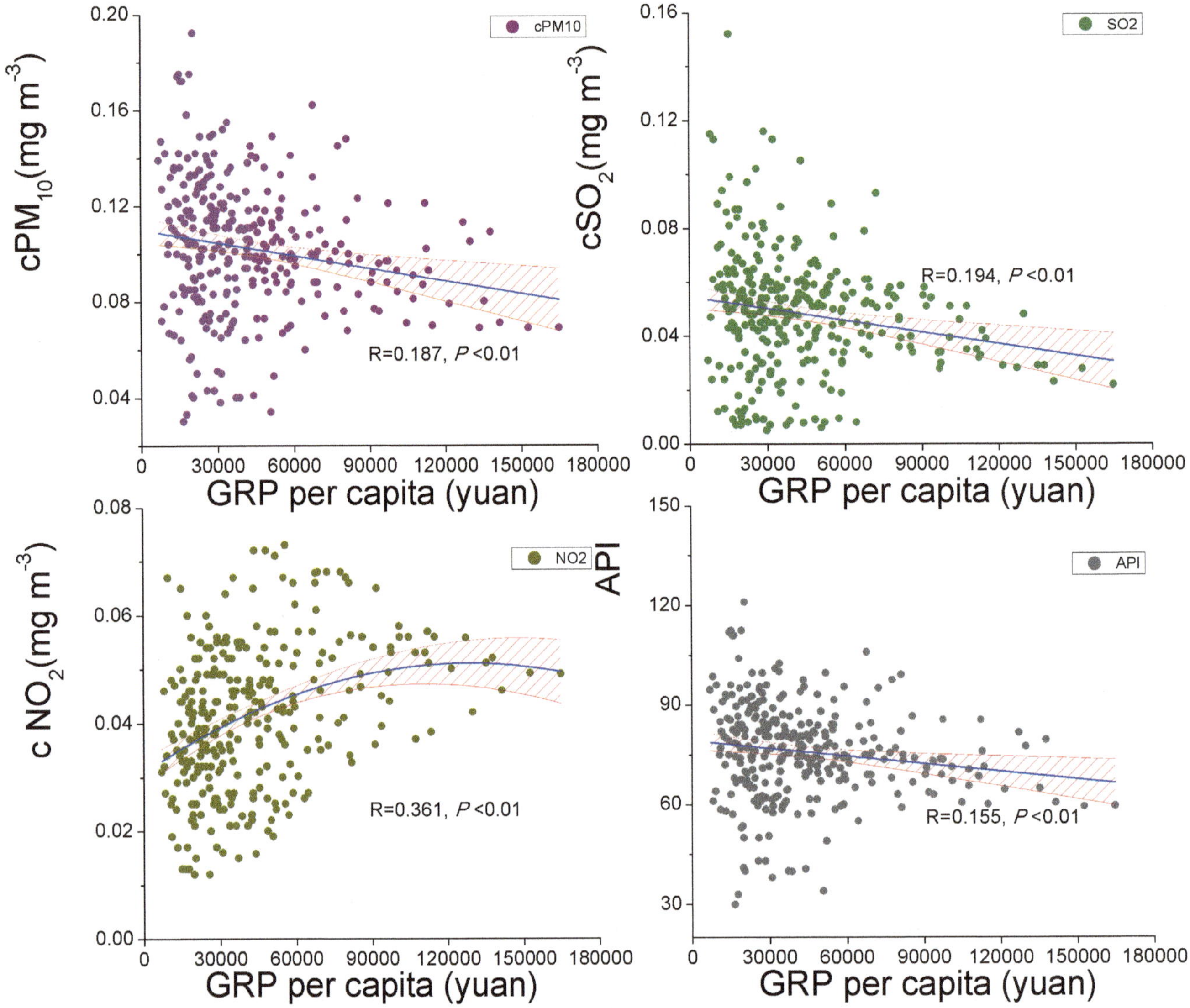

Figure 3. Regression curves between GRP per capita and air pollutant index (PM_{10}, SO_2, NO_2, API) in all the provincial capital cities during 2003–2012. The blue line is the regression line and the pink area the 95% confidence limits.

dollars. Consequently, the relationship curve of SO_2 per capita and GRP per capita appeared a declining trend [53]. Besides, the relationship curve of one air pollutant varied from another for its particular features. Carbon emissions like CO_2 was found to increase at ever-decreasing rates, with predicted peaks beyond reasonable income level because of its cross-border externalities which result in no sufficient incentives to urge countries to regulate emission [18,54–55]. These findings remind us to view the relationship of economy and air pollution with consideration of the time period and specific air pollutants [52,56–57].

In the national scale, our results showed a negative relationship between GRP per capita and PM_{10}, SO_2 or API while inverted-U shape relationship with NO_2 (Fig. 3). The EKC did exist for Chinese cities because the concentration of PM_{10} and SO_2 stop rising from mid-1990s [58]. However, the turning point of EKC for air pollutants seemed to vary with the place, or, with the economical level. Taking SO_2 as an example, the turning point of EKC was approximately 20 thousand yuan in Changsha-Zhuzhou-Xiangtan Urban Agglomeration [59], but 37 thousand in Beijing [60]. Peng & Bao [61] reported a national EKC knee point of around 36 thousand yuan, close to that of 30 thousand yuan claimed by Li et al [62]. Though we lacked data about SO_2 concentration before 2003, our analysis made an estimated turning point of less than 30 thousand yuan, also consistent with other results.

Such EKC pattern was probably caused by the following: (1) The structure of Chinese economy has changed from energy-intensive heavy industry to a more market-oriented service-based economy [62], which, with its lower environmental damage [25], helped China in ameliorating the environment rather than aggravating pollution. Furthermore, in order to stay competitive, firms are keen on investing new and improved technology to enhance cost effectiveness. One of the most significant consequences of this trend is an improvement in resource use efficiency within industrial sector which cut the industrial energy intensity by 50 percent during 1990s [62]. (2) Citizens' environmental awareness is improved. As Chinese people get richer and more educated, they become more concerned about the ambient

Table 2. Regression for concentration of PM_{10}, SO_2, NO_2, API and GRP per capita (panel data of all provincial cities).

	Model summary							
	Described	Regression					Coefficients T test	
Region	Relationship	Model	n	R	SE	Sig.	Independent variable	Constant
	PM10	Linear		0.187	0.027	0.001**	0.001**	0.000**
	&	Quadratic	305	0.187	0.027	0.005**	0.267	0.000**
	GDP per capita	Cubic		0.204	0.027	0.005**	0.076	0.000**
	SO2	Linear		0.194	0.021	0.001**	0.001**	0.000**
	&	Quadratic	305	0.194	0.022	0.003**	0.439	0.000**
All	GDP per capita	Cubic		0.213	0.021	0.003**	0.084	0.000**
Provincial								
City	NO2	Linear		0.344	0.012	0.000**	0.000**	0.000**
	&	Quadratic	305	0.361	0.012	0.000**	0.000**	0.000**
	GDP per capita	Cubic		0.363	0.012	0.000**	0.361	0.000**
	API	Linear		0.155	14.327	0.007**	0.007**	0.000**
	&	Quadratic	305	0.155	14.260	0.026*	0.346	0.000**
	GDP per capital	Cubic		0.179	14.277	0.021*	0.071	0.000**

* $P<0.05$; ** $P<0.01$.

Table 3. Regression for concentration of PM_{10}, SO_2, NO_2, API and GRP per capita.

	Model summary							
	Described	Regression					Coefficients T test	
Region	Relationship	Model	n	R	SE	Sig.	Independent variable	Constant
	PM_{10}	Linear		0.078	0.030	0.441		
	&	Quadratic	100	0.192	0.030	0.163		
	GDP per capita	Cubic		0.192	0.030	0.305		
	SO_2	Linear		0.096	0.022	0.343		
	&	Quadratic	100	0.228	0.022	0.076		
East	GDP per capita	Cubic		0.228	0.022	0.161		
	NO_2	Linear		0.383	0.014	0.000^{**}	0.000^{**}	0.000^{**}
	&	Quadratic	100	0.497	0.013	0.000^{**}	0.000^{**}	0.000^{**}
	GDP per capita	Cubic		0.495	0.013	0.000^{**}	0.016^{*}	0.164
	API	Linear		0.283	6.517	0.129		
	&	Quadratic	100	0.285	6.633	0.318		
	GDP per capita	Cubic		0.296	6.737	0.490		
	PM_{10}	Linear		0.449	0.018	0.000^{**}	0.000^{**}	0.000^{**}
	&	Quadratic	60	0.465	0.018	0.001^{**}	0.050^{*}	0.000^{**}
	GDP per capita	Cubic		0.474	0.018	0.002^{**}	0.870	0.000^{**}
	SO_2	Linear		0.297	0.019	0.021^{*}	0.021^{*}	0.000^{**}
	&	Quadratic	60	0.332	0.019	0.036^{*}	0.063	0.000^{**}
Central	GDP per capita	Cubic		0.391	0.019	0.082		
	NO_2	Linear		0.386	0.010	0.002^{**}	0.002^{**}	0.000^{**}
	&	Quadratic	60	0.386	0.010	0.010^{**}	0.464	0.000^{**}
	GDP per capita	Cubic		0.405	0.010	0.018^{*}	0.244	0.116
	API	Linear		0.449	9.204	0.000^{**}	0.000^{**}	0.000^{**}
	&	Quadratic	60	0.465	9.917	0.001^{**}	0.050^{*}	0.000^{**}
	GDP per capita	Cubic		0.474	9.230	0.002^{**}	0.871	0.000^{**}
	PM_{10}	Linear		0.283	0.013	0.129		
	&	Quadratic	30	0.285	0.013	0.318		
	GDP per capita	Cubic		0.296	0.013	0.490		
	SO_2	Linear		0.463	0.012	0.010^{**}	0.010^{**}	0.000^{**}
	&	Quadratic	30	0.463	0.013	0.038^{*}	0.642	0.013^{*}
Northeast	GDP per capita	Cubic		0.465	0.013	0.091		
	NO_2	Linear		0.333	0.009	0.072		

Table 3. Cont.

		Model summary						
	Described	Regression					Coefficients T test	
Region	Relationship	Model	n	R	SE	Sig.	Independent variable	Constant
	&	Quadratic	30	0.334	0.010	0.202		
	GDP per capita	Cubic		0.348	0.010	0.332		
	API	Linear		0.283	6.517	0.129		
	&	Quadratic	30	0.285	6.633	0.318		
	GDP per capita	Cubic		0.296	6.737	0.490		
	PM_{10}	Linear		0.149	0.030	0.113		
	&	Quadratic	115	0.161	0.031	0.230		
	GDP per capita	Cubic		0.184	0.031	0.227		
	SO_2	Linear		0.169	0.023	0.071		
	&	Quadratic	115	0.222	0.022	0.059		
West	GDP per capita	Cubic		0.225	0.022	0.122		
	NO_2	Linear		0.211	0.012	0.024*	0.024*	0.000**
	&	Quadratic	115	0.233	0.012	0.044*	0.087	0.000**
	GDP per capita	Cubic		0.251	0.012	0.065		
	API	Linear		0.152	15.321	0.104		
	&	Quadratic	115	0.166	15.354	0.208		
	GDP per capita	Cubic		0.190	15.356	0.250		

Only significant P-values of T test are listed.
* $P<0.05$; ** $P<0.01$.

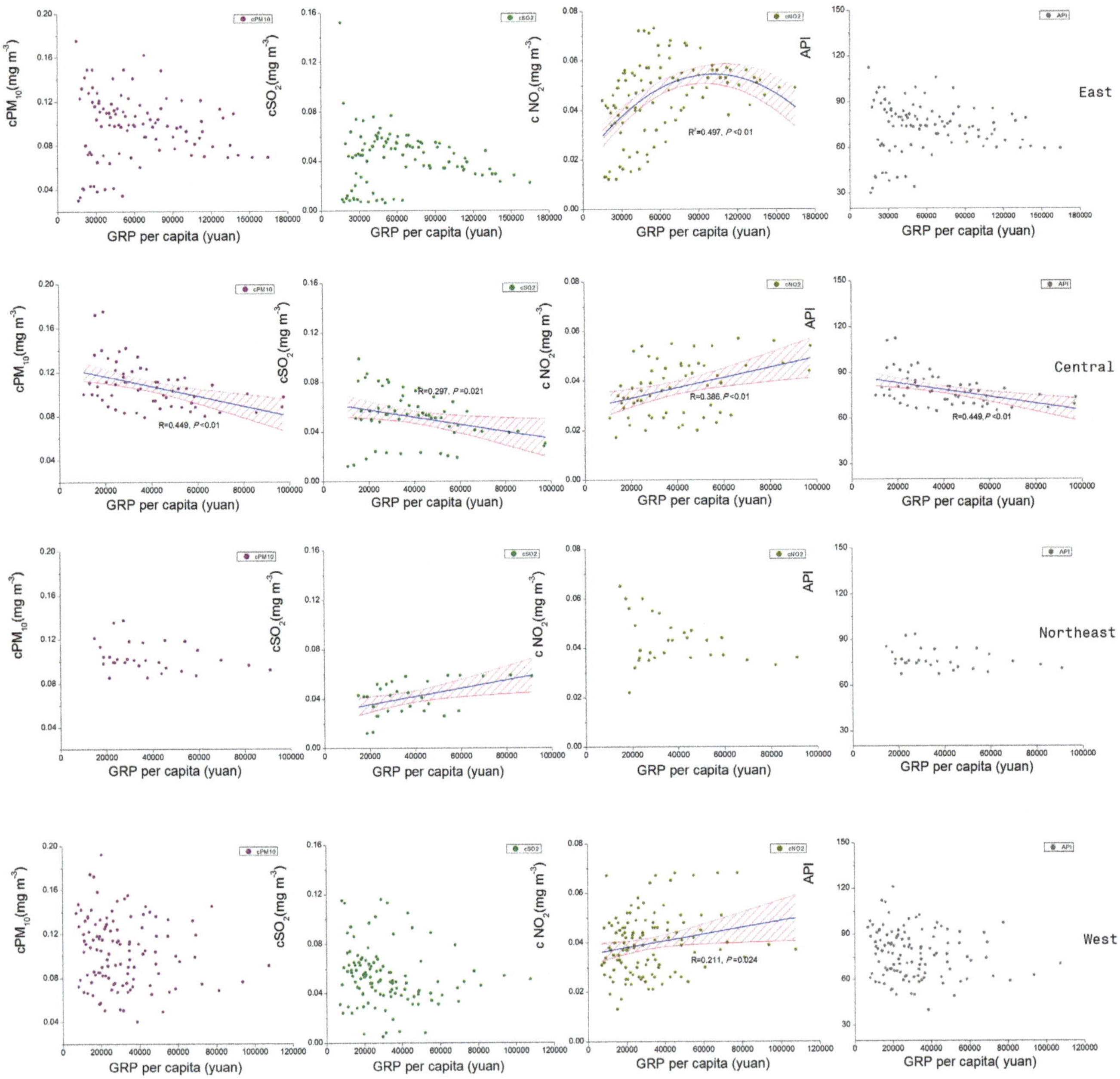

Figure 4. Regression curves between GRP per capita and air pollutant index (PM_{10}, SO_2, NO_2, API) in four economic regions during 2003–2012. The blue line is the regression line and the pink area the 95% confidence limits.

environment they dwell [63–65]]. At this time, their behaviors to protect environment and striving for more governmental support to do so contribute to the emergence of EKC [64,66]. For example, a gigantic demonstration against production of p-xylene in Dalian on August 14th, 2011 [67] reflected the strong demand for better living environment. (3) Regulatory policy for environment protection has been established and effectively implemented, which is another important factor to spur EKC [68-69]. In China, the first law against air pollution took into effect in 1987 and was amended in 1995 and 2000. The environment-friendly measures conducted by the government also provide significant support to air quality. The central government, for example, adopted drastic new pollution control measures for town and village industrial enterprises (TVIEs) and closed 65,000 high-pollution TVIEs in a national campaign in 1996. Therefore, with strict and effective regulatory measures, air pollutants such as SO_2, soot and industrial fugitive dust began to decrease since mid-1990s [57]. It's believed that the environment will continue to improve with Chinese central government policies making efforts to promote ecological progress [70]. However, the rise of the SO_2 concentration in the Northeastern China simultaneous with the enhancement of civil revenues in this area (Fig. 4) might be a result of the policy of "revitalizing the old Northeast industry" by the Central Government [71].

The emergence of the knee point of EKC in the developed cities of Eastern China may be a result of well-implemented environmental policies and high investment in pollution control. But in the meantime, the Central and Northeast regions did not show a

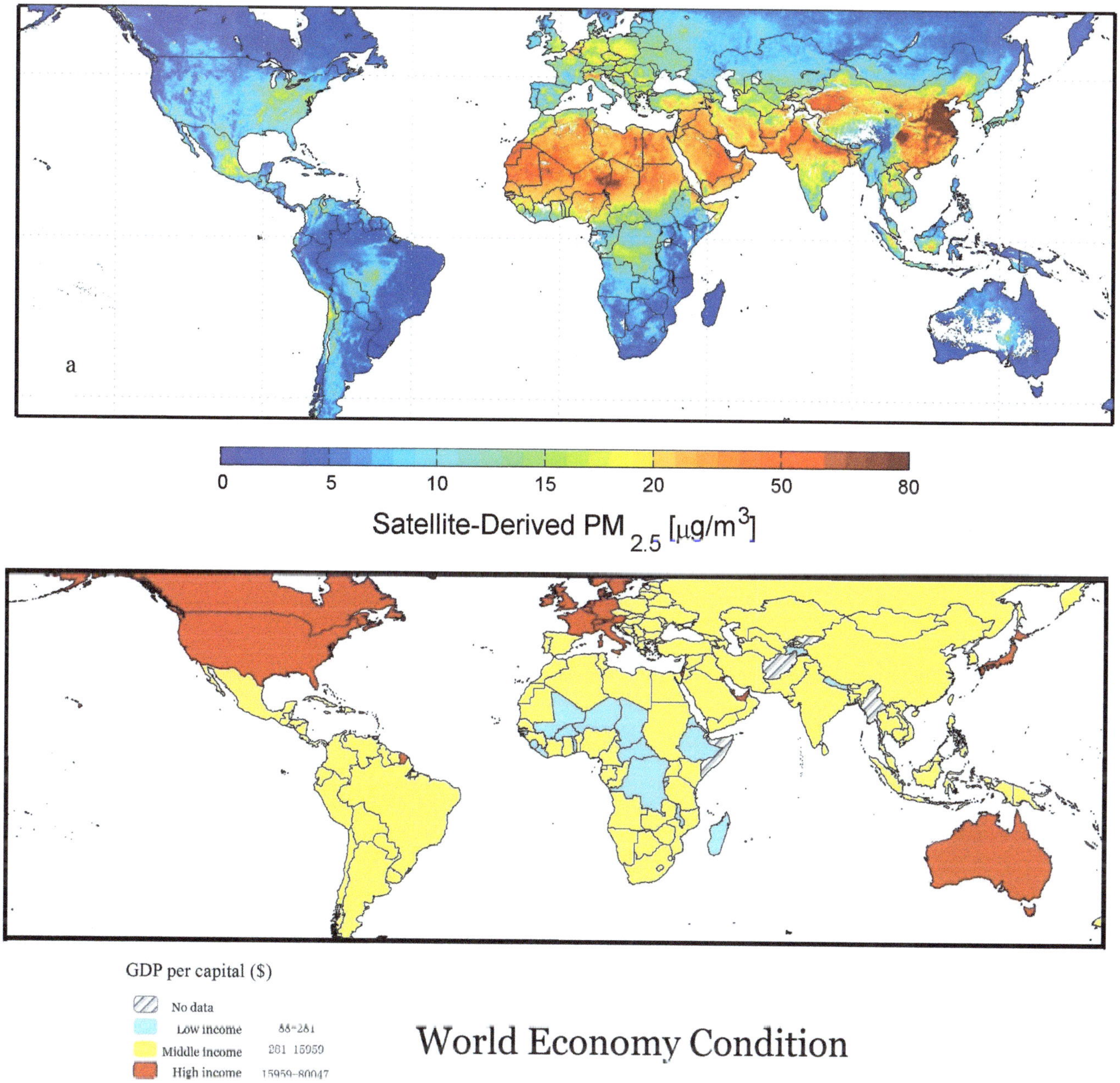

Figure 5. Maps of world $PM_{2.5}$ ($\mu g\ m^{-3}$) and GRP per capita ($) during 2001 to 2006. (a) $PM_{2.5}$, downloaded from NASA website and reproduced with permission from its authors and publisher (van Donkelaar et al., 2010); (b) GRP per capital, derived from the World Development Indicators of the World Bank (http://data.worldbank.org/country).

downward trend of NO_2 concentration (Fig. 4), which was probably attributable to the growing impact of vehicular emissions [51,72]. NO_2 is one of the dominant components in vehicle exhaust [73]. The ever-increasing civil vehicles, particularly the surge of vehicles in the cities after 2000, probably emit enough NO_2 to compensate the decrease of the pollutant from technical advancement of the industries [74]. It would be difficult to decrease NO_2 concentration in most cities if civil vehicles continue to increase in the near future, despite the controlling measures already taken [75].

Since API is a simple and generalized indicator, its variation can reflect the general trend of air pollutants. The negative linear trend of API and GRP per capita in the Central region (Fig. 4) was probably attributable to the overall decreasing trend of the three categories of air pollutants (Fig. 3) [76].

It is worth noticing air pollutants were not significantly related to GRP per capita in any of the four economic regions. Since the classification criterion of the four economic regions was not only the economic development level but also including geographical location, variance of economic levels within a region might have obscured the relationship between air pollutants and GRP per capita. Some detailed classification is needed to improve the accuracy of analysis.

Comparative qualitative analysis of the world also illustrated the existence of EKC and pointed out the developmental status of China in the world scale (Fig. 5 and Table 4). Two comparative

Table 4. Concentration of PM_{10} in cities of different continents.

Continent	Country(time period)	Mean PM_{10} concentration (mg m^{-3})	Scale	Reference
Asia	China(2003–2010)	0.1056±0.0259	National	This study
	Japan (2007–2008)	0.0151±0.0078	Yokohama	[81]
	India (1998–1999,2002)	0.2317±0.0815	New Delhi	[82]
Africa	South Africa (winter of 1997)	0.0933±0.0188	National	[83]
	Tanzania (2005)	0.0510±0.0210	National	[84]
	Guinea (2004)	0.1453±0.1092	Conakry	[85]
South America	Brazil (2008)	0.064±0.0190	São Paulo	[86]
	Argentina (2008)	0.0470±0.0120	Buenos Aires	
	Columbia (2008)	0.0640±0.0490	Bogotá	
Europe	(1992–2009)	0.0306±0.0084	Continental	[87]
	Netherlands (1985–2008)	0.0180	Rotterdam	[88]
	Greece (1999–2000)	0.0755±0.0275	Athens,	[89]
	German (2002–2005)	0.0663±0.0105	National	[90]
North America	US (1992–2009)	0.0276±0.0081	National	[87]
	Canada (1993–2009)	0.0155±0.0052	National	
Oceania	Australia (1998–2001)	0.0175±0.0018	National	[91]
	New Zealand (1999–2007)	0.0299±0.0132	National	[92]

"World Map" in Fig. 5 identified the existence of EKC at the global level. Emerging economics like China and India have the highest concentration of $PM_{2.5}$ (Fig. 5) and PM_{10} (Table 4), mainly resulting from fossil fuel and biomass burning in order to meet the energy demand for rapid economic growth [7]. Developed countries in the Europe, Oceania and North America, however, have lower concentration of particulate matter, probably because of the following two reasons: (1) The developed countries have already accomplished the transition of industrialization which is currently taking place in developing countries like China [77]. (2) The developed countries possess more environment-friendly technology to enhance energy efficiency and reduce pollution [78,79]. As for the undeveloped countries, most of them in Africa, the high concentration of pollutants are mainly caused by the Sahara Desert, which brings them seasonal dry, dust-laden wind known as Harmattan [80]. Of course there are other undeveloped countries with low level of pollutants, simply because they are still at the very early stage of economic development.

Conclusions

The quadratic relationship between the concentration of NO_2 and the output per capita of the secondary or tertiary industry, as well as the negative correlation between the concentration of PM_{10}, SO_2 and industry output per capita, indicate the declining trend of the pollutant concentrations with the improvement of energy efficiency and implementation of environment protection policies. With technology innovation and modulation in industries together with policy implementation from 1990s, ratios of pollutants to GRP in the 31 provincial capitals in mainland China shows a downward trend. Such negative or invert-U quadratic relationship curve between air pollutants and GRP per capita verifies the existence of EKC in China.

Acknowledgments

We thank Mingxu Li for doing a favor with transcribing data in order. The authors give special thanks to Ms. Wan Xiong for her editing and valuable comments on the manuscript.

Author Contributions

Conceived and designed the experiments: HC YPL QAZ CHP. Analyzed the data: YPL. Contributed reagents/materials/analysis tools: YPL YZY GY YZ. Wrote the paper: YPL GY.

References

1. Davis DL (2002) A look back at the London smog of 1952 and the half century since. Environmental health perspectives 110: A734.
2. Akimoto H (2003) Global air quality and pollution. Science 302: 1716–1719.
3. Kopp RE, Mauzerall DL (2010) Assessing the climatic benefits of black carbon mitigation. Proc Natl Acad Sci U S A 107: 11703–11708.
4. Ramanathan V, Feng Y (2009) Air pollution, greenhouse gases and climate change: Global and regional perspectives. Atmospheric Environment 43: 37–50.
5. Clarke A, Kapustin V (2010) Hemispheric Aerosol Vertical Profiles: Anthropogenic Impacts on Optical Depth and Cloud Nuclei. Science 329: 1488–1492.
6. Ramanathan V, Chung C, Kim D, Bettge T, Buja L, et al. (2005) Atmospheric brown clouds: Impacts on South Asian climate and hydrological cycle. Proc Natl Acad Sci U S A 102: 5326–5333.
7. Allen RJ, Sherwood SC, Norris JR, Zender CS (2012) Recent Northern Hemisphere tropical expansion primarily driven by black carbon and tropospheric ozone. Nature 485: 350–U393.
8. Bollasina MA, Ming Y, Ramaswamy V (2011) Anthropogenic aerosols and the weakening of the South Asian summer monsoon. Science 334: 502–505.

9. Booth BB, Dunstone NJ, Halloran PR, Andrews T, Bellouin N (2012) Aerosols implicated as a prime driver of twentieth-century North Atlantic climate variability. Nature 484: 228–232.
10. Menon S, Hansen J, Nazarenko L, Luo YF (2002) Climate effects of black carbon aerosols in China and India. Science 297: 2250–2253.
11. Silva RA, West JJ, Zhang Y, Anenberg SC, Lamarque J-F, et al. (2013) Global premature mortality due to anthropogenic outdoor air pollution and the contribution of past climate change. Environmental Research Letters 8: 034005.
12. Dadvand P, de Nazelle A, Triguero-Mas M, Schembari A, Cirach M, et al. (2012) Surrounding greenness and exposure to air pollution during pregnancy: an analysis of personal monitoring data. Environmental health perspectives 120: 1286.
13. Tonne C, Wilkinson P (2013) Long-term exposure to air pollution is associated with survival following acute coronary syndrome. European heart journal 34: 1306–1311.
14. Chen Y, Ebenstein A, Greenstone M, Li H (2013) Evidence on the impact of sustained exposure to air pollution on life expectancy from China's Huai River policy. Proceedings of the National Academy of Sciences.
15. Crandall RW (1983) Controlling industrial pollution: The economics and politics of clean air.
16. Seaton A, Godden D, MacNee W, Donaldson K (1995) Particulate air pollution and acute health effects. The Lancet 345: 176–178.
17. Stieb DM, Judek S, Burnett RT (2002) Meta-analysis of time-series studies of air pollution and mortality: effects of gases and particles and the influence of cause of death, age, and season. Journal of the Air & Waste Management Association 52: 470–484.
18. Galeotti M (2003) Economic Development and Environmental Protection. Fondazione Eni Enrico Mattei 2003:89.
19. Carson RT (2010) The environmental Kuznets curve: seeking empirical regularity and theoretical structure. Review of Environmental Economics and Policy 4: 3–23.
20. Selden TM, Song D (1994) Environmental quality and development: is there a Kuznets curve for air pollution emissions? Journal of Environmental Economics and management 27: 147–162.
21. Shafik N (1994) Economic development and environmental quality: an econometric analysis. Oxford Economic Papers: 757–773.
22. Cole MA, Rayner AJ, Bates JM (1997) The environmental Kuznets curve: an empirical analysis. Environment and development economics 2: 401–416.
23. De Bruyn SM, van den Bergh JC, Opschoor JB (1998) Economic growth and emissions: reconsidering the empirical basis of environmental Kuznets curves. Ecological Economics 25: 161–175.
24. Grossman GM, Krueger AB (1991) Environmental impacts of a North American free trade agreement. National Bureau of Economic Research.
25. Panayotou T (1993) Empirical tests and policy analysis of environmental degradation at different stages of economic development. International Labour Organization.
26. Wang Y-C (2011) Short-and Long-run Environmental Kuznets Curve: Case Studies of Sulfur Emissions in OECD Countries International Journal of Economic Research 9: 1–18.
27. Akbostancı E, Türüt-Aşık S, Tunç Gİ (2009) The relationship between income and environment in Turkey: Is there an environmental Kuznets curve? Energy Policy 37: 861–867.
28. Fodha M, Zaghdoud O (2010) Economic growth and pollutant emissions in Tunisia: An empirical analysis of the environmental Kuznets curve. Energy Policy 38: 1150–1156.
29. De Groot HL, Withagen CA, Minliang Z (2004) Dynamics of China's regional development and pollution: an investigation into the Environmental Kuznets Curve. Environment and development economics 9: 507–537.
30. Jalil A, Mahmud SF (2009) Environment Kuznets curve for CO_2emissions: A cointegration analysis for China. Energy Policy 37: 5167–5172.
31. Vincent JR (1997) Testing for environmental Kuznets curves within a developing country. Environment and development economics 2: 417–431.
32. Song T, Zheng T, Tong L (2008) An empirical test of the environmental Kuznets curve in China: A panel cointegration approach. China Economic Review 19: 381–392.
33. Ren W-X, Xue B, Zhang L, Ma Z-X, Geng Y (2013) Spatiotemporal variations of air pollution index in China's megacities. Chinese Journal of Ecology 32 (10):2788–2796, in Chinese with english abstract.
34. GAO B-L (2009) Relationship Between Environmental Pollution and Economic Growth in Jiangsu Province. POLLUTION CONTROL TECHNOLOGY 22(6):36–39, in Chinese with english abstract.
35. Li H-Y, Li J, Zhang D-H, Chen S-S (2013) The Environmental Kuznets Curve in Xi'an. ARID ZONE RESEARCH 30(3): 556–562, in Chinese with english abstract.
36. Shen J (2006) A simultaneous estimation of Environmental Kuznets Curve: Evidence from China. China Economic Review 17: 383–394.
37. Hu M-X, Hu H, Liu Z (2005) Air Environmental Kuznets Curve (EKC) Characteristics in Wuhan City. Journal of Wuhan Polytechnic University 24(1):72–75, in Chinese with english abstract.
38. He K, Huo H, Zhang Q (2002) Urban air pollution in China: current status, characteristics, and progress. Annual Review of Energy and the Environment 27: 397–431.
39. Ministry of Environmental Protection of the People's Republic of China (2012) http://kjs.mep.gov.cn/hjbhbz/bzwb/dqhjbh/jcgfffbz/201203/t20120302_224166.htm.
40. Wikipedia. (2013). http://en.wikipedia.org/wiki/List_of_regions_of_China.
41. Agras J, Chapman D (1999) A dynamic approach to the Environmental Kuznets Curve hypothesis. Ecological Economics 28: 267–277.
42. LI R-e, ZHANG H-j (2008) An empirical analysis on the regional discrepancy and tendency of EKC in China (1981–2004). Journal of Xi'an Jiaotong University (Social Sciences) 4: 007.
43. Kampa M, Castanas E (2008) Human health effects of air pollution. Environmental Pollution 151: 362–367.
44. Gustafsson Ö, Kruså M, Zencak Z, Sheesley RJ, Granat L, et al. (2009) Brown clouds over South Asia: biomass or fossil fuel combustion? Science 323: 495–498.
45. Zhou Y, Wu Y, Yang L, Fu L, He K, et al. (2010) The impact of transportation control measures on emission reductions during the 2008 Olympic Games in Beijing, China. Atmospheric Environment 44: 285–293.
46. Kim Oanh NT, Ly BT, Tipayarom D, Manandhar BR, Prapat P, et al. (2011) Characterization of particulate matter emission from open burning of rice straw. Atmospheric Environment 45: 493–502.
47. Sinton JE, Fridley DG (2000) What goes up: recent trends in China's energy consumption. Energy Policy 28: 671–687.
48. Hu J-L, Wang S-C (2006) Total-factor energy efficiency of regions in China. Energy Policy 34: 3206–3217.
49. Dhakal S (2009) Urban energy use and carbon emissions from cities in China and policy implications. Energy Policy 37: 4208–4219.
50. Fan J (2006) Industrial Agglomeration and Difference of Regional Labor Productivity: Chinese Evidence with International Comparison [J]. Economic Research Journal 11: 72–81.
51. Li L, Chen C, Xie S, Huang C, Cheng Z, et al. (2010) Energy demand and carbon emissions under different development scenarios for Shanghai, China. Energy Policy 38: 4797–4807.
52. Levinson A (2002) The ups and downs of the environmental Kuznets curve. Recent Advances in Environmental Economics Edward Elgar, Cheltenham: 119–139.
53. Markandya A, Golub A, Pedroso-Galinato S (2006) Empirical analysis of national income and SO_2 emissions in selected European countries. Environmental and resource economics 35: 221–257.
54. Lieb CM (2004) The environmental Kuznets curve and flow versus stock pollution: the neglect of future damages. Environmental and resource economics 29: 483–506.
55. Miah D, Masum FH (2010) Global observation of EKC hypothesis for CO_2, SO_2 and NO_2 emission: A policy understanding for climate change mitigation in Bangladesh. Energy Policy 38: 4643–4651.
56. Figueroa B, Pastén C (2009) Country specific environmental Kuznets curves: A random coefficient approach applied to high-income countries. Estudios de Economía 36: 5–32.
57. Harris PG (2006) Environmental Perspectives and Behavior in China Synopsis and Bibliography. Environment and Behavior 38: 5–21.
58. Zeng H-X, Qin D-L, Luo F, Zhang H, Luo Y-P (2010) Study on the Relationship between Enironmetal Pollution and Economic Growth in the Changsha - Zhuzhou - Xiangtan Urban Agglomera tion. Ecological Economy 1:347–350, in Chinese with english abstract.
59. Zhou Y-M, Huang S-P (2010) Research on Relationship between Economic Growth and Environmental Pollution: an Empirical Analysis Based on Panel Data of Beijing. In Chinese with english abstract.
60. Peng S-J, Bao Q (2006) Economic Growth and Environmental Pollution: An Empirical Test for the Environmental Kuznets Curve Hypothesis in China. Research on Financial and Economic Issues 8: 3–17.
61. Li R-P, Wang G-S, Wang A-J, Luo J-H, Geng N (2010). Factor Analysis of SO2 Emission Trend in Typical Industrialized Countries and Its Revelation to China. Acta Geoscientica Sinica 31(5):749–758, in Chinese with english abstract.
62. World Bank (2001) China: air, land, and water: environmental priorities for a new millennium. Washington, DC: World Bank.
63. Liu J, Diamond J (2005) China's environment in a globalizing world. Nature 435: 1179–1186.
64. Guo X, Marinova D (2011) Environmental awareness in China: Facilitating the greening of the economy.
65. Martens I S (2006) Public participation with Chinese characteristics: citizen consumers in China's environmental management. Environmental politics 15: 211–230.
66. Xie L (2011) China's environmental activism in the age of globalization. Asian Politics & Policy 3: 207–224.
67. BBC News. (2011) China protest closes toxic chemical plant in Dalian. http://www.bbc.co.uk/news/world-asia-pacific-14520438
68. He J, Wang H (2012) Economic structure, development policy and environmental quality: An empirical analysis of environmental Kuznets curves with Chinese municipal data. Ecological Economics 76: 49–59.
69. Kijima M, Nishide K, Ohyama A (2011) EKC-type transitions and environmental policy under pollutant uncertainty and cost irreversibility. Journal of Economic Dynamics and Control 35: 746–763.
70. Xinhua News Agency. (2012) Full text of Hu Jintao's report at 18th Party Congress. http://news.xinhuanet.com/english/special/18cpcnc/2012-11/17/c_131981259.htm

71. Zhang P, Ma Y, Liu W, Chen Q (2004) New Urbanization Strategy for Revitalizing the Tradianal Industrial Base of Northeast China. Acta Geographica Sinica: S1.
72. World Bank. (1997). Clear Water, blue skies: China's environmental pollution in the new century. Washington, DC: World Bank.
73. Brugge D, Durant JL, Rioux C (2007) Near-highway pollutants in motor vehicle exhaust: a review of epidemiologic evidence of cardiac and pulmonary health risks. Environmental Health 6: 23.
74. National Bureau of Statistics. (2011) http://www.stats.gov.cn/tjsj/ndsj/2011/indexch.htm.
75. Li X-F, Zhang M-J, Wang S-J, Zhao A-F, Ma Q (2012) Variation Characteristics and Influencing Factors of Air Pollution Index in China. Environmental Science 33(6):1936-1943, in Chinese with english abstract.
76. Wang L, Jang C, Zhang Y, Wang K, Zhang Q, et al. (2010) Assessment of air quality benefits from national air pollution control policies in China. Part II: Evaluation of air quality predictions and air quality benefits assessment. Atmospheric Environment 44: 3449-3457.
77. Great Lakes Invitational Conference Association (GLICA). (2008) Promoting Sustainable Industry in the Developing World. http://www.glica.org/topics/show/53
78. Dechezleprêtre A, Glachant M, Haščič I, Johnstone N, Ménière Y (2011) Invention and transfer of climate change–mitigation technologies: a global analysis. Review of Environmental Economics and Policy 5: 109–130.
79. Popp D (2011) International technology transfer, climate change, and the clean development mechanism. Review of Environmental Economics and Policy 5: 131–152.
80. Afeti G, Resch F (2000) Physical characteristics of Saharan dust near the Gulf of Guinea. Atmospheric Environment 34: 1273–1279.
81. Khan MF, Shirasuna Y, Hirano K, Masunaga S (2010) Characterization of PM2.5, PM2.5–10 and PM> 10 in ambient air, Yokohama, Japan. Atmospheric Research 96: 159–172.
82. Monkkonen P, Uma R, Srinivasan D, Koponen IK, Lehtinen KEJ, et al. (2004) Relationship and variations of aerosol number and PM10 mass concentrations in a highly polluted urban environment - New Delhi, India. Atmospheric Environment 38: 425–433.
83. Engelbrecht JP, Swanepoel L, Chow JC, Watson JG, Egami RT (2001) PM2.5 and PM10 concentrations from the Qalabotjha low-smoke fuels macro-scale experiment in South Africa. Environmental Monitoring and Assessment 69: 1–15.
84. Mkoma SL, Maenhaut W, Chi XG, Wang W, Raes N (2009) Characterisation of PM10 atmospheric aerosols for the wet season 2005 at two sites in East Africa. Atmospheric Environment 43: 631–639.
85. Weinstein JP, Hedges SR, Kimbrough S (2010) Characterization and aerosol mass balance of PM2.5 and PM10 collected in Conakry, Guinea during the 2004 Harmattan period. Chemosphere 78: 980–988.
86. Vasconcellos PC, Souza DZ, Avila SG, Araujo MP, Naoto E, et al. (2011) Comparative study of the atmospheric chemical composition of three South American cities. Atmospheric Environment 45: 5770–5777.
87. Wang KC, Dickinson RE, Su L, Trenberth KE (2012) Contrasting trends of mass and optical properties of aerosols over the Northern Hemisphere from 1992 to 2011. Atmospheric Chemistry and Physics 12: 9387–9398.
88. Keuken M, Zandveld P, van den Elshout S, Janssen NAH, Hoek G (2011) Air quality and health impact of PM10 and EC in the city of Rotterdam, the Netherlands in 1985-2008. Atmospheric Environment 45: 5294–5301.
89. Chaloulakou A, Kassomenos P, Spyrellis N, Demokritou P, Koutrakis P (2003) Measurements of PM10 and PM2.5 particle concentrations in Athens, Greece. Atmospheric Environment 37: 649–660.
90. Buns C, Klemm O, Wurzler S, Hebbinghaus H, Steckelbach I, et al. (2012) Comparison of four years of air pollution data with a mesoscale model. Atmospheric Research 118: 404–417.
91. Barnett AG, Williams GM, Schwartz J, Neller AH, Best TL, et al. (2005) Air pollution and child respiratory health - A case-crossover study in Australia and new Zealand. American Journal of Respiratory and Critical Care Medicine 171: 1272–1278.
92. Trompetter WJ, Davy PK, Markwitz A (2010) Influence of environmental conditions on carbonaceous particle concentrations within New Zealand. Journal of Aerosol Science 41: 134–142.

16

The Relationship between Particulate Pollution Levels in Australian Cities, Meteorology, and Landscape Fire Activity Detected from MODIS Hotspots

Owen F. Price[1]*, Grant J. Williamson[2], Sarah B. Henderson[3], Fay Johnston[2,4], David M. J. S. Bowman[2]

1 School of Biological Sciences, University of Wollongong, Wollongong, New South Wales, Australia, **2** School of Plant Sciences, University of Tasmania, Hobart, Tasmania, Australia, **3** Environmental Health Services, British Columbia Centre for Disease Control, Vancouver, British Columbia, Canada, **4** Menzies Research Institute, University of Tasmania, Hobart, Tasmania, Australia

Abstract

Smoke from bushfires is an emerging issue for fire managers because of increasing evidence for its public health effects. Development of forecasting models to predict future pollution levels based on the relationship between bushfire activity and current pollution levels would be a useful management tool. As a first step, we use daily thermal anomalies detected by the MODIS Active Fire Product (referred to as "hotspots"), pollution concentrations, and meteorological data for the years 2002 to 2008, to examine the statistical relationship between fire activity in the landscapes and pollution levels around Perth and Sydney, two large Australian cities. Resultant models were statistically significant, but differed in their goodness of fit and the distance at which the strength of the relationship was strongest. For Sydney, a univariate model for hotspot activity within 100 km explained 24% of variation in pollution levels, and the best model including atmospheric variables explained 56% of variation. For Perth, the best radius was 400 km, explaining only 7% of variation, while the model including atmospheric variables explained 31% of the variation. Pollution was higher when the atmosphere was more stable and in the presence of on-shore winds, whereas there was no effect of wind blowing from the fires toward the pollution monitors. Our analysis shows there is a good prospect for developing region-specific forecasting tools combining hotspot fire activity with meteorological data.

Editor: Alex J. Cannon, Pacific Climate Impacts Consortium, Canada

Funding: This work was supported by the ARC (Australian Research Council) Linkage Grant LP0882048 (www.arc.gov.au), the Rural Fire Service of NSW, and a grant from the Tasmanian Environment Fund Grant. The funders had no role in study design, data collection and analysis, decision to publish, or preparation of the manuscript.

Competing Interests: The authors have declared that no competing interests exist.

* E-mail: oprice@uow.edu.au

Introduction

An increasingly important issue of fire management revolves around the health impacts of smoke pollution. Smoke is a complex mixture of particulate and gaseous pollutants [1] that has been associated with a wide range of adverse health outcomes [2]. Smoke from bushfires can travel vast distances to affect towns and cities far from its original source [3,4]. Bushfire smoke has been clearly associated with exacerbation of respiratory illnesses, increased respiratory hospital admissions, and visits to emergency departments [5]. The effect of bushfire smoke on other health outcomes such as cardiovascular morbidity and mortality has been less extensively researched. Of six studies into smoke-related particulate-matter mortality, three found an association [6,7,8], while associations with cardiovascular disease have rarely been reported [2]. However, the weight of evidence suggests that smoke particles elicit toxicological effects similar to those of particles from urban pollution (eg motor vehicle emmissions) [9,10,11], and the association between urban particles and respiratory and cardiovascular morbidity and mortality is well established [12].

Fire managers are required to make many decisions based on assessments of risks related to many variables (ecological, property, human safety and health), and each decision may impact other management variables. One of the greatest trade-offs is the use of prescribed burning to limit the extent and intensity of uncontrolled wildfires. For example, the 2009 Victorian Bushfire Royal Commission recommends increasing the level of burning of public bush land from 1% per year currently to 5% [13]. Most recent wildfire enquiries in Australia have made similar recommendations [14,15]. In the USA, there is mounting pressure to increase prescribed burning rates to counter the increasing area burnt by wildland fires [16,17,18]. Recent evidence suggests that such increases to prescribed fire will actually increase the total area burnt, because several hectares of prescribed fire are required for each hectare of wildfire area reduction [19,20]. Such a marked increase in prescribed burning may have adverse impacts on urban airsheds, counteracting the fact that prescribed fires have less extensive smoke plumes than wildfires [21].

Resolving this management trade-off associated with prescribed burning requires understanding of the risks associated with smoke pollution, and tools for reliably predicting where smoke is likely to travel. With appropriate warning, susceptible people can minimize their exposure by taking appropriate precautions. In addition, managers could use predictive models to decide whether or not the atmospheric conditions suit a prescribed fire, because they have a legal responsibility to avoid causing exceedances of the

Australian air quality standards for particulate matter. The regulatory standard for PM_{10} (particulate matter <10 μm in aerodynamic diameter, in μg/m^3) is currently 50 μg/m^3 while the advisory reporting standard for $PM_{2.5}$ (particulate matter <2.5 μm in diameter, in μg/m^3) is 25 μg/m^3 [22]. The standards for ambient air pollution are currently being reviewed and it is possible that the $PM_{2.5}$ standard will be upgraded from an advisory to a regulatory standard. As $PM_{2.5}$ is the major component of particular matter in bushfire smoke, prescribed and wild fires could become an important cause of failure to achieve the regulatory standards [23].

Most methods for tracking smoke from bushfires are based on atmospheric dispersion modelling, which is extremely data and resource intensive. At present, such models are routinely used for smoke pollution forecasting in North America [24], but not in Australia [25]. Evaluation of these models has been limited [26,27], but the correlation between predicted and actual pollution concentrations at monitoring sites is moderate (between 0.35 and 0.71 [26,28]), as is the spatial overlap between predicted and observed plumes (average 12% [29]). Ultimately, they are physical models rather than empirical models. We believe there is merit in investigating the independent empirical relationship between observed fire activity and measured pollution for two reasons. First, empirical models quantify the direct link between cause and effect, rather than relying on physical models and all of their necessary assumptions. Second, the insights gained may be used to improve computationally demanding physical models.

Remote sensing has revolutionised the analysis of landscape fire activity, and satellite data may also be useful for measuring smoke transport. The Moderate Resolution Imaging Spectroradiometer (MODIS) instruments aboard the Terra and Aqua satellites (operated by the US National Aeronautics and Space Administration) provide a wide range of fire-relevant information. In the active fire product (MOD14 from Terra and MYD14 from Aqua), thermal anomalies (referred to herein as "hotspots") are recorded at a nadir resolution of 1 km for each satellite overpass (occurring globally at approximately 02:00, 10:00, 14:00 and 22:00 local time) [30], and the Fire Radiative Power (FRP) of each hotspot is measured. This attribute provides valuable information about the rate at which a fire is generating energy, and has been directly correlated with its aerosol emissions [31,32]. Analytic use of the FRP measurement provides a unique opportunity to better examine the air quality impacts of bushfire smoke.

Here we investigate the relationship between daily pollution levels and the daily spatial pattern of hotspot activity in the landscapes surrounding two large Australian cities: Sydney and Perth. We also investigate how atmospheric conditions affect the relationship between pollution and hotspots, because smoke from fires will only move towards cities if winds are blowing from the fire to the city, and the smoke remains in the lower levels of the atmosphere. Specifically we examine: (1) whether there is a detectable link between fire activity as measured by MODIS FRP and particulate matter concentrations in cities; (2) the spatial and temporal zone of influence of bushfire activity on city pollution; (3) the potential for FRP to be combined with weather variables for smoke pollution forecasting; and (4) the use of FRP as a tool to filter the contribution of biomass smoke to records of air pollution in urban airsheds.

Materials and Methods

Study Areas

Sydney is a city of 4.0 million people, lying in a highly developed coastal lowland plain,surrounded by dissected sandstone tablelands (Figure 1a, Figure 2). The native vegetation in the tablelands is largely intact and is dominated by a diverse dry sclerophyll eucalypt forest [33], with a total area of approximately 20,000 km^2. The climate is warm and temperate, and the rainfall total of 1200 mm is evenly distributed through the year (Bureau of Meteorology data). Approximately 5% of the forest is burnt by unplanned fires each year, and another 1% is burnt by prescribed fires [20]. Other flammable vegetation types (grasslands, shrublands and woodlands) are minor components and mostly further from Sydney than the forests.

Perth is the capital City of Western Australia (WA), with a population of 1.4 million people in a metropolitan area of 5300 km^2 (Figure 2). The climate is Mediterranean, with cool wet winters and hot dry summers that are conducive to bushfires. The surrounding forest has a similar total area to Sydney (approximately 20,000 km^2), but is much more scattered (Figure 1b). The forests are also dry sclerophyll and the communities are mostly dominated by one of two commercial species, namely jarrah (Eucalyptus marginata) and karri (Eucalyptus diversicolor) [34]. In the Warren region, typical of the forested part of WA, approximately 7% of the forest is burnt each year by prescribed fire and 1.5% by wildfire [19].

Data

Pollution records were obtained from monitoring stations in each of the cities. For Sydney the monitor is at Richmond, 50 km inland (−33.618, 150.746, New South Wales (NSW) Department of Environment and Climate Change, http://www.environment.nsw.gov.au/AQMS) and for Perth it is at Duncraig, 2 km inland (−31.826, 115.783, WA Department of Environment and Conservation, WA Air Quality Data). Daily peak $PM_{2.5}$ data was available for both cities from the beginning of 2002 to the end of 2007 (n = 2191). For Perth, the mean daily peak value across the year is approximately 8 μg/m^3, and for Sydney it is about 10 μg/m^3 (Figure 3). Exceedance of the 25 μg/m^3 24-hour air quality guideline occurs in less than 5% of days in all seasons for both cities. Exceedances are equally likely in all seasons in Perth, but for Sydney they are more frequent in summer, and these are most likely due to bushfires. The ambient level of pollution is mostly due to fossil fuel combustion [35], but other sources, including dust storms, cause some of the exceedances [36].

The daytime detects from the MODIS active fire data from Aqua and Terra were acquired from the Fire Information for Resource Management System (FIRMS – http://maps.geog.umd.edu/firms/) for 2002 through 2007. Note that Collection 4 values were used, meaning that they were not multiplied by pixel area as in Collection 5. As such, this study follows the methods developed by Ichoku and Kaufman (2005) [37] and Henderson *et al.* (2010) [38] for the use of MODIS Collection 4 data for smoke detection. For each day, all of the daytime hotspots within 500 km were identified, and their distance from and direction to the pollution monitors were calculated, and their FRP and distance-weighted FRP (FRP/distance^2) values were recorded. The distance weighting was included because we expect smoke concentrations to diminish in proportion to the square of the distance from the source.

The daily weather record was obtained for the closest weather station from each city from the Bureau of Meteorology: Sydney Airport (BOM Station number 066037) and Perth Airport (BOM Station number 009021), which are approximately 50 km and 25 km away from the pollution monitors, respectively. We derived daily values of mean wind speed, whether the wind was onshore or not (wind angle <=90), and Forest Fire Danger Index (FFDI). The FFDI is a function of wind speed, temperature, humidity and

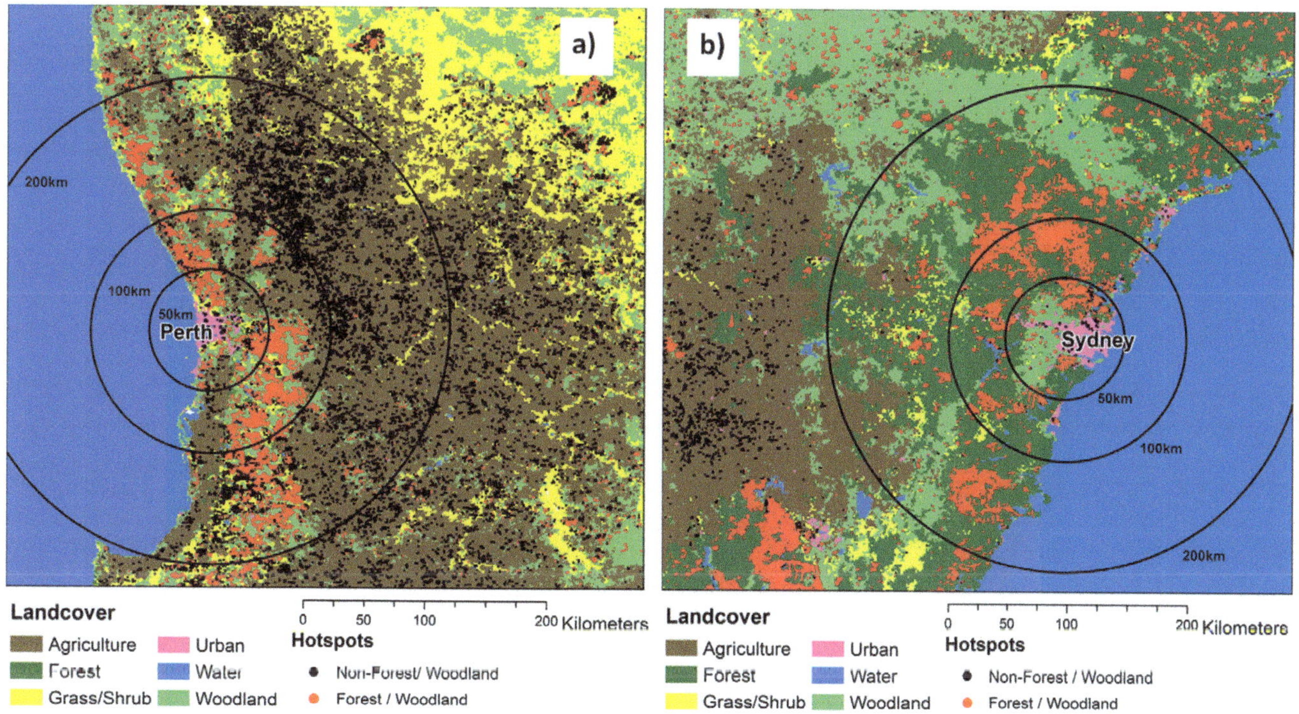

Figure 1. Map of the two study landscapes: a) Sydney; b) Perth. Dominant vegetation types are shaded and hotspots during the study are superimposed: forest/woodland hotspots in red; and non forest/woodland hotspots in black. Distances from the pollution monitor are shown as concentric rings.

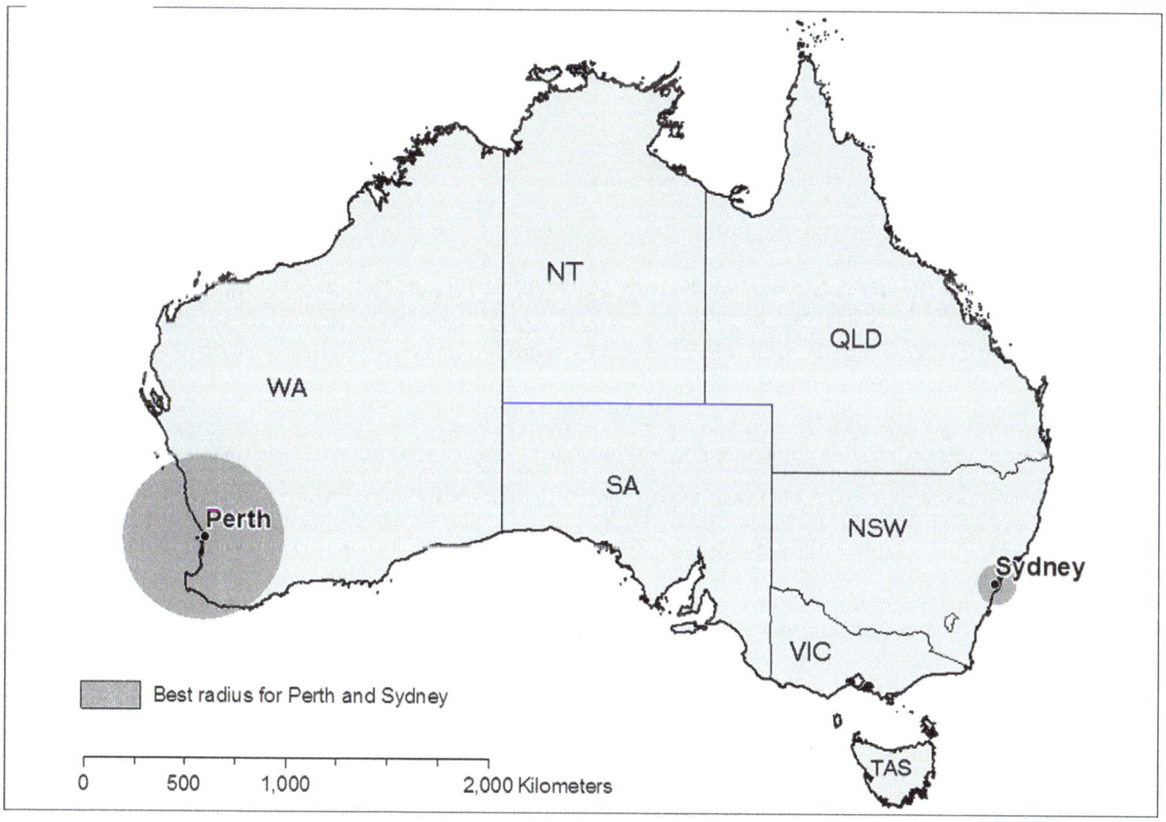

Figure 2. Location of Sydney and Perth, with the best radius also shown.

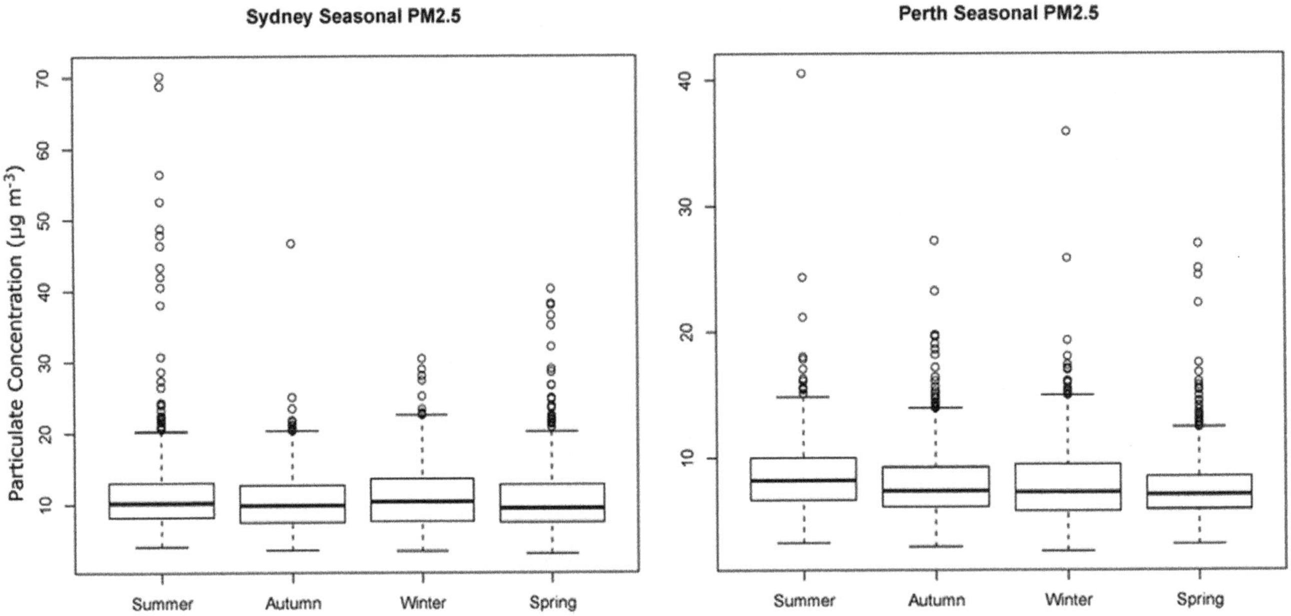

Figure 3. Seasonal $PM_{2.5}$ distribution for a) Sydney; b) Perth. For each season the mean, standard deviation, 95th percentile and individual outliers are plotted. The y axis is the maximum daily $PM_{2.5}$ (Particulate Pollution <2.5 µm in size, in $\mu g/m^3$).

a drought factor, and is routinely used to predict bushfire risk in Australia [39]. In addition, upper atmospheric weather data were obtained from the same weather stations. These were used to calculate the Continuous Haines Index of atmospheric stability, using temperature and dewpoint values at 700 and 850 hPa [40], as well as the wind direction and wind speed at the surface and 700 hPa. These data are recorded several times per day, and through preliminary analysis we found that the early morning record (6 am) provided the best relationships with pollution levels. We also calculated the wind-angle relative to the dominant direction of hotspots using both the station and upper atmospheric wind directions. That is, the mean bearing between hotspots and the pollution monitor was calculated, and this was subtracted from the wind-direction so that a value of 0 represents the situation where the monitor is directly downwind from the mean centre of hotspots, and 180 where the monitor is upwind. For this calculation, all hotspots within 100 km of Sydney and 400 km of Perth were used, because these were found to be the distances over which hotspots influenced pollution levels (see below).

Lag Effects and Duration of Pollution Events

If fire activity is to be used in a pollution forecasting tool, then it is necessary that the fire activity from one day is related to the pollution level from the following day. That means either it must take one day for the smoke to travel to the city, or otherwise the pollution event must last more than one day. It is also possible that this lag effect lasts more than one day. To explore the nature of lag effects, we constructed generalised linear models of $PM_{2.5}$ (dependent) against FRP values (predictor) for different lag periods from 1 to 5 days. Three different lagged FRP measures were tested at each lag duration: the lag day on its own (Lag FRP), all days from the lag to the day before the current day (Combined Forecast FRP) and all days including the current day (Combined FRP). The models were structured with a normal error distribution and identity link function. The raw values of $PM_{2.5}$ and the residuals of the model were inspected for normality, and the goodness of fit was compared among the variables using the proportion of variance explained (adjusted r^2). To explore how long smoke may persist, we calculated the lengths of periods when the respective 95th percentile values were exceeded for each of FRP and $PM_{2.5}$. This generated two frequency distributions of duration. If we define an "event" as any day above the 95th percentile among all days, the sum of the number of events and their durations will be approximately 109 (5% of the total number of days). For both of these analyses, the distance-weighted FRP within the best-fitting radius around the pollution monitor was used (see below, which for Sydney was 100 km and for Perth 400 km).

Radius

There is a limit to the distance at which smoke from fires will spread to a city. Wise (2008) [4] acknowledged this in his study comparing pollution with fire activity in Texas, by using a 45 km cut-off, though he did not know the actual extent of the influence. To investigate how distance influenced pollution, we constructed generalised linear models of the relationship between FRP (predictor) and $PM_{2.5}$ (dependent) using subsets of the hotspots at increasing radii from the monitor: at 25, 50, 75, 100, 150, 200, 300, 400 and 500 km. These were inclusive subsets, meaning that all of the hotspots closer than each radius were included. As with the lag analysis, the models used a normal error distribution, and the best interval was identified using the proportion of variance captured by the model (adjusted r^2). For each radius, we also determined whether the r^2 could be improved by log-transforming the FRP, $PM_{2.5}$, or both variables. For all analyses, we used the mean FRP from the current and previous day (Combined FRP with lag = 1, which we now term FRP_{01}), to take account of residual smoke from the previous day's fires. The lag analysis (above) identified this to be the best FRP variable for predicting $PM_{2.5}$. Also, days without fires on both days (zero FRP_{01}) were excluded, giving a final sample size of 1845 days for Sydney and 1687 for Perth. This was to reduce the 'noise' in the data because pollution from other sources still occurred on those days, but the cause was unlikely to have been bushfire. Also, from an operational perspective, a bushfire smoke prediction model is only relevant during bushfire events, and removing zero FRP_{01} days also reduced skewness in the FRP_{01} data. Exploratory analysis revealed that removing zero FRP_{01} days did not change the form of the relationships in the models but improved their goodness of fit.

Best Model

A model was developed to maximise the explained variation in pollution levels for each city. The variables available for this process were FRP (either single day or FRP_{01}), $PM_{2.5}$ from yesterday (PM_{lag}), and all of weather variables: FFDI, Haines index, wind speed, direction and angle relative to the hotspots. PM_{lag} was also included purely to improve the predictive ability of the model. The best model was identified using a model selection process, examining all of combinations of variables and selecting the one with the lowest Akaike Information Criteria (AIC) value [41]. Supported alternative models (those with $\Delta AIC < 2$) were also identified. From the best model, all two-way interactions between FRP and the other variables were tested and retained if they reduced the AIC. The importance of each variable in the final model was assessed by calculating the proportion of explained variation attributable to each variable [42]. The modelling was repeated using the log of FRP, using the log of $PM_{2.5}$ as the dependent variable and using all days for the sample (i.e. not restricted to days with fire activity, which is the default).

Evaluation

A database of historical causes for extreme pollution events had previously been compiled for six cities, including Perth and Sydney [36]. This was done by matching the days where pollution exceeded the 95th percentile of all values against a variety of media information and empirical data sources. Probable causes could be indentified for 67% of events and, of those, 94% were confirmed as bushfire days. The raw FRP_{01} data were validated against this database by matching days that were known bushfire days and $PM_{2.5}$ peaks to 95th percentile peaks in the FRP_{01}.

Results

During the study period (2002 to 2007) 1845 and 1687 days had both pollution measures and fire hotspots in Sydney and Perth, respectively. In Sydney, the hotspots occurred in a wide arc of forest surrounding the city from north, west, and south (Figure 1a). Major, damaging fires occurred in the 2002/3 and 2006/7 fire seasons. A fire rose diagram for unweighted hotspots (showing the number of hotspots in each 10° compass sector, centred on the pollution monitor) is dominated by the fires that burnt the Australian Alps and Canberra in January 2003 (300–500 km to the south-west of Sydney) (Figure 4a). The distance-weighted fire rose shows a much more even spread of fire directions, with the fires from 2002/3 and 2006/7 prominent, but with fires from land in all directions. In Perth, hotspots were more dispersed than in Sydney and the fires were smaller (Figure 1b). Consequently, the fire roses were similar for the weighted and unweighted distances (Figure 4b). Notice the occurrence of hotspots in the direction of the ocean for Sydney, which is due to the inland position of the pollution monitor.

In Sydney, the explanatory power of Lag FRP decreased with the length of the lag: r^2 decreased from 0.17 for the current day to 0.04 for 4 days previously (Figure 5a). However, the best explanatory power was obtained with the Combined FRP with 1 day lag (FRP_{01}, $r^2 = 0.23$) and was also high for Combined FRP with 2 and 3 day lags. For the Combined Forecast FRP (excluding the current day), the explanatory power was highest for a 3 day lag (0.19). Perth showed a similar pattern, with the Combined FRP with 1 day lag giving the best explanatory power, though this was low ($r^2 = 0.84$, Figure 5b). For both cities, the best FRP variable to use for modelling is the current day + previous day (FRP_{01}).

Analyses on the radius of influence showed a peak in explanatory power at 100 km for Sydney ($r^2 = 0.24$, Table 1). The peak for Perth was at 400 km with a lower explanatory power ($r^2 = 0.08$). None of the log-transformations improved the fit of the relationships.

There were 69 exceedance events for $PM_{2.5}$ in Sydney, of which 52% lasted a single day. Only four events lasted four days or more (Figure 6). The two longest events were associated with major fire events in the Sydney region (Christmas 2001 fires and early December 2002). Two other four-day events were both during winter, and were not verified fire events (11th July 2002 and 7th June 2005). In Perth, there were more events with shorter durations (n = 80, 75% lasted a single day), and only two lasted four days. These were both in May (26th May 2004 and 9th May 2006), and were not verified fire events. The exceedances in FRP showed similar patterns to the $PM_{2.5}$ and similar differences between the cities. In Sydney there were 48 events, of which 54% lasted a single day, but ten lasted four days or more. In Perth there were 75 events, of which 75% lasted a single day, and only three were four days or more.

The best statistical model for Sydney included positive effects for FRP_{01}, FFDI, C-Haines, PM_{lag} and with on-shore winds and a negative effect of wind-speed. $PM25_{lag}$ had the strongest effect (see the importance value, Table 2a), followed by FRP_{01}. There were also negative interactions between FRP_{01} and wind-speed,and FRP_{01} and $PM25_{lag}$, but these were weak effects (Table 2, Figure 7). The model explained 56% of variance in $PM_{2.5}$ ($r^2 = 0.562$) and there were no supported alternatives. When the analysis was repeated without $PM25_{lag}$, the final model was similar, but explained 45% of variation. Using log($PM_{2.5}$) as the dependent variable or including all days also resulted in similar models and explanatory power ($r^2 = 0.561$, 0.569 respectively). Using the log of FRP as a predictor variable gave weaker results than the untransformed FRP variables.

The best model for Perth was similar to Sydney, but explained less variation ($r^2 = 0.310$, Table 2b). There were positive effects of FFDI, C-Haines and $PM25_{lag}$ and on-shore wind and a negative effect of wind speed. The FRP_{01} effect was weak, and there was an interaction between FRP_{01} and C-Haines, which meant that FRP had a positive relationship with $PM_{2.5}$ which became slightly negative when the C-Haines index was less than 1.3 (on about 31% of days). There were no supported alternative models. $PM25_{lag}$ was the most important variable, followed by C-Haines. Without $PM25_{lag}$, the best model explained only 19% of variance. Using log($PM_{2.5}$) as the dependent variable, including all days, or using log of FRP as the predictor variable all resulted in similar models with slightly lower explanatory power ($r^2 = 0.294$, 0.308, 0.306 respectively).

There were 54 validated bushfire events in Sydney with matching $PM_{2.5}$ peaks, and 74% of these corresponded with peaks in the raw FRP_{01} value and 61% corresponded with peaks in the model predictions. For Perth there were 23 events and 35% corresponded with the raw FRP_{01} and 43% with the model predictions.

Discussion

We have demonstrated that there is a relationship between fire activity detectable from satellites and pollution in urban centres. Moreover, there are windows of spatial and temporal influence of

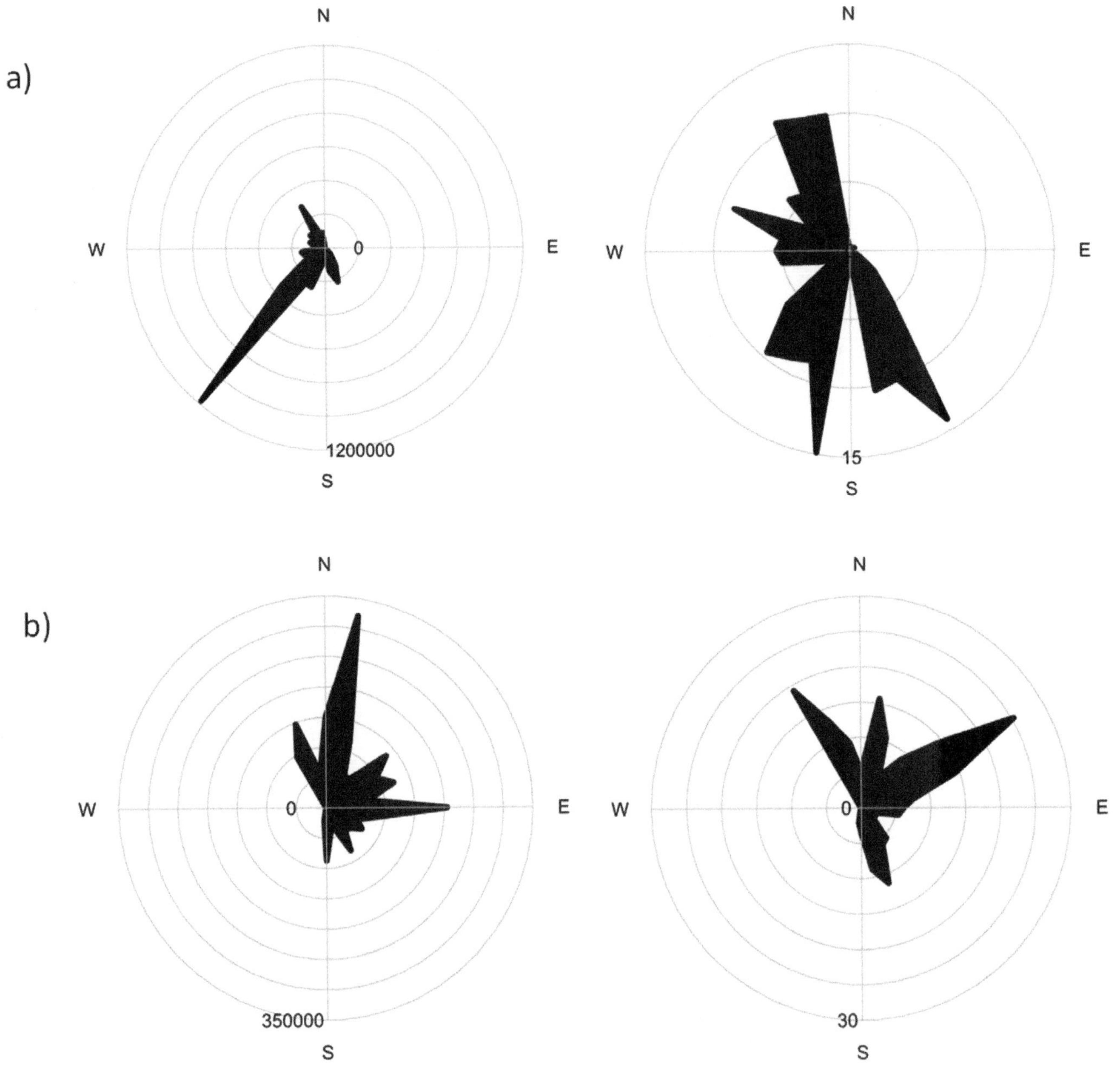

Figure 4. Unweighted (left) and weighted (right) hotspot rose. a) Sydney; b) Perth. The diagram shows the sum of FRP values (raw or weighted as FRP/distance^2) in 10° sectors around the compass. The scale is in FRP units.

fire activity on pollution. For Sydney, the radius of influence is relatively small (100 km) and the relationship is strong, while for Perth the radius is much larger (400 km) and the relationship is weak. The explanatory power of the models was improved by including weather and pollution history variables, to the extent that more than half of the variance in pollution could be explained for Sydney. The weather relationships revealed higher pollution when the atmosphere was stable, when wind speed was low and when the wind in the upper atmosphere was blowing on-shore. Importantly, wind flow from the fires to the monitor was not present in the models. This is presumably because on-shore winds that trap in-situ pollution in the city basins are more important to the overall level of pollution than winds transporting smoke from bushfires, even if days without fire activity are excluded. This trapping effect is probably why stable air and low wind speed also lead to more pollution.

The differences between the cities have a logical explanation. Sydney is bounded on three sides by extensive forests (approximately 20,000 km^2) at a distance of about 100 km. Beyond this is an extensive agricultural belt with very little fire activity. The mean area burnt in these forests is approximately 5% per year, of which more than 80% is unplanned fire [20]. Generally, the wind associated with bushfire events blows from an arc between northwest and southwest, and so blows smoke into the city. The forests around Perth are much more scattered, comprising small patches spread across a larger region. A higher proportion of the forests are burnt each year (8.5%) [19], but the regime is dominated by prescribed fires (7% compared to 1.5% for

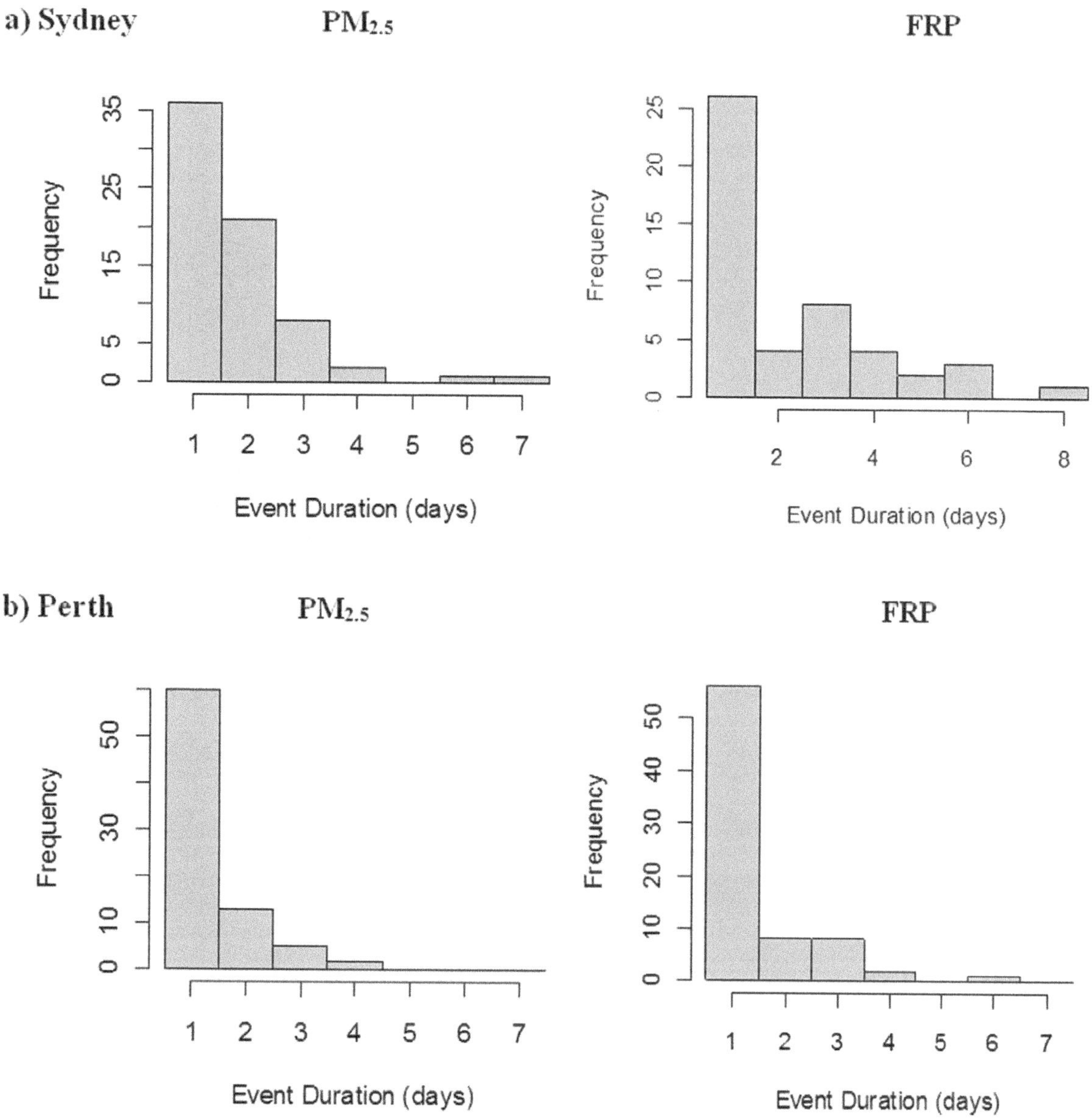

Figure 5. The goodness of fit between FRP and $PM_{2.5}$ at increasing lag durations for Sydney and Perth. For each lag, three measures were compared: A) the lag day on its own (Lag FRP); B) all days including the current (Combined FRP); and C) all days from the lag to the day before the current one (Combined Forecast FRP).

Table 1. Strength of model for various hotspot radii for Sydney and Perth.

Radius (km)	Sydney r^2	Perth r^2
50	0.122	0.050
75	0.161	0.046
100	0.243	0.044
150	0.183	0.045
200	0.194	0.062
300	0.181	0.072
400	0.126	0.075
500	0.112	0.062

unplanned fire). The closeness of the forests and the more extensive unplanned fire explain why the model has a smaller radius and better goodness of fit for Sydney than for Perth. These differences between cities mean that a forecast tool must be based on a model specific to each city.

Our empirical modelling method shows some potential for developing a forecasting tool for bushfire planning. For both cities, the best single predictor of daily pollution is the previous day's pollution, which may be of use to health authorities. Nevertheless, the FRP and weather variables also contribute substantially to the predictive power. For Sydney, the predictive model explained most of the variation in pollution and three quarters of the known bushfire caused pollution days were identified as peaks in the FRP data. For Perth, the previous days FRP is actually a better predictor than the current day, albeit with low predictive power. Our method may be more efficient and accurate for identifying

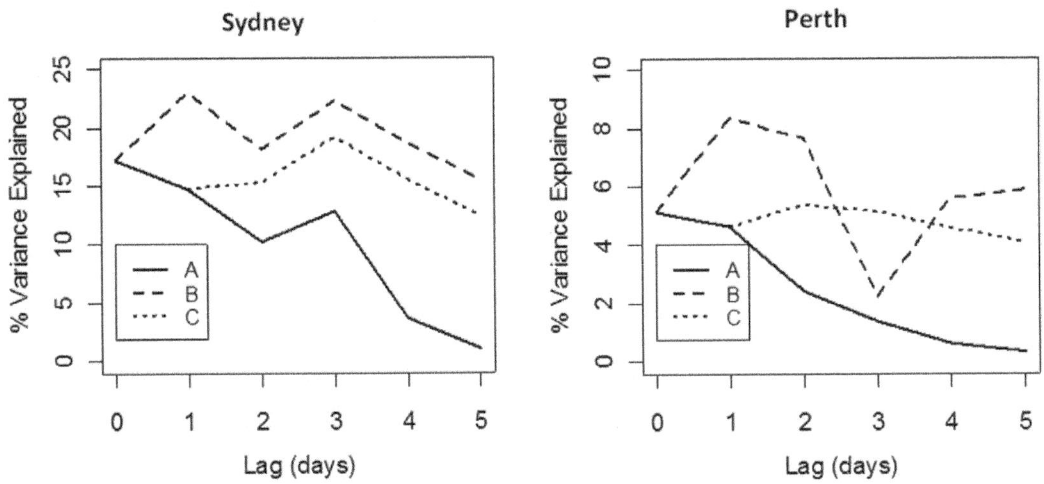

Figure 6. Frequency distributions of event durations for $PM_{2.5}$ and FRP exceedances. a) Sydney and b) Perth. Values for $PM_{2.5}$ and FRP are the 95th percentiles.

historical fires than other methods available that require laborious searches of print media and government reports [43].

Improvements could be made to the forecasting ability by addressing some of the limitations of the data. The most important of these is the lack of identification of non-bushfire sources of

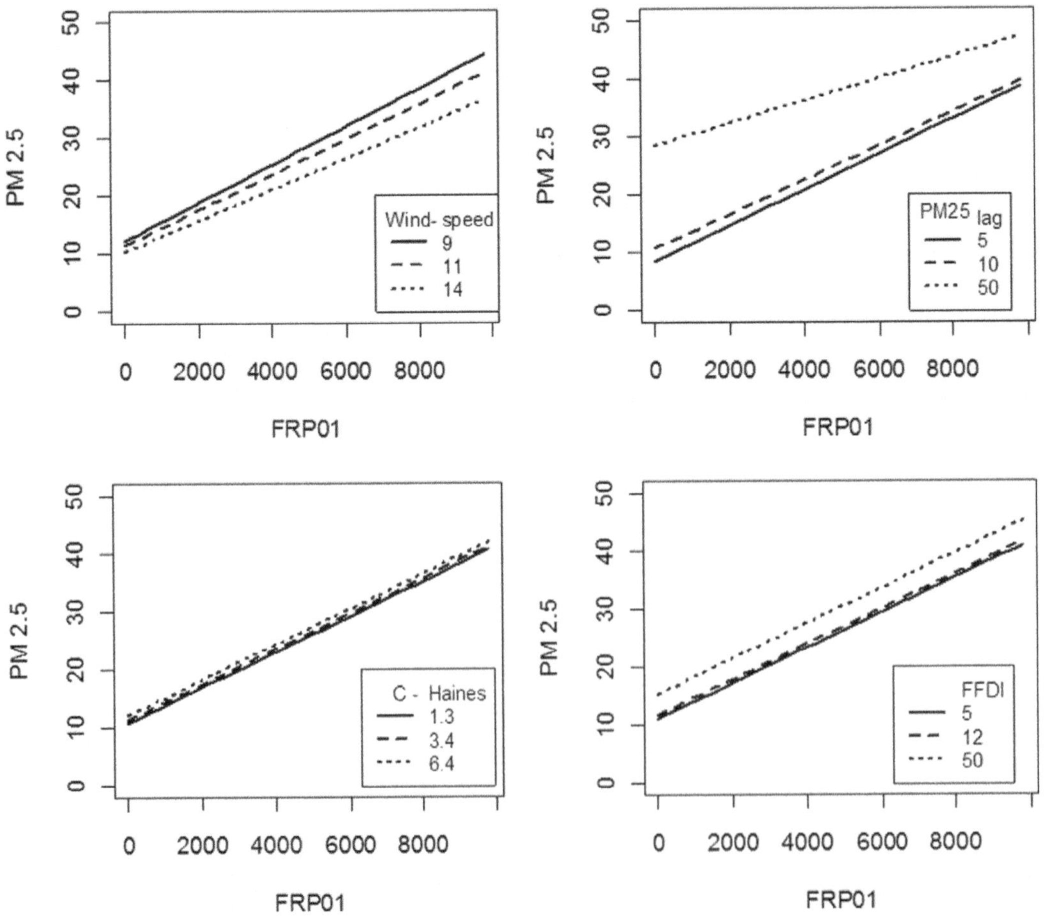

Figure 7. Predictions from the best model for Sydney. The plots give the relationship between PM_{25} and FRP_{01} at different levels of wind $speed_{sta}$, $PM25_{lag}$, C-Haines and FFDI. For wind $speed_{sta}$ and C-Haines these levels are the quartiles (25, median, 75) while for $PM25_{lag}$ and FFDI they are set levels (i.e. 50 represents $PM_{2.5}$ exceedance and also a fire danger of Severe). In each plot, all other variables are held at their median values.

Table 2. Final model tables.

Sydney, $r^2 = 0.562$, n = 1845.					
	Estimate	**Std. Err**	**t**	**P**	**Importance**
(Intercept)	8.165	0.371	21.988	<0.001	
FFDI	0.094	0.015	6.210	<0.001	0.053
FRP_{01}	0.0047	0.0004	12.473	<0.001	0.236
Wind-$speed_{sta}$	−0.369	0.021	−17.316	<0.001	0.129
C-Haines	0.262	0.026	9.932	<0.001	0.103
On-$shore_{700}$	0.592	0.234	2.532	0.011	0.015
$PM25_{lag}$	0.444	0.019	23.266	<0.001	0.428
FRP_{01}:Wind-$speed_{sta}$	−1.22e-4	2.50e-5	−4.874	<0.001	0.020
FRP_{01}:$PM25_{lag}$	−2.57e-5	8.78e-6	−2.926	0.003	0.016
Perth, $r^2 = 0.310$, n = 1667.					
	Estimate	**Std. Err**	**t**	**P**	**Importance**
(Intercept)	6.083	0.312	19.510	<2e-16	
FFDI	0.029	0.009	3.071	0.002	0.033
FRP_{01}	2.80e-5	8.27e-5	−0.338	0.735	0.107
Wind-$speed_{sta}$	−0.146	0.014	−10.339	<2e-16	0.148
C-Haines	0.129	0.018	7.102	0.000	0.205
On-$shore_{700}$	0.441	0.159	2.779	0.006	0.012
$PM25_{lag}$	0.364	0.021	17.333	<2e-16	0.483
FRP_{01}:C-Haines	3.33e-5	1.00e-5	3.316	0.001	0.013

Importance refers to Relative Importance: the proportion of explained variance attributable to each variable (sum to 1.0).

pollution, since these are the principle driver of pollution levels on most days. Pollution from these sources also varies and our analysis suggests that they respond to weather conditions, perhaps more than smoke does. If non-smoke pollution days could be filtered from the data then the accuracy of the models would be greatly increased. A more explicit representation of atmospheric transport such as a two or three-dimensional wind-field between the source and monitor would also improve the relationship. Many fires, especially larger ones, inject smoke high into the atmosphere which may reduce pollution on the ground [44]. Since there is actually a relationship between FRP and injection height [44], this effect could be directly incorporated into a predictive model and may be more successful than the use of an atmospheric stability index such as C-Haines. Ultimately, a hybrid model using daily snapshots of FRP and a three-dimensional atmospheric transport model may be the most fruitful approach.

There are other limitations that affect the accuracy of the models. The raw FRP values used here do not account for differences in vegetation type. Vegetation correction factors have been applied elsewhere to FRP values [37], but none have been developed for Australia. The episodic nature of major bushfire events means that the six year time series used here only captured a small number of events (54 events reported in the media for Sydney and 23 for Perth), and these varied greatly in magnitude. The analysis would be improved with a longer time series. Sydney experiences extreme fire seasons (where >10% of the forest area burns) between once and twice per decade [20], so it may be that several decades of data are required to determine the nature of the relationship between FRP and pollution levels.

Our results are also relevant to the debate over the appropriate level of prescribed fire treatment. Prescribed fire is imposed as a means of reducing the area and impact of wildfire, and has been suggested as one means of preventing increased wildfire activity due to climate change. However, in many Australian landscapes several hectares of prescribed fire are required to reduce the area of wildfire by one hectare. For example, around Sydney 3–4 ha of prescribed fire are required per hectare reduction in wildfire area [20,45], and the forests around Perth, 6 ha are required [19]. However, prescribed fires consume less biomass per hectare than wildfires because they are more patchy [46], are usually confined to the surface fuels and leave larger proportion of fuel unburned. It has been shown for Southern Australia that smoke plumes from prescribed fires blow over cities, though on average these are five times smaller than plumes from wildfires [21]. Further examination of the contribution of prescribed and wildfires to smoke is needed before the likely impact of increased prescribed burning could be accurately predicted.

In conclusion, this study has identified a clear relationship between remotely sensed fire activity and pollution levels in two Australian cities. It has also identified the distance over which fires tend to supply smoke to the cities, and some of the challenges to developing a useful operational tool. With some improvements, models developed using this method could become useful in many fire-prone regions around the world for forecasting smoke pollution. They may be used as an adjunct to smoke-plume modelling or may eventually prove to be more useful. Even as they are now, the method can be used for rapid retrospective identification of bushfire pollution days, in contrast to the labour intensive methods used by Johnston et al. [36]. With further separation of FRP into wild and prescribed fires, our approach could also provide a new tool for understanding the comparative smoke pollution exposure to humans from fires of both types.

Acknowledgments

We would like to thank two reviewers for their suggestions.

Author Contributions

Conceived and designed the experiments: DMJSB OFP SBH FJ. Performed the experiments: GW. Analyzed the data: GW OFP. Wrote the paper: OFP DMJSB GW SBH FJ.

References

1. De Vos A, Reisen F, Cook A, Devine B, Weinstein P (2009) Respiratory Irritants in Australian Bushfire Smoke: Air Toxics Sampling in a Smoke Chamber and During Prescribed Burns. Archives of Environmental Contamination and Toxicology 56: 380–388.
2. Naeher LP, Brauer M, Lipsett M, Zelikoff JT, Simpson CD, et al. (2007) Woodsmoke Health Effects: A Review. Inhalation Toxicology 19: 67–106.
3. Sapkota A, Symons JM, Kleissl J, Wang L, Parlange MB, et al. (2005) Impact of the 2002 Canadian forest fires on particulate matter air quality in Baltimore City. Environmental Science & Technology 39: 24–32.
4. Wise EK (2008) Meteorologically influenced wildfire impacts on urban particulate matter and visibility in Tucson, Arizona, USA. International Journal of Wildland Fire 17: 214–223.
5. Dennekamp M, Abramson MJ (2011) The effects of bushfire smoke on respiratory health. Respirology 16: 198–209.
6. Johnston F, Hanigan I, Henderson S, Morgan G, Bowman D (2011) Extreme air pollution events from bushfires and dust storms and their association with mortality in Sydney, Australia 1994–2007. Environmental Research 111: 811–816.
7. Sastry N (2002) Forest fires, air pollution, and mortality in Southeast Asia. Demography 39: 1–23.

8. Jayachandran S (2009) Air Quality and Early-Life Mortality Evidence from Indonesia's Wildfires. Journal of Human Resources 44: 916–954.
9. Jalava PI, Salonen RO, Halinen AI, Penttinen P, Pennanen AS, et al. (2006) In vitro inflammatory and cytotoxic effects of size-segregated particulate samples collected during long-range transport of wildfire smoke to Helsinki. Toxicology and Applied Pharmacology 215: 341–353.
10. Mazzoli-Rocha F, Magalhaes CB, Malm O, Saldiva PHN, Zin WA, et al. (2008) Comparative respiratory toxicity of particles produced by traffic and sugar cane burning. Environmental Research 108: 35–41.
11. Danielsen PH, Loft S, Kocbach A, Schwarze PE, Moller P (2009) Oxidative damage to DNA and repair induced by Norwegian wood smoke particles in human A549 and THP-1 cell lines. Mutation Research-Genetic Toxicology and Environmental Mutagenesis 674: 116–122.
12. Pope CA, Dockery DW (2006) Health effects of fine particulate air pollution: Lines that connect. Journal of the Air & Waste Management Association 56: 709–742.
13. Teague B, Mcleod R, Pascoe S (2010) 2009 Victorian bushfires Royal Commission final report. Melbourne: Parliament of Victoria.
14. Anon (2002) Report on inquiry into 2001/2002 bushfires. Sydney: Joint Select Committee on Bushfires, Parliament, NSW. 130 p.
15. Esplin B, Gill M, Enright N (2003) Report of the Inquiry into the 2002–03 Victorian Bushfires. Melbourne: Department of Premier and Cabinet. 117 p.
16. Minnich RA, Chou YH (1997) Wildland fire patch dynamics in the chaparral of southern California and northern Baja California. International Journal of Wildland Fire 7: 221–248.
17. Snider G, Daugherty PJ, Wood D (2006) The irrationality of continued fire suppression: An avoided cost analysis of fire hazard reduction treatments versus no treatment. Journal of Forestry 104: 431–437.
18. Donovan GH, Brown TC (2007) Be careful what you wish for: the legacy of Smokey Bear. Frontiers in Ecology and the Environment 5: 73–79.
19. Boer MM, Sadler RJ, Wittkuhn R, McCaw L, Grierson PF (2009) Long-term impacts of prescribed burning on regional extent and incidence of wildfires – evidence from fifty years of active fire management in SW Australian forests. Forest Ecology and Management 259: 132–142.
20. Price OF, Bradstock R (2011) The influence of weather and fuel management on the annual extent of unplanned fires in the Sydney region of Australia. International Journal of Wildland Fire 20: 142–151.
21. Williamson GJ, Price OF, Henderson SB, Bowman DMJS (in press) Satellite-based comparison of fire intensity and smoke emissions from prescribed and wildfires in south-eastern Australia. International Journal of Wildland Fire.
22. NEHPC (2003) NEPC, National Environment Protection Measure (Ambient Air Quality) as amended. National Environment and Heritage Protection Council.
23. NEHPC (2009) Review of the National Environment Protection (Ambient Air Quality) Measure. Discussion Paper on Air Quality Standards. National Environment and Heritage Protection Council.
24. Larkin NK, O'Neill SM, Solomon R, Raffuse S, Strand T, et al. (2009) The BlueSky smoke modeling framework. International Journal of Wildland Fire 18: 906–920.
25. Johnston F, Williamson GJ, Bowman D (2010) A review of approaches to monitoring smoke from vegetation fires for public health. Air Quality and Climate Change 44: 17–21.
26. Henderson SB, Burkholder B, Jackson PL, Brauer M, Ichoku C (2008) Use of MODIS products to simplify and evaluate a forest fire plume dispersion model for PM10 exposure assessment. Atmospheric Environment 42: 8524–8532.
27. O'Neill S, Larkin N, Hoadley J, Mills G, Vaughan J, et al. (2008) Regional real-time smoke prediction systems. In: A Bytnerowicz MA, A Riebau, C Andersen, editor. Wildland Fires and Air Pollution. Amsterdam: Elsevier. 499–534.
28. Stein AF, Rolph GD, Draxler RR, Stunder B, Ruminski M (2009) Verification of the NOAA Smoke Forecasting System: Model Sensitivity to the Injection Height. Weather and Forecasting 24: 379–394.
29. Rolph GD, Draxler RR, Stein AF, Taylor A, Ruminski MG, et al. (2009) Description and Verification of the NOAA Smoke Forecasting System: The 2007 Fire Season. Weather and Forecasting 24: 361–378.
30. Justice CO, Giglio L, Korontzi S, Owens J, Morisette JT, et al. (2002) The MODIS fire products. Remote Sensing of Environment 83: 244–262.
31. Wooster MJ, Roberts G, Perry GLW, Kaufman YJ (2005) Retrieval of biomass combustion rates and totals from fire radiative power observations: FRP derivation and calibration relationships between biomass consumption and fire radiative energy release. Journal of Geophysical Research 110.
32. Ichoku C, Kaufman YJ (2005) A method to derive smoke emission rates from MODIS fire radiative energy measurements. IEEE Transactions on Geoscience and Remote Sensing 43: 26–36.
33. Keith DA (2004) Ocean shores to desert dunes: The native vegetation of New South Wales and the ACT. Hurstville, NSW: Department of Environment and Conservation.
34. Stoneman GL (2007) 'Ecological forestry' and eucalypt forests managed for wood production in south-western Australia. Biological Conservation 137: 558–566.
35. Water NDoECCa (2010) Air Emissions Inventory for the Greater Metropolitan Region in NSW. Sydney.
36. Johnston FH, Hanigan IC, Henderson SB, Morgan GG, Portner T, et al. (2011) Creating an integrated historical record of extreme bushfire smoke events in 3 Australian cities from 1994 to 2007. Journal of the Air and Waste Management Association 61: 390–398.
37. Ichoku C, Kaufman YJ (2005) A method to derive smoke emission rates from MODIS fire radiative energy measurements. Ieee Transactions on Geoscience and Remote Sensing 43: 2636–2649.
38. Henderson SB, Ichoku C, Burkholder BJ, Brauer M, Jackson PL (2010) The validity and utility of MODIS data for simple estimation of area burned and aerosols emitted by wildfire events. International Journal of Wildland Fire 19: 844–852.
39. Bradstock RA, Gill AM, Kenny BJ, Scott J (1998) Bushfire risk at the urban interface estimated from historical weather records: consequences for the use of prescribed fire in the Sydney region of south-eastern Australia. Journal of Environmental Management 52: 259–271.
40. Mills G, McCaw L (2010) Atmospheric Stability Environments and Fire Weather in Australia – extending the Haines Index. Melbourne: Centre for Australian Weather and Climate Research. 151 p.
41. Burnham KP, Anderson DR (2002) Springer-Verlag. 353 p.
42. Grömping U (2006) Relative Importance for Linear Regression in R: The Package relaimpo. Journal of Statistical Software 17: 1–27.
43. Morgan G, Sheppeard V, Khalaj B, Ayyar A, Lincoln D, et al. (2010) Effects of Bushfire Smoke on Daily Mortality and Hospital Admissions in Sydney, Australia. Epidemiology 21: 47–55.
44. Amiridis V, Giannakaki E, Balis DS, Gerasopoulos E, Pytharoulis I, et al. (2010) Smoke injection heights from agricultural burning in Eastern Europe as seen by CALIPSO. Atmospheric Chemistry and Physics 10: 11567–11576.
45. Bradstock RA, Cary GJ, Davies I, Lindenmayer DB, Price O, et al. (in press) Wildfires, fuel treatment and risk mitigation in Australian eucalypt forests: insights from landscape-scale simulation. Journal of Environmental Management.
46. Penman TD, Kavanagh RP, Binns DL, Melick DR (2007) Patchiness of prescribed burns in dry sclerophyll eucalypt forests in South-eastern Australia. Forest Ecology and Management 252: 24–32.

17

Composition of PM Affects Acute Vascular Inflammatory and Coagulative Markers - The RAPTES Project

Maciej Strak[1,2], Gerard Hoek[2], Krystal J. Godri[3,4], Ilse Gosens[1], Ian S. Mudway[3], René van Oerle[5], Henri M. H. Spronk[5], Flemming R. Cassee[1,2], Erik Lebret[1,2], Frank J. Kelly[3], Roy M. Harrison[4,6], Bert Brunekreef[2,7], Maaike Steenhof[2], Nicole A. H. Janssen[1]*

1 Centre for Environmental Health (MGO), National Institute for Public Health and the Environment (RIVM), Bilthoven, The Netherlands, **2** Division of Environmental Epidemiology and Division of Toxicology, Institute for Risk Assessment Sciences (IRAS), Utrecht University, Utrecht, The Netherlands, **3** Environmental Research Group, MRC-HPA Centre for Environmental Health, King's College London, London, United Kingdom, **4** Division of Environmental Health and Risk Management, School of Geography, Earth & Environmental Sciences, University of Birmingham, Birmingham, United Kingdom, **5** Laboratory for Clinical Thrombosis and Haemostasis, Department of Internal Medicine, Cardiovascular Research Institute Maastricht, Maastricht University Medical Center, Maastricht, The Netherlands, **6** Department of Environmental Sciences / Center of Excellence in Environmental Studies, King Abdulaziz University, Jeddah, Saudi Arabia, **7** Julius Centre for Health Sciences and Primary Care, University Medical Centre Utrecht, Utrecht, The Netherlands

Abstract

Background: Exposure to ambient particulate matter (PM) has been associated with adverse cardiovascular effects in epidemiological studies. Current knowledge of independent effects of individual PM characteristics remains limited.

Methods: Using a semi-experimental design we investigated which PM characteristics were consistently associated with blood biomarkers believed to be predictive of the risk of cardiovascular events. We exposed healthy adult volunteers at 5 different locations chosen to provide PM exposure contrasts with reduced correlations among PM characteristics. Each of the 31 volunteers was exposed for 5 h, exercising intermittently, 3–7 times at different sites from March to October 2009. Extensive on-site exposure characterization included measurements of PM mass and number concentration, elemental- (EC) and organic carbon (OC), trace metals, sulfate, nitrate, and PM oxidative potential (OP). Before and 2 h and 18 h after exposure we measured acute vascular blood biomarkers - C-reactive protein, fibrinogen, platelet counts, von Willebrand Factor, and tissue plasminogen activator/plasminogen activator inhibitor-1 complex. We used two-pollutant models to assess which PM characteristics were most consistently associated with the measured biomarkers.

Results and Conclusion: We found OC, nitrate and sulfate to be most consistently associated with different biomarkers of acute cardiovascular risk. Associations with PM mass concentrations and OP were less consistent, whereas other measured components of the air pollution mixture, including PNC, EC, trace metals and NO_2, were not associated with the biomarkers after adjusting for other pollutants.

Editor: Stephania Ann Cormier, Louisiana State University Health Sciences Center, United States of America

Funding: The RAPTES project was funded by the RIVM Strategic Research Program (S630002). The funders had no role in study design, data collection and analysis, decision to publish, or preparation of the manuscript.

Competing Interests: The authors have declared that no competing interests exist.

* E-mail: nicole.janssen@rivm.nl

Introduction

Exposure to ambient particulate matter (PM) has been associated with cardiovascular health effects in epidemiological studies [1,2]. These effects have been linked with exposure to ambient PM, mostly with concentrations of $PM_{2.5}$ (particulate matter less than 2.5 μm in aerodynamic diameter; expressed in $\mu g/m^3$). However, PM composition is variable in time and space and is strongly dependent on (local) sources of emission [3]. This heterogeneity in physiochemical composition has been shown to affect PM toxicity [4]. PM characteristics, like particle number concentration (expressed in particles/cm^3), surface area, content of transition metals, organics, sulfate and nitrate, and biological components (e.g., endotoxin) have been proposed as possibly being responsible for observed adverse health effects. Recently, oxidative potential (OP) of PM has been suggested as a promising integrated and biologically meaningful measure to predict human health effects related to the PM mixture [5]. However, current knowledge of health effects of individual PM characteristics remains limited [2,6,7].

Current generally accepted hypotheses on the biological mechanisms of cardiovascular effects of PM exposure include the *indirect* pathway, where inhaled particles provoke lung cells to release pro-inflammatory mediators (e.g., cytokines, reactive oxygen species) that are consequently released into the circulation, and the *direct* pathway, with translocation of particles (i.e., ultrafines) or their constituents (e.g., soluble metals, organics) from the lungs into the circulation through the alveolar-capillary barrier [2]. Both pathways would cause systemic inflammation and (potentially) oxidative stress.

Over the recent years, several biomarkers of acute vascular response have been identified as contributors to the risk of cardiovascular disease, namely: C-reactive protein (CRP) as an acute-phase reactant and a sensitive systemic marker of inflammation and tissue damage [8]; fibrinogen as an acute-phase reactant in systemic inflammation [8] and involvement in blood coagulation [9]; platelets for their role in hemostasis and systemic inflammation [2,10]; von Willebrand Factor (vWF) as an important factor in hemostasis and possible indicator of endothelial cell activation or damage [11]; and the complex between tissue plasminogen activator and plasminogen activator inhibitor-1 (tPA/PAI-1 complex) indicating impaired fibrinolytic activity [12]. Though not consistent across studies, these markers were associated with (typically 24-h average) ambient air pollution [2].

Often the limitations in study design are an obstacle in investigating the independent health effects of individual PM characteristics. In classical epidemiological studies a frequent limitation in disentangling independent health contributions is the high correlation between air pollutants [7], as well as pollutant-specific measurement error related to non-representativeness of a central monitoring site for personal exposure assessment. Those problems are avoided in controlled exposure studies in laboratory settings, but here the limitations often are individual air pollutants or defined mixtures (e.g., diesel engine exhaust) with high and constant concentrations, which may not be reflective of variable real-world exposures. The aim of the RAPTES project ("Risk of Airborne Particles: a Toxicological-Epidemiological Hybrid Study") was to assess the independent contribution of specific PM characteristics to various health outcomes. We studied those by a short-term, semi-experimental exposure of healthy volunteers to ambient PM at selected real-world locations. At those locations we previously established substantial differences in PM characteristics [13] and investigated PM toxicity *in vitro* [14], as well as acute respiratory changes in human volunteers [15]. The current paper focuses on the associations of specific PM characteristics, including OP, with acute vascular changes in healthy human volunteers. We hypothesized that the changes would have a stronger and more consistent relationship with OP of PM, as a measure of integrated biological response of the total PM mixture, than with other measured characteristics.

Materials and Methods

Study Design

The RAPTES study design has been explained previously [15]. Briefly, we used a semi-experimental rather than a pure observational design to reduce exposure measurement error while still exposing the volunteers to real-world conditions. We exposed 31 healthy human volunteers to ambient PM at five locations in the Netherlands: an underground train station, a continuous and a stop-and-go traffic location, a farm and an urban background site. The locations all had different source characteristics, which resulted in increased exposure contrasts and reduced correlations between PM characteristics compared to a study conducted at a single site [13]. A detailed characterization of PM air pollution was performed on-site during exposure over 30 sampling days between March and October 2009. The participants were exposed at the five sampling locations multiple times, one site at a time, and each time pre- and post-exposure measurements were made to assess changes in biomarkers of acute vascular effects. Participants directly inhaled ambient air at each location; no air concentration or inhalation devices were used. Exposure contrast for the epidemiological analysis is derived from contrasts between sites and (temporal) variation within sites. Exposure always started around 9:00–9:30 and lasted for 5 h (Figure S1). Each hour participants cycled for 20 minutes on a stationary bicycle to increase the experimental dose and obtain a clearer health response, needed to disentangle the effects of different pollutants. The activity and 5-h exposure period was also selected to provide a contrast with the background exposure at the campus. To avoid potential carry-over effects from previous exposures, an individual's visits to the sites were separated by a minimum of 14 days.

Study Population

Volunteers were healthy, young, non-smoking students of Utrecht University, the Netherlands. We chose participants living at the campus to equalize and minimize exposure to traffic-related air pollution prior to the experimental exposure. Before the study, each participant attended a check-up by a physician to obtain medical clearance for participation. The study was approved by the ethics committee at University Medical Center Utrecht and written informed consent was provided by all participants.

Exposure Measurements

We previously described in detail the measurements of exposure to air pollution on location and the instrumentation used [13]. Briefly, we determined mass concentrations and absorbance of PM_{10} and $PM_{2.5}$ samples. The mass concentration of $PM_{2.5-10}$ was calculated as the difference between PM_{10} and $PM_{2.5}$. PM_{10} samples were analyzed for endotoxin content. Particle number concentration (PNC) between 0.007 µm and 3 µm was measured using a condensation particle counter. $PM_{2.5-10}$ and $PM_{2.5}$ samples collected with a high volume sampler were measured for the concentrations of elemental (EC) and organic carbon (OC), trace metals, e.g., iron (Fe), copper (Cu), nickel (Ni), vanadium (V) (both water-soluble and "total" acid-extracted fractions), and secondary inorganics (nitrate, sulfate). Gaseous pollutants concentrations (O_3, NO_2, NO_X) were measured using real-time monitors. $PM_{2.5-10}$, $PM_{2.5-0.18}$ and $PM_{0.18}$ samples were collected using the Micro-Orifice cascade Impactor (MOUDI, model M100-NR, MSP Corporation, Shoreview, MN, USA) and their OP was determined *in vitro* by the extent of ascorbate (OP^{AA}) and reduced glutathione (OP^{GSH}) depletion [16] and the sum of both metrics is presented as OP^{TOTAL}, expressed per m^3. We report absorbance, nitrate and sulfate in the fine fraction, whereas the individual PM fraction concentrations of OP and trace metals were aggregated.

To minimize exposure during transport of participants between the collection point and the sampling locations, we equipped a minibus with a custom-made cabin air filter. To estimate the traffic-related air pollution during transport, we measured the PNC in the minibus during each commute, as described previously [15].

Clinical Measurements

Venous blood was collected into K_2-EDTA, 3.2% (w/v) citrate and serum tubes (BD) through vena punction using 21 gauge needles before exposure (t = 0), 2 h after exposure (t = 9), and the next morning (t = 25) at the collection point located at the Utrecht University campus. Short after the collection (at t = 0 and t = 25) or the next morning (refrigerated after t = 9) EDTA plasma samples were transported to an external, commercial laboratory that performed the complete blood cell counts. Serum for analysis of high-sensitivity CRP (hs-CRP) and citrate plasma for analysis of fibrinogen, vWF and tPA/PAI-1 complex were stored at −80°C and analyzed at the end of the sampling campaign. Hs-CRP was measured in serum using CRP Vario turbidimetric immunoassay (Sentinel, Milan, Italy) with clinical chemistry analyzer (ARCHITECT; Abbott Laboratories, Abbott Park, IL, USA); platelets were analyzed as part of complete blood cell count using an automated hematology analyzer (CELL-DYN; Abbott Laboratories, Abbott Park, IL, USA); fibrinogen was measured on a Behring Coagulation System (BCS) with Multifibren U kit (Siemens Healthcare Diagnostics, Marburg, Germany) using a modification of the Clauss method; vWF antigen was measured on a BCS with vWF:ag immunoturbidimetric determination kit (Siemens Healthcare Diagnostics, Marburg, Germany); tPA/PAI-1 complex was measured using Technoclone tPA/PAI-1 Complex ELISA Reagent Kit (Technoclone, Vienna, Austria) according to the manufacturers instructions. The coefficients of variation (CV) for hs-CRP, fibrinogen, vWF and tPA/PAI-1 complex were <1.25%, <3%, <2% and <11%, respectively.

At t = 0 and t = 25 participants filled in a questionnaire reporting e.g., time spent in traffic, medication use and respiratory symptoms during the last 24 h.

Statistical Analyses

We analyzed the associations between air pollution concentrations measured during exposure at the sampling locations and the difference in acute vascular markers between post- (t = 9, t = 25) and pre-exposure (t = 0) for each sampling day using mixed linear regression. We used mixed models to account for the influence of repeated observations per subject (using compound symmetry of the residuals). Pre- and post-exposure levels of biomarkers were log-transformed to stabilize outliers. The associations between vWF and PM characteristics were driven by one obvious extreme outlier which influence was not diminished by log-transformation. Since the outlier decreased several otherwise stronger and significant estimates, we decided to report the vWF associations from the models where this outlier was excluded. As independent variables, we used the 5-h average concentrations of air pollutants measured at the locations. We first specified single-pollutant models. Then, to disentangle the individual effects of different pollutants, we specified two-pollutant models with all possible combinations of measured pollutants. We specified two-pollutant models because, due to the experimental nature of the study, exposure measurement error is much less important compared to pure observational studies in determining which of the two exposure variables has the strongest association with health. Further, we did not perform a factor analysis to group correlated pollutants because of an insufficient amount of observations available for such analysis (30 days).

The two-pollutant models are our main models, as in these the independent associations of particle components are assessed. We report first the results for single-pollutant models, to be able to compare with previous studies. If we only showed two-pollutant model results with e.g. no significant associations with $PM_{2.5}$ then we would not be able to distinguish between the scenario of no associations with $PM_{2.5}$ at all (in single and two-pollutant models) and the scenario of significant association with $PM_{2.5}$ (in single-pollutant model) that was explained by a specific component when included in two-pollutant model.

Effect estimates and confidence intervals (CI) are presented as percentage change over a study population mean of the baseline (t = 0) values expressed per changes in interquartile ranges (IQR) for the outdoor locations. We largely focus on the results of two-pollutant models. We did not interpret models with Spearman's R larger than 0.7 between two pollutants as those were considered highly correlated. Inclusion of highly correlated predictors in regression models may result in unstable regression coefficients (co-linearity). Since we defined a large number of models, we focus on the consistency of significant associations rather than single isolated significant associations. Due to a substantial difference in some exposures between outdoor locations and the underground location [13], we also analyzed data separately for the outdoor sites.

As potential confounders in the analysis, we adjusted for temperature and relative humidity measured at the location during sampling, and season (before or after the start date of the calendar summer), as in our analysis of respiratory effects [15]. We additionally adjusted for use of oral contraceptives and the use of oral contraceptives on the sampling day or a day before, since those may affect our biomarkers [17,18]. The impact of influential observations on the regression results was assessed by comparing effect estimates with and without 1% of observations with the highest Cook's Distance value. Data analyses were carried out using SAS 9.2 (SAS Institute, Cary, NC).

Results

We obtained a total of 170 observations from 31 participants (Table 1). Participants were exposed between three and seven times at different sites, and each person visited the underground train station at least once – in total we obtained 45 observations at the underground and 28–37 at the other sites. Baseline markers mean levels agreed closely with a recent Dutch study in healthy adults [19].

Concentrations of Air Pollutants

Nearly every PM characteristic, especially levels of Fe, Cu and OP^{TOTAL}, had substantially higher levels at the underground train station (Table S1 in Supporting Information). The highest PNC was at the continuous traffic site. Endotoxin levels were the highest at the farm. Variability within sites was smaller than between sites.

Correlations between air pollution concentrations are shown in Table S2 and we discussed them more extensively in our previous publication [13]. Briefly, we found PM_{10} and $PM_{2.5}$ to be highly correlated with each other, as well as with absorbance, EC, OC, trace metals, OP^{TOTAL}, but less so with PNC. O_3 showed a strong negative correlation with several PM characteristics. The high correlations decreased considerably after we excluded the measurements from the underground train station. Due to this exclusion, we observed substantially stronger correlations between PNC and EC or absorbance (whereas the correlations between

Table 1. Population characteristics and baseline (t = 0) levels of blood biomarkers.

Characteristic	Value
Age (years)	22 (19–26)
Sex, n (%)	
Female	21 (68%)
Male	10 (32%)
Body Mass Index (kg/m²)	22.3 (17.0–32.0)
Hs-CRP (mg/L)	1.00 (0.10–14.48)
Fibrinogen (g/L)	3.02 (1.43–5.19)
Platelet counts (10⁹/L)	267.79 (130.00–416.00)
vWF (% of normal)	89.41 (37.72–199.64)
tPA/PAI-1 complex (ng/mL)	2.81 (0.08–27.39)

Unless otherwise stated, values are means (ranges), or geometric means (ranges) for the biomarkers.
N = 170, except vWF (N = 169) and tPA/PAI-1 complex (N = 168).

PNC and PM_{10} or $PM_{2.5}$ remained low). Overall, both in the "all sites" and the "outdoor-only sites" dataset, the correlations between several PM characteristics were sufficiently low to investigate their independent effects on human health in two-pollutant models.

Associations 2 h after Exposure

In the single-pollutant models, hs-CRP (Table S4), fibrinogen (Table S5) and tPA/PAI-1 complex (Table S6) were not associated with any of the PM characteristics. There was a negative association of platelet counts with nitrogen oxides in the outdoor-only dataset (Table S7). NO_2 and OC (fine) were positively associated with vWF in the complete dataset and the latter association was also present in the outdoor-only models (Table S8).

Generally, we did not observe strong and consistent associations 2 h after exposure, therefore we did not analyze the two-pollutant models for this time point.

Associations in the Following Morning

Air pollution and C-reactive protein. In single-pollutant models, an increase in hs-CRP was associated with an increase in PM mass concentrations, OP, OC (fine) and transition metals, and negatively associated with O_3 (Table S4). Those associations were not present in the outdoor-only data subset.

In two-pollutant models, none of the PM characteristics remained significant after including each of the other co-pollutants (Table 2 and Table S9). We use the term "consistent" if a component is (borderline) significantly associated after adjusting for all other components (shown in the row per component) and the effect estimates do not change substantially compared to the single-pollutant estimate. However, OP and OC (fine) were fairly consistently associated with increase in hs-CRP after inclusion of most other PM characteristics (Table 2 and Table S9). OC (fine) had (borderline) significant associations in 16 of the models shown in the OC (fine) row and effect estimates did not deviate much (4.6% to 6.9%) from the single-pollutant estimate (6.7%). In contrast, e.g. PM_{10} remained significant in seven models but the effect estimate became negative after adjusting for Fe and Cu with a borderline significant association in the single-pollutant model, whereas the associations with $PM_{2.5}$ (also borderline significant in the single-pollutant model) decreased and became non-significant after adjusting for EC, some trace metals and gaseous pollutants (Table S9). In the outdoor-only models no consistent associations were found (Table S13).

Air pollution and fibrinogen. Increases in nitrate, sulfate, OC (coarse) and water-soluble Cu were associated with increases in fibrinogen in single-pollutant models (Table S5). Positive associations with nitrate and sulfate were also found in the outdoor-only models.

Nitrate and sulfate were most consistently associated with fibrinogen in two-pollutant models (Table 3 and Table S10). In the outdoor-only subset, the associations with both sulfate and nitrate were consistently robust, but decreased somewhat after adjusting for each other (and sulfate after adjusting for OP^{AA}) (Table S14).

The associations with sulfate decreased and were non-significant (particularly in the complete dataset) after excluding 1% observations with the highest Cook's Distance.

Air pollution and platelets counts. In single-pollutant models there were positive associations of platelet counts with OC (coarse) and nitrate and negative associations with nitrogen oxides (Table S7). The positive OC (coarse) and nitrate associations were also present in the outdoor-only models, along with positive associations with PM_{10} and $PM_{2.5}$.

In two-pollutant models, a consistent and robust positive association was found for platelet counts with OC (coarse) (Table 4 and Table S11). A fairly consistent positive association was also found for nitrate. Negative associations with NO_X remained, except for adjusting for PNC when NO_X became somewhat less strong and less significant. In the outdoor-only set, none of the PM characteristics remained significant after adjusting for all other pollutants. Fairly consistent associations of a similar magnitude, though less robust than in the complete dataset, were found for OC (coarse), nitrate, PM_{10} and $PM_{2.5}$, with the two former being more stable (Table S15). The negative associations with NO_X found in the outdoor-only dataset were less consistent than in the complete dataset.

Air pollution and von Willebrand Factor. Changes in vWF in single-pollutant models were positively associated with PM mass concentrations, EC (coarse), OC (fine), OP^{GSH} and OP^{TOTAL} (Table S8). In the outdoor subset, only $PM_{2.5}$ and OC (fine) were positively associated.

In the two-pollutant models, no single component was consistently associated with vWF (Table 5 and Table S12). Associations with OC (fine) were most consistent, whereas the associations with PM mass concentrations were not consistent after adjusting for a range of components (Table S12). In the outdoor-only subset, the association with OC (fine) was consistent and robust and only decreased somewhat after adjusting for OP^{AA} (Table S16). We also saw a fairly consistent positive association with $PM_{2.5}$.

Air pollution and tPA/PAI-1 complex. In single-pollutant models, PM mass concentrations, EC, absorbance, total Fe and nitrogen oxides were positively associated with increase in the tPA/PAI-1 complex, whereas O_3 had an opposite association (Table S6). The latter one is the only association that was also present in the outdoor-only models.

In the two-pollutant models, no single component was consistently associated with increase in the tPA/PAI-1 complex (Table S17). Strong consistent negative associations with O_3 were found, possibly due to high negative correlation with EC. In the outdoor-sites models, O_3 was similarly strong, though the estimates after adjusting for absorbance and EC (coarse) lost statistical significance (Table S18). The associations with water-soluble Ni were driven by 1% of influential observations with the highest Cook's Distance and were not present anymore after those were removed.

We log-transformed the tPA/PAI-1 complex levels for consistency with other endpoints, but even though the transformation did not affect normality, it resulted in disappearance of fairly consistent associations with absorbance, EC, OP^{AA} and NO_2 seen in the "all sites" dataset before log-transformation (Table S19).

Sensitivity Analyses

Exposure of participants to PNC during transport was not associated with changes in acute vascular markers investigated. Inclusion of in-transport exposure did not affect the reported associations with our experimental exposures (data not shown).

Deletion of 1% observations with the highest Cook's Distances generally did not affect the effect estimates, except for the aforesaid associations with vWF and other associations mentioned in the text.

Further adjustment for infections/colds at baseline, did not affect the reported associations with air pollution (data not shown).

Table 2. Associations between exposure to air pollution and percentage changes (25 h post-pre) in hs-CRP (all sites).

	IQR	Single-pollutant associations	Two-pollutant associations									
			PM_{10}	$PM_{2.5}$	PNC	EC(F)	OC(F)	OC(C)	NO_3^{-a}	SO_4^{2-a}	OP^{TOTAL}	NO_2
PM_{10}	13.50	0.74* (−0.01, 1.50)		*0.19 (−2.80, 3.26)*	0.79** (0.03, 1.56)	1.46* (−0.24, 3.19)	*0.36 (−0.57, 1.30)*	*1.01* (−0.12, 2.14)*	0.72* (−0.05, 1.48)	0.77** (0.02, 1.53)	*−0.71 (−2.80, 1.43)*	0.63 (−0.14, 1.41)
$PM_{2.5}$	11.54	1.74* (0.00, 3.51)	*1.32 (−5.49, 8.62)*		1.81** (0.07, 3.59)	2.54 (−0.64, 5.83)	*0.76 (−1.53, 3.10)*	2.46* (−0.15, 5.14)	1.71* (−0.13, 3.58)	1.74* (0.00, 3.51)	*0.13 (−3.48, 3.89)*	1.45 (−0.36, 3.28)
PNC	32,906	−4.31 (−14.35, 6.92)	−5.75 (−15.66, 5.32)	−5.23 (−15.15, 5.85)		−9.91 (−20.56, 2.17)	−4.89 (−14.78, 6.16)	−2.23 (−12.51, 9.26)	−3.46 (−14.05, 8.44)	−3.49 (−13.91, 8.19)	−8.95 (−18.84, 2.13)	−11.23* (−21.75, 0.71)
EC(F)	4.35	4.23 (−2.05, 10.91)	−6.30 (−18.50, 7.72)	−3.30 (−13.63, 8.28)	7.24* (−0.07, 15.08)		1.30 (−5.46, 8.54)	3.82 (−3.50, 11.70)	4.53 (−1.80, 11.27)	4.98 (−1.47, 11.86)	*−8.79 (−19.03, 2.74)*	2.63 (−4.01, 9.73)
OC(F)	1.82	6.68** (0.81, 12.89)	*5.00 (−2.14, 12.67)*	*4.95 (−2.69, 13.20)*	6.81** (0.92, 13.04)	6.13* (−0.40, 13.08)		6.94** (0.60, 13.69)	6.94** (0.52, 13.78)	6.48** (0.56, 12.74)	5.52 (−1.93, 13.54)	5.69* (−0.48, 12.24)
OC(C)	0.79	1.64 (−2.17, 5.60)	*−2.18 (−7.65, 3.63)*	−2.40 (−7.87, 3.40)	1.61 (−2.22, 5.58)	0.36 (−4.11, 5.05)	−0.20 (−4.25, 4.03)		1.11 (−3.06, 5.45)	1.72 (−2.10, 5.68)	*−0.79 (−5.55, 4.21)*	1.18 (−2.61, 5.12)
NO_3^{-a}	5.19	1.89 (−3.21, 7.26)	1.21 (−3.89, 6.57)	0.30 (−4.96, 5.86)	1.41 (−3.90, 7.01)	2.28 (−2.86, 7.69)	−0.54 (−5.92, 5.15)	1.71 (−3.81, 7.55)		0.97 (−5.42, 7.79)	3.01 (−2.11, 8.39)	0.79 (−4.44, 6.30)
SO_4^{2-a}	2.99	2.51 (−3.59, 9.00)	3.07 (−3.06, 9.58)	2.53 (−3.55, 8.98)	2.05 (−4.21, 8.71)	3.53 (−2.75, 10.22)	1.64 (−4.41, 8.08)	2.92 (−3.13, 9.35)	1.79 (−5.87, 10.07)		3.05 (−2.96, 9.43)	1.94 (−4.14, 8.41)
OP^{TOTAL}	38.71	0.28** (0.05, 0.52)	*0.51 (−0.20, 1.23)*	*0.27 (−0.27, 0.81)*	0.31** (0.07, 0.55)	*0.61** (0.13, 1.08)*	0.13 (−0.18, 0.45)	*0.30* (−0.02, 0.61)*	0.27** (0.04, 0.51)	0.29** (0.05, 0.53)		0.27** (0.03, 0.51)
NO_2	10.54	10.81 (−2.30, 25.67)	8.11 (−4.99, 23.02)	7.56 (−5.65, 22.61)	18.64** (2.70, 37.05)	8.60 (−5.24, 24.46)	6.33 (−6.94, 21.50)	12.79* (−0.59, 27.96)	10.21 (−3.37, 25.70)	10.29 (−2.87, 25.23)	4.44 (−8.30, 18.96)	

[a]measured in $PM_{2.5}$.

*p<0.10, ** p<0.05.

"C" denotes the coarse, whereas "F" the fine PM fraction.

Italics indicate Spearman's R above 0.7. In each row effect estimates (95% confidence intervals) for the indicated pollutant are presented. All models are adjusted for the use of oral contraceptives and the use of oral contraceptives on the sampling day or the day before, temperature, relative humidity, season and adjustment pollutant (indicated in the header of each column). For example, in the single-pollutant model PM_{10} has the effect estimate of 0.74 (−0.01, 1.50), whereas after adjusting for PNC, the effect estimate is 0.79 (0.03, 1.56). Estimates are percentage increases above population-average baseline expressed per outdoor-sites IQR. N = 170, except all models including OP where N = 153 and all models including EC (C) and OC (C) where N = 166.

Table 3. Associations between exposure to air pollution and percentage changes (25 h post-pre) in fibrinogen (all sites).

	IQR	Single-pollutant associations	Two-pollutant associations									
			PM_{10}	$PM_{2.5}$	PNC	EC(F)	OC(F)	OC(C)	NO_3^{-a}	SO_4^{2-a}	OP^{TOTAL}	NO_2
PM_{10}	13.50	0.10 (−0.04, 0.25)		*−0.18 (−0.75, 0.40)*	0.11 (−0.03, 0.26)	0.30* (−0.02, 0.62)	*0.07 (−0.11, 0.25)*	*0.02 (−0.17, 0.21)*	0.08 (−0.06, 0.23)	0.11 (−0.03, 0.25)	*0.06 (−0.37, 0.50)*	0.11 (−0.04, 0.25)
$PM_{2.5}$	11.54	0.27 (−0.06, 0.60)	*0.68 (−0.65, 2.02)*		0.28* (−0.04, 0.62)	0.59* (−0.01, 1.19)	*0.22 (−0.22, 0.66)*	0.14 (−0.30, 0.57)	0.18 (−0.16, 0.53)	0.26 (−0.06, 0.59)	*0.33 (−0.42, 1.08)*	0.28 (−0.06, 0.63)
PNC	32,906	−0.92 (−2.98, 1.19)	−1.12 (−3.19, 0.99)	−1.06 (−3.11, 1.05)		−1.62 (−3.95, 0.76)	−0.99 (−3.05, 1.12)	−0.93 (−2.75, 0.92)	−0.35 (−2.50, 1.85)	−0.46 (−2.55, 1.69)	−1.27 (−3.56, 1.07)	−1.40 (−3.77, 1.04)
EC(F)	4.35	0.40 (−0.78, 1.59)	−1.82 (−4.38, 0.80)	−1.36 (−3.45, 0.77)	0.84 (−0.51, 2.21)		0.06 (−1.26, 1.39)	−0.23 (−1.45, 0.99)	0.52 (−0.66, 1.71)	0.66 (−0.53, 1.86)	*−1.33 (−3.70, 1.10)*	0.40 (−0.88, 1.70)
OC(F)	1.82	0.73 (−0.36, 1.82)	*0.41 (−0.94, 1.77)*	*0.24 (−1.20, 1.70)*	0.75 (−0.34, 1.85)	0.70 (−0.51, 1.94)		0.47 (−0.56, 1.51)	0.34 (−0.84, 1.53)	0.56 (−0.53, 1.65)	0.66 (−0.84, 2.18)	0.77 (−0.39, 1.94)
OC(C)	0.79	0.64* (0.00, 1.29)	*0.56 (−0.41, 1.54)*	0.41 (−0.56, 1.40)	0.62* (−0.02, 1.27)	0.72* (−0.05, 1.49)	0.51 (−0.19, 1.22)		0.39 (−0.32, 1.10)	0.64** (0.01, 1.27)	*0.71 (−0.17, 1.59)*	0.64* (−0.01, 1.30)
NO_3-^a	5.19	0.98** (0.01, 1.96)	0.90* (−0.08, 1.89)	0.81 (−0.22, 1.84)	0.94* (−0.08, 1.96)	1.03** (0.05, 2.01)	0.86 (−0.21, 1.94)	0.77 (−0.18, 1.72)		0.54 (−0.69, 1.78)	0.98* (−0.06, 2.02)	1.03** (0.02, 2.06)
$SO_4^2-^a$	2.99	1.33** (0.11, 2.57)	1.40** (0.17, 2.64)	1.31** (0.08, 2.54)	1.27** (0.02, 2.55)	1.47** (0.22, 2.73)	1.23* (−0.01, 2.49)	1.34** (0.28, 2.41)	0.90 (−0.66, 2.49)		1.21* (−0.07, 2.51)	1.34** (0.10, 2.59)
OP^{TOTAL}	38.71	0.04 (−0.01, 0.08)	*0.02 (−0.13, 0.16)*	*−0.01 (−0.12, 0.10)*	0.04 (−0.01, 0.09)	*0.08 (−0.02, 0.18)*	0.02 (−0.05, 0.08)	*0.01 (−0.05, 0.06)*	0.03 (−0.02, 0.08)	0.04 (−0.01, 0.08)		0.04 (−0.01, 0.09)
NO_2	10.54	0.29 (−2.09, 2.73)	−0.11 (−2.55, 2.38)	−0.29 (−2.75, 2.24)	1.10 (−1.67, 3.95)	−0.02 (−2.59, 2.62)	−0.28 (−2.81, 2.30)	−0.01 (−2.12, 2.15)	−0.42 (−2.86, 2.09)	−0.05 (−2.42, 2.37)	−0.35 (−2.96, 2.33)	

For explanation see Table 2.

Table 4. Associations between exposure to air pollution and percentage changes (25 h post-pre) in platelet counts (all sites).

	IQR	Single-pollutant associations	Two-pollutant associations									
			PM_{10}	$PM_{2.5}$	PNC	EC(F)	OC(F)	OC(C)	NO_3^{-a}	SO_4^{2-a}	OP^{TOTAL}	NO_2
PM_{10}	13.50	0.04 (−0.07, 0.14)		0.[illegible]2 (−0.31, 0.5[illegible])	0.05 (−0.06, 0.16)	0.23* (0.00, 0.47)	0.07 (−0.06, 0.20)	−0.[illegible]7** (−0.31, −0.02)	0.03 (−0.08, 0.13)	0.04 (−0.06, 0.15)	0.17 (−0.14, 0.48)	0.07 (−0.04, 0.17)
$PM_{2.5}$	11.54	0.07 (−0.17, 0.32)	−0.19 (−1.17, 0.80)		0.09 (−0.15, 0.34)	0.30 (−0.14, 0.75)	0.17 (−0.16, 0.49)	−0.39** (−0.72, −0.05)	0.00 (−0.25, 0.26)	0.08 (−0.17, 0.32)	0.08 (−0.45, 0.62)	0.15 (−0.10, 0.40)
PNC	32,906	−1.15 (−2.69, 0.40)	−1.26 (−2.80, 0.32)	−1.21 (−2.75, 0.36)		−1.44 (−3.20, 0.35)	−1.15 (−2.69, 0.41)	−1.[illegible]5 (−2.68, 0.40)	−0.75 (−2.36, 0.88)	−1.09 (−2.67, 0.51)	−0.76 (−2.42, 0.94)	−0.51 (−2.29, 1.30)
EC(F)	4.35	−0.08 (−0.95, 0.80)	−1.77* (−3.66, 0.16)	−0.98 (−2.55, 0.61)	0.32 (−0.68, 1.33)		−0.03 (−0.99, 0.95)	−1.06** (−2.02, −0.08)	0.03 (−0.84, 0.90)	−0.02 (−0.91, 0.88)	−0.58 (−2.26, 1.14)	0.29 (−0.65, 1.23)
OC(F)	1.82	−0.12 (−0.91, 0.69)	−0.43 (−1.41, 0.56)	−0.47 (−1.51, 0.59)	−0.10 (−0.90, 0.70)	−0.11 (−0.99, 0.79)		−0.65 (−1.47, 0.18)	−0.50 (−1.35, 0.36)	−0.14 (−0.95, 0.67)	−0.31 (−1.37, 0.75)	0.11 (−0.72, 0.95)
OC(C)	0.79	0.79** (0.25, 1.33)	1.[illegible]4** (0.56, 2.23)	1.44** (0.67, 2.21)	0.78** (0.24, 1.32)	1.16** (0.53, 1.79)	0.96** (0.38, 1.54)		0.63** (0.04, 1.23)	0.79** (0.25, 1.34)	1.06** (0.36, 1.77)	0.84** (0.30, 1.37)
NO_3^{-a}	5.19	0.74** (0.02, 1.47)	0.[illegible]2* (−0.01, 1.46)	0.74* (−0.02 1.51)	0.64* (−0.12, 1.41)	0.75** (0.02, 1.48)	0.92** (0.14, 1.71)	0.52 (−0.26, 1.31)		0.90* (0.01, 1.81)	0.37 (−0.39, 1.13)	0.98** (0.25, 1.73)
SO_4^{2-a}	2.99	0.29 (−0.58, 1.17)	0.[illegible]3 (−0.55, 1.21)	0.30 (−0.57 1.[illegible]8)	0.18 (−0.71, 1.07)	0.29 (−0.60, 1.19)	0.31 (−0.57, 1.19)	0.41 (−0.44, 1.26)	−0.33 (−1.39, 0.74)		0.02 (−0.87, 0.91)	0.39 (−0.48, 1.26)
OP^{TOTAL}	38.71	0.02 (−0.02, 0.05)	−0.04 (−0.14, 0.06)	0.01 (−0.07, 0.08)	0.02 (−0.02, 0.05)	0.04 (−0.03, 0.10)	0.02 (−0.02, 0.07)	−0.03 (−0.07, 0.02)	0.01 (−0.02, 0.05)	0.02 (−0.02, 0.05)		0.02 (−0.02, 0.05)
NO_2	10.54	−1.69* (−3.40, 0.06)	−[illegible].93** (−3.68, −0.14)	−1.98** (−3.75, −0.17)	−1.40 (−3.39, 0.64)	−1.91** (−3.76, −0.02)	−1.76* (−3.55, 0.07)	−1.[illegible]3** (−3.31, −0.11)	−2.28** (−4.01, −0.52)	−1.77** (−3.50, −0.02)	−1.24 (−3.09, 0.64)	

For explanation see Table 2.

Table 5. Associations between exposure to air pollution and percentage changes (25 h post-pre) in von Willebrand Factor (all sites).

	IQR	Single-pollutant associations	Two-pollutant associations									
			PM_{10}	$PM_{2.5}$	PNC	EC(F)	OC(F)	OC(C)	NO_3^{-a}	SO_4^{2-a}	OP^{TOTAL}	NO_2
PM_{10}	13.50	0.22** (0.02, 0.41)		*−0.32 (−1.09, 0.46)*	0.22** (0.02, 0.42)	0.31 (−0.12, 0.75)	*−0.04 (−0.27, 0.20)*	*0.07 (−0.22, 0.37)*	0.21** (0.01, 0.41)	0.13 (−0.07, 0.33)	*0.02 (−0.55, 0.60)*	0.20* (−0.01, 0.40)
$PM_{2.5}$	11.54	0.58** (0.12, 1.03)	*1.29 (−0.51, 3.12)*		0.58** (0.12, 1.03)	0.78* (−0.04, 1.60)	*−0.06 (−0.64, 0.53)*	0.32 (−0.36, 1.00)	0.55** (0.07, 1.03)	0.35 (−0.11, 0.81)	*0.34 (−0.65, 1.34)*	0.53** (0.06, 1.00)
PNC	32,906	−0.04 (−2.80, 2.80)	0.16 (−2.70, 3.09)	0.28 (−2.56, 3.20)		−0.40 (−3.53, 2.83)	−0.13 (−2.87, 2.69)	0.87 (−2.05, 3.87)	0.40 (−2.48, 3.38)	0.37 (−2.60, 3.42)	−1.40 (−4.47, 1.78)	−0.73 (−3.88, 2.52)
EC(F)	4.35	0.32 (−1.27, 1.94)	−0.91 (−4.42, 2.73)	−0.87 (−3.73, 2.08)	0.44 (−1.37, 2.28)		−0.40 (−2.14, 1.36)	0.06 (−1.83, 2.00)	1.54* (−0.12, 3.22)	0.88 (−0.81, 2.59)	*−2.13 (−5.21, 1.04)*	0.12 (−1.58, 1.85)
OC(F)	1.82	1.81** (0.26, 3.38)	*1.51 (−0.29, 3.34)*	*1.46 (−0.47, 3.44)*	1.34* (−0.11, 2.81)	1.50* (−0.12, 3.13)		1.68** (0.05, 3.33)	1.69* (0.01, 3.41)	1.24 (−0.29, 2.80)	0.98 (−1.03, 3.02)	1.30* (−0.23, 2.86)
OC(C)	0.79	0.48 (−0.53, 1.50)	*−0.05 (−1.55, 1.46)*	−0.29 (−1.78, 1.22)	0.50 (−0.52, 1.53)	0.20 (−0.97, 1.38)	0.03 (−1.05, 1.13)		−0.02 (−1.09, 1.07)	0.13 (−0.89, 1.17)	*0.56 (−0.78, 1.92)*	0.17 (−0.81, 1.16)
$NO_3^{-\,a}$	5.19	0.66 (−0.63, 1.97)	0.59 (−0.74, 1.95)	0.29 (−1.10, 1.70)	0.72 (−0.63, 2.09)	0.92 (−0.42, 2.28)	0.25 (−1.21, 1.74)	0.73 (−0.70, 2.19)		0.55 (−1.19, 2.33)	0.87 (−0.54, 2.30)	0.57 (−0.78, 1.94)
SO_4^{2-a}	2.99	0.93 (−0.91, 2.80)	0.99 (−0.86, 2.88)	0.88 (−0.97, 2.76)	0.97 (−0.91, 2.89)	1.11 (−0.78, 3.03)	0.67 (−1.19, 2.56)	1.00 (−0.86, 2.88)	0.46 (−1.89, 2.87)		1.04 (−0.85, 2.97)	0.80 (−1.07, 2.70)
OP^{TOTAL}	38.71	0.06* (−0.01, 0.12)	*0.01 (−0.18, 0.21)*	*−0.02 (−0.17, 0.12)*	0.06* (0.00, 0.13)	*0.10 (−0.03, 0.22)*	0.03 (−0.06, 0.12)	*−0.01 (−0.09, 0.08)*	0.06 (−0.01, 0.12)	0.03 (−0.03, 0.10)		0.06* (−0.01, 0.12)
NO_2	10.54	1.19 (−2.00, 4.49)	1.79 (−1.56, 5.26)	1.52 (−1.86, 5.01)	1.61 (−2.08, 5.45)	1.10 (−2.33, 4.66)	0.25 (−3.09, 3.70)	1.41 (−1.86, 4.78)	0.78 (−2.55, 4.22)	1.48 (−1.90, 4.98)	0.53 (−3.03, 4.21)	

For explanation see Table 2.

Discussion

We investigated the acute vascular effects in a panel of healthy, young volunteers semi-experimentally exposed for 5 h to ambient air pollution with contrasting PM characteristics. Most associations were observed in the next morning rather than 2 h after exposure. In two-pollutant models we found associations of OC (fine) and OP with hs-CRP; sulfate and nitrate with fibrinogen; OC (coarse) and nitrate with platelet counts; and OC (fine) with vWF. We also observed unexpected negative associations with NO_2/NO_X (platelet counts) and O_3 (tPA/PAI-1 complex).

Our study substantially adds to the existing limited evidence of the effects of particulate matter air pollution on the biomarkers of cardiovascular risk. Compared to previous epidemiological studies which often use 24-hour average exposures, this study shows that five-hour exposure with exercise is sufficient to elicit biological responses. The semi-experimental design of our study is a strength when compared to observational studies as bias due to exposure measurement error and confounding is reduced. This study further adds to the very scarce evidence of the effects of different particle components. We have documented that associations with specific components (OC, nitrate, sulfate) were stronger than with widely used, legally regulated, PM mass metrics.

PM Characteristics and Components

In single-pollutant models many significant associations of PM mass and composition with acute vascular biomarkers were observed. Those were, however, not robust against adjustment for co-pollutants. In two-pollutant models, with the exception of fairly consistent associations in the "outdoor sites" models with platelets and von Willebrand Factor, the associations with PM mass concentrations generally disappeared. This supports the notion that particle composition affects the biological response to PM. We found little ($PM_{2.5}$) or no effect ($PM_{2.5-10}$, PNC) of PM fractions defined on the basis of particle size. PNC has been associated with cardiovascular events in time series studies, but not consistently [2]. A concern in observational studies is that PNC exposure measurement error is larger than for $PM_{2.5}$. This however does not apply to our study as we measured exposure on-site.

We also observed no associations with either total or water-soluble trace metals, or EC. Contrary to the results of the respiratory analysis [15], we observed no positive associations with PNC and NO_2 in the current analysis. Similarly to our respiratory analysis [15], and contrary to our hypothesis, OP did not display a strong and consistent association with acute vascular effects - in the two-pollutant models associations with OP were only fairly consistent with changes in hs-CRP. The absence of associations of the examined endpoints with OP may reflect the fact that the assay employed only examined the intrinsic potential of the particles to drive oxidation reactions in an acellular model: reflecting their content of redox-active transition metals and quinones. As PM can elicit oxidative stress through alternative pathways upon interaction with airway cells only a fraction of their in vivo activity can be accounted for by this assay. The OP values themselves also reflect, as with the transition metal concentrations, an aggregate (OP^{TOTAL}) of the observed values in coarse ($OP^{PM2.5-10}$), fine ($OP^{PM2.5}$) and (quasi)ultrafine ($OP^{PM0.18}$) fractions. On average, $OP^{PM2.5-10}$ contributed in 30% to the OP^{TOTAL} (26% at the underground location and 31% at the outdoor locations), $OP^{PM2.5}$ contributed 40% to the OP^{TOTAL} (30% underground, 44% outdoor sites), and $OP^{PM0.18}$ contributed 31% to the OP^{TOTAL} (45% underground, 24% outdoor sites). There may be further merit therefore in exploring these parameters in the individual size fraction. In the current analysis we restricted ourselves to PM_{10}, to limit the number of exposure variables under examination and thus the possibility of observing spurious interactions by chance.

The components consistently seen in the current analysis were OC, either in coarse or fine PM fraction, nitrate and sulfate, which were seen both in the "all sites" and the "outdoor-only sites" datasets. OC and other organic components of PM (e.g., polycyclic aromatic hydrocarbons, (semi)quinones) may be relevant for cardiovascular diseases [2], with two recent US studies reporting associations of organic carbon with cardiovascular hospital admissions [20] and mortality [21]. Sulfate and nitrate may not have direct effects - a review of the biological effects of secondary inorganic aerosols in controlled exposure studies showed the effects due to acidity of inorganic components rather than presence of single specific anions [22]. Since ambient PM in the Netherlands is less acidic due to neutralization by ambient ammonia [22], the sulfate and nitrate are unlikely to have a direct causal effect. However, a recent American Heart Association Scientific Statement did not completely exclude a direct role of particle sulfate in cardiovascular events [2] and a large epidemiological literature supports associations of sulfate with hospital admissions and mortality [23–26]. As the associations found for sulfate and nitrate with acute vascular biomarkers were robust against adjustment for a large series of co-pollutants, it remains unclear from our data what these associations represent. Sulfate and nitrate are likely indicators for other secondary aerosol components that may be biologically active.

Timing of Effects

Different from our findings in the respiratory analysis [15], where the associations were the strongest immediately after and 2 h after exposure, in the current analysis the associations with acute vascular markers were mostly seen in the morning following the exposure. This difference seems reasonable as systemic inflammation would need more time to develop. The lack of associations is also consistent with the relatively long half-lives of CRP and especially fibrinogen. However, the literature regarding the timing of the effects is somewhat inconsistent. Some experimental studies in human volunteers reported increases in platelet count [27] and activation [28], thrombus formation [28], CRP and tPA [29] between 2 and 6 h after exposure to diesel exhaust or concentrated ambient ultrafine particles. However, no changes in CRP, fibrinogen, platelets or vWF were observed either 6 h after exposure of commuters to traffic-related air pollution [19], or immediately after experimental exposure to concentrated (quasi)ultrafine particles ($PM_{0.16}$) [30]. When the more delayed effects were investigated, experimental diesel exhaust exposure led to a decrease in PAI-1 levels 22 h after exposure [31]; exposure to concentrated ambient particles was associated with elevated fibrinogen 24 h later [32]; and exposure of healthy volunteers to subway air pollution was associated with increase in fibrinogen 14 h later [33]. However, exposure of healthy volunteers to road-tunnel air pollution was not associated with changes in fibrinogen and PAI-1 14 h later [34]; and no changes in CRP, fibrinogen, platelets, vWF, tPA or PAI-1 were observed 18 h after exposure to concentrated $PM_{0.16}$ [30].

It has to be noted that all of these studies used 1–2 h exposure periods which could potentially be too short to observe a marked increase in acute vascular biomarkers. A study on cardiovascular effects in healthy highway patrol officers by Riediker et al. showed that 9 h in-vehicle exposure to traffic-related $PM_{2.5}$ was associated with increase in CRP and vWF 10-14 h after exposure [35]. When investigating the components of $PM_{2.5}$, Cu and its compounds were associated with a decrease in PAI-1 [36]. However, none of the discussed studies measured the PM components with which we found robust associations; they also did not take a systematic two-pollutant approach.

Effect estimates for OC, nitrate and sulfate expressed per IQR were small (1–2% changes) and in this healthy adult population probably do not reflect an adverse effect. In patients with preexisting disease, changes in these markers might however elicit cardiovascular events. Our study therefore adds to the plausibility of observed associations of ambient air pollution with cardiovascular events.

Strengths and Limitations

The use of a semi-experimental design in the current investigation allowed us to define two-pollutant models to investigate independent effects of a large number of individual PM characteristics. Since we performed air pollution characterization on site during exposure of volunteers, the exposure measurement error was largely due to instrumental errors and likely negligible compared to observational studies relying on central site monitoring. Instrumental precision of measurements was in general between 5–10% (Table S3), which is very low compared to the range of measured concentrations. This means that difference in instrumental precision was not a likely explanation for stronger associations with specific pollutants. In our design, we also reduced correlations between PM characteristics by performing repeated measurements at multiple locations with different source characteristics. Despite that, some correlations remained too high to interpret two-pollutant models and disentangle independent health effects of single PM characteristics. Since we specified a large number of models to investigate all possible combinations of air pollutants measured, we potentially faced a problem of chance finding in our results. That is why in our interpretation of the results we focused on the consistency of (significant) associations rather than individual significant associations present.

In conclusion, OC, nitrate and sulfate were most consistently associated with different biomarkers of acute cardiovascular risk. Associations with PM mass concentrations and OP were less consistent, whereas other measured components of the air pollution mixture, including PNC, EC, trace metals and NO_2, were not associated with the biomarkers after adjusting for other pollutants.

Supporting Information

Figure S1 Timeline of a typical sampling day in the RAPTES project.

Table S1 Geometric means and minimum-maximum of 5-hour average air pollution concentrations.

Table S2 Spearman's correlation coefficients between PM characteristics.

Table S3 Precision and limits of detection (LOD) for 5-hour sampling periods.

Table S4 Adjusted associations between exposure to air pollution and percentage changes (post-pre) in hs-CRP.

Table S5 Adjusted associations between exposure to air pollution and percentage changes (post-pre) in fibrinogen.

Table S6 Adjusted associations between exposure to air pollution and percentage changes (post-pre) in tPA/PAI-1 complex.

Table S7 Adjusted associations between exposure to air pollution and percentage changes (post-pre) in platelet counts.

Table S8 Adjusted associations between exposure to air pollution and percentage changes (post-pre) in von Willebrand Factor.

Table S9 Two-pollutant models of associations between exposure to air pollution and percentage changes (25 h post-pre) in hs-CRP (all sites).

Table S10 Two-pollutant models of associations between exposure to air pollution and percentage changes (25 h post-pre) in fibrinogen (all sites).

Table S11 Two-pollutant models of associations between exposure to air pollution and percentage changes (25 h post-pre) in platelet counts (all sites).

Table S12 Two-pollutant models of associations between exposure to air pollution and percentage changes (25 h post-pre) in von Willebrand Factor (all sites).

Table S13 Two-pollutant models of associations between exposure to air pollution and percentage changes (25 h post-pre) in hs-CRP (outdoor sites).

Table S14 Two-pollutant models of associations between exposure to air pollution and percentage changes (25 h post-pre) in fibrinogen (outdoor sites).

Table S15 Two-pollutant models of associations between exposure to air pollution and percentage changes (25 h post-pre) in platelet counts (outdoor sites).

Table S16 Two-pollutant models of associations between exposure to air pollution and percentage changes (25 h post-pre) in von Willebrand Factor (outdoor sites).

Table S17 Two-pollutant models of associations between exposure to air pollution and percentage changes (25 h post-pre) in tPA/PAI-1 complex (all sites).

Table S18 Two-pollutant models of associations between exposure to air pollution and percentage changes (25 h post-pre) in tPA/PAI-1 complex (outdoor sites).

Table S19 Two-pollutant models of associations between exposure to air pollution and percentage changes

(25 h post-pre) in tPA/PAI-1 complex (not log-transformed; all sites).

Acknowledgments

We are grateful to all study participants; John Boere, Paul Fokkens, Daan Leseman, Lise van den Burg, Veerle Huijgen, Maartje Kleintjes, Marja Meijerink and Jet Musters, for their excellent support in data collection; Eef van Otterloo for his help with participant's recruitment, and Miriam Groothoff, for medical supervision.

References

1. Brunekreef B, Holgate ST (2002) Air pollution and health. Lancet 360: 1233–1242.
2. Brook RD, Rajagopalan S, Pope III CA, Brook JR, Bhatnagar A, et al. (2010) Particulate matter air pollution and cardiovascular disease: An update to the scientific statement from the American Heart Association. Circulation 121: 2331–2378.
3. Solomon PA, Sioutas C (2008) Continuous and semicontinuous monitoring techniques for particulate matter mass and chemical components: a synthesis of findings from EPA's Particulate Matter Supersites Program and related studies. J Air Waste Manag Assoc 58: 164–195.
4. Valavanidis A, Fiotakis K, Vlachogianni T (2008) Airborne particulate matter and human health: toxicological assessment and importance of size and composition of particles for oxidative damage and carcinogenic mechanisms. J Environ Sci Health C Environ Carcinog Ecotoxicol Rev 26: 339–362.
5. Ayres JG, Borm P, Cassee FR, Castranova V, Donaldson K, et al. (2008) Evaluating the toxicity of airborne particulate matter and nanoparticles by measuring oxidative stress potential–a workshop report and consensus statement. Inhal Toxicol 20: 75–99.
6. WHO (2007) Health relevance of particulate matter from various sources. Report on a WHO workshop Bonn, Germany 26–27 March 2007. Copenhagen, Denmark: World Health Organization Europe. p.
7. Brunekreef B (2010) The Color of Smoke. Epidemiology 21: 903–904.
8. Pearson TA, Mensah GA, Alexander RW, Anderson JL, Cannon III RO, et al. (2003) Markers of Inflammation and Cardiovascular Disease: Application to Clinical and Public Health Practice: A Statement for Healthcare Professionals From the Centers for Disease Control and Prevention and the American Heart Association. Circulation 107: 499–511.
9. Kannel WB, Wolf PA, Castelli WP, D'Agostino RB (1987) Fibrinogen and risk of cardiovascular disease. J Am Med Assoc 258: 1183–1186.
10. Wagner DD (2005) New Links Between Inflammation and Thrombosis. Arterioscler Thromb Vasc Biol 25: 1321–1324.
11. Spiel AO, Gilbert JC, Jilma B (2008) Von Willebrand Factor in Cardiovascular Disease: Focus on Acute Coronary Syndromes. Circulation 117: 1449–1459.
12. Wiman B, Andersson T, Hallqvist J, Reuterwall C, Ahlbom A, et al. (2000) Plasma levels of tissue plasminogen activator/plasminogen activator inhibitor-1 complex and von Willebrand factor are significant risk markers for recurrent myocardial infarction in the Stockholm Heart Epidemiology Program (SHEEP) study. Arterioscler Thromb Vasc Biol 20: 2019–2023.
13. Strak M, Steenhof M, Godri KJ, Gosens I, Mudway IS, et al. (2011) Variation in Characteristics of Ambient Particulate Matter at Eight Locations in the Netherlands - The RAPTES Project. Atmos Environ 45: 4442–4453.
14. Steenhof M, Gosens I, Strak M, Godri KJ, Hoek G, et al. (2011) In vitro toxicity of particulate matter (PM) collected at different sites in the Netherlands is associated with PM composition, size fraction and oxidative potential - the RAPTES project. Part Fibre Toxicol 8.
15. Strak M, Janssen NAH, Godri KJ, Gosens I, Mudway IS, et al. (2012) Respiratory Health Effects of Airborne Particulate Matter: The Role of Particle Size, Composition and Oxidative Potential - The RAPTES Project. Environ Health Persp 120: 1183–1189.
16. Godri KJ, Duggan ST, Fuller GW, Baker T, Green D, et al. (2010) Particulate Matter Oxidative Potential from Waste Transfer Station Activity. Environ Health Persp 118: 493–498.
17. Krintus M, Sypniewska G, Kuligowska-Prusinska M (2010) Effect of second and third generation oral contraceptives on C-reactive protein, lipids and apolipoproteins in young, non-obese, non-smoking apparently healthy women. Clin Biochem 43: 626–628.

Author Contributions

Performed chemical and oxidative potential analyses: KJG. Assisted in and co-supervised oxidative potential analyses: ISM. Assisted in biomarkers measurements: RvO. Supervised biomarkers measurements: HMHS. Supervised oxidative potential analyses: FJK. Supervised chemical analyses: RMH. Project leader of RAPTES: NAHJ. Read and approved the manuscript: M. Strak GH KJG IG FRC ISM RvO HMHS FJK RMH BB EL M. Steenhof NAHJ. Conceived and designed the experiments: M. Strak GH IG FRC ISM FJK RMH BB EL M. Steenhof NAHJ. Performed the experiments: M. Strak M. Steenhof. Analyzed the data: M. Strak GH NAHJ. Wrote the paper: M. Strak GH BB EL NAHJ.

18. Trigg DE, Wood MG, Kouides PA, Kadir RA (2011) Hormonal Influences on Hemostasis in Women. Semin Thromb Hemost 37: 77–86.
19. Zuurbier M, Hoek G, Oldenwening M, Meliefste K, Krop E, et al. (2011) In-Traffic Air Pollution Exposure and CC16, Blood Coagulation, and Inflammation Markers in Healthy Adults. Environ Health Persp 119: 1384–1389.
20. Peng RD, Bell ML, Geyh AS, McDermott A, Zeger SL, et al. (2009) Emergency Admissions for Cardiovascular and Respiratory Diseases and the Chemical Composition of Fine Particle Air Pollution. Environ Health Persp 117: 957–963.
21. Ostro B, Feng W-Y, Broadwin R, Green S, Lipsett M (2007) The Effects of Components of Fine Particulate Air Pollution on Mortality in California: Results from CALFINE. Environ Health Persp 115: 13–19.
22. Schlesinger RB, Cassee F (2003) Atmospheric secondary inorganic particulate matter: the toxicological perspective as a basis for health effects risk assessment. Inhal Toxicol 15: 197–235.
23. Krewski D, Burnett RT, Goldberg MS, Hoover K, Siemiatycki J, et al. (2000) Reanalysis of the Harvard Six Cities Study and the American Cancer Society Study of Particulate Air Pollution and Mortality. Boston, MA: Health Effects Institute. p.
24. Burnett RT, Dales R, Krewski D, Vincent R, Dann T, et al. (1995) Associations between Ambient Particulate Sulfate and Admissions to Ontario Hospitals for Cardiac and Respiratory Diseases. Am J Epidemiol 142: 15–22.
25. Pope III CA, Burnett RT, Thun MJ, Calle EE, Krewski D, et al. (2002) Lung cancer, cardiopulmonary mortality, and long-term exposure to fine particulate air pollution. J Am Med Assoc 287: 1132–1141.
26. Pope III CA, Rodermund DL, Gee MM (2007) Mortality Effects of a Copper Smelter Strike and Reduced Ambient Sulfate Particulate Matter Air Pollution. Environ Health Persp 115: 679–683.
27. Salvi S, Blomberg A, Rudell B, Kelly F, Sandström T, et al. (1999) Acute inflammatory responses in the airways and peripheral blood after short-term exposure to diesel exhaust in healthy human volunteers. Am J Respir Crit Care Med 159: 702–709.
28. Lucking AJ, Lundback M, Mills NL, Faratian D, Barath SL, et al. (2008) Diesel exhaust inhalation increases thrombus formation in man. Eur Heart J 29: 3043–3051.
29. Gong H, Linn WS, Clark KW, Anderson KR, Sioutas C, et al. (2008) Exposures of healthy and asthmatic volunteers to concentrated ambient ultrafine particles in Los Angeles. Inhal Toxicol 20: 533–545.
30. Samet JM, Rappold A, Graff D, Cascio WE, Berntsen JH, et al. (2009) Concentrated Ambient Ultrafine Particle Exposure Induces Cardiac Changes in Young Healthy Volunteers. Am J Respir Crit Care Med 179: 1034–1042.
31. Carlsten C, Kaufman JD, Peretz A, Trenga CA, Sheppard L, et al. (2007) Coagulation markers in healthy human subjects exposed to diesel exhaust. Thromb Res 120: 849–855.
32. Ghio AJ, Hall A, Bassett MA, Cascio WE, Devlin RB (2003) Exposure to concentrated ambient air particles alters hematologic indices in humans. Inhal Toxicol 15: 1465–1478.
33. Klepczynska Nystrom A, Svartengren M, Grunewald J, Pousette C, Rodin I, et al. (2010) Health effects of a subway environment in healthy volunteers. Eur Respir J 36: 240–248.
34. Larsson B-M, Sehlstedt M, Grunewald J, Sköld CM, Lundin A, et al. (2007) Road tunnel air pollution induces bronchoalveolar inflammation in healthy subjects. Eur Respir J 29: 699–705.
35. Riediker M, Cascio WE, Griggs TR, Herbst MC, Bromberg PA, et al. (2004) Particulate Matter Exposure in Cars Is Associated with Cardiovascular Effects in Healthy Young Men. Am J Respir Crit Care Med 169: 934–940.
36. Riediker M (2007) Cardiovascular Effects of Fine Particulate Matter Components in Highway Patrol Officers. Inhal Toxicol 19: 99–105.

18

Association between Ambient Temperature and Acute Myocardial Infarction Hospitalisations in Gothenburg, Sweden: 1985–2010

Janine Wichmann[1]*, Annika Rosengren[2], Karin Sjöberg[3], Lars Barregard[1], Gerd Sallsten[1]

1 Department of Occupational and Environmental Medicine, Sahlgrenska University Hospital and University of Gothenburg, Gothenborg, Sweden, **2** Department of Molecular and Clinical Medicine, Sahlgrenska University Hospital and University of Gothenburg, Gothenborg, Sweden, **3** IVL Swedish Environmental Research Institute, Gothenborg, Sweden

Abstract

Cardiovascular disease (CVD) is the number one cause of death globally and evidence is steadily increasing on the role of non-traditional risk factors such as meteorology and air pollution. Nevertheless, many research gaps remain, such as the association between these non-traditional risk factors and subtypes of CVD, such as acute myocardial infarction (AMI). The objective of this study was to investigate the association between daily ambient temperature and AMI hospitalisations using a case-crossover design in Gothenburg, Sweden (1985–2010). A secondary analysis was also performed for out-of-hospital ischemic heart disease (IHD) deaths. Susceptible groups by age and sex were explored. The entire year as well as the warm (April–September) and cold periods (October–March) were considered. In total 28 215 AMI hospitalisations (of 22 475 people) and 21 082 out-of-hospital IHD deaths occurred during the 26-year study period. A linear exposure-response corresponding to a 3% and 7% decrease in AMI hospitalisations was observed for an inter-quartile range (IQR) increase in the 2-day cumulative average of temperature during the entire year (11°C) and the warm period (6°C), respectively, with and without adjustment for PM_{10}, NO_2, NO_x or O_3. No heat waves occurred during the warm period. No evidence of an association in the cold period nor any association between temperature and IHD deaths in the entire year, warm or cold periods - with and without adjusting for PM_{10}, NO_2, NO_x or O_3 was found. No susceptible groups, based on age or sex, were identified either. The inverse association between temperature and AMI hospitalisations (entire year and warm period) in Gothenburg is in accordance with the majority of the few other studies that investigated this subtype of CVD.

Editor: Bruno Caramelli, University of Sao Paulo, Brazil

Funding: The Swedish Research Council Formas funded the study. The funders had no role in study design, data collection and analysis, decision to publish, or preparation of the manuscript.

Competing Interests: The authors have declared that no competing interests exist.

* E-mail: gerd.sallsten@amm.gu.se

Introduction

Cardiovascular disease (CVD) is the number one cause of death globally and also in European countries, such as Sweden [1]. Ischemic heart disease (IHD) is responsible for 61% of the 2 million CVD deaths in Europe annually, and stroke accounts for most of the rest. Acute myocardial infarction (AMI) is the most important manifestation of ischemic heart disease.

The biggest risk factors for CVD are age, sex, smoking, low physical activity, increased waist circumference, diabetes and hypertension [2–4]. Evidence is increasing on the effects of non-traditional risk factors, such as air pollution and weather (e.g. temperature) changes on CVD mortality and morbidity, specifically as short-term risk factors [4–6]. Nevertheless, fewer studies focused on subtypes of CVD, such as AMI hospitalisations.

The objective of this study was to investigate the association between daily ambient temperature and acute myocardial infarction (AMI) hospitalisations in a case-crossover design in Gothenburg, Sweden (1985–2010). A secondary aim was to study associations with out-of-hospital ischemic heart disease (IHD) deaths. Susceptible groups by age and sex were explored. We studied the entire year as well as warm (April–September) and cold periods (October–March). We addressed some of the limitations of the previous studies: investigated confounding by ambient air pollution (PM_{10}, NO_2, NO_x and O_3), examined the lag and nonlinear effects of temperature and relative humidity, explored effect modification by age and sex by stratification and interaction models, and compared the results of the case-crossover models to those of the generalised additive Poisson time-series models.

Methods

Population and register data

Sweden has a publicly financed health care system with hospital care available to all citizens at low cost. Swedish hospitals register principal and contributory discharge diagnoses for all patients in the national hospital discharge register. For the purpose of the present study, data from the national hospital discharge register with coverage of all hospitalisations in Gothenburg since 1970 was used. The national cause of death register, based on diagnoses from death certificates, is complete from 1961.

Major coronary events and related case-fatality. The population for the present study includes all cases of AMI or fatal IHD events registered in Gothenburg from 1 January 1985 to 31

December 2010. The study period was determined by the availability of air pollution data. The International Classification of Diseases (ICD) version 9 (ICD-9) was used from 1987 to 1996 and version 10 (ICD 10) from 1997 and onwards.

Hospitalisation for an AMI was defined as a discharge (dead or alive) with a principal diagnosis of ICD 8–9 410 (until 1996); ICD 10 I21. Only emergency hospitalisations to any of the five hospitals within Gothenburg (about 1.7 to 5.8 km from the urban background station) were included in the study. In case AMI hospitalisations for the same person occurred at more than one hospital (i.e. transferrals) on a specific day, then only the first hospitalisation was retained in the study. AMI hospitalisations that occurred within 28 days after a previous AMI hospitalisation were excluded (3 068 admissions) as readmissions following discharge for AMI are quite high [7].

Out-of-hospital death was defined as a death from IHD (defined as ICD 8–9 410–414 (until 1996); ICD 10 I20–I25) in a person who was not hospitalised, as defined by not having a recorded admission on the day of death. Only deaths that occurred in Gothenburg were included in the study. The location of death was based on the home address of the deceased, but a minority might have died somewhere else.

Air pollution data

Meteorological and air pollution data, except ground-level ozone (O_3), were measured at the urban background monitoring station by the Gothenburg Environment and Health Department. The urban background monitoring station is located on the roof of a 25 m high building in the centre of Gothenburg with typical weekday traffic flows of about 65 000 vehicles/day in several directions (within 300 m) and minimal direct contribution from local pollution sources, in accordance with World Health Organisation (WHO) guidelines [8]. O_3 was measured by IVL Swedish Environmental Research Institute on behalf of the Swedish Environmental Protection Agency and within the national air quality monitoring network at a rural background site located in Råö, 45 km south of the city centre.

Air pollution data included measurements of PM_{10} (tapered element oscillating microbalance instrument), nitrogen dioxides (NO_2 and NO_x) (chemiluminescence instrument) and O_3 (UV absorption instrument). Please see Table S1 for the technical names of the various air pollution measurement instruments used during the study period. Temperature and relative humidity were measured with the HMP45a probe (Vaisala, Helsinki). Daily averages (midnight to midnight) data were applied in the statistical analyses. Missing values were not imputed. For 437 days the PM_{10} levels were lower than 5 $\mu g.m^{-3}$ and were set as 5 $\mu g.m^{-3}$.

Ethics

Registry based health outcome data were applied in this study and all identifying variables were deleted. The study adheres to the standards of the Swedish Data Protection Agency. It was approved by the institutional review board of the University if Gothenburg which waived the need for written informed consent.

Statistical analysis

The time-stratified case-crossover design was applied to investigate the association between temperature and AMI hospitalisations and IHD deaths. The case-crossover design was developed as a variant of the case-control design to study the effects of transient exposures on emergency events, comparing each person's exposure in a time period just prior to a case-defining event with person's exposure at other times [9]. Hereby, control on all measured and unmeasured personal characteristics that do not vary over a short time period is accomplished. If in addition, the control days are chosen close to the event day, personal characteristics that vary slowly over time (e.g. body mass index, smoking status) are also controlled by matching. A time-stratified approach was applied to select the control days, defining the day of hospitalisations or death as the case day and the same day of the other weeks in the same month and year as control days. With this approach even very strong confounding of exposure by seasonal patterns is controlled by design [10–12]. The data were analysed using conditional logistic regression analysis (PROC PHREG in SAS 9.2, SAS Institute, Cary, NC).

Lag0 (same day of exposure as day of hospitalisation or death) and lag1 (day prior to day of hospitalisation or death) were investigated, as well as the 2-day cumulative average (CA2) (mean of lag0–1). The values of CA2 were set as missing if any of the values of lag0 or lag1 were missing. Control days for lag1 and CA2 were defined as for lag0.

All models were adjusted for public holidays (as a binary variable). Previous studies reported a linear relationship between the air pollutants and AMI hospitalisations and CVD deaths [13–25]. The pollutants were therefore adjusted for as linear terms, one pollutant at a time.

Although intra-individual factors cannot be examined due to the nature of the case-crossover design where each person is his/her own control, inter-individual variation using an interaction term between the susceptibility variable and an exposure variable in the conditional logistic regression model yields the possibility to detect a p-value for interaction and when significant the subgroup specific estimates are valid. Susceptibility was therefore investigated in stratified analyses by sex and age (<75 and ≥75 years), followed by models with interaction terms.

Odds ratios (OR) and the 95% confidence intervals (CI) were calculated per inter-quartile range (IQR) increase in temperature and pollutant levels, which provide the magnitude-of-risk estimates that are comparable across the exposure variables. The results are presented as the percent excess risk in AMI hospitalisations or IHD deaths per IQR increase in a pollutant using the following calculation: $(\exp^{(\beta \times IQR)} - 1) \times 100\%$, where β is the model estimate. For analysis of a given lagged exposure, a case was automatically dropped if exposure and meteorological data were not available for the case and at least one control day.

Sensitivity analyses were applied. The linearity of the relationship between AMI hospitalisations and the CA2 of temperature and the CA2 of relative humidity was confirmed in generalised additive Poisson time-series regression models (GAM) with the use of the *gam* procedure, *mgcv* package in R statistical software (R Development Core Team, 2010). We did not investigate this for the IHD deaths as no clear association with temperature was observed. The GAM design is the most suitable to explore the shape of an association. Its usefulness lies in the possibility of incorporating variables in a non-parametric way using smooth functions such as loess or spline, therefore avoiding the need to assume the shape of the association and later trying to reproduce it by means of an approximate functional expression. It also offers the opportunity to compare the results to that of the case-crossover analysis, i.e. investigate the robustness of the associations using different epidemiological study designs. Smoothing splines of calendar time with degrees of freedom per year (df/year) that varied from 0 to 5 were used to control for long-term trend and seasonality. The sum of the partial autocorrelation coefficients was minimised at 2.5 to 2.6 df/year, depending whether models were adjusted for PM_{10} or NO_2 (CA2 lag) or none. The GAM models were also adjusted for day of the week and public holidays. Models were run with linear and non-linear terms of temperature and relative humidity, the latter as a smoothing spline function with 3 df. No difference in the results was observed by applying 4 or 5 df. We investigated whether the non-linear terms of temperature and relative humidity improved the models by conducting log-likelihood ratio tests. We decided not to use splines for temperature and relative humidity, as the splines were insignificant and did not add value to the models (Figure S1). As a linear dose-response was observed when data of the entire year were applied, GAM models were not run by the warm or cold period.

Results

Table 1 indicates the characteristics of the 28 215 AMI hospitalisations (of 22 475 people) and 21 082 out-of-hospital IHD deaths during the study period. Of the 22 475 people who had AMI hospitalisations, 4 415 died at a later date out-of-hospital due to IHD. The mean age for AMI hospitalisations and out-of-hospital IHD deaths was 74 and 80 years, respectively. More hospitalisations and deaths occurred among men, those older than 75 years and during the cold period. Figures S2 and S3 illustrate the time-series of the daily number of AMI hospitalisations and IHD deaths during the study period. The number of AMI hospitalisations and the IHD deaths varied from 0 to 23 and 0 to 11 per day, respectively.

Table 2 provides an overview of the daily temperature, relative humidity and air pollution data. As expected, temperature levels were on average higher during the warm period (April–September). PM_{10} levels were on average similar during the warm and cold periods (October–March). Relative humidity, NO_2 and NO_x levels were on average higher during the cold period, whilst this was the case for O_3 levels during the warm period. The daily WHO and EU air quality limits (50 $\mu g.m^{-3}$) for the measured PM_{10} levels were exceeded on 60 days at the urban background level during 1990–2010 (Figure S4) [8,26]. On average, the temperature, relative humidity and air pollution levels on the case days were not significantly different from those on the control days (Table 2). No heat waves occurred during the 26-year study period. The Swedish Meteorological and Hydrological Institute defines a heat wave as those days when the daily maximum temperature is greater than 25°C for 5 days consecutively [27].

Table 3 display the Spearman correlations between the daily temperature, relative humidity and air pollution levels during the warm and cold periods. Temperature had an inverse correlation with NO_2 and NO_x during both periods - stronger during the cold period. Temperature had a stronger correlation with PM_{10} and O_3 during the warm period. Relative humidity had an inverse correlation with temperature and the pollutants during both periods, except for temperature during the cold period. O_3 had an inverse association with NO_2 and NO_x - the strongest during the cold period. PM_{10} had a stronger correlation with the other pollutants during the warm period.

Figure 1 and Figure S5 illustrate the % change in the AMI hospitalisations per IQR increase in the lags of temperature and the pollutants, respectively, after adjusting for public holidays and relative humidity in single pollutant models. The same lag of the pollutants, temperature and relative humidity was included in each model. None of the pollutants were significantly associated with AMI hospitalisations in the entire year, warm or cold period (Figure S5). However, the pollutants attenuated the association between temperature and AMI hospitalisations. In the entire year, an IQR increase in the CA2 of temperature was associated with an insignificant decrease of 4% (95% CI: −1%; 9%) in AMI hospitalisations without adjusting for any pollutant or adjusting for O_3 (Figure 1). After adjusting for PM_{10}, the association strengthened to −6% (95% CI: 0%; −12%) in the entire year. Adjusting for NO_2 and NO_x resulted in a protective association of 5%, albeit not statistically significant.

In the warm period, the inverse association between the CA2 of temperature and AMI hospitalisations was slightly stronger than in the entire year (5% vs. 4%), without adjusting for any pollutant (Figure 1). Adjusting for O_3 resulted in a weaker inverse association, and adjusting for PM_{10}, NO_2 and NO_x strengthened the inverse association somewhat to 5–7%. No association

Table 1. Characteristics of the acute myocardial infarction hospital admissions and ischemic heart disease deaths in Gothenburg.

	Acute myocardial infarction hospital admissions during 1 January 1985 to 31 December 2010		Ischemic heart disease deaths during 1 January 1987 to 31 December 2010	
	No. cases	%	No. cases	%
Total	28215	100.0	21082	100.0
Sex				
Male	16627	58.9	11013	52.2
Female	11588	41.1	10069	47.8
Age				
23*–35 years	62	0.2	20	0.1
36–55 years	2459	8.7	659	3.1
56–65 years	3951	14.0	1488	7.1
66–75 years	7390	26.2	3913	18.6
76–84 years	9122	32.3	7012	33.3
85–102* years	5231	18.5	7990	37.9
Period				
Warm	13381	47.4	10149	48.1
Cold	14834	52.6	10933	51.9

*25 years and 108 years for IHD deaths.

Table 2. Descriptive statistics for daily meteorological and air pollutant levels (lag0) in Gothenburg (1 January 1985–31 December 2010).

	No. days missing data	Mean	SD	Range	Percentiles 25th	50th	75th	IQR	AMI admissions Difference between case days and mean control days (95% CI)[a]	IHD deaths Difference between case days and mean control days (95% CI)[a]
All year (9496 days)										
Temperature (°C)	146	8.5	7.2	−22.0–26.2	3.3	8.4	14.4	11.1	0.0 (−0.1–0.0)	0.0 (0.0–0.1)
Relative humidity (%)	637	77.4	12.5	15.0–99.9	69.8	79.4	87.1	17.3	0.0 (−0.2–0.1)	0.1 (0.0–0.3)
PM_{10} (μg/m³)	2004	15.9	9.4	5–78.1	9.6	14.0	20.0	10.4	0.0 (−0.1–0.2)	−0.1 (−0.2–0.1)
NO_x (μg/m³)	739	58.2	71.2	4.4–1322.6	23.0	37.2	64.7	41.7	0.2 (−0.8–1.2)	−1.1 (−2.2– −0.1)
NO_2 (μg/m³)	739	27.2	13.7	1.1–130.5	17.6	24.6	34.0	16.4	0.0 (−0.2–0.2)	−0.2 (−0.4–0.0)
O_3 (μg/m³)	253	60.5	20.5	1.0–174.8	47.0	61.5	73.9	26.9	0.0 (−0.2–0.2)	0.1 (−0.1–0.3)
Warm period (4758 days)										
Temperature (°C)	47	13.8	4.6	−1.1–26.2	10.9	14.3	16.9	6.1	0.0 (−0.1–0.0)	0.0 (0.0–0.1)
Relative humidity (%)	304	72.2	12.3	32.0–99.0	64.7	73.5	81.2	16.5	0.0 (−0.3–0.2)	−0.1 (−0.3–0.2)
PM_{10} (μg/m³)	973	15.3	8.9	5–76.0	9.5	13.5	19.0	9.5	−0.1 (−0.2–0.1)	0.0 (−0.2–0.2)
NO_x (μg/m³)	435	42.5	35.1	4.4–520.6	20.6	31.6	51.6	31.0	−0.1 (−0.8–0.6)	−0.8 (−1.5–0.0)
NO_2 (μg/m³)	435	24.0	11.2	1.1–102.3	15.8	21.9	30.1	14.3	0.0 (−0.2–0.2)	−0.2 (−0.4–0.0)
O_3 (μg/m³)	90	71.3	17.0	14.5–174.8	60.3	70.8	81.6	21.2	−0.1 (−0.3–0.2)	0.2 (−0.2–0.5)
Cold period (4738 days)										
Temperature (°C)	99	3.1	5.0	−22.0–16.3	0.2	3.5	6.4	6.2	0.0 (−0.1–0.0)	0.0 (0.0–0.1)
Relative humidity (%)	333	82.7	10.3	15.0–99.9	77.4	84.8	90.1	12.7	0.0 (−0.2–0.1)	0.3 (0.1–0.5)
PM_{10} (μg/m³)	1031	16.4	9.8	5–78.1	9.7	14.6	20.8	11.1	0.1 (−0.1–0.3)	−0.2 (−0.4–0.0)
NO_x (μg/m³)	304	73.5	90.9	6.1–1322.6	26.7	44.7	81.5	54.8	0.5 (−1.2–2.2)	−1.5 (−3.3–0.3)
NO_2 (μg/m³)	304	30.3	15.0	5.0–130.5	19.7	27.4	37.8	18.1	0.0 (−0.2–0.3)	−0.2 (−0.5–0.1)
O_3 (μg/m³)	163	49.5	17.9	1.0–112.4	37.5	50.4	62.3	24.8	0.0 (−0.3–0.3)	0.0 (−0.3–0.3)

SD: Standard deviation.
IQR: Interquartile range.
[a]Differences between case days and control days are calculated by subtracting the average of the level on the associated control days from the case day. The average of these differences for the 28215 acute myocardial infarction admissions and 21082 ischemic heart disease deaths is then calculated.

Table 3. Spearman correlation coefficients between exposure variables (daily lag0) in Gothenburg during the warm and cold periods (1 January 1985–31 December 2010).

Warm period	Temp	PM_{10}	NO_x	NO_2	O_3
Rel. hum	−0.161	−0.157	−0.144	−0.137	−0.315
	4454[a]	3781	4122	4122	4381
Temp		0.109	−0.098	−0.080	0.102
		3781	4319	4319	4623
PM_{10}			0.120	0.234	0.364
			3555	3555	3777
NO_x				0.920	−0.247
				4323	4249
NO_2					−0.092
					4249
Cold period	**Temp**	**PM_{10}**	**NO_x**	**NO_2**	**O_3**
Rel. hum	0.183	−0.170	−0.126	−0.121	−0.502
	4405[a]	3664	4221	4221	4355
Temp		0.054[b]	−0.353	−0.348	0.032[b]
		3701	4428	4428	4480
PM_{10}			0.042[b]	0.117	0.120
			3587	3587	3672
NO_x				0.913	−0.279
				4434	4284
NO_2					−0.162
					4284

[a]Number of days less than 4758 in the warm and 4738 in the cold period due to missing exposure data.
[b]p-value<0.05, otherwise p-value<0.0001.

between the CA2 of temperature and AMI hospitalisations was noticeable in the cold period.

There was no association between the CA2 of temperature and IHD deaths in the entire year, warm or cold periods, with and without adjusting for any pollutant (Figure S6). Neither were there any significant associations between the pollutants and IHD deaths (Figure S7).

Figure 2 illustrates that stronger associations were observed between the CA2 of temperature and AMI hospitalisations for men and those older than 75 years (after adjustment for PM_{10}), but the interaction terms were insignificant. No susceptible groups were identified in the warm period either, after adjusting for PM_{10} or NO_2.

Inverse associations between temperature and AMI hospitalisations of similar effect size were in general observed in the case-crossover and GAM analyses (Table S2), with or without adjusting for PM_{10} or NO_2.

Discussion

In this case-crossover study from Gothenburg, Sweden, we observed a linear exposure-response relation corresponding to a 3% and 7% decrease in AMI hospitalisations for an IQR increase in the CA2 of temperature during the entire year (11°C) and the warm period (6°C), respectively, with and without adjustment for PM_{10}, NO_2, NO_x or O_3. This translates into a 1–5% decrease in AMI hospitalisations per 1°C increase in the CA2 of temperature. We found no evidence of an association in the cold period or any association between temperature and IHD deaths in the entire year, warm or cold periods, with and without adjusting for any pollutant. No susceptible groups were identified either.

A review summarised the evidence of the association between temperature indices (e.g. daily mean, minimum and maximum temperature, diurnal temperature range, 3-hr maximum apparent temperature) and AMI hospitalisations [28]. Barnett et al concluded in a review that there is no single temperature index that is superior to others [29]. The observed effects in our study occurring within a day or two following exposure is compatible with other studies [6,28]. The significant inverse association between daily mean temperature and AMI hospitalisations in Gothenburg (entire year and warm period) is in accordance with eight of 11 previous studies that reported significant effect estimates per unit change in a temperature index (i.e. not by categories of a temperature index) [13,16,17,19–21,23,25]. These 11 studies were conducted in Europe, the USA and Asia. The strongest inverse association was observed for women in a time-series study from Trapani, Sicily (−9% per 1°C increase in minimum temperature) [19]. One study reported an insignificant inverse association between AMI hospitalisations in California and an increase in the daily apparent temperature (warm period) [22]. Two studies observed either an inverse or a positive association (not necessarily significant) between AMI hospitalisations with an increase in a temperature index, depending on whether the temperature index is minimum or maximum temperature and the location of the study (Los Angeles, San Francisco or Sacramento, USA) [15,24]. The strongest positive association was observed for women (55–69 years) in Sacramento, California, USA (8% per 1°C increase in minimum temperature) [15]. Two of the 11 studies

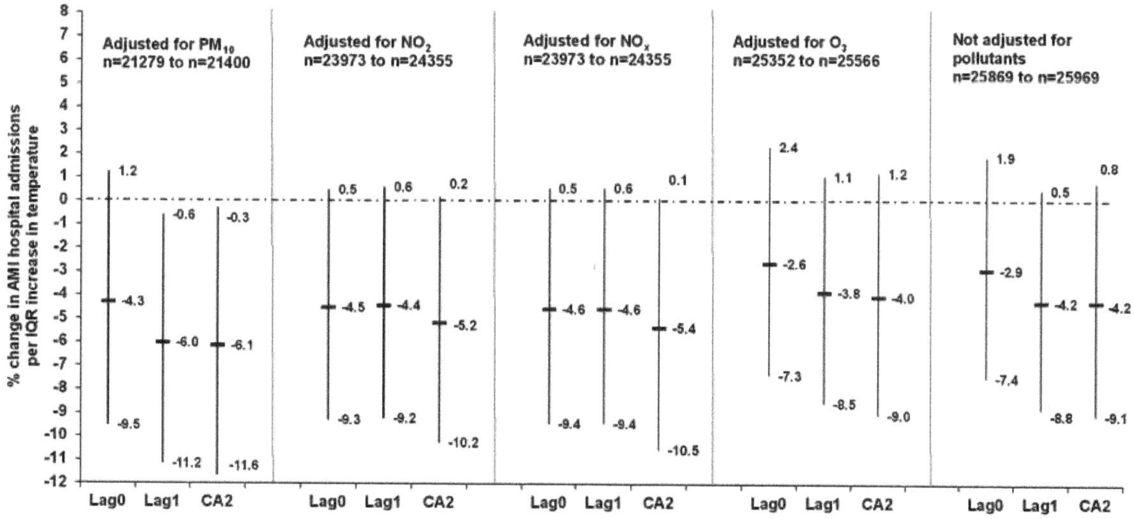
% change in AMI hospital admissions per IQR increase in temperature
Adjusted for PM_{10} n=21279 to n=21400
1.2 -4.3 -9.5 -0.6 -6.0 -11.2 -0.3 -6.1 -11.6
Adjusted for NO_2 n=23973 to n=24355
0.5 -4.5 -9.3 0.6 -4.4 -9.2 0.2 -5.2 -10.2
Adjusted for NO_x n=23973 to n=24355
0.5 -4.6 -9.4 0.6 -4.6 -9.4 0.1 -5.4 -10.5
Adjusted for O_3 n=25352 to n=25566
2.4 -2.6 -7.3 1.1 -3.8 -8.5 1.2 -4.0 -9.0
Not adjusted for pollutants n=25869 to n=25969
1.9 -2.9 -7.4 0.5 -4.2 -8.8 0.8 -4.2 -9.1
Lag0 Lag1 CA2

(a)

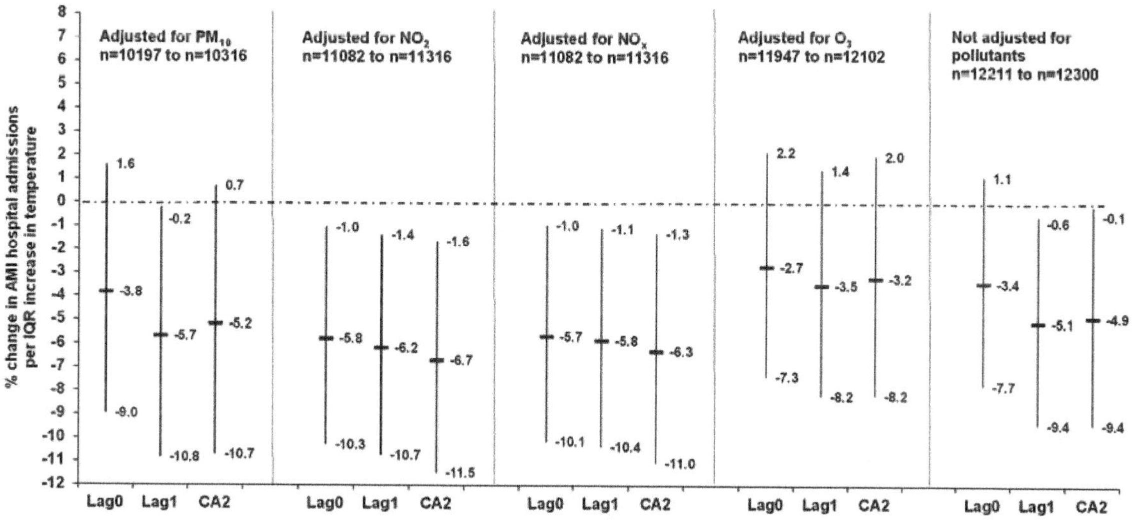
% change in AMI hospital admissions per IQR increase in temperature
Adjusted for PM_{10} n=10197 to n=10316
1.6 -3.8 -9.0 -0.2 -5.7 -10.8 0.7 -5.2 -10.7
Adjusted for NO_2 n=11082 to n=11316
-1.0 -5.8 -10.3 -1.4 -6.2 -10.7 -1.6 -6.7 -11.5
Adjusted for NO_x n=11082 to n=11316
-1.0 -5.7 -10.1 -1.1 -5.8 -10.4 -1.3 -6.3 -11.0
Adjusted for O_3 n=11947 to n=12102
2.2 -2.7 -7.3 1.4 -3.5 -8.2 2.0 -3.2 -8.2
Not adjusted for pollutants n=12211 to n=12300
1.1 -3.4 -7.7 -0.6 -5.1 -9.4 -0.1 -4.9 -9.4
Lag0 Lag1 CA2

(b)

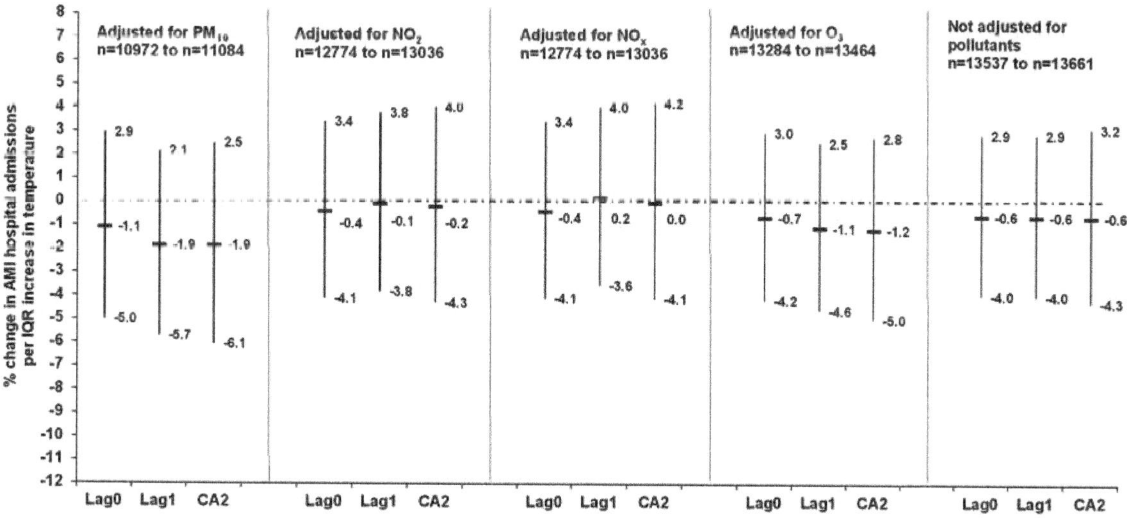
% change in AMI hospital admissions per IQR increase in temperature
Adjusted for PM_{10} n=10972 to n=11084
2.9 -1.1 -5.0 2.1 -1.9 -5.7 2.5 -1.9 -6.1
Adjusted for NO_2 n=12774 to n=13036
3.4 -0.4 -4.1 3.8 -0.1 -3.8 4.0 -0.2 -4.3
Adjusted for NO_x n=12774 to n=13036
3.4 -0.4 -4.1 4.0 0.2 -3.6 4.2 0.0 -4.1
Adjusted for O_3 n=13284 to n=13464
3.0 -0.7 -4.2 2.5 -1.1 -4.6 2.8 -1.2 -5.0
Not adjusted for pollutants n=13537 to n=13661
2.9 -0.6 -4.0 2.9 -0.6 -4.0 3.2 -0.6 -4.3
Lag0 Lag1 CA2

(c)

Figure 1. Association between temperature and acute myocardial infarction hospital admissions in Gothenburg, expressed as percentage increase in risk (%) and 95% confidence intervals per inter-quartile increase in daily lag0, lag1 and 2-day cumulative average during (a) the entire year, (b) warm period (April–September) and (c) cold period (October–March). Models adjusted for a single pollutant (same lag as temperature), relative humidity (same lag as temperature) and public holidays. Number of cases (n) used in the models is less than the original number due to missing exposure data.

focused on the warm period (non-heat wave days) [22,24] and one study investigated both warm and cold periods [25], whilst the rest investigated the entire year. The temperature ranged from −15°C in Lille, France to 38°C in California, USA. Although numerous ecological studies have shown a U-shape association between temperature and all types of CVD mortality or morbidity combined (i.e. all ICD10 codes I or all ICD9 codes 390–459) [30–34], most of the 11 studies on AMI confirmed a linear

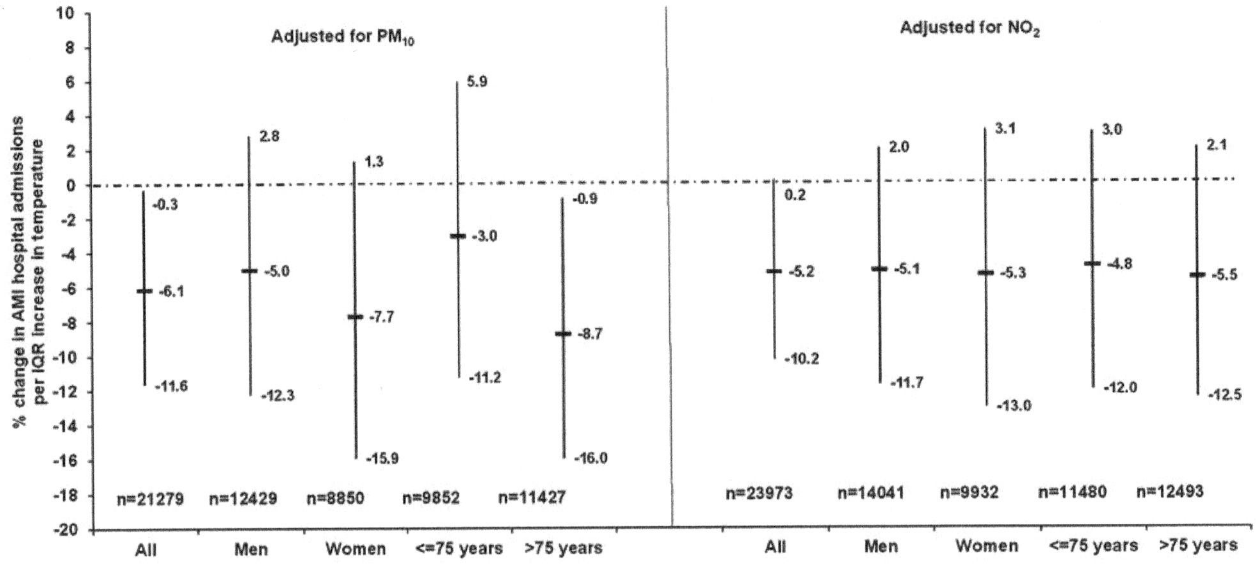

(a)

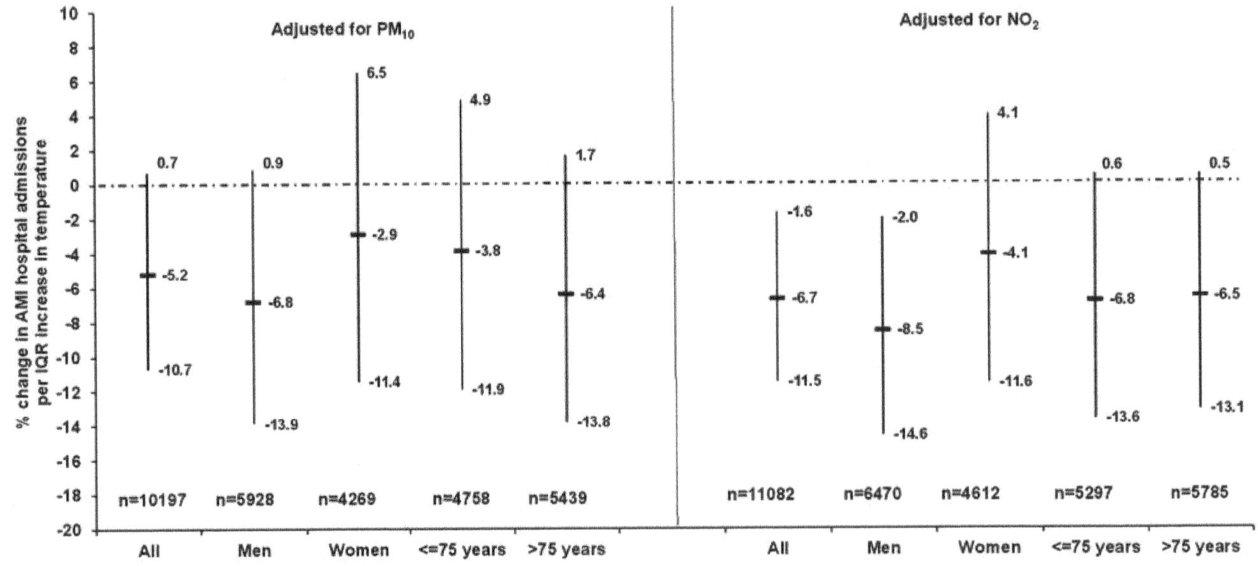

(b)

Figure 2. Association between temperature and acute myocardial infarction hospital admissions in Gothenburg by sex and age, expressed as percentage increase in risk (%) and 95% confidence intervals per inter-quartile increase in daily 2-day cumulative average during (a) the entire year and (b) warm period (April–September). Models adjusted for a single pollutant (same lag as temperature), relative humidity (same lag as temperature) and public holidays. Number of cases (n) used in the models is less than the original number due to missing exposure data p-values for the interaction term of CA2 of temperature and sex/age group were not significant (>0.05).

association. Only four of the studies investigated confounding by air pollution [20–22,25].

Of the aforementioned studies, few investigated susceptible groups. Those that did, reported associations of the stratified analyses, but did not explicitly mention whether the interaction term between the susceptible variable and that of temperature was significant or not.

Apart from the general lack of investigating confounding by air pollution, other factors may explain the heterogeneity of risk estimates. These factors are related to the statistical methods used (different lags selected, lack of control for confounding by day of the week, public holidays and seasonality), the demographical profile of the study population (e.g. age, sex, socio-economic status), the efficiency and accessibility of the health system, diagnostic criteria of AMI, and type of hospitalisation (planned or emergency). Gothenburg had lower PM_{10} levels than the studies from Copenhagen, Denmark and Augsburg, Germany [20,25]. Higher and lower O_3 levels occurred in the city compared to those in Augsburg, Germany and California, USA, respectively [20,22]. Slightly higher NO_2 levels were observed in Gothenburg compared to those from Copenhagen, Denmark [25]. Other reasons for heterogeneity may be related to different PM composition, different indoor and outdoor air pollution sources, climate, behaviour and building traditions, which may influence indoor infiltration of outdoor air pollution and exposure to outdoor temperature [35].

Seasonal variation in CVD mortality and morbidity has been reported in the scientific literature since the early 1950s [36]. Some of this variation is dependent on temperature, as indicated in 11 other studies and ours, after controlling for seasons. Although we observed more AMI hospitalisations in the cold period, we did not observe any association between temperature and AMI hospitalisations in the cold period. A possible reason for the latter may be due to the widespread use of central heating in well insulated buildings in Sweden [35]. Most AMI hospitalisations also occurred among people older than 65 years and some of them may have restricted mobility and remain mostly indoors. Nevertheless, a study from Norway found that mean room temperature in the living room, kitchen and bathroom is about 20°C throughout the year in households of the elderly, but the bedroom temperature varies with the outdoor temperature as people sleep with a window slightly opened [37]. This behaviour is also common in Sweden. One likely mechanism behind the lower AMI morbidity and CVD mortality in warm seasons is vasodilation, resulting in more favourable cardiac hemodynamics [38]. Although the physiological mechanisms resulting from temperature changes to CVD outcomes that manifest at the clinical and public health scales (e.g. hospitalisations and deaths) are still not completely understood [28], vasoconstriction following a drop in ambient temperature is a likely mechanism. Other physiological markers along with blood pressure have a clear seasonal variation (increases in the colder months): red and white blood cell counts, blood viscosity, plasma cholesterol, coagulation factors (e.g. plasma fibrinogen, plasminogen activator inhibitor type 1) and inflammatory markers (e.g. IL-6, C-reactive protein) during winter [28].

The inverse association between outdoor temperature and AMI hospitalisations during the entire year in our study was driven by the inverse association in the warm period. This poses the question of the biological plausibility of such an association. Cold exposure studies (i.e. human volunteer studies) have also observed an increase in the afore-mentioned physiological factors with a decrease in temperature independent of season [39]. It is therefore possible that changes in these physiological markers for CVD may affect hospitalisations due to AMI, independent of season. Some of these physiological markers (e.g. blood pressure, heart rate, cardiac output and endothelial function) have a circadian rhythm [40]. It should be noted that the climate in Gothenburg in April to September is not hot; 25th and 75th percentiles 11 and 17°C with frequent rain and no heat waves. We therefore postulate that the physiological mechanisms behind the seasonality for AMI (higher risk during winter) may apply also in the "warm" season in countries with a mild climate in summer. In addition, people display different behaviour in diet, activity, housing (open windows), psychosocial factors and mood disorders on cold and warm days, and general well-being may be attenuated in cold and rainy days during summer [41,42].

Our secondary aim was to examine associations between temperature and IHD deaths, but we did not find any such associations. The reason for this could be that IHD deaths that occurred out-of-hospital may have a lower diagnostic validity than the hospitalised AMI cases of the present study. Asymptomatic AMIs are not uncommon. Potentially, cold temperatures could precipitate chest pain symptoms, leading to hospitalisation and recognition of AMI, rather than influencing mechanisms, such as plaque rupture. However, in the absence of more detailed information on the episodes involved, this remains highly conjectural.

Although the focus of this study is on temperature, the lack of an association between AMI hospitalisations and IHD deaths with PM_{10}, NO_2 and O_3 does warrant some discussion. Please see Text S1 for this discussion.

Advantages of our study include accurate meteorological, air pollution, AMI hospitalisation and IHD mortality data. Data from the Swedish Hospital Registry has been shown to have high validity for a diagnosis of MI [43]. Autopsy rates for persons dying outside hospital from IHD are high for younger but not older people [44]. Some disease misclassification is possible, but it is unlikely to be related to temperature. Our study period of 26 years is longer than the other studies that applied GAM analyses, which had study periods of 1–15 years [13,15,16,18,20,21,23,25]. Unlike our study, most studies do not report the actual number of cases or days included in their case-crossover or GAM analyses, due to missing exposure data.

Another advantage of our study is that unlike most of the other studies, we investigated confounding by air pollution. Moreover, our results were robust regarding the association between temperature and AMI hospitalisations in the case-crossover and GAM analyses.

As with all ecological epidemiology study designs, our study has a disadvantage of exposure misclassification, i.e. the assumption that the ambient temperature, humidity and air pollution measured in the city are the same across Gothenburg. The exposure error resulting from using ambient temperature and air pollution as a surrogate for personal exposure can potentially lead to bias in the estimated association, and this can be more pronounced among the elderly and other frail groups who generally spend most of their time indoors.

Other limitations are the inability to adjust for $PM_{2.5}$ (data available from 2006) and the lack of chemical composition of PM_{10} [45].

A third limitation is that information on effect modifiers, e.g. the use of medications, preexisting CVD or comorbidities [46–48], was not available in our study. Such effect modifiers may bias the association between the air pollutants and AMI hospital admissions in either direction.

In conclusion, our results support the notion that moderate increases in temperature are associated with a decrease in AMI hospitalisations in the entire year and warm period in a setting with no heat waves. This association (assumed to be causal) is complex and depends on the specific health outcome (death or hospitalisation), population characteristics (age, sex, socio-economic status), exposure conditions and the efficiency of the health care system, which all vary with time [49]. The International Panel on Climate Change stressed that many similar studies on temperature and health cannot be extrapolated infinitely into the future without considering major uncertainties regarding changes in populations, the rate and intensity of projected climate change and adaptation [50].

Supporting Information

Figure S1 Smoothed relationship (expressed as the model estimate) between acute myocardial infarction hospital admissions and a unit increase in daily 2-day cumulative average of (a) temperature and (b) relative humidity in Gothenburg, Sweden during the entire year (1985–2010).

Figure S2 Time-series of acute myocardial infarction hospital admissions in Gothenburg, Sweden (1 January 1985–31 December 2010).

Figure S3 Time-series of out-of-hospital ischemic heart disease deaths in Gothenburg, Sweden (1 January 1987–31 December 2010).

Figure S4 Time-series of PM_{10} levels in Gothenburg, Sweden (1 January 1985–31 December 2010).

Figure S5 Association between air pollutants and acute myocardial infarction hospital admissions in Gothenburg, expressed as percentage increase in risk (%) and 95% confidence intervals per inter-quartile increase in daily lag0, lag1 and 2-day cumulative average during (a) the entire year, (b) warm period (April−September) and (c) cold period (October−March).

Figure S6 Association between temperature and out-of-hospital ischemic heart disease deaths in Gothenburg, expressed as percentage increase in risk (%) and 95% confidence intervals per inter-quartile increase in daily lag0, lag1 and 2-day cumulative average during (a) the entire year, (b) warm period (April−September) and (c) cold period (October−March).

Figure S7 Association between air pollutants and out-of-hospital ischemic heart disease deaths in Gothenburg, expressed as percentage increase in risk (%) and 95% confidence intervals per inter-quartile increase in daily lag0, lag1 and 2-day cumulative average during (a) the entire year, (b) warm period (April−September) and (c) cold period (October−March).

Table S1 Technical names of the various air pollution measurement instruments used in Gothenburg, Sweden during 1985–2010.

Table S2 Association between temperature and acute myocardial infarction hospitalisations in Gothenburg, expressed as percentage increase in risk (%) and 95% confidence intervals per inter-quartile increase in the 2-day cumulative average (11°C).

Text S1 Association between acute myocardial infarction hospitalisations and out-of-hospital ischemic heart disease deaths, and PM_{10}, NO_2 and O_3 in Gothenburg, Sweden during 1985–2010.

Acknowledgments

The authors would like to thank Georgios Lappas (Department of Emergency and Cardiovascular Medicine, Sahlgrenska Academy, University of Gothenburg), Peter Molnár, Eva Andersson, Sarah Gustavsson (Occupational and Environmental Medicine, Sahlgrenska University Hospital & University of Gothenburg), Marie Haeger-Eugensson and Lin Tang (IVL Swedish Environmental Research Institute, Gothenburg) for data management and/or discussions regarding the study. Gratitude is also expressed towards the Gothenburg Environment and Health Department for supplying the meteorological and PM_{10}, NO_x and NO_2 data.

Author Contributions

Conceived and designed the experiments: JW GS LB. Performed the experiments: JW. Analyzed the data: JW. Contributed reagents/materials/analysis tools: AR KS. Wrote the paper: JW AR KS LB GS. Cleaned and contributed the acute myocardial infarction hospital admission and ischemic heart disease mortality data: AR. Provided and ensured the quality of the air pollution and meteorological data: KS.

References

1. Rayner M, Allender S, Scarborough P (2009) Cardiovascular disease in europe. Eur J Cardiovasc Prev Rehabil 16(suppl 2): S43–S47.
2. Yusuf S, Hawken S, Ounpuu S, Dans T, Avezum A, et al. (2004) Effect of potentially modifiable risk factors associated with myocardial infarction in 52 countries (the INTERHEART study): case-control study. Lancet 364: 937–952.
3. Lloyd-Jones DM, Leip EP, Larson MG, D'Agostino RB, Beiser A, et al. (2006) Prediction of lifetime risk for cardiovascular disease by risk factor burden at 50 years of age. Circulation 113: 791–798.
4. Brook RD, Rajagopalan S, Pope CA 3rd, Brook JR, BhatnagarA, et al. (2010) Particulate matter air pollution and cardiovascular disease: An update to the scientific statement from the American Heart Association. Circulation 121(21): 2331–2378.
5. Mustafic H, Jabre P, Caussin C, Murad MH, Escolano S, et al (2012) Main air pollutants and myocardial infarction: a systematic review and meta-analysis. JAMA 307(7): 713–721.
6. Turner LR, Barnett AG, Connell D, Tong S (2012) Ambient temperature and cardiorespiratory morbidity: a systematic review and meta-analysis. Epidemiology 23(4): 594–606.
7. Kociol RD, Lopes RD, Clare R, Thomas L, Mehta RH, et al. (2012) International variation in and factors associated with hospital readmission after myocardial infarction. JAMA 307(1): 66–74.
8. World Health Organisation: Air Quality Guideline Global Update 2005 (2006) Available: http://www.who.int/phe/health_topics/outdoorair_aqg/en. Accessed 14 November 2012.
9. Maclure M (1991) The case-crossover design: a method for studying transient effects on the risk of acute events. Am J Epidemiol 133(2): 144–153.
10. Bateson TF, Schwartz J (1999) Control for seasonal variation and time trend in case-crossover studies of acute effects of environmental exposures. Epidemiology 10(5):539–544.
11. Bateson TF, Schwartz J (2001) Selection bias and confounding in case-crossover analyses of environmental time-series data. Epidemiology 12(6): 654–661.
12. Levy D, Lumley T, Sheppard L, Kaufman J, Checkoway H (2001) Referent selection in casecrossover analyses of acute health effects of air pollution. Epidemiology 12(2): 186–192.
13. Danet S, Richard F, Montaye M, Beauchant S, Lemaire B, et al. (1999) Unhealthy effects of atmospheric temperature and pressure on the occurrence of myocardial infarction and coronary deaths. A 10-year survey: the Lille-World

Health Organization MONICA project (Monitoring trends and determinants in cardiovascular disease). Circulation 100(1): E1–7.
14. Anderson HR, Atkinson RW, Peacock JL, Marston L, Konstantinou K (2004) Meta-Analysis of Time Series Studies and Panel Studies of Particulate Matter (PM) and Ozone (O_3). World Health Organization: Copenhagen, Denmark. Available: http://www.euro.who.int/__data/assets/pdf_file/0004/74731/e82792.pdf. Accessed 14 November 2012.
15. Ebi KL, Exuzides KA, Lau E, Kelsh M, Barnston A (2004) Weather changes associated with hospitalizations for cardiovascular diseases and stroke in California, 1983–1998. Int J Biometeorol 49(1): 48–58.
16. Panagiotakos DB, Chrysohoou C, Pitsavos C, Nastos P, Anadiotis A, et al. (2004) Climatological variations in daily hospital admissions for acute coronary syndromes. Int J Cardiol 94(2–3): 229–233.
17. Misailidou M, Pitsavos C, Panagiotakos DB, Chrysohoou C, Stefanadis C (2006) Short-term effects of atmospheric temperature and humidity on morbidity from acute coronary syndromes in free of air pollution rural Greece. Eur J Cardiovasc Prev Rehabil.13(5): 846–848.
18. Anderson HR, Atkinson RW, Bremner SA, Carrington J, Peacock J (2007) Quantitative systematic review of short term associations between ambient air pollution (particulate matter, ozone, nitrogen dioxide, sulphur dioxide and carbon monoxide), and mortality and morbidity. Report to Department of Health revised following first review.Available: http://www.dh.gov.uk/en/Publicationsandstatistics/Publications/PublicationsPolicyAndGuidance/DH_121200. Accessed 14 November 2012.
19. Abrignani MG, Corrao S, Biondo GB, Renda N, Braschi A, et al. (2009) Influence of climatic variables on acute myocardial infarction hospital admissions. Int J Cardiol 137(2): 123–129.
20. Wolf K, Schneider A, Breitner S, von Klot S, Meisinger C, et al. (2009) Air temperature and the occurrence of myocardial infarction in Augsburg, Germany. Circulation 120(9):735–742.
21. Bhaskaran K, Hajat S, Haines A, Herrett E, Wilkinson P, et al. (2010) Short term effects of temperature on risk of myocardial infarction in England and Wales: time series regression analysis of the Myocardial Ischaemia National Audit Project (MINAP) registry. BMJ 341:c3823. doi: 10.1136/bmj.c3823.
22. Green RS, Basu R, Malig B, Broadwin R, Kim JJ, et al. (2010) The effect of temperature on hospital admissions in nine California counties. Int J Public Health 55: 113–121.
23. Lee JH, Chae SC, Yang DH, Park HS, Cho Y, et al. (2010) Influence of weather on daily hospital admissions for acute myocardial infarction (from the Korea Acute Myocardial Infarction Registry). Int J Cardiol 144(1): 16–21.
24. Ostro B, Rauch S, Green R, Malig B, Basu R (2010) The effects of temperature and use of air conditioning on hospitalizations. Am J Epidemiol 172(9): 1053–1061.
25. Wichmann J, Ketzel M, Ellerman T, Loft S (2012) Apparent temperature and acute myocardial infarction hospital admissions in Copenhagen, Denmark. Environ Health 11: 19. doi:10.1186/1476-069X-11-19.
26. Directive 2008/50/EC of the European Parliament and of the Council of 21 May 2008 on ambient air quality and cleaner air for Europe. Available at: http://eur-lex.europa.eu/LexUriServ/LexUriServ.do?uri = OJ:L:2008:152:0001:0044:EN:PDF.Accessed 19 December 2012.
27. Swedish Meteorological and Hydrological Institute (2012) Available: http://www.smhi.se/polopoly_fs/1.16889!webbFaktablad_49.pdf. Accessed 20 December 2012.
28. Bhaskaran K, Hajat S, Haines A, Herrett E, Wilkinson P, et al. (2009). Effects of ambient temperature on the incidence of myocardial infarction. Heart 95(21): 1760–1769.
29. Barnett AG, Tong S, Clements ACA (2010) What measure of temperature is the best predictor of mortality? Environ Res 110: 604–611.
30. Eurowinter Group (1997) Cold exposure and winter mortality from ischaemic heart disease, cerebrovascular disease, respiratory disease, and all causes in warm and cold regions of Europe. Lancet 349: 1341–1346.
31. Analitis A, Katsouyanni K, Biggeri A, Baccini M, Forsberg B, et al. (2008) Effects of cold weather on mortality: results from 15 European cities within the PHEWE project. Am J Epidemiol 168(12): 1397–1408.
32. Kovats RS, Hajat S (2008) Heat stress and public health: a critical review. Annu Rev Public Health 29: 41–55.
33. Basu R (2009) High ambient temperature and mortality: a review of epidemiologic studies from 2001 to 2008. Environ Health 8(40), Online firstdoi: 10.1186/1476-069X-8-40.
34. Michelozzi P, Accetta G, De Sario M, D'Ippoliti D, Marino C, et al. (2009) High temperature and hospitalizations for cardiovascular and respiratory causes in 12 European cities. Am J Respir Crit Care Med 179(5): 383–389.
35. Hänninen O, Hoek G, Mallone S, Chellini E, Katsouyanni K, et al. (2011) Seasonal patterns of outdoor PM infiltration into indoor environments: review and meta-analysis of available studies from different climatological zones in Europe. Air Quality, Atmosphere & Health 4(3-4): 221–233.
36. Teng HC, Heyer HE (1955) The relationship between sudden changes in weather and the occurrence of acute myocardial infarction. Am Heart J 49(1): 9–20.
37. Bøkenes L, Mercer JB, MacEvilly S, Andrews JF, Bolle R (2009) Annual variations in indoor climate in the homes of elderly persons living in Dublin, Ireland and Tromsø, Norway. Eur J Public Health 21(4): 526–531.
38. Elwood PC, Beswick A, O'Brien JR, Renaud S, Fifield R, et al. (1993) Temperature and risk factors for ischaemic heart disease in the Caerphilly prospective study. Br Heart J 70: 520–523.
39. Schäuble CL, Hampel R, Breitner S, Rückerl R, Phipps R, et al. (2012) Short-term effects of air temperature on blood markers of coagulation and inflammation in potentially susceptible individuals. Occup Environ Med 69(9):670–678. doi: 10.1136/oemed-2011-100469.
40. Scheer FA, Hu K, Evoniuk H, Kelly EE, Malhotra A, et al. (2010) Impact of the human circadian system, exercise, and their interaction on cardiovascular function. Proc Natl Acad Sci USA 107(47): 20541–20546.
41. Sher L (1999) Seasonal distribution of myocardial infarction and seasonal mood changes. J Am Coll Cardiol 33: 2088–2089.
42. Matthews C, Freedson P, Hebert J, Stanek EJ 3rd, Merriam PA, et al. (2001). Seasonal variation in household, occupational, and leisure time physical activity: longitudinal analyses from the Seasonal Variation of Blood Cholesterol Study. Am J Epidemiol 53: 172–183.
43. Ludvigsson JF, Andersson E, Ekbom A, Feychting M, Kim JL, et al. (2011) External review and validation of the Swedish national inpatient register. BMC Public Health 11:450. doi: 10.1186/1471-2458-11-450.
44. Dudas K, Lappas G, Stewart S, Rosengren A (2011) Trends in out-of-hospital deaths due to coronary heart disease in Sweden (1991 to 2006). Circulation 123: 46–52.
45. Bhaskaran K, Hajat S, Haines A, Herrett E, Wilkinson P, et al. (2009) Effects of air pollution on the incidence of myocardial infarction. Heart 95(21): 1746–1759.
46. Miller KA, Siscovick DS, Sheppard L, Shepherd K, Sullivan JH, et al. (2007) Long-term exposure to air pollution and incidence of cardiovascular events in women. N Engl J Med 356: 447–458.
47. O'Neill MS, Veves A, Sarnat JA, Zanobetti A, Gold DR, et al. (2007) Air pollution and inflammation in type 2 diabetes: a mechanism for susceptibility. Occup Environ Med 64: 373–379.
48. Park SK, O'Neill MS, Wright RO, Hu H, Vokonas PS, et al. (2006) HFE genotype, particulate air pollution, and heart rate variability: a gene-environment interaction. Circulation 114: 2798–2805.
49. Lucas RM, McMichael AJ (2005) Association or causation: evaluating links between "environment and disease". Bull World Health Organ 83(10): 792–795.
50. World Meteorological Organisation. Contribution of Working Group II to the Fourth Assessment Report of the Intergovernmental Panel on Climate Change (2007) Parry ML, Canziani OF, Palutikof JP, van der Linden PJ, Hanson CE (eds). Cambridge University Press, Cambridge, United Kingdom and New York, NY, USA. Available: http://www.ipcc-wg2.gov/publications/AR4/index.html. Accessed 14 November 2012.

19

Spatial Interpolation of Fine Particulate Matter Concentrations Using the Shortest Wind-Field Path Distance

Longxiang Li[1], Jianhua Gong[2], Jieping Zhou[1]*

1 Institute of Remote Sensing and Digital Earth, Chinese Academy of Sciences, Olympic Science & Technology Park of CAS, Beijing, China, **2** Institute of Remote Sensing and Digital Earth, Chinese Academy of Sciences, Olympic Science & Technology Park of CAS, Beijing, China and Zhejiang-CAS Application Center for Geoinformatics, Jiashan, Zhejiang, China

Abstract

Effective assessments of air-pollution exposure depend on the ability to accurately predict pollutant concentrations at unmonitored locations, which can be achieved through spatial interpolation. However, most interpolation approaches currently in use are based on the Euclidean distance, which cannot account for the complex nonlinear features displayed by air-pollution distributions in the wind-field. In this study, an interpolation method based on the shortest path distance is developed to characterize the impact of complex urban wind-field on the distribution of the particulate matter concentration. In this method, the wind-field is incorporated by first interpolating the observed wind-field from a meteorological-station network, then using this continuous wind-field to construct a cost surface based on Gaussian dispersion model and calculating the shortest wind-field path distances between locations, and finally replacing the Euclidean distances typically used in Inverse Distance Weighting (IDW) with the shortest wind-field path distances. This proposed methodology is used to generate daily and hourly estimation surfaces for the particulate matter concentration in the urban area of Beijing in May 2013. This study demonstrates that wind-fields can be incorporated into an interpolation framework using the shortest wind-field path distance, which leads to a remarkable improvement in both the prediction accuracy and the visual reproduction of the *wind-flow effect*, both of which are of great importance for the assessment of the effects of pollutants on human health.

Editor: Qinghua Sun, The Ohio State University, United States of America

Funding: This work was partially supported by the Plans of National Sci-Tech Major Special Item 2014ZX10003002, the National Science Foundation of China 41301437, the Foundation of the State Key Laboratory of Remote Sensing Sciences in China and the National Science Foundation of China 41371387. The funders had no role in study design, data collection and analysis, decision to publish, or preparation of the manuscript.

Competing Interests: The authors have declared that no competing interests exist.

* E-mail: jpzhou@irsa.ac.cn

Introduction

Public health studies of air-pollution exposure require accurate predictions of concentrations at unmonitored locations to minimize the misclassification of exposure levels [1]. Recent studies have reported that intra-urban-scale variations in air-pollution concentration may exceed the differences between cities [2,3], suggesting the potential importance of predicting air-pollution at fine spatial scales [4,5]. Correspondingly, estimates of exposure to pollutants on small temporal scales are necessary to study the short-term or acute impacts of air-pollution [6]. Spatial interpolation models, landuse regression (LUR) models, remote-sensing-based models and diffusion models are robust tools for intra-urban air-pollution prediction [7,8]. However, spatial interpolation techniques, which generate concentration surfaces from *in situ* observations, are preferred for the estimation of real-time concentrations when data availability and software and hardware costs are taken into account [3,9,10].

Wind is a key meteorological factor that has major impacts on the movement and distribution of air pollutants in a region. When the wind-speed is relatively high, local wind-field exert substantial influence on the horizontal transport of air-pollution; this phenomenon is known as the *wind-flow effect* [11,12]. For example, areas downwind of highways are more heavily exposed to traffic-related pollutants than are upwind areas. This effect illustrates the necessity of incorporating wind-field into spatial interpolation. In a number of recent studies, the consideration of a negative correlation between air-pollution concentration and wind-speed has led to the application of the wind-speed as an auxiliary variable in multi-variable interpolation methods [9,13,14]. Although there have been several attempts to incorporate long-term, large-scale wind-fields into corresponding air-pollution estimations, short-term, small-scale wind-fields have not been extensively used for this purpose, because no direct numerical relations exist between the angle of the wind-direction and the concentration level in such cases. As a result, these approaches fail to capture the expected short-term effects of the wind flow.

By including the wind-fields indirectly, some regression-based methods are able to capture the complex features of pollutant distributions [2,3,14]. A recent study assessed the use of the wind-direction in LUR to improve predictions of nitrogen dioxide levels in Toronto-Hamilton area [11,15]. This method shows great potential, as it quantifies the influence of the wind-direction with the downwind distances from highways. However, real-time air-pollution assessment using this model is economically infeasible because of the cost of collecting sufficiently diverse data sets. Therefore, one objective of the present study is to incorporate wind-fields directly into interpolation frameworks.

Most interpolation techniques depend on Euclidean or straight-line distances to compute spatial dependency. However, the complex features of certain spatial phenomena impede the ability to obtain accurate dependency descriptions using Euclidean distances [16,17]. An appropriate non-Euclidean distance may outperform the Euclidean distance in determining such types of spatial dependency and in capturing complex features [18]. The shortest path distance (SPD) is an important subclass of non-Euclidean distance and has exhibited great potential in diverse interpolation studies. The hydrological distance, a derivative of the SPD, has been used to characterize the spatial configurations, connectivity and directionality of the water temperatures and chemical pollutants in stream networks [17,19–22]. Accounting for geological anisotropy has led to the interpolation of deposits in conjunction with shortest anisotropy path distances [23]. Along-road continuity has been described in carbon dioxide estimations after replacing the Euclidean distance with the SPD [24]. The inclusion of the effect of topographical factors in simulations of the genetic dispersion path has led to the development of the concept of effective distance, which is the length of a virtual movement route. Although the quantitation of these factors remains unclear, this metric exhibits a greater correlation with genetic variance than does the straight-line distance and has been used to characterize the nonlinear features of genetic dispersal [25]. Road-network connectivity has also been incorporated into the interpolations of urban travel speeds using the approximate road-network distance, another derivative of the SPD [26]. The works listed above are important references for the methodology presented in this paper. However, the characteristics of the wind-flow effect differ from those investigated in these previous works. Current SPD techniques are insufficient to successfully capture such features. To address this shortcoming, another derivation of SPD, the shortest wind-field path distance (SWPD), is proposed to determine the spatial dependency effected by a wind-field and is exploited to integrate wind-fields into interpolation frameworks.

In this study, a new interpolation method based on the SPD is developed to describe the influence of the wind-field. This technique is then applied to generate concentration surfaces for fine particulate matter of less than 2.5μm in diameter ($PM_{2.5}$) on the experimental dates in the study area. Comparisons are performed between this technique and the conventional methodology to illustrate the improvements achieved by including the wind-field.

Materials and Methods

Data collection and processing

Study area and context. Beijing, the capital of the People's Republic of China, is an international metropolis and has experienced a rapid increase in urban population, energy consumption and vehicle numbers over the past several decades [27]. An urban area inside the surrounding ring road (Fig. 1) was selected as the study area (approximately 30×30 km) because of the relatively dense air-pollution monitoring network that is present in this area. Six non-consecutive days in May 2013 with major air-pollution in terms of PM2.5 and daily wind-speeds above 1.5 m/s were selected as the experimental dates because no accurate wind-direction measurements were available for low-speed wind conditions.

Observed PM2.5 concentration. To improve air-pollution monitoring, a network of 35 automated stations has been established by the Beijing Environmental Protection Bureau (BJEPB). Each station measures hourly $PM_{2.5}$ concentrations and releases real-time data to the public through the Beijing Municipal Environmental Monitoring Center (www.bjmemc.com.cn). 13 urban sampling sites compose a dense urban monitoring subsystem across the study area (Fig. 1). This monitoring network enables the detection of real-time, small-area variations in the $PM_{2.5}$ concentration. We collected an experimental data set from all 13 sites for six selected dates. The concentration data are given in units of$\mu g/m^3$. Daily average concentrations were calculated from the hourly data.

Observed wind-fields. Hourly wind-field observations were obtained through the Chinese Meteorological Data Service Platform (cdc.bjmb.gov.cn). The measurements were collected over a network of 16 weather stations throughout the study area operated by the Beijing Meteorological Bureau (Fig. 1). The daily average wind-speed and wind-direction were calculated from the hourly real-time data. Influenced by the complex urban morphology, the urban wind-fields exhibit dramatic small-scale variations that cannot be captured by model-simulated fields with overlarge grid sizes [11].

Methodology

At the heart of the proposed method is the shortest-path analysis. An appropriate simulation of air-pollution movement using this method entails the construction of a cost surface, which lays the foundation for the shortest-path analysis. After the SWPD between every pair of unmonitored locations and sampling locations is obtained, this distance metric is used to determine the spatial dependency. Inverse Distance Weighting (IDW) in conjunction with SWPD is then implemented to calculate the concentration surface. Therefore, this method consists of generating a continuous wind-field, then implementing the shortest-path analysis and finally creating the estimation surface using SWPD-based IDW.

Generation of continuous wind-field. The creation of the cost surface for shortest-path analysis first requires the generation of continuous wind-field. This process also involves spatial interpolation. Unlike other scalar weather variables, a wind-field is a vector quantity whose interpolation is unique in meteorology. Typically, one wind-vector is decomposed into two Cartesian wind components (an east-west component and a north-south component). Each component is then interpolated separately into a corresponding surface using multiquadric (MQ) radial basis functions (RBF) [28]. The wind-field is then constructed backward from the two Cartesian-component surfaces using trigonometry (Fig. 2). This methodology has been widely used to interpolate diverse vector-type data since its proposal and is considered to be a robust approach for various meteorological studies [28]. In this study, a continuous wind-field is established using a grid size of 0.5 km, which is an appropriate resolution for urban air-pollution research [29].

Shortest-path analysis. The shortest-path analysis includes two stages. First, a continuous wind-field is modeled onto a cost surface that depicts the movement cost between adjacent cells. Second, a shortest-path algorithm is implemented to acquire the SWPDs between locations.

Creating a cost surface using wind-field data. The movement of PM2.5 from one location to another may be facilitated or impeded by the local wind-field. A cost surface must be well defined based on the properties of the air-pollution movement to ensure that each shortest path acquired represents

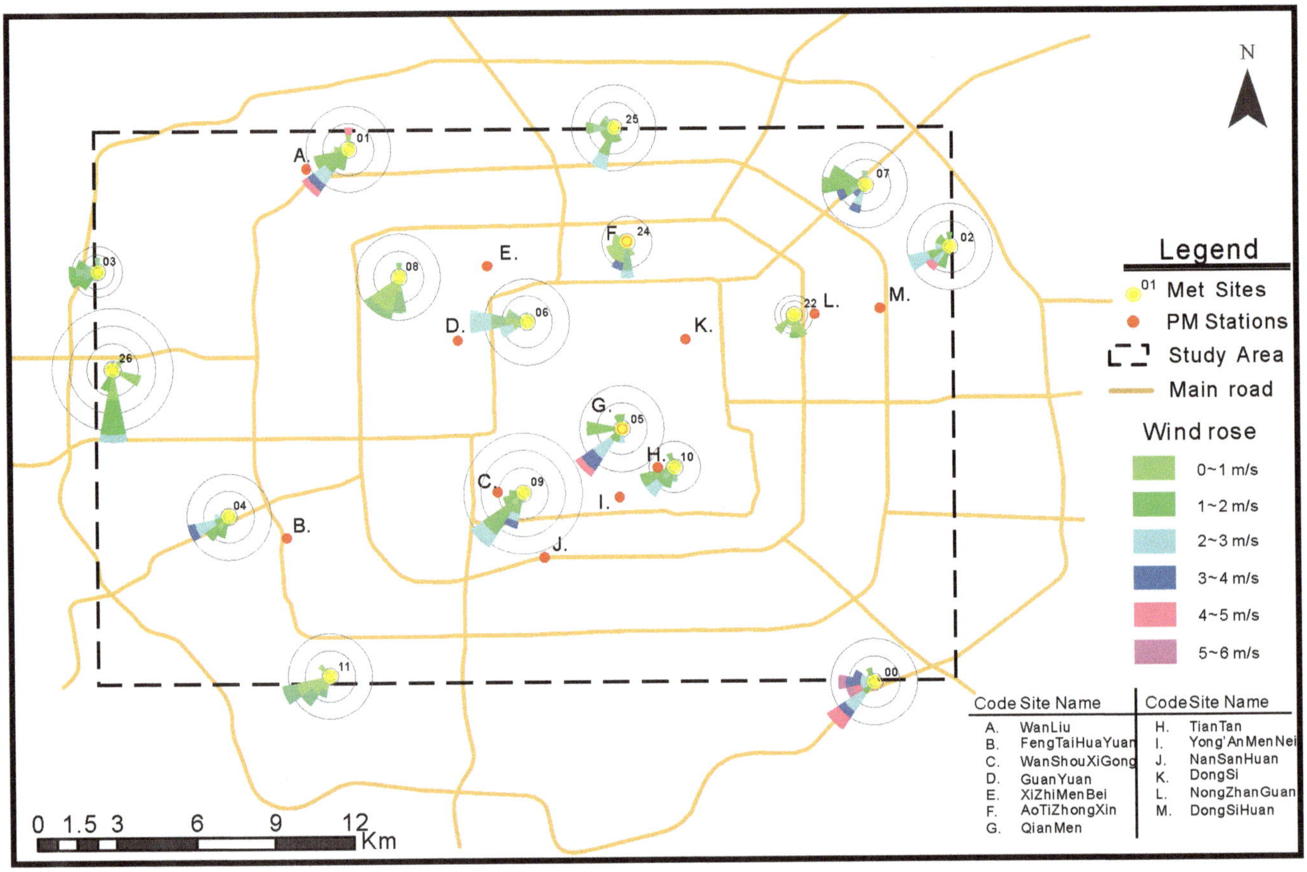

Figure 1. Study area and locations of $PM_{2.5}$ monitoring stations and meteorological sites with daily wind roses for May 18th.

the true movement trajectories and thus reveals the path along which the two locations are related.

The grid-based representation of field data is sufficient to depict the cost of traversing each cell but is incapable of representing the movement cost associated with not only the distance between cells but also the relative positions of adjacent cells. An alternate methodology is to reform the grid raster (Fig. 3A) into a graph (Fig. 3C) with pixels as vertices and virtual connecting lines as edges, where each edge has an associated cost value that indicates the cost of traveling along this edge.

The calculation of the edge cost that depicts the movement difficulty between adjacent cells is performed based on Gaussian dispersion model, which is the standard model for the study of the transport of airborne contaminants under the influence of wind-field. This model simulates a cross section of the air-pollution dispersion and assumes that both the horizontal and vertical concentration distributions are normal [30,31]. The basic formula for this model can be written as follows:

$$C_0(x,y,z,u)=\frac{Q}{\pi u\sigma_y\sigma_z}\exp(-\frac{y^2}{2\sigma_y^2}-\frac{z^2}{2\sigma_z^2}) \quad (1)$$

where $C_0(x,y,z,u)$ is the concentration at ground level, x is the downwind distance, y is the horizontal distance between the point of interest and the centerline, z is the height of the emission source, u is the horizontal wind-speed, σ_y is the standard deviation of horizontal dispersion and σ_z is the standard deviation of vertical dispersion.

The Cartesian coordinates (x and y) in the model can be transformed into polar coordinates (γ andθ) as shown below [32]:

$$C_0(\gamma,\theta,z,u)=\frac{Q}{\pi u\sigma_\theta\sigma_z}\exp(-\frac{z^2}{2\sigma_z^2}-\frac{\gamma^2\theta^2}{2\sigma_\theta^2}-\Omega\frac{\theta^2}{2}) \quad (2)$$

where Ω is a correction term given by

$$\Omega=\alpha(\gamma)+\beta(\gamma)\frac{z^2}{\sigma_z^2} \quad (3)$$

with

$$\alpha(\gamma)=-2+\frac{p\gamma}{a+\gamma}+\frac{q\gamma}{a+\gamma} \quad (4)$$

$$\beta(\gamma)=1-\frac{q\gamma}{a+\gamma} \quad (5)$$

where a, $p$and $q$are dispersion parameters that depend on the atmospheric stability. Further information can be found in the cited references. When the focus is placed only on horizontal diffusion and downwind advection, the formula can be written as

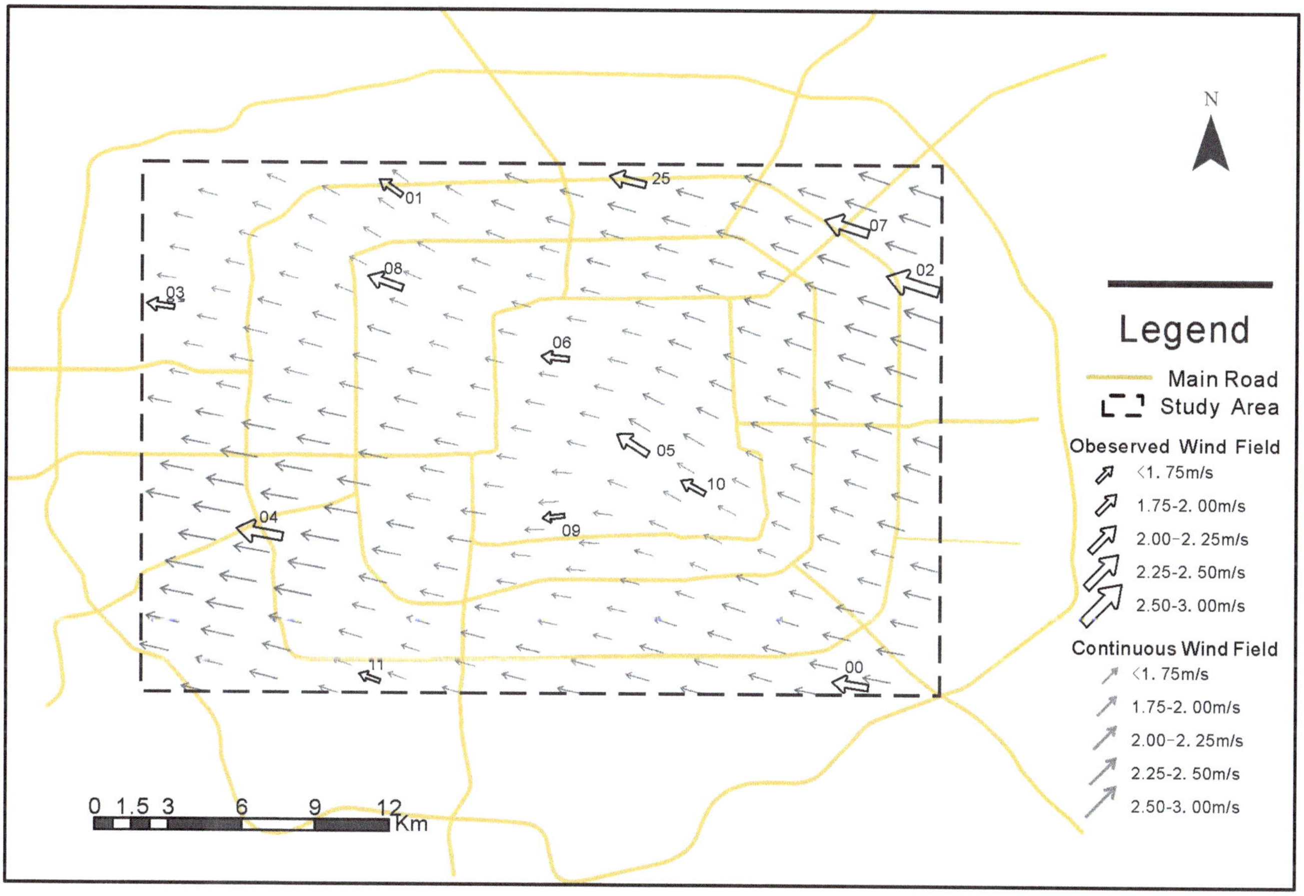

Figure 2. Observed wind-field and continuous wind-field generated from the monitoring data of May 28th.

$$C_0(\gamma,\theta,u) = \frac{Q}{\pi u \sigma_\theta}\exp\left(-\frac{\gamma^2\theta^2}{2\sigma_\theta^2} - \Omega\frac{\theta^2}{2}\right) \tag{6}$$

which indicates that the concentration will decrease with increasing downwind distance (γ) and azimuth (θ), both of which are used to determine the movement cost in the following steps. However, this formula is too complicated to be directly used in calculation, and the following simplified version is applied instead:

$$Cost(E_{AB}) = [\mathbb{F}(D_A,D_M) + \mathbb{F}(D_B,D_M)] \times L_{AB} \tag{7}$$

where A and B are two points in the wind-field, $E_{AB}$is the edge connecting these two points, D_A and D_B are the wind-directions at these two points and D_M is the direction of E_{AB}, namely, the potential movement direction. L_{AB} is the length of E_{AB}, which is functionally equivalent toγ. The function $\mathbb{F}$ is used to calculate the azimuth which is functionally identical toθ. It is important to note that the Gaussian dispersion model is designed to simulate the diffusion of contaminants from definite sources and cannot be directly applied in this study because there are no stable emission sources. However, this model illustrates the origin of the movement cost, which serves as the foundation of this section.

Implementing shortest-path analysis. Based on the establishment of the cost surface, shortest-path algorithms can be used to calculate the paths with the minimum accumulated movement cost between point pairs, indicating the most likely path along which the two points are related (Fig. 3d). In this study, the classic Dijkstra algorithm [33] is employed. The shortest paths and their associated costs are computed as the output of this step. However, SWPD is not measured as a summation of movement costs but as the total length of the shortest path segments between location pairs.

Interpolation based on SWPD. After the SWPD between each pair of prediction locations and measurements is acquired, SWPD-based interpolation can be used to generate estimation surfaces. IDW in conjunction with the SWPD was selected as the technique used to incorporate wind-field in this study. Although the total sample size in this study is small, the density of the observation network permits the estimation of small-area variations in the air-pollution concentration using the proposed method [34]. The feasibility of the method is confirmed by the low mean squared error and mean absolute error in the following sections.

The IDW approach aims to predict the pollutant concentration at a given location based on a weighted average of the measurements obtained at surrounding stations. As a direct application of Tobler's First Law (TFL), the relations between the point of interest and the nearby stations are determined by the distances between them. The method takes the following form:

$$z^*(u) = \sum_{\alpha=1}^{n} \lambda_\alpha z(u_\alpha) \tag{8}$$

where$z^*(u)$is the estimate at location$u$, $z(u_\alpha)$is the measurement at

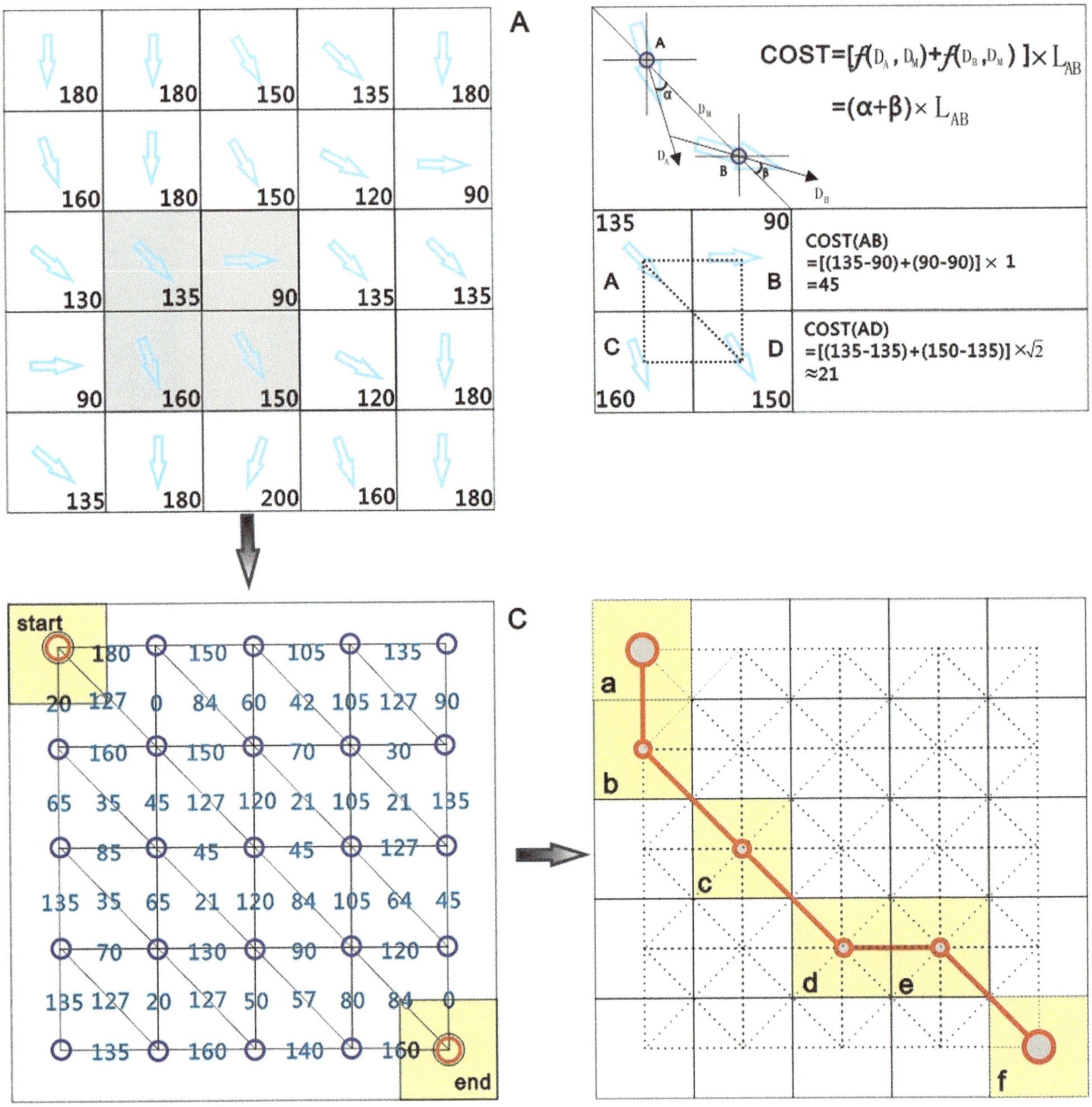

Figure 3. Steps for calculating SWPDs using a wind-field. (A) Grid-based representation of a wind-field. (B) Computing the cost associated with the edge between adjacent cells. (C) Reforming a grid-based wind-field into a graph. (D) SWPD calculated between starting point A and ending point F.

locationu_α, n is the number of stations used for the estimation and$\lambda_\alpha$is the interpolation weight of the measurement atu_α. The calculation of$\lambda_\alpha$takes the following form:

$$\lambda_\alpha = \frac{1/D_\alpha^\omega}{\sum_{\alpha=1}^{n} 1/D_\alpha^\omega} \quad (9)$$

where $D_\alpha$is the distance between the monitoring station numbered$\alpha$and the point of interest; ω is the exponent of the distance and is set to 2 by default. The incorporation of the SWPD can be achieved by replacing the distances with the SWPD values, as demonstrated below:

$$\lambda_\alpha = \frac{1/SWPD_\alpha^{\omega}}{\sum_{\alpha=1}^{n} 1/SWPD_\alpha^{\omega}} \quad (10)$$

where$SWPD_\alpha$is the shortest path distance in the wind-field between the monitoring station numbered$\alpha$and the point of interest.

Results

Here, comparisons are made on various temporal scales between IDW based on the SWPD (IDWS), as proposed in this paper, and IDW based on the Euclidean distance (IDWE). First, a cross-validation by "leaving one out" is performed to assess the estimation accuracy, as described below. Second, the abilities of the two methods to visually reproduce the wind-flow effect in the interpolation results are also compared. The method used in cross-validation involves temporarily removing one $PM_{2.5}$ measurement from the data set and then predicting the concentration at this location based on the remaining measurements using the same methodology.

The three comparison criteria below are used to assess the performance of the interpolations:

$$MSE = \frac{1}{n}\sum_{\alpha=1}^{n}[z(u_\alpha) - z^*(u_\alpha)]^2 \tag{11}$$

$$MAE = \frac{1}{n}\sum_{\alpha=1}^{n}|z(u_\alpha) - z^*(u_\alpha)| \tag{12}$$

$$MRE = \frac{1}{n}\sum_{\alpha=1}^{n}\frac{z(u_\alpha) - z^*(u_\alpha)}{z(u_\alpha)} \tag{13}$$

The mean squared error (MSE) measures the average squared difference between the removed true $PM_{2.5}$ measurement $z(u_\alpha)$ and its estimate $z^*(u_\alpha)$. The mean absolute error (MAE) measures the average absolute difference between $z(u_\alpha)$ and $z^*(u_\alpha)$. The mean relative error (MRE) measures the average relative deviation between $z(u_\alpha)$ and $z^*(u_\alpha)$. In the case of reasonably accurate estimation, the values of all three statistics should be close to zero [26,34,35].

Interpolation of the daily $PM_{2.5}$ concentration

Daily estimation surfaces for the $PM_{2.5}$ concentration on the experimental dates were calculated using IDWS and IDWE. The local prediction is improved when wind-fields are incorporated, as evidenced by the average decrease of 15.66% in the MSE, the average decrease of 6.46% in the MAE and the evident decline in the MRE obtained for the IDWS estimation compared to the IDWE estimation (Table 1). The spatial distributions of the relative errors are presented in Table 2.

In addition to the benefit of lowering these three statistics, improvements are evident when the estimation surfaces obtained using two methods are compared visually (Fig. 4). The two methods produce different distributions when the $PM_{2.5}$ value measured at a single location is much higher than those measured at surrounding stations. In the results obtained using IDWS, greater continuity is apparent on the downwind side of the downtown area and there is a shorter dispersion distance on the upwind side, as would be expected from the wind-flow effect (Fig. 4A). IDWE methods always produce an eye-shaped pattern in such cases, which is commonly considered to be a major shortcoming of this interpolation method (Fig. 4B). Under the complex local wind-field northwest of the urban area, the results of the interpolation method proposed in this paper also exhibit an

Table 1. Comparison of the interpolation accuracies achieved using MSE, MAE and MRE based on a cross-validation analysis.

Date	$PM_{2.5}$ ($\mu g/m^3$)	wind speed (m/s)	wind direction(°)	MSE of IDWS	MSE of IDWE	MAE of IDWS	MAE of IDWE	MRE of IDWS	MRE of IDWE	Improvement in MSE (%)	Improvement in MAE (%)
4th May	92.2	2.19	206.2	49.33	62.51	5.81	6.24	−0.36%	−1.21%	21.07%	7.40%
18th May	118.8	1.96	194.2	105.86	129.71	8.69	8.91	0.93%	2.53%	18.39%	2.53%
21st May	110.8	1.65	93.4	695.81	781.08	18.54	20.99	1.48%	5.03%	10.92%	13.21%
26th May	106.9	1.71	101.9	85.78	90.23	8.17	8.09	−0.27%	0.41%	4.93%	−0.98%
28th May	17.7	1.99	285.3	96.71	113.19	8.22	9.09	0.22%	0.35%	14.56%	10.58%
29th May	78.8	1.53	251.7	13.83	18.22	2.82	2.99	−1.63%	−2.59%	24.11%	6.03%

Table 2. Distributions of the relative errors of the two methods for the daily $PM_{2.5}$ estimation on 21st May.

PM station Number	Estimation of IDWS	Estimation of IDWE	Measured value	Relative error of IDWS	Relative error of IDWE
A	100.9	100.2	129	−21.78%	−22.33%
B	116.8	115.9	100	16.80%	15.90%
C	115.7	121.6	104	11.25%	16.92%
D	113.7	112.6	95	19.68%	18.53%
E	100.7	106.1	104	−3.17%	2.02%
F	106.2	107.6	98	8.37%	9.80%
G	101.3	103.2	179	−43.41%	−42.35%
H	123.8	136.3	98	26.33%	39.08%
I	111.2	117.5	112	−0.71%	4.91%
J	111.9	113.6	118	−5.17%	−3.73%
K	111	118.4	100	11.00%	18.40%
L	105.6	112.5	91	16.04%	23.63%
M	94.9	95.6	113	−16.02%	−15.40%

accordingly complex anisotropy (Fig. 4A). However, the estimation surface obtained using IDWE fails to capture this feature (Fig. 4B).

Interpolation of the hourly $PM_{2.5}$ concentration

Here, we consider the process of interpolating the $PM_{2.5}$ concentration on a smaller temporal scale. Although IDWS outperforms IDWE on most experimental dates, the improvement in the MSE on the 26th of May is only 4.93%, and the MAE of IDWS is larger than that of IDWE (Table 1). On May 26th, the $PM_{2.5}$ concentration increased gradually during the diurnal hours, reached a peak at approximately 4 pm and then decreased dramatically because of the washout caused by a moderate rainfall event that offset the impact of increasing traffic volume during the evening rush hour (Fig. 5). The hourly measured $PM_{2.5}$ concentrations were interpolated from 6 am to 8 pm. Twelve of the 15 experimental hours exhibit smaller MSE values in the IDWS estimation than in the IDWE estimation, whereas the remaining three hours exhibit larger MSE values, suggesting that the incorporation of the wind-fields had a negative influence on the interpolation accuracy during these three hours. The hours immediately preceding and immediately following these three hours also exhibit a limited improvement of less than 10%. Moreover, the improvement-ratio curve of the MAE follows a similar trend. Thus, two valleys appear in the curves: at noon and at sunset.

The prevailing wind-direction was approximately 65° NE in the morning (Fig. 6A) and changed to 120° SE in the afternoon (Fig. 6C). From 11 am to 1 pm, corresponding to the first valley in both improvement-ratio curves, the prevailing direction experienced dramatic variations from northeast to southeast (Fig. 6B), which limited the accuracy of the wind-field measurements. Because modest direction errors on the order of 10 degrees can lead to large errors in the estimation of air-pollution trajectories, the effectiveness of the proposed methodology no longer holds in cases of strongly varying wind-direction [36]. As is also indicated by the meteorological data set, the study area experienced rainfall from 4 pm to 6 pm, corresponding to the second valley. Because

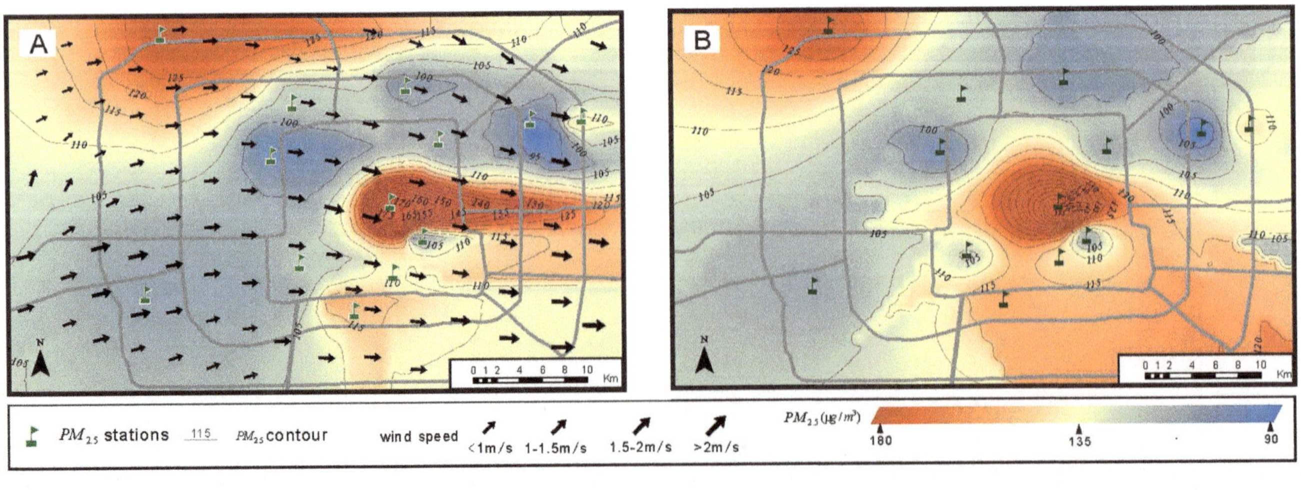

Figure 4. Comparison between the daily $PM_{2.5}$ (May 21st) estimations obtained using IDW based on the SWPD and the Euclidean distance. (A)Interpolation results obtained using IDWS. (B) Interpolation results obtained using IDWE.

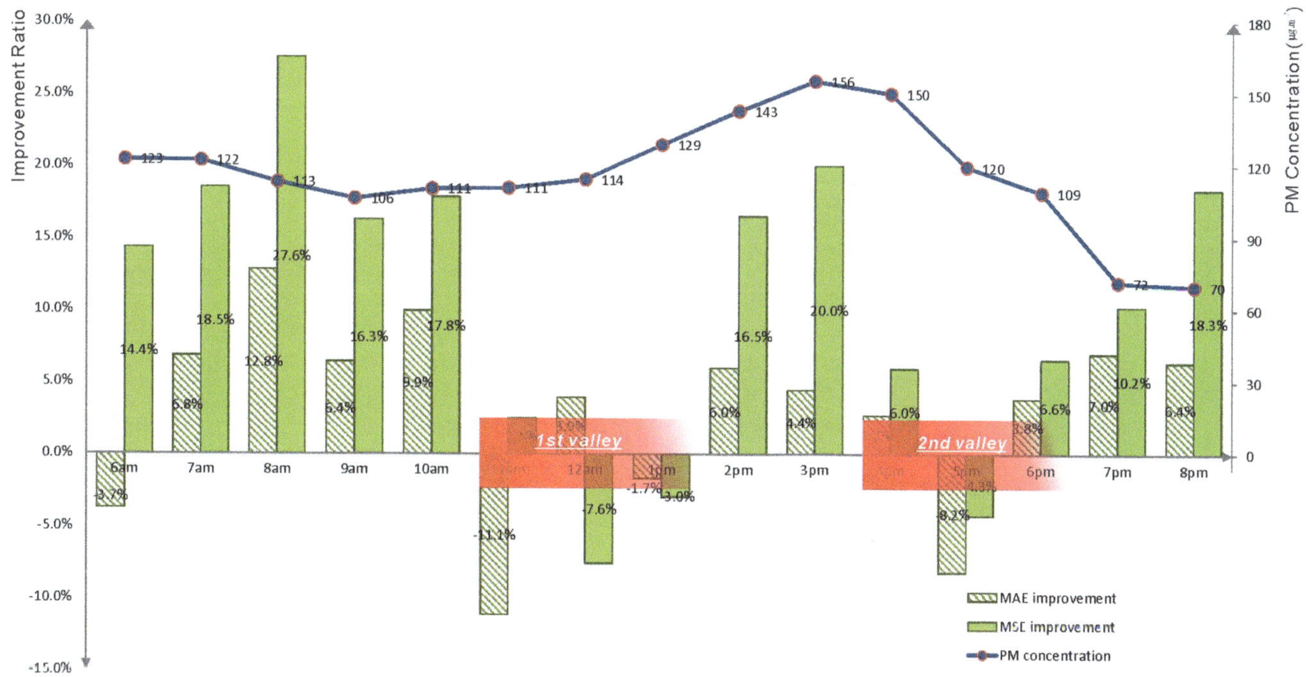

Figure 5. Temporal variation in the hourly $PM_{2.5}$ concentration and the improvement ratio of IDWS with respect to IDWE on May 26th.

precipitation accelerates the deposition of particulate matter, the transport effects of the wind-field were reduced (Fig. 6D). However, under weather conditions with fewer or weaker variations, the methodology proposed in this paper yielded better results on an hourly temporal scale than did the conventional interpolation method.

Discussion

This study demonstrates the potential of incorporating wind-field into interpolation using the IDWS approach. In addition to minimizing estimation errors, a major advantage of this approach stems from its ability to reproduce complex nonlinear features caused by the wind-flow effect. This capability deserves further investigation for its potential use in studies of air-pollution and the negative health effects thereof. As shown in Fig. 4A, the asymmetric distribution of $PM_{2.5}$ on the two sides of the downtown area suggests that residents living east of downtown were exposed to higher concentrations, whereas those living to the west were protected by the wind. By contrast, the symmetric distribution predicted by IDWE (Fig. 4B) may overestimate the $PM_{2.5}$ exposure of upwind residents. Furthermore, the estimation surface obtained using IDWS exhibits greater downwind continuity. Ignoring the wind-flow effect will lead to the underestimation of downwind dispersion distance and the overestimation of dispersion distances in other directions. The IDWS technique also enables the modeling of smaller-scale variations, which can reduce prediction uncertainties in exposure assessments.

As demonstrated in the previous sections, the proposed method produced relatively inaccurate estimates on certain experimental dates. A combination of the dramatically changing wind-direction at noon and the precipitation that occurred at sunset led to poor performance on May 26th. The daily $PM_{2.5}$ on May 20th was also interpolated, and the improvement in the MSE with respect to the results obtained using IDWE was only 3.31%. Although the daily average wind-speed was greater than 1.5 m/s on May 20th, the major pollutant was dust particles caused by blowing sand, which was assumed to be the major source of the prediction uncertainty. These results suggest the need for careful evaluation of the specific weather conditions prior to including the wind-field using IDWS.

The basic version of IDW was applied to test the feasibility of the novel distance metric proposed in this study. Neither the problem of the influence radius nor the problem of zero distance was considered. Some variants of the classic method may be used to verify the validity of the SWPD or even to achieve more accurate estimations. Although it offers a number of advantages, IDW always achieves poorer performance than kriging or other more sophisticated interpolation methods. Thus, now that the effectiveness of the SWPD has been demonstrated and the metric has been shown to offer improved accuracy and realistic visual representation, a need exists to combine this distance metric with a more robust technique to obtain prediction surfaces with higher accuracy [23].

The incorporation of secondary information, instead of relying solely on station measurements, enables the estimation surfaces obtained to reflect localized variations and thus improves the predictive capacity of the analysis [34]. Most currently used auxiliary data are scalar, and little weight is given to vector data. One objective of this paper is to propose a methodology for incorporating vector-type secondary information into interpolation. Now that the feasibility of this methodology has been confirmed, a more general method that is capable of including both scalar-type data and vector-type data is desirable.

The primary intent of this study was to verify the effectiveness of the proposed method, so little attention was directed toward improving the computing efficiency. All algorithms were implemented using C++ with no optimization, and the visualization was performed using ArcGIS (ESRI). At the current stage of

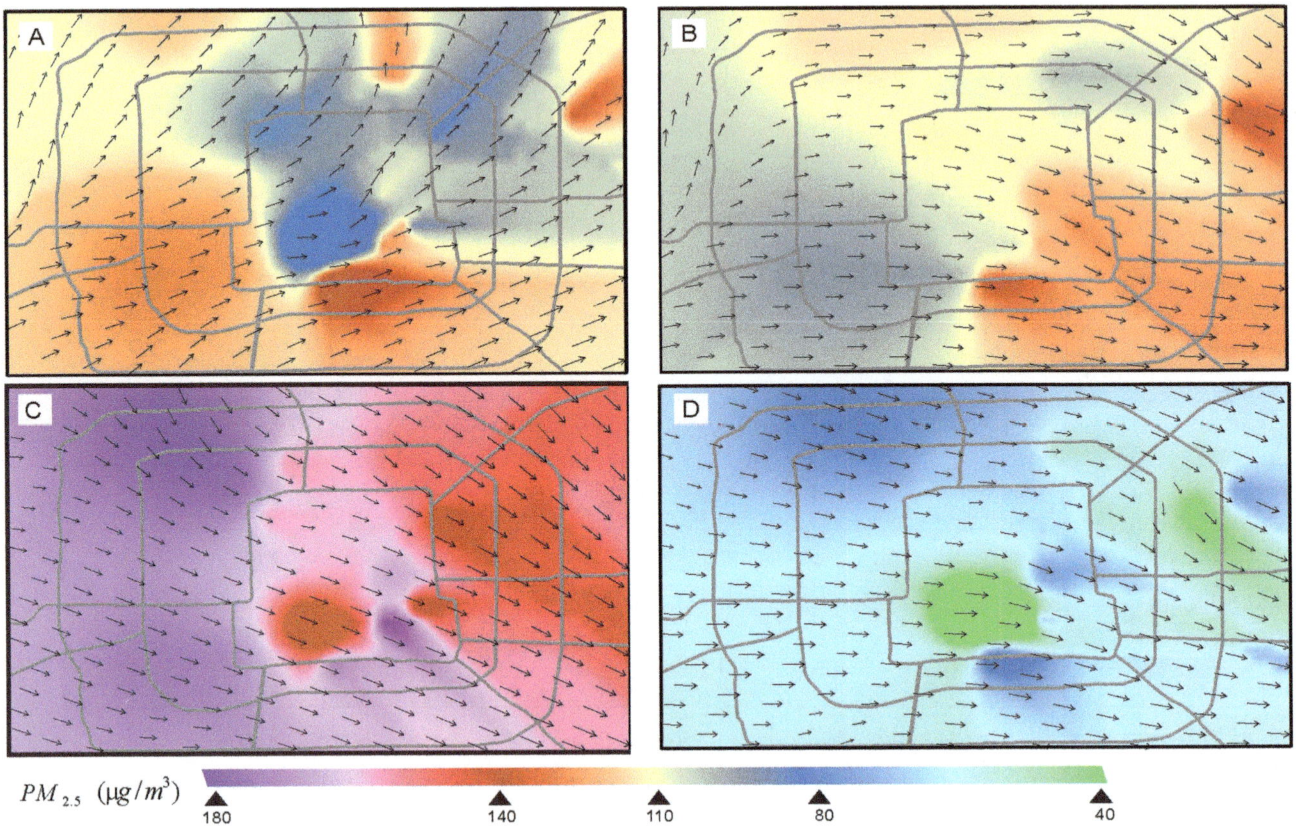

Figure 6. Interpolation results of the hourly $PM_{2.5}$ concentrations on May 26th at various times: (A) 8 am, (B) 12 pm, (C) 4 pm and (D) 8 pm.

development, the time required to generate one estimation surface for the study area is approximately 10 minutes. Further research is necessary to accelerate the calculation and to allow this technique to be used for real-time estimation.

The results of cross-validation and visual assessment have demonstrated that including wind-field in the interpolation of the $PM_{2.5}$ concentration improves the predictive performance. However, the experiments were conducted only in the core area of Beijing during six selected days in May. The method should be assessed over much longer monitoring periods spanning all four seasons to confirm its year-round effectiveness. Additional measurements are also required to confirm the usefulness of the SWPD or even to discover a better distance metric. The interpolation of other types of air-pollution, such as nitrogen dioxide and coarse particulate matter, should also be performed to verify that the model has general applicability. Furthermore, wind-fields with higher resolution may have the potential to improve the predictive capability of the technique and deserve further research.

Conclusions

Three major conclusions can be drawn from this study:

(1) Wind-fields are of great importance to studies of the negative effects of airborne pollutants. Incorporating wind-fields into the spatial interpolation of air-pollution distributions serves to enhance the predictive capability of such interpolation.

(2) The shortest wind-field path distance (SWPD) shows great potential for determining the spatial dependence and enables SWPD-based interpolations to capture complex features of air-pollution distributions with higher accuracy than methods based on the Euclidean distance.

(3) The workflow proposed in this paper, which consists of wind-field generation, shortest-path analysis and IDW in conjunction with the SWPD, has been demonstrated to be a robust technique for predicting urban-scale $PM_{2.5}$ concentrations.

Acknowledgments

We thank Professor Dongping Ming, Professor Jinfeng Wang and Professor Tao Pei for their advice. We also thank Zhuoqin Song for her assistance.

Author Contributions

Conceived and designed the experiments: LL JG. Performed the experiments: LL JZ. Analyzed the data: LL JG. Contributed reagents/materials/analysis tools: LL JZ. Wrote the paper: LL JZ.

References

1. Ryan PH, LeMasters GK (2007) A review of land-use regression models for characterizing intraurban air pollution exposure. Inhal Toxicol 19 Suppl 1: 127–133.
2. Lee SJ, Serre ML, van Donkelaar A, Martin RV, Burnett RT, et al. (2012) Comparison of geostatistical interpolation and remote sensing techniques for estimating long-term exposure to ambient PM2.5 concentrations across the continental United States. Environ Health Perspect 120: 1727–1732.
3. Jerrett M, Arain A, Kanaroglou P, Beckerman B, Potoglou D, et al. (2004) A review and evaluation of intraurban air pollution exposure models. Journal of Exposure Science and Environmental Epidemiology 15: 185–204.
4. Liao D, Peuquet DJ, Duan Y, Whitsel EA, Dou J, et al. (2006) GIS approaches for the estimation of residential-level ambient PM concentrations. Environmental health perspectives 114: 1374.
5. Ensor KB, Raun LH, Persse D (2013) A Case-Crossover Analysis of Out-of-Hospital Cardiac Arrest and Air PollutionClinical Perspective. Circulation 127: 1192–1199.
6. Sullivan J, Sheppard L, Schreuder A, Ishikawa N, Siscovick D, et al. (2005) Relation between short-term fine-particulate matter exposure and onset of myocardial infarction. Epidemiology 16: 41–48.
7. Bowman KW (2013) Toward the next generation of air quality monitoring: Ozone. Atmospheric Environment 80: 571–583.
8. Engel-Cox J, Kim Oanh NT, van Donkelaar A, Martin RV, Zell E (2013) Toward the next generation of air quality monitoring: Particulate Matter. Atmospheric Environment 80: 584–590.
9. Pearce JL, Rathbun SL, Aguilar-Villalobos M, Naeher LP (2009) Characterizing the spatiotemporal variability of PM2.5 in Cusco, Peru using kriging with external drift. Atmospheric Environment 43: 2060–2069.
10. Janssen S, Dumont G, Fierens F, Mensink C (2008) Spatial interpolation of air pollution measurements using CORINE land cover data. Atmospheric Environment 42: 4884–4903.
11. Arain MA, Blair R, Finkelstein N, Brook JR, Sahsuvaroglu T, et al. (2007) The use of wind fields in a land use regression model to predict air pollution concentrations for health exposure studies. Atmospheric Environment 41: 3453–3464.
12. Seaman NL (2000) Meteorological modeling for air-quality assessments. Atmospheric Environment 34: 2231–2259.
13. Liu Y, Paciorek CJ, Koutrakis P (2009) Estimating regional spatial and temporal variability of PM2. 5 concentrations using satellite data, meteorology, and land use information. Environmental Health Perspectives 117: 886.
14. Tian J, Chen D (2010) A semi-empirical model for predicting hourly ground-level fine particulate matter (PM2.5) concentration in southern Ontario from satellite remote sensing and ground-based meteorological measurements. Remote Sensing of Environment 114: 221–229.
15. Hoek G, Beelen R, de Hoogh K, Vienneau D, Gulliver J, et al. (2008) A review of land-use regression models to assess spatial variation of outdoor air pollution. Atmospheric Environment 42: 7561–7578.
16. Curriero F (2006) On the Use of Non-Euclidean Distance Measures in Geostatistics. Mathematical Geology 38: 907–926.
17. Greenberg JA, Rueda C, Hestir EL, Santos MJ, Ustin SL (2011) Least cost distance analysis for spatial interpolation. Computers & Geosciences 37: 272–276.
18. Boisvert JB, Manchuk JG, Deutsch CV (2009) Kriging in the Presence of Locally Varying Anisotropy Using Non-Euclidean Distances. Mathematical Geosciences 41: 585–601.
19. Peterson E, Merton A, Theobald D, Urquhart NS (2006) Patterns of Spatial Autocorrelation in Stream Water Chemistry. Environmental Monitoring and Assessment 121: 569–594.
20. Peterson EE, Theobald DM, Ver Hoef JM (2007) Geostatistical modelling on stream networks: developing valid covariance matrices based on hydrologic distance and stream flow. Freshwater Biology 52: 267–279.
21. Hoef JV, Peterson E, Theobald D (2006) Spatial statistical models that use flow and stream distance. Environmental and Ecological Statistics 13: 449–464.
22. Gardner B, Sullivan PJ, Lembo JAJ (2003) Predicting stream temperatures: geostatistical model comparison using alternative distance metrics. Canadian Journal of Fisheries and Aquatic Sciences 60: 344–351.
23. Boisvert JB, Deutsch CV (2011) Programs for kriging and sequential Gaussian simulation with locally varying anisotropy using non-Euclidean distances. Computers & Geosciences 37: 495–510.
24. Boisvert J, Deutsch C (2011) Modeling locally varying anisotropy of CO2 emissions in the United States. Stochastic Environmental Research and Risk Assessment 25: 1077–1084.
25. Broquet T, Ray N, Petit E, Fryxell J, Burel F (2006) Genetic isolation by distance and landscape connectivity in the American marten (Martes americana). Landscape Ecology 21: 877–889.
26. Zou H, Yue Y, Li Q, Yeh AG (2012) An improved distance metric for the interpolation of link-based traffic data using kriging: a case study of a large-scale urban road network. International Journal of Geographical Information Science 26: 667–689.
27. Zhang A, Qi Q, Jiang L, Zhou F, Wang J (2013) Population Exposure to PM2. 5 in the Urban Area of Beijing. PloS one 8: e63486.
28. Goodin WR, McRae GJ, Seinfeld JH (1980) An objective analysis technique for constructing three-dimensional urban-scale wind fields. Journal of Applied Meteorology 19: 98–108.
29. Kanaroglou PS, Jerrett M, Morrison J, Beckerman B, Arain MA, et al. (2005) Establishing an air pollution monitoring network for intra-urban population exposure assessment: A location-allocation approach. Atmospheric Environment 39: 2399–2409.
30. Turner DB (1973) Workbook of atmospheric dispersion estimates: US Government Printing Office.
31. de Mesnard L (2013) Pollution models and inverse distance weighting: Some critical remarks. Computers & Geosciences 52: 459–469.
32. Green A, Singhal R, Venkateswar R (1980) Analytic extensions of the Gaussian plume model. Journal of the Air Pollution Control Association 30: 773–776.
33. Dijkstra EW (1959) A note on two problems in connexion with graphs. Numerische mathematik 1: 269–271.
34. Goovaerts P (2000) Geostatistical approaches for incorporating elevation into the spatial interpolation of rainfall. Journal of Hydrology 228: 113–129.
35. Picard RR, Cook RD (1984) Cross-validation of regression models. Journal of the American Statistical Association 79: 575–583.
36. Warner TT, Fizz RR, Seaman NL (1983) A Comparison of Two Types of Atmospheric Transport Models—Use of Observed Winds Versus Dynamically Predicted Winds. Journal of Applied Meteorology 22: 394–406.

Estimation of Short-Term Effects of Air Pollution on Stroke Hospital Admissions in Wuhan, China

Hao Xiang[1,2], Kristen J. Mertz[3], Vincent C. Arena[4], Luann L. Brink[3], Xiaohui Xu[5], Yongyi Bi[2,6], Evelyn O. Talbott[3]*

1 Department of Epidemiology and Biostatistics, School of Public Health, Wuhan University, Wuhan, Hubei, China, **2** Wuhan University Global Health Institute, Wuhan University, Wuhan, Hubei, China, **3** Department of Epidemiology, Graduate School of Public Health, University of Pittsburgh, Pittsburgh, Pennsylvania, United States of America, **4** Department of Biostatistics, Graduate School of Public Health, University of Pittsburgh, Pittsburgh, Pennsylvania, United States of America, **5** Department of Epidemiology, College of Public Health and Health Professions, College of Medicine, University of Florida, Gainesville, Florida, United States of America, **6** Department of Labor and Environmental Hygiene, School of Public Health, Wuhan University, Wuhan, Hubei, China

Abstract

Background and Objective: High concentrations of air pollutants have been linked to increased incidence of stroke in North America and Europe but not yet assessed in mainland China. The aim of this study is to evaluate the association between stroke hospitalization and short-term elevation of air pollutants in Wuhan, China.

Methods: Daily mean NO_2, SO_2 and PM_{10} levels, temperature and humidity were obtained from 2006 through 2008. Data on stroke hospitalizations (ICD 10: I60–I69) at four hospitals in Wuhan were obtained for the same period. A time-stratified case-crossover design was performed by season (April-September and October-March) to assess effects of pollutants on stroke hospital admissions.

Results: Pollution levels were higher in October-March with averages of 136.1 $\mu g/m^3$ for PM_{10}, 63.6 $\mu g/m^3$ for NO_2 and 71.0 $\mu g/m^3$ for SO_2 than in April-September when averages were 102.0 $\mu g/m^3$, 41.7 $\mu g/m^3$ and 41.7 $\mu g/m^3$, respectively ($p<.001$). During the cold season, every 10 $\mu g/m^3$ increase in NO_2 was associated with a 2.9% (95%C.I. 1.2%–4.6%) increase in stroke admissions on the same day. Every 10 ug/m^3 increase in PM_{10} daily concentration was significantly associated with an approximate 1% (95% C.I. 0.1%–1.4%) increase in stroke hospitalization. A two-pollutant model indicated that NO_2 was associated with stroke admissions when controlling for PM_{10}. During the warm season, no significant associations were noted for any of the pollutants.

Conclusions: Exposure to NO_2 is significantly associated with stroke hospitalizations during the cold season in Wuhan, China when pollution levels are 50% greater than in the warm season. Larger and multi-center studies in Chinese cities are warranted to validate our findings.

Editor: Michelle L. Block, Virginia Commonwealth University, United States of America

Funding: These authors have no support or funding to report.

Competing Interests: The authors have declared that no competing interests exist.

* E-mail: eot1@pitt.edu

Introduction

Numerous epidemiological studies have demonstrated increased risk for acute myocardial infarction (MI), coronary heart disease (CHD) and other cardiovascular mortality associated with acute exposure to ambient air pollutants [1–4]. More recently, studies have focused on morbidity with use of hospital admission data to assess the short-term effects of air pollution on stroke; however, the results of these studies are inconsistent. Some studies have found associations between air pollution and stroke hospital admissions [5–11], while other studies have reported no significant links [12–16]. Moreover, most studies have been conducted in North American and European cities [5–7,9,13,14,17] with only limited information from Asia [8,12,18].

Wuhan is the capital of Hubei Province and the most populous city in Central China. There are approximately 10 million people in the metropolitan area, with approximately 6 million in the center city within an area of 201 km^2. Wuhan's climate is subtropical with abundant rainfall and very high humidity in the spring and summer. Wuhan's major industries include optic-electronics, automobile manufacturing, steel manufacturing, chemical plants and power plants; it is also a major transportation hub.The major sources of air pollution in the city are motor vehicles and the use of coal for domestic cooking, heating, and industrial processes [19,20].

Qian et al's study in Wuhan showed that every 10 $\mu g/m^3$ increase in daily PM_{10} concentration was significantly associated with an increase in stroke mortality (0.44%) on the same day [20]; however, the effect of air pollution on stroke hospital admissions has not been assessed in Wuhan, which has a hotter climate and higher levels of pollutants than most places studied to date. The objective of this study is to assess the associations between hospitalizations for stroke and short-term elevations in particulate matter with diameter <10 μm (PM_{10}), nitrogen dioxide (NO_2) and sulfur dioxide (SO_2), in Wuhan, China. The research covers a 3-year period from 2006 through 2008.

Materials and Methods

Ethics Statement

This study was approved by the Institutional Review Boards of Wuhan University and the University of Pittsburgh. Because the data were analyzed anonymously, names of participants were not essential, and we did not collect the names of participant. Informed consent from human participant was not necessary in this study.

Study Population

Hospital admission data for this analysis were collected from four major medical centers with a total of 10,000 beds in the center city because there is no unified hospital data collection system for all hospitals in Wuhan, We obtained electronic data for all hospital admissions for acute strokes (International Classification of Diseases, 10th Revision: I60–I69) from January 1, 2006 through December 31, 2008, including admission date, age, sex, and discharge diagnostic code (ICD-10).

Exposure Information

Nine monitoring stations operated by the Wuhan Environmental Monitoring Center in the metropolitan area measure levels of air pollution. The monitoring program has been certified by the US Environment Protection Agency (EPA). We obtained the mean daily Air Pollution Index (API) of PM_{10}, NO_2 and SO_2 for the 9 monitoring stations for the period January 1, 2006 through December 31, 2008. API reporting system, based on health effects, is introduced for a consistent comparison among the pollution levels by different air pollutants. In China, the API system has been applied in cities from 1997 [21]. Other similar index systems are also used in other countries, for example, AQI (Air Quality Index) in Canada, IMECA (Air Quality Metropolitan Index) in Mexico and PSI (Pollutant Standards Index) in Singapore. According to the breakpoints of air pollutant and the transform formula between API and air pollutant concentrations. We converted the API values to concentration levels using standard formula [22]. Meteorological data, including mean daily temperature and relative humidity, were obtained from the Wuhan Meteorological Administration.

Data Analysis

A time-stratified case-crossover study design was used to evaluate the association between air pollution and the number of stroke admissions. This method is a special case-control design used to explore the effects of transient exposures on acute events, in which every case serves as his or her own control [22,23]. We compared the air pollution exposure on the current day (lag 0) of the stroke onset to exposure on the same weekdays during the same month as the lag 0 day. Because previous studies indicated that increased hospital admissions were associated with high air pollution levels up to 2 days before the event [24], we also considered pollution concentrations one or two days before stroke admission (lag1, lag2) and the accumulated exposure over 3 days (mean of lag 0 through lag 2). This design is useful for controlling for seasonality, time trends, and other potential confounders [25].

We performed multivariate conditional logistic regression modeling to obtain estimates of the odds ratios (ORs) and 95% confidence intervals (CIs) for the effect of PM_{10}, NO_2 and SO_2 on stroke hospitalization, controlling for meteorological covariates including daily mean temperature and relative humidity. For understanding the mean effect of air pollution during lag 0–2 day, we calculated lag 0–2 with available data, we reported effect estimates as the percentage of change in the rate of hospitalizations associated with a 10 $\mu g/m^3$ increase in mean daily pollutant levels. Subgroup analysis is based on age group (≥65 years and <65 years), sex (male and female), categories of stroke (hemorrhagic stroke: I61 and ischemic stroke: I63) and seasons (warm: April-September and cold: October-March). All of the reported p-values are based on 2-sided tests at the 0.05 level. Analyses were performed using SAS Version 9 and the C-CAT package [26].

Results

We observed 10663 hospital admissions with a primary discharge diagnosis of stroke from four major hospitals, of which 51% (n = 5422) were coded as ischemic, 21% (n = 2225) as hemorrhagic, and 28% (n = 3016) as other. Age on admission ranged from 0 to 98 years with a mean of 64(S.D.14) years, and 51% (n = 5480) of those admitted were equal or older than 65 years. Approximately 61% of cases were male.

A summary of air pollutants and meteorological parameters is shown in Table 1. During the 1096 days of our study period, daily mean temperatures fluctuated between −2.7 and 33.6°C. The overall mean concentrations of pollutants were 119±53 $\mu g/m^3$ for PM_{10}, 53±21 $\mu g/m^3$ for NO_2, and 56±31 $\mu g/m^3$ for SO_2. There were significant differences(P<0.001) in pollution levels by season with the April through September means of 102.0 $\mu g/m^3$ for PM_{10}, 41.7 $\mu g/m^3$ for NO_2 and 41.7 $\mu g/m^3$ for SO_2; and the October through March means of 136.1 $\mu g/m^3$, 63.6 $\mu g/m^3$ and 71.0 $\mu g/m^3$, respectively. (Table 1).

As shown in Table 2, none of the single-pollutant models with lag 0, lag 1 and lag 2 were associated with hospital admissions for stroke, ischemic stroke or hemorrhagic stroke with all seasons combined. Subgroup analysis by age and sex found no significant associations.

The associations between air pollutants and stroke hospitalization by season are presented in Table 3. For the cold season, significant associations were found between stroke hospital admissions and levels of PM_{10}, and NO_2. Every 10 $\mu g/m^3$ increase in NO_2 was associated with a 2.9% (95%C.I. 1.2%–4.6%) increase in stroke events on the same day and a 2.1% increase 2 days later. In addition, each 10 $\mu g/m^3$ increase in PM_{10} daily concentration was significantly associated with an approximate 0.7% (95% C.I. 0.1%–1.4%) increase in stroke hospitalization the next day. SO_2 levels had no significant effect on stroke hospitalization (Table 3). For the warm season, a time when the pollution levels were lower, no significant associations with stroke hospitalization were noted for any of the pollutants controlling for temperature and humidity.

A two-pollutant model was examined to determine which individual pollutant might influence stroke occurrence independently of the effect of the other (Table 4).

SO_2 was not included in this model because it was not significantly associated with hospital admissions for stroke. For the cold season, there was no significant association between PM_{10} and stroke admissions after adjusting for NO_2; however, the effect of NO_2 remained significant after adjusting for PM_{10}.

Discussion

This is one of the few investigations of the association between short-term health effects and air pollution in mainland China. Single-pollutant models, when adjusted for temperature and humidity, indicated a significant effect from NO_2 and PM_{10} on

Table 1. Ambient air pollution concentrations and meteorological parameters in Wuhan, 2006–2008.

	Mean	Min.	Q1*	Median	Q3*	Max.	Apr.-Sep. Median (n = 549 days)	Oct.-Mar. Median (n = 547 days)
Pollutants								
PM_{10}($\mu g/m^3$)	119	35	78	110	148	350	94†	132
NO_2($\mu g/m^3$)	53	18	35	48	66	154	38†	66
SO_2($\mu g/m^3$)	56	8	33	50	72	267	38†	61
Meteorology								
Temperature (°C)	18.2	−2.7	9.9	19.6	26.3	33.6	26.2	10.0
Humidity (%)	69.6	21.0	61.0	70.5	78.0	97.0	68.0	73.0

*Q1, first quartile value; Q3, third quartile value.
†There are significant differences in PM_{10}, NO_2, SO_2 concentration in summer/fall and winter/spring, Wilcoxon test $P<0.001$.

stroke hospital admissions during the cold season. The two-pollutant model showed that the effect of NO_2 remained significant after adjusting for PM_{10}.

It should be noted that hospital admissions for hemorrhagic and ischemic strokes, when analyzed separately, were not associated with levels of any air pollutants in single-pollutant models in our study. Given that there were only 2, 225 hemorrhagic and 5, 422 ischemic stroke admissions during the three–year period, we believe that there may be a lack of statistical power to identify such a risk.

Seasonal effects between air pollution and hospital admissions have been reported frequently, and various explanations have been postulated. In the present study, we found that during the warm season, no significant associations were obtained. In contrast, for the cold season of October through March, stroke admissions and PM_{10} and NO_2 levels were significantly associated. One explanation is that the concentrations of pollutants during the warm season were much lower than in the cold season; the association between air pollution and stroke hospitalization is stronger when pollution levels are higher.

There are three main reasons for higher air pollution levels in the cold season in Wuhan. First, the amount of rainfall decreases significantly in the cold season. The average precipitation in November, December, and January were 2.04, 1.02, 1.71 inches respectively, in contrast to the average rainfall in May, June, and July of 6.47, 8.86, 7.49 inches. Rainfall plays an important role in chemical reactions with air pollutants, diluting their effects. [27] Second, Wuhan was almost controlled by stagnant air and there is less wind in winter than in summer, which went against pollutants diffusion. The formation of inversion layer also made air suspended particles hard to diffuse to high altitude but close to the ground. Third, coal emissions are higher on cold days because of the greater use of coal for electricity and heating. Under extreme conditions, in relative high temprature and low humidity days in winter, air pollutants could be transferred into secondary

Table 2. Association of all strokes, ischemic and hemorrhagic strokes and pollutant concentrations by lag days for 10 $\mu g/m^3$ increases in pollutant levels adjusted for temperature and humidity, Wuhan, 2006–2008.

	Stroke N = 10663		Ischemic Stroke N = 5422		Hemorrhagic Stroke N = 2225	
Lag days	O.R. (95%C.I.)	*P* Value	O.R. (95%C.I.)	*P* Value	O.R. (95%C.I.)	*P* Value
PM_{10}						
0	0.997(0.989,1.005)	0.4350	0.999(0.992,1.006)	0.7394	0.993(0.982,1.004)	0.2265
1	0.997(0.990,1.005)	0.5189	1.000(0.993,1.007)	0.8932	0.998(0.987,1.008)	0.6485
2	0.996(0.988,1.004)	0.3134	1.007(0.992,1.005)	0.6358	0.999(0.989,1.009)	0.8663
0–2	0.995(0.985,1.005)	0.3048	1.005(0.990,1.007)	0.7016	0.995(0.983,1.008)	0.4706
NO_2						
0	0.997(0.977,1.018)	0.7902	1.002(0.984,1.021)	0.8267	0.989(0.961,1.018)	0.4421
1	0.994(0.973,1.015)	0.5703	1.003(0.969,1.006)	0.1849	1.007(0.978,1.036)	0.6403
2	1.010(0.989,1.031)	0.3566	1.000(0.982,1.019)	0.9704	1.004(0.976,1.032)	0.7990
0–2	1.004(0.977,1.031)	0.7703	1.004(0.973,1.018)	0.6672	1.000(0.965,1.035)	0.9841
SO_2						
0	1.001(0.986,1.016)	0.8792	0.996(0.983,1.009)	0.5549	0.996(0.976,1.016)	0.6773
1	0.992(0.978,1.006)	0.2824	0.994(0.982,1.007)	0.3740	1.003(0.984,1.022)	0.7943
2	0.991(0.977,1.005)	0.1989	0.996(0.984,1.009)	0.5825	0.991(0.973,1.011)	0.3784
0–2	0.991(0.972,1.010)	0.3587	0.993(0.977,1.009)	0.3922	0.995(0.971,1.019)	0.6658

Table 3. The association of stroke admissions and air pollutants in cold (winter and spring) and warm (summer and fall) seasons for 10 μg/m^3 increases in pollutant level, adjusted for temperature and humidity, Wuhan, 2006–2008.

Lag Days	Cold seasons N = 5341		Warm seasons N = 5322	
	O.R. (95%C.I.)	*P* Value	O.R. (95%C.I.)	*P* Value
PM_{10}				
0	1.006(0.999,1.013)	0.0875	1.000(0.999,1.000)	0.3174
1	1.007(1.001,1.014)	0.0300	0.995(0.988,1.002)	0.1442
2	1.005(0.999,1.012)	0.0830	0.996(0.989,1.002)	0.1860
0–2	1.009(1.001,1.016)	0.0256	0.994(0.985,1.002)	0.1555
NO_2				
0	1.029(1.012,1.046)	0.0007	0.998(0.980,1.016)	0.8207
1	1.012(0.995,1.029)	0.1620	0.983(0.965,1.001)	0.0681
2	1.021(1.005,1.037)	0.0096	0.996(0.978,1.014)	0.6584
0–2	1.031(1.011,1.052)	0.0025	0.990(0.967,1.014)	0.4096
SO_2				
0	1.004(0.993,1.015)	0.4677	0.997(0.984,1.010)	0.6519
1	0.996(0.986,1.007)	0.4929	0.988(0.976,1.001)	0.0621
2	0.997(0.986,1.007)	0.5258	0.994(0.981,1.006)	0.3251
0–2	0.997(0.983,1.011)	0.6844	0.987(0.970,1.004)	0.1307

Table 4. Relationship between stroke and PM_{10} and NO_2 concentrations in a two-pollutant model in the cold season for 10 μg/m^3 increases in pollutant level, adjusted for temperature and humidity.

Pollutant	O.R. (95%CI)	P
PM_{10}		
0	0.995(0.986,1.005)	0.3305
1	1.008(0.999,1.017)	0.0931
2	0.998(0.989,1.008)	0.7333
0–2	0.999(0.987,1.011)	0.8945
NO_2		
0	1.037(1.013,1.061)	0.0021
1	0.997(0.973,1.021)	0.7955
2	1.024(1.000,1.049)	0.0496
0–2	1.033(1.001,1.065)	0.0412

aerosol and lead to more serious pollution, even formed haze or smog.

It is widely accepted that there is a significant association between mortality and PM_{10} levels. [5,6,8] Recent studies showed that air particles maybe a possible causal factor in stroke mortality [28] [29,30]. Several potential mechanisms have been proposed. First, it has been hypothesized that particulate exposure can induce an acute systemic inflammatory response with an increased number of circulating neutrophils [28], provoking inflammation and increased blood coagulation. [29] Second, particles may affect the autonomic nervous system resulting in increased risk for heart rate variability. [30] Third, particles can increase concentrations of fibrinogen [31] and enhance peripheral arterial thrombosis [32], which will promote stroke. Exposure to cold days has also been proved to increase plasma viscosity and serum cholesterol levels, which is also regarded as a stimulus to stroke [33].

In this study, PM_{10} was significantly associated with stroke hospitalizations in the cold season in a single pollutant model, but not after adjusting for NO_2. The reason why our result is inconsistent with the past studies may be due to the problem of co-linearity between air pollutants (the pearson correlation between PM_{10} and NO_2 is 0.75, see supplement Table S1), it was difficult to separate the effect of individual pollutants on stroke admission. Haidong Kan et al's study showed that in single air pollutant models, each 10 μg/m^3 increase in PM_{10} and NO_2 corresponded to 1.008 and 1.029 relative risks of stroke mortality, respectively, but in multiple pollutant models, there were no significant associations [34].

The importance of NO_2 as a cause of increased stroke admissions is not sufficiently understood. We cannot rule out the possibility that NO_2 acts, at least in part, as a surrogate for some other unmeasured end product of reaction sequences initiated by NO_2.

The strengths of our study include the fact that data were derived from the electronic data of four major medical centers, which provided relative accurate information and allowed us to stratify the effect of air pollution by age, sex and stroke type. A further strength of our study was the availability of data on PM_{10}, SO_2 and NO_2 levels from Wuhan Environmental Monitoring Center, which could be linked for the first time to stroke admissions in Wuhan city. Finally, the case-crossover studies design control for confounders by design rather than by modeling [21,35]. As a result, time-invariant variables such as sex, age and family history do not act as confounders. In addition, by using the same weekdays as the onset day as controls could avoid possible confounding from the effects of day of the week [22,36].

We do acknowledge that our study has several limitations. First, there is no unified hospital data collection system for all hospitals in Wuhan, and it is not possible to obtain data on all the stroke admissions in the city during the study period. However, the catchment area of these four hospitals is relatively stable. Second, we used the mean air pollutant concentrations from nine monitoring stations to estimate individuals' exposure because the air pollutants are highly correlated in 9 monitors. However, errors in exposure estimation resulting from the differences between the population-average exposure and local ambient levels cannot be avoided. Such exposure misclassification may lead to inaccurate results, as reported in previous studies [37,38]. Third, it should also be noted that Wuhan has a subtropical humid monsoon climate, and this fact may restrict the generalization of these findings to other locations with different meteorological characteristics. However, this would not affect the study's internal validity.

In summary, our study shows that there is an association between exposure to air pollution and hospital admissions for stroke during the cold season in Wuhan, China. Although these findings support the possibility that there are acute pathogenetic processes in the cerebrovascular system induced by air pollution, the mechanism is still not clear and toxicological analyses are expected in further studies.

Acknowledgments

We thank Dr. Jiming Fang of Department of Medicine, University of Toronto and Institute for Clinical Evaluative Sciences for his thoughtful comments.

Author Contributions

Conceived and designed the experiments: EOT YB. Performed the experiments: HX YB KJM. Analyzed the data: HX KJM LLB VCA XX. Contributed reagents/materials/analysis tools: VCA XX LLB KJM EOT. Wrote the paper: HX KJM VCA XX EOT LLB.

References

1. Peters A, Dockery DW, Muller JE, Mittleman MA (2001) Increased particulate air pollution and the triggering of myocardial infarction. Circulation 103: 2810–2815.
2. D'Ippoliti D, Forastiere F, Ancona C, Agabiti N, Fusco D, et al. (2003) Air pollution and myocardial infarction in Rome: a case-crossover analysis. Epidemiology 14: 528–535.
3. Rich DQ, Schwartz J, Mittleman MA, Link M, Luttmann-Gibson H, et al. (2005) Association of short-term ambient air pollution concentrations and ventricular arrhythmias. Am J Epidemiol 161: 1123–1132.
4. Wellenius GA, Bateson TF, Mittleman MA, Schwartz J (2005) Particulate air pollution and the rate of hospitalization for congestive heart failure among medicare beneficiaries in Pittsburgh, Pennsylvania. Am J Epidemiol 161: 1030–1036.
5. Wordley J, Walters S, Ayres JG (1997) Short term variations in hospital admissions and mortality and particulate air pollution. Occup Environ Med 54: 108–116.
6. Ponka A, Virtanen M (1996) Low-level air pollution and hospital admissions for cardiac and cerebrovascular diseases in Helsinki. Am J Public Health 86: 1273–1280.
7. Wellenius GA, Schwartz J, Mittleman MA (2005) Air pollution and hospital admissions for ischemic and hemorrhagic stroke among medicare beneficiaries. Stroke 36: 2549–2553.
8. Tsai SS (2003) Evidence for an Association Between Air Pollution and Daily Stroke Admissions in Kaohsiung, Taiwan. Stroke 34: 2612–2616.
9. Ballester F, Tenias JM, Perez-Hoyos S (2001) Air pollution and emergency hospital admissions for cardiovascular diseases in Valencia, Spain. J Epidemiol Community Health 55: 57–65.
10. Moolgavkar SH (2000) Air pollution and hospital admissions for diseases of the circulatory system in three U.S. metropolitan areas. J Air Waste Manag Assoc 50: 1199–1206.
11. Linn WS, Szlachcic Y, Gong H, Jr., Kinney PL, Berhane KT (2000) Air pollution and daily hospital admissions in metropolitan Los Angeles. Environ Health Perspect 108: 427–434.
12. Wong TW, Lau TS, Yu TS, Neller A, Wong SL, et al. (1999) Air pollution and hospital admissions for respiratory and cardiovascular diseases in Hong Kong. Occup Environ Med 56: 679–683.
13. Le Tertre A, Medina S, Samoli E, Forsberg B, Michelozzi P, et al. (2002) Short-term effects of particulate air pollution on cardiovascular diseases in eight European cities. J Epidemiol Community Health 56: 773–779.
14. Poloniecki JD, Atkinson RW, de Leon AP, Anderson HR (1997) Daily time series for cardiovascular hospital admissions and previous day's air pollution in London, UK. Occup Environ Med 54: 535–540.
15. Burnett RT, Smith-Doiron M, Stieb D, Cakmak S, Brook JR (1999) Effects of particulate and gaseous air pollution on cardiorespiratory hospitalizations. Arch Environ Health 54: 130–139.
16. Sunyer J, Ballester F, Tertre AL, Atkinson R, Ayres JG, et al. (2003) The association of daily sulfur dioxide air pollution levels with hospital admissions for cardiovascular diseases in Europe (The Aphea-II study). Eur Heart J 24: 752–760.
17. Arena VC, Mazumdar S, Zborowski JV, Talbott EO, He S, et al. (2006) A Retrospective Investigation of PM10 in Ambient Air and Cardiopulmonary Hospital Admissions in Allegheny County, Pennsylvania: 1995–2000. Journal of Occupational and Environmental Medicine 48: 38–47.
18. Chan CC (2005) Urban air pollution and emergency admissions for cerebrovascular diseases in Taipei, Taiwan. European Heart Journal 27: 1238–1244.
19. Ali R, Zhao H (2008) Wuhan, China and Pittsburgh, USA: Urban Environmental Health Past, Present, and Future. EcoHealth 5: 159–166.
20. Qian Z, He Q, Lin H, Kong L, Liao D, et al. (2007) Association of daily cause-specific mortality with ambient particle air pollution in Wuhan, China. Environmental Research 105: 380–389.
21. Carracedo-Martinez E, Taracido M, Tobias A, Saez M, Figueiras A (2010) Case-crossover analysis of air pollution health effects: a systematic review of methodology and application. Environ Health Perspect 118: 1173–1182.
22. Levy D, Lumley T, Sheppard L, Kaufman J, Checkoway H (2001) Referent selection in case-crossover analyses of acute health effects of air pollution. Epidemiology 12: 186–192.
23. Tong S, Wang XY, Guo Y (2012) Assessing the short-term effects of heatwaves on mortality and morbidity in brisbane, australia: comparison of case-crossover and time series analyses. PLoS One 7: e37500.
24. Katsouyanni K, Touloumi G, Spix C, Schwartz J, Balducci F, et al. (1997) Short-term effects of ambient sulphur dioxide and particulate matter on mortality in 12 European cities: results from time series data from the APHEA project. Air Pollution and Health: a European Approach. BMJ 314: 1658–1663.
25. Wichmann J, Andersen Z, Ketzel M, Ellermann T, Loft S (2011) Apparent temperature and cause-specific emergency hospital admissions in Greater Copenhagen, Denmark. PLoS One 6: e22904.
26. Abraham JH, Gold DR, Dockery DW, Ryan L, Park JH, et al. (2005) Within-home versus between-home variability of house dust endotoxin in a birth cohort. Environ Health Perspect 113: 1516–1521.
27. Yang CY, Chen YS, Yang CH, Ho SC (2004) Relationship between ambient air pollution and hospital admissions for cardiovascular diseases in kaohsiung, taiwan. J Toxicol Environ Health A 67: 483–493.
28. van Eeden SF, Hogg JC (2002) Systemic inflammatory response induced by particulate matter air pollution: the importance of bone-marrow stimulation. J Toxicol Environ Health A 65: 1597–1613.
29. Seaton A, MacNee W, Donaldson K, Godden D (1995) Particulate air pollution and acute health effects. Lancet 345: 176–178.
30. Fang SC, Cassidy A, Christiani DC (2010) A systematic review of occupational exposure to particulate matter and cardiovascular disease. Int J Environ Res Public Health 7: 1773–1806.
31. Schwartz J (2001) Air pollution and blood markers of cardiovascular risk. Environ Health Perspect 109 Suppl 3: 405–409.
32. Nemmar A, Hoet PH, Dinsdale D, Vermylen J, Hoylaerts MF, et al. (2003) Diesel exhaust particles in lung acutely enhance experimental peripheral thrombosis. Circulation 107: 1202–1208.
33. Hong YC (2002) Air Pollution: A New Risk Factor in Ischemic Stroke Mortality. Stroke 33: 2165–2169.
34. Kan H, Jia J, Chen B (2003) Acute stroke mortality and air pollution: new evidence from Shanghai, China. J Occup Health 45: 321–323.
35. Janes H, Sheppard L, Lumley T (2005) Case-crossover analyses of air pollution exposure data: referent selection strategies and their implications for bias. Epidemiology 16: 717–726.
36. Maclure M (1991) The case-crossover design: a method for studying transient effects on the risk of acute events. Am J Epidemiol 133: 144–153.
37. Zeger SL, Thomas D, Dominici F, Samet JM, Schwartz J, et al. (2000) Exposure measurement error in time-series studies of air pollution: concepts and consequences. Environ Health Perspect 108: 419–426.
38. Zeka A, Schwartz J (2004) Estimating the independent effects of multiple pollutants in the presence of measurement error: an application of a measurement-error-resistant technique. Environ Health Perspect 112: 1686–1690.

21

Effectiveness of Low Emission Zones: Large Scale Analysis of Changes in Environmental NO_2, NO and NO_x Concentrations in 17 German Cities

Peter Morfeld[1,2*], David A. Groneberg[3], Michael F. Spallek[3,4]

1 Institute for Occupational Epidemiology and Risk Assessment (IERA) of Evonik Industries, Essen, Germany, **2** Institute and Policlinic for Occupational Medicine, Environmental Medicine and Preventive Research, University of Cologne, Cologne, Germany, **3** Institute of Occupational Medicine, Social Medicine and Environmental Medicine, Goethe-University, Frankfurt am Main, Germany, **4** European Research Group on Environment and Health in the Transport Sector (EUGT), Berlin, Germany

Abstract

Background: Low Emission Zones (LEZs) are areas where the most polluting vehicles are restricted from entering. The effectiveness of LEZs to lower ambient exposures is under debate. This study focused on LEZs that restricted cars of Euro 1 standard without appropriate retrofitting systems from entering and estimated LEZ effects on NO_2, NO, and NO_x (=NO_2+NO).

Methods: Continuous half-hour and diffuse sampler 4-week average NO_2, NO, and NO_x concentrations measured inside and outside LEZs in 17 German cities of 6 federal states (2005–2009) were analysed as matched quadruplets (two pairs of simultaneously measured index values inside LEZ and reference values outside LEZ, one pair measured before and one after introducing LEZs with time differences that equal multiples of 364 days) by multiple linear and log-linear fixed-effects regression modelling (covariables: e.g., wind velocity, amount of precipitation, height of inversion base, school holidays, truck-free periods). Additionally, the continuous half-hour data was collapsed into 4-week averages and pooled with the diffuse sampler data to perform joint analysis.

Results: More than 3,000,000 quadruplets of continuous measurements (half-hour averages) were identified at 38 index and 45 reference stations. Pooling with diffuse sampler data from 15 index and 10 reference stations lead to more than 4,000 quadruplets for joint analyses of 4-week averages. Mean LEZ effects on NO_2, NO, and NO_x concentrations (reductions) were estimated to be at most -2 µg/m^3 (or -4%). The 4-week averages of NO_2 concentrations at index stations after LEZ introduction were 55 µg/m^3 (median and mean values) or 82 µg/m^3 (95th percentile).

Conclusions: This is the first study investigating comprehensively the effectiveness of LEZs to reduce NO_2, NO, and NO_x concentrations controlling for most relevant potential confounders. Our analyses indicate that there is a statistically significant, but rather small reduction of NO_2, NO, and NO_x concentrations associated with LEZs.

Editor: Qinghua Sun, The Ohio State University, United States of America

Funding: The study was funded by the European Research Group on Environment and Health in the Transport Sector; www.eugt.org. The funder supported data collection and preparation of the manuscript.

Competing Interests: PD Dr. Peter Morfeld and Prof Dr. David Groneberg are members of the scientific advisory group of the European Research Group on Environment and Health in the Transport Sector (EUGT, www.eugt.org). PD Dr. Spallek is managing director of EUGT e.V. EUGT performs and supports research on emissions, immissions and health effects related to the transport sector.

* Email: Peter.Morfeld@evonik.com

Introduction

Low Emission Zones (LEZs) are areas or roads where the most polluting vehicles are restricted from entering. They are currently introduced in 13 European countries [1]. In Europe, vehicle emissions are classified by the so-called "Euro Standards" with a current range from Euro 1 to Euro 6 regarding the technical features of the vehicles which are fixed in several EU-Directives for passenger cars and heavy-duty trucks [e.g., 2]. Basically, this means that vehicles are restricted in relation to their Euro emission level. The configuration of LEZs is extremely different and heterogeneous in Europe, for example in Italy, where the entry standards, the subsistent regulations and the daily duration of LEZ conditions differ substantially from town to town. However, most LEZs in Europe operate 24 hours a day, 365 days a year [see 1].

One of the most developed applications is found in Germany. Low emission zones have been introduced in Germany since 2008 in different stages, resulting in meanwhile 48 LEZs with restrictions for pollutant groups 2 or 3 in 11 Federal states by the end of January 2014 [3]. In this study we analysed the effect of introducing the "LEZ of pollutant group 1" which restricts from entering Diesel cars of an European emission standard below Euro 2 without particulate reduction system and gasoline cars of an European emission standard below Euro 1 without appropriate exhaust gas catalytic converters [3].

Traffic emissions are considered to be a relevant source of air pollution [4] and LEZs are believed to be the most effective measure that cities can take to reduce vehicle-induced air pollution problems in their area [5–7]. The emissions that are aimed to be reduced by LEZs are mainly fine particles like PM10 or smaller [8–12]. The effectiveness of LEZs to reduce traffic-related exposures is still under debate [13] and there is an open discussion in the public about the "outcome" and cost-benefit ratio of LEZs [14–16]. Most of the published information refers to particulate matter.

Additionally, nitrogen dioxide is discussed to be a major traffic-related pollutant as well as an epidemiologic marker of air quality and related adverse health effects [17–21]. On the other hand, a systematic literature review showed only moderate evidence for adverse health effects at a long-term exposure below an annual mean of 40 $\mu g/m^3$ NO2 [22].

According to EU rules [23,24] limits were additionally imposed for NO2 and are enforced in Germany since 2010: 200 $\mu g/m^3$ as an 1 hour average (acceptable: 18 excursions/year) and 40 $\mu g/m^3$ as an annual average. Values were and are in excess: about 69% of all stations near to traffic showed annual averages higher than 40 $\mu g/m^3$ in Germany [7,25]. This non-compliance is not restricted to Germany but the European limit value for NO2 is exceeded in many European cities [26–28]. The LEZ concept was extended and it was assumed that LEZs are an effective measure not only to lower PM10 dust levels but also to reduce NO2 concentrations [6,29]. There are indications that LEZs may indeed reduce NOx concentrations effectively [30–32], but ozone has to be considered a confounder in NO2 measurements [e. g. 33], and the gases NO and NO2 rapidly interconvert, too [34]. Furthermore, national emission ceilings were defined for NOx, i.e., the sum of NO2 and NO [35]. Thus, there is interest in the impact of LEZs on concentrations of NO and NOx also [36].

However, a scientific proof of the LEZ concept targeting at NO_2, NO, and NO_x is still missing. In order to test the views of legislators and researchers that LEZs are effective measures to reduce nitrogen oxide concentrations [29,37], this study focused on the potential effects of LEZs on ambient concentrations of NO_2, NO, and NO_x in LEZ areas of 17 German cities.

We reported on the effect of LEZs on PM_{10} concentrations elsewhere [32].

Methods

Target parameters

The aim of the study was to analyse the effectiveness of German LEZs (as many as eligible) to lower NO_2, NO, and NO_x ($=NO_2+$ NO) concentrations. The first analysis series of NO_2, NO, and NO_x were based on continuous half-hour measurement data of NO_2 and NO. Second, measurement data for NO_2 and NO concentrations collected by diffuse samplers and determined over longer sampling periods were available. These data were allocated to 4-week periods. Third, we collapsed the half-hour measurement data to four-week averages and pooled these collapsed continuous data and the diffuse sampler data to perform joint analyses over 4-week periods. The original NO_2 and NO measurements were performed by the Environmental State Institutions in Germany (Landesumweltämter). A federal data base [38] reports on the applied measurement procedures.

Measuring procedure

Two measuring procedures were applied: continuous measurement devices (chemiluminescence), data stored as half-hour averages and diffuse samplers (Palmes tubes, chromatography), data stored as long-term averages over weeks. The chemiluminescence method relies on the reaction of NO with O3: NO+ O3→NO2*+O2. Chemiluminescence is generated in the range of 600 nm to 3,000 nm when the excited molecules return to the ground state. The light intensity is proportional to the concentration of NO molecules. A deoxidation converter is used to reduce NO2 to NO. Thus, the NO2 concentration is determined as the difference between the NOx concentration measured when the sample gas is directed through a deoxidation converter and the NO concentration measured when the gas is not run through the converter. The diffuse samplers were Palmes type tubes modified with a glass frit as turbulence barrier. In these passive samplers molecules diffuse because of a concentration gradient through an intake opening with a defined cross-section along a fixed diffusion path to a sampling medium by which they are adsorbed. This process is described by Fick's first diffusion law. The chemical analysis is done by chromatography. [More details on both methods may be found in 39,40–43].

Period of investigation

The period of investigation was from 2005 until the end of 2009 (31 December 2009), starting at least from the introduction of the individual LEZ minus the length of the respective LEZ phase (or earlier if restrictions of truck traffic were enforced before the introduction of the LEZ).

Low Emission Zones

There were 34 German active LEZs until the end of 2009 and 774 monitoring stations in use. With introduction of these LEZs, as a main effect, only those diesel vehicles with an exhaust emission standard better than Euro 1 (with sticker) were allowed to enter the zone. In principle, the German "LEZ of pollutant group 1" restricts from entering

- Diesel passenger cars, trucks and buses of an European emission standard below Euro 2 without particulate reduction system, and
- Gasoline passenger cars, trucks and buses of an European emission standard below Euro 1 without appropriate exhaust gas catalytic converters.

Local authorities can set up exception permits especially for light duty vehicles, trucks and buses due to local necessities [3].

According to protocol LEZs were included into the study if and only if

- monitoring stations existed, that operated before and after the LEZ introduction and measured inside the LEZ area (*index stations*) and
- monitoring stations existed, that operated before and after the LEZ introduction and measured outside the LEZ area – in a circle around the centre with a radius of about 25 km – and if

outside the city area, than in no other LEZ (*reference stations*) and

- these monitoring stations measured NO_2 or NO (continuous measurements or diffuse samplers).

(For the terminology and the use of index and reference values in comparisons if exposures levels see Rothman et al. [44])

Seventeen cities with LEZs in 6 German Federal states could be included into the study (Baden-Württemberg: Herrenberg, Ludwigsburg, Mannheim, Reutlingen, Stuttgart, Tübingen; Bavaria: Augsburg, Munich; Berlin: Berlin; Hesse: Frankfurt; Lower Saxony: Hannover; North Rhine-Westphalia: Dortmund, Duisburg, Düsseldorf, Essen, Cologne, Wuppertal). Figure 1 shows all active 34 German active LEZs in December 2009 and the 17 LEZs included for study. File S1 entails maps of all LEZs eligible for study with all index and reference stations marked (Figure S1 in S1 to Figure S19 in S1).

In total, these 17 LEZs, eligible for study, contained 108 eligible monitoring stations with 53 index stations and 55 reference stations. The data base constructed from transferred data encompassed a total of 9,517,911 data lines which were used as input to analysis. An overview is given in Table 1.

Data analysis

The data set structured for analysis consisted of matched quadruplets. A matched quadruplet comprises four pairwise corresponding measurement values consisting of two index- and two reference values. One index value and the simultaneously measured reference value were obtained during the active LEZ period, the other pair of values was obtained before introducing the LEZ. The pairs of values had a 364 days difference in time of or a multiple of 364 days, hence keeping the season, day of the week and time of day constant within the quadruplets. The allocation of reference stations to index stations was done pairwise, i.e., quadruplets were constructed by the data of one index station and allocating to it all appropriate reference stations with their data without a prior collapsing ("collapsing" is a technical term widely used in statistics describing the summary of a table in marginal, http://www.stata.com/manuals13/dcollapse.pdf). The method has been described in detail before [45] and is a refined approach in comparison to other analytical strategies [46]. The analysis plan was critically reviewed by a chair of statistics.

The quadruplets were analysed by the "difference score method in the two period case" [47]: Differences in index values were regressed on differences in reference values while other data were taken into account as covariates in fixed-effect regression analyses. Two types of models were fitted: a linear (additive) model and a log-linear (multiplicative) model. The difference of the index concentration data was used as the response variable in the linear model. The log of this response variable was entered into the log-linear regression model after applying an appropriate positive offset calculated from the data [48]. The two model types differ in the assumption on how covariables may influence the index station concentration data: on an additive scale or on a multiplicative scale [49,50].

The following covariables were taken into account in the basic fixed effects regression analyses: differences at reference stations in $\mu g/m^3$ (to control e.g. for large-scale meteorological changes and seasonal effects), baseline data at reference stations in $\mu g/m^3$ (to control for time-dependent effects of reference data, Allison [47], and baseline data at index stations in $\mu g/m^3$ (to control for "regression to the mean" [51]. This structure defines the basic regression approach. The covariables were entered into the log-linear (multiplicative) models after adding an appropriate offset if indicated [48] and then taking logs.

The following equation describes the analysis of matched quadruples in the basic fixed-effect linear ("additive") regression model [47]

$$\Delta\, x_{mdh} = E + \sum_{k=1}^{Z} E_k \cdot z_k + b_x \cdot x_{0mdh,cent} + b_{\Delta r} \cdot \Delta r_{zdh} + b_r \cdot r_{z0dh,cent} + \varepsilon.$$

$\Delta\, x_{mdh}$ describes the difference of the index station data at monitoring station m between days d and d-364 (= day d+1 in the year before), always at time (hour) h, i.e., x_{1mdh} - x_{0mdh} (compare Figure 2). $x_{0mdh,cent}$ denotes the baseline value at station m on day d at time h, centred at the mean of all baseline values at station m. The terms Δr_{dh} and $r_{z0dh,cent}$ are the corresponding reference value data. The coefficient of major interest is the intercept of the regression model because it estimates the LEZ effect: E measures the mean effect across all LEZs, E+E_k the mean effect in zone k, $1 \leq k \leq Z$. The coefficient b_x accounts for "regression to the mean", $b_{\Delta r}$ for the bias in annual levels (e.g., changed meteorological conditions), b_r for a time-dependent effect of reference values and ε is the residual error of the concentration difference at the index stations. The second model type had the same structure but used logs of the terms ("log-linear", "multiplicative"). An appropriate small offset was added to avoid undefined logarithms [48].

The equation of the basic fixed-effect linear ("additive") regression model can be justified as follows (to keep the notation simple we suppress the time index: we write, eg, x_{z0} instead of x_{z0dh}).

Let us start with assuming an ideal hypothetical situation: measurements are without any distortions and random errors, no covariates are operating. In that case we will measure the index concentration at time point 0 before the LEZ was introduced as a constant value c_0 at all index stations of the LEZ. After the introduction of the LEZ we will measure at time point 1 at the same index station the constant value c_1. The effectiveness of the zone is simply $E = c_1 - c_0$.

But even if there are no biases, random errors and no covariates we do not expect to see the same E for all LEZs. The effectiveness may depend on characteristics of the zone k like the area of the LEZ, A_k (e.g., we may expect a larger effect if the LEZ area is larger). The concentration at time point 1 may be written more appropriately as $c_1 + f \times A_k$ (with a multiplicative coefficient f mapping the effect of the area into the concentration scale). The effect of zone k can be described as $E = c_1 + f \times A_k - c_0$. Note that A_k operates as an effect modifier.

We can take account of differences between the zones without referring to a specific characteristic of LEZ k, like the area: We may describe the effect of zone k in more abstract terms as E+E_k [E+E_k means E+$E_k \times z_k$ with a multiplicative indicator z_k, that takes the value 1 for zone k and zero otherwise, $1 \leq k \leq Z$]. E_k ist the specific effect offset of LEZ k in comparison to the overall mean E of the LEZ effects. It is simple to extend the notation to cover different baseline concentrations for the different LEZs. Thus, $E + E_k = c_1 + (c_{z1} - c_1) - [c_0 + (c_{z0} - c_0)] = c_{z1} - c_{z0}$, i.e., the effect of zone k is the difference between the zone specific measurement values after (c_{z1}) and before (c_{z0}) the introduction of the LEZ k (c_1 and c_0 now denote the averages of the concentrations across all LEZs under study).

Still, the approach is not very realistic. We should take into account background variations of the intensities, resulting from e.g. large area changes of the concentrations. These large area variations are reflected in the values r_{z0} and r_{z1} at the reference

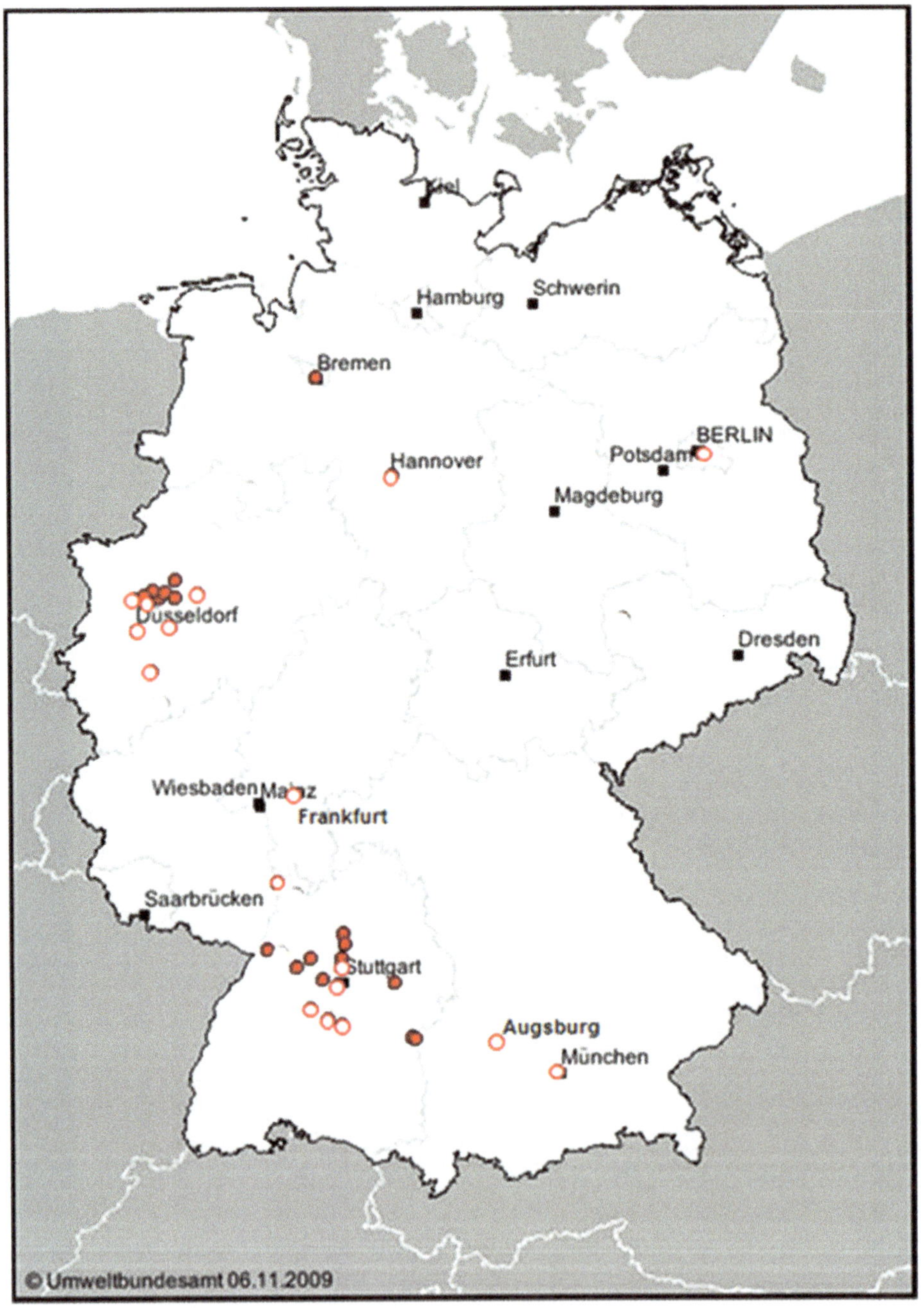

Figure 1. Active and investigated LEZs in Germany, December 2009. The 17 LEZs included into the study are marked with open red circles. The 17 LEZs that were active but were excluded according to protocol are indicated by full red circles. Capital cities of the Federal states are shown by black squares. This is a modification of the map as published at URL www.umweltbundesamt.de/umweltzonen. Date of access: 6th November 2009.

stations. Despite all efforts to measure the concentrations as precisely as possible we always will have random errors ε_1 and ε_0. In this extended approach the measurement values for zone k before introducing the LEZ are $x_{z0}=c_0+(c_{z0}-c_0)+g\times r_{z0}+\varepsilon_0$ and $x_{z1}=c_1+(c_{z1}-c_1)+g\times r_{z1}+\varepsilon_1$ after the introduction. The factor g measures how strong the reference values do influence the index values. It follows that $x_{z1}-x_{z0}=E+E_k+g^*\,(r_{z1}-r_{z0})+(\varepsilon_1-\varepsilon_0)$. With $\Delta\, x_z=x_{z1}-x_{z0}$, $\Delta\, r_z=r_{z1}-r_{z0}$, $\varepsilon=\varepsilon_1-\varepsilon_0$ and $b_{\Delta r}=g$ we yield the major part of the equation of the basic fixed-effect linear ("additive") regression model. Note that we substituted z by m which means that we apply the approach in a refined way to every index monitoring station m. We have demonstrated above that a potential confounder/adjuster, like the concentration at a reference station, enters the equation in terms of the difference of the values across time (e.g., $\Delta\, r_z=r_{z1}-r_{z0}$). And we have seen that the model can be extended by potential modifiers of the LEZ effect ("interaction terms"), like the area, by adding terms like $f\times A_k$ (in contrast to adjusters not as a difference in time). We will now explain in more detail why we included the effect modifying variables x_{z0} and r_{z0} additionally.

Altman and Bland [52] and Bland and Altman [53] suggested including the mean value x_{zm} of the concentrations at index station

Table 1. Overview of low emission zones (LEZs) in German cities included for analysis: 17 LEZs in 6 federal states.

LEZ				Measurement Stations Suitable For Analysis									
		Data Since	Introduction of LEZ	NO_2		NO		NO_x		Area LEZ/km²	Area City/km²	Population LEZ	Population City
Federal State	City			Index	Reference	Index	Reference	Index	Reference			2007	2011
Baden-Württemberg[a]	Herrenberg	2006	01Jan2009	1	1[d]	1	1[d]	1	1[d]	17	64.71	15000	29935
	Ludwigsburg	2005	01Mar2008	2	1[d]	2	1[d]	2	1[d]	30	43.33	55000	86939
	Mannheim	2005	01Mar2008	2	2	2	2	2	2	7.5	144.96	93900	291458
	Reutlingen	2004	01Mar2008	1	2[d]	1	2[d]	1	2[d]	4.5	87.06	24500	110084
	Stuttgart	2004	01Mar2008	5	2[d]	5	2[d]	5	2[d]	207	207	590000	591015
	Tübingen	2005	01Mar2008	1	1[d]	1	1[d]	1	1[d]	11.9	108.12	64000	83248
Total Baden-W[a]				12	9	12	9	12	9	277.9	655.18	842400	1192679
Bavaria[a]	Augsburg	2007	01Jul2009	3	1	3	1	3	1	5.2	146.93	40000	272699
	Munich	2006	01Oct2008	5	3	5	3	5	3	44	310.71	431000	1388308
Total Bavaria[a]				8	4	8	4	8	4	49.2	457.64	471000	1661007
Berlin "BLUME"[a]	Berlin	2004	01Jan2008	5	9	5	9	5	9	88	891.85	1100000	3375222
Hesse[a]	Frankfurt a.M.	2003	01Oct2008	1	3	1	3	1	3	110	248.31	550000	687775
Lower Saxony[a]	Hannover	2005	01Jan2008	1	4	1	4	1	4	50	204.14	218000	514137
North Rhine-Westphalia[a]	Dortmund	2006	01Jan2008	2	4[d]	2	4[d]	2	4[d]	19.1	280.71	ca. 158000	572087
	Duisburg	2006	01Oct2008	2	1	2	1	2	1	ca. 50	232.83	ca. 376500	486816
	Düsseldorf	2006	15Feb2009	1	3	1	3	1	3	13.8	217.41	36500	593682
	Essen	2006	01Oct2008	3	2[d]	3	2[d]	3	2[d]	84	210.34	ca. 291000	566862
	Cologne	2006	01Jan2008	2	4	2	4	2	4	16	405.17	130000	1023373
	Wuppertal	2006	01Feb2009	1	2[d]	1	2[d]	1	2[d]	35.3	168.39	194000	342885
Total North Rhine-W[a]				11	16	11	16	11	16	218.2	1514.85	1186000	3585705
Total (contin) [a]				38	45	38	45	38	45	793.3	3971.97	4367400	11016525
Berlin "RUBIS"[b]	Berlin	2004	01Jan2008	8	9	8	9	8	9	88	891.85	1100000	3375222
North Rhine-Westphalia[b]	Essen	2006	01Oct 2008	7	1	-	-	-	-	84	210.34	ca.291000	566862
Total (diffuse)[b]				15	10	8	9	8	9	172	1102.19	1391000	3942084
Total (overall)[c]				53	55	46	54	46	54	965.3	5074.16	5758400	14958609

Number of index and reference stations per gaseous component (NO_2, NO, NO_x) suitable for analysis of the first tier of LEZs until Dec 31, 2009.
[a]continuous measurements, half-hour concentration averages.
[b]diffuse samplers, longer sampling periods (over weeks).
[c]all data combined (continuous, diffuse), adjusted to concentration averages over 4-weeks periods.
[d]reference station used more than once in comparisons to index stations.

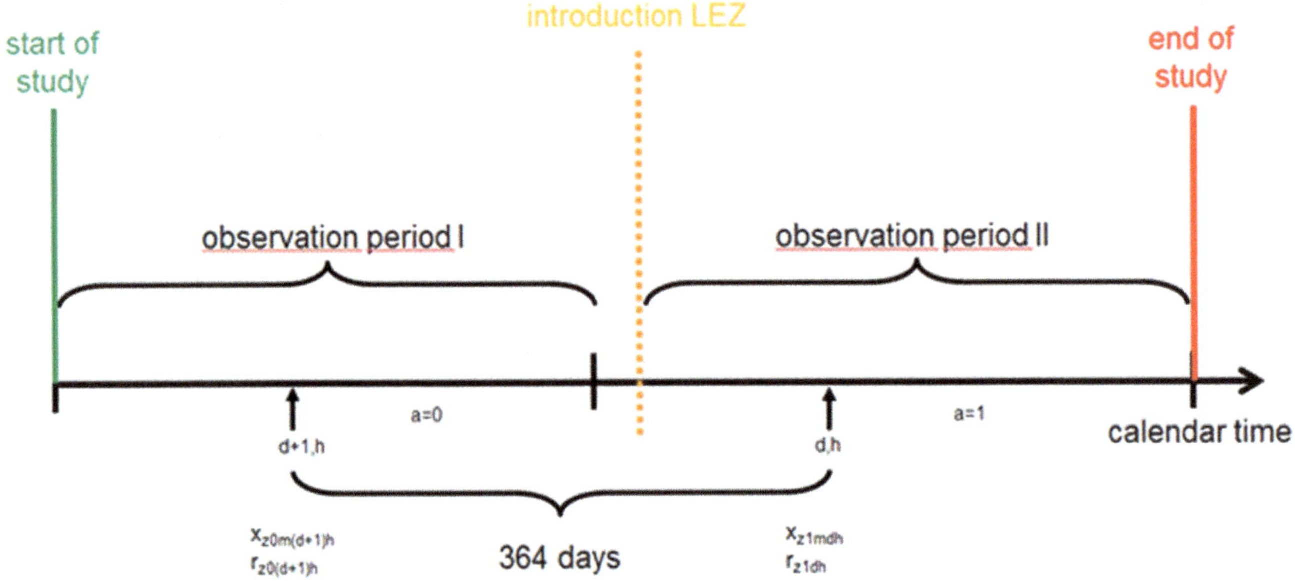

Figure 2. Index (x_{zamdh}) and reference concentration (r_{zadh}) at index measurement station m and LZ z in observation period II with active LEZ (a = 1) and in observation period I with inactive LEZ (a = 0): matched quadruplets consisting of two index measurement values and two reference measurement values. The time difference between compared measurement values at day d of the year in period 2 and d+1 of the year in period 1, always with starting time h, is not a full year but 364 days to keep the weekday constant.

m as another covariate: this allows the difference Δx_{zm} to depend on the average concentration at the station. This inclusion of x_{zm} operates again a distortion due to "regression to the mean" [54]. This phenomenon is inevitably complicating longitudinal comparisons. Baseline values that are very high due to random errors will probably not be reproduced but lower values will be measured, and this is so even if the null hypotheses of no effect is true [51,55–57]. A better strategy to correct for this potential distortion is to include $x_{z0m,cent}$, i.e., the baseline values at the index station [58,59]. Including additionally $r_{z0,\ cent}$ was exercised in Allison [47], p. 10. This approach allows for a flexible adjustment of the annual level bias because we get rid of the assumption of a time-invariant effect of the reference station values on the index stations values. The covariates $x_{z0m,\ cent}$ and $r_{z0,\ cent}$ are centered on the mean of the values of each measuring station so that the terms E und E_k can be interpreted without further transformations.

Since the impact of meteorological conditions is extremely relevant [e.g., 60], the following data were collected to be used in addition to the reference station data to control for distortions due to meteorological changes. We took over the height of the inversion base H in m, the wind velocity V in m/s, the amount of precipitation P in mm/h from the PAREST project for all investigated measurement stations and half hours in the follow-up period [61–65].

We extended the basic regression models to the regression model 1 approach by adjusting additionally for the change of the three meteorological variables at the index stations According to the box model of meteorology [66] the differences were calculated on the additive scale after transforming the variables into 1/H, 1/(V+0.1 m/s), and 1/(P+0.1 mm/h). The smallest unit of scale was 0.1 throughout, hence this value was used as an offset to avoid divisions by zero [48]. Differences were determined on the multiplicative scale after taking logs of these terms [48]. We adjusted for the time span (in years) between measurements considered within a quadruplet in order to adjust for trends in concentration levels before the LEZ was introduced. In multiplicative models log of the time span was used after applying an appropriate offset [48].

In regression model 2 approach, the following time-dependent binary indicators were additionally adjusted for: period of school holidays (yes/no), period of environmental bonus paid (yes/no) and periods when trucks were not allowed to enter the area where the measurement station was located (yes/no). In Germany a bonus was paid to car owners between January 14, 2009 and November 2, 2009 if they bought a new car with a reduced exhaust emission (http://www.bafa.de/bafa/de/wirtschaftsfoerderung/umweltpraemie/index.html). These binary indicators were entered also into the extended log-linear (multiplicative) models.

This statistical approach was successfully validated in advance to the study in an analysis of simulated data from FU Berlin [67]. The simulated data was produced by the PAREST project [61,63, www.parest.de]. The major aim of this project was the identification of emission reducing strategies by simulation. Transport and distribution models were developed and applied, the so called REM-CALGRID approach [68–70]. The model was applied to the city of Munich, and simulated half-hour PM_{10} data were generated for each of the five index and three reference stations (see Figure S10 in File S1). Data of the year 2005 were simulated twice, with and without adding an LEZ effect (the value of the imprinted effect was unknown to the analyzing working group). 280,320 data lines were transferred. The simulated PM_{10} concentrations were analyzed and showed a mean value of about 21 $\mu g/m^3$ at the index stations and 18 $\mu g/m^3$ at the reference stations. The basic additive regression model estimated an LEZ effect of $-0.130\ \mu g/m^3$, the multiplicative model a relative change of -0.7%. The PAREST research report [63, www.parest.de] described the LEZ effect that was imprinted: PM_{10} mean values are reduced in Munich city by at most 0.2 $\mu g/m^3$ or at most 1%.

Additive and multiplicative regression models were fitted to subsets of the data to perform sensitivity analyses: continuous measurement data, continuous measurement data collapsed to four-week averages, diffuse sampler data, pooled diffuse sampler and collapsed continuous data, always with and without excluding times with restrictions of truck traffic; quadruplets produced by index traffic stations only. The pooled continuous and diffuse sampler measurement data determined for four weeks periods was of major interest in this study because the annual average is the most critical endpoint to consider (see Introduction section) and these data cover both types of measurement data. Because annual data are generally too coarse for LEZ effect estimation, we followed-up on averages over about a month. The additive and multiplicative regression models analysing these sets of data were specified with three sets of covariables as described above. All basic models, the models evaluating continuous data collapsed to 4-week averages and the models fitted to single index stations were not used for statistical testing of the LEZ effects. Tests for effects measured by NO_2, NO, and NO_x quadruplets were not considered as independent. According to this structure, we evaluated 2*3*2*2*2 = 48 statistical tests for each of the three endpoints. Due to this multiple testing scenario we applied an adapted significance level of 5%/50 = 0.1% ["family wise error rate", 71].

We fitted additionally explorative models that estimated the size of the LEZ effect at each index station enrolled. In addition, we estimated mean effects of the LEZs across the Federal states. The results of these exploratory analyses were mainly used for internal discussions of the project steering committee (see Acknowledgement).

All regression models used robust estimators of coefficient variances. All data analyses were performed using Stata 11 [72] on a 64-bit PC.

Results

NO_2 - continuous measurements

The basic data consisted of 6,412,864 data lines leading to 3,038,781 quadruplets of continuous NO_2 measurement (half-hour averages) from 6 Federal states and 17 LEZs with 38 index stations and 45 reference stations. Table 2 gives an overview of the distributions observed: on average, NO_2 concentrations were between 50 $\mu g/m^3$ and 52 $\mu g/m^3$ at the index stations and between 26 $\mu g/m^3$ and 27 $\mu g/m^3$ at the reference stations. The differences at the stations varied substantially in a range of hundreds of $\mu g/m^3$ upwards and downwards. A comparison of mean and median differences at index and reference stations indicated a crude LEZ effect estimate of about -1 $\mu g/m^3$. In the linear model 1 the absolute effect estimate was similar: -1.11 $\mu g/m^3$ (Table 3). The model 1 results showed a time-dependent impact of reference station data, a pronounced "regression to the mean", a clear influence of the three meteorological variables (independently from the crude adjustment by reference station data), and a downward trend of concentrations before the LEZs were introduced. The direction of impact of the meteorological variables was as expected: the smaller H, V, or P the larger the index NO2 concentrations. In linear model 2 the LEZ effect estimate was slightly more pronounced: -1.85 μg/m3. In the log-linear model 1 (multiplicative approach), the relative effect estimate was 0.979, i.e., a reduction of 2.1% was found (Table 4). The estimated impact of covariables agreed with the finding in the corresponding linear model. When applying regression model 2 the relative LEZ effect estimate was 0.961, i.e., the reduction was estimated to be 3.9%.

NO_2 - pooled continuous and diffuse sampler measurement data

6,133 data lines and 4,095 quadruplets of NO_2 pooled continuous and diffuse sampler measurement data (averaging period: four weeks) were examined from 17 LEZs with 53 index stations and 55 reference stations. A crude comparison based on the observed distributions revealed a LEZ effect of about -0.2 $\mu g/m^3$ to -0.6 $\mu g/m^3$ (Table 5). Using the linear model 1 approach the absolute effect estimate was -0.826 $\mu g/m^3$ (Table 6). The meteorological variables showed no substantial impact due to the long averaging period. Model 2 estimated the LEZ effect as -1.73 $\mu g/m^3$. The log-linear modelling led to a relative effect of 0.980 (Table 7, model 1) or 0.961 (model 2). Table S1 in File S1 provides a detailed overview of the results when fitting a series of models to analyse the NO_2 measurements. LEZ effect estimates were about -1 $\mu g/m^3$ to -2 $\mu g/m^3$ (additive models) or -2% to -4% (multiplicative models).

NO - pooled continuous and diffuse sampler measurement data

A total of 5,790 data lines from 17 LEZs with 46 index stations and 54 reference stations were available to analyse pooled continuous and diffuse sampler NO measurement data. A descriptive analysis of the 4,005 quadruplets indicated a LEZ effect of about 0 $\mu g/m^3$ to -1 $\mu g/m^3$ (Table 8). Using the additive approach the absolute effect estimate was -1.13 $\mu g/m^3$ in model 1 (Table 9). When the model specification 2 was applied the LEZ effect estimate changed the sign: +0.38 $\mu g/m^3$, i.e., no reduction was indicated in this extended model type. The log-linear regression model of type 1 yielded a relative effect estimate of 0.968 (Table S2 in File S1). The direction of the estimated relative effect changed when model 2 was applied: +1.20.

NO_x - pooled continuous and diffuse sampler measurement data

The analysis of pooled continuous and diffuse sampler NO_x measurement data was performed using 4,005 quadruplets that originated from a set of 5,790 data lines generated by 46 index stations and 54 reference stations of 17 LEZs. According to the distributions of differences a crudely estimated LEZ effect (based on averages or medians) was present of about -0.2 $\mu g/m^3$ to -1.3 $\mu g/m^3$ (Table S3 in File S1). Adjusting for covariables in linear model 1 returned an absolute effect estimate of -1.74 $\mu g/m^3$ (Table S4 in File S1). The adjustment for further covariables (regression model 2) led to an effect estimate of -0.89 $\mu g/m^3$. When the log-linear model 1 was used (Table S5 in File S1), a relative LEZ effect of 0.976 was found. The adjustment for additional covariates (model 2) led to a change in direction: the relative effect was estimated as 1.048.

Summary of Results for NO_2, NO, and NO_x

Table 10 gives an overview of the findings for NO_2, NO, and NO_x. The mean concentration levels at the index stations were about 50 $\mu g/m^3$ for NO_2 and for NO, thus, about 100 $\mu g/m^3$ for NO_x. Model 1 analyses showed reductions of the concentrations after introducing the LEZs. Although small, all effect estimates were statistically significant at the 0.1% level. Model 1 estimates based on an additive structure gave compatible findings to the log-linear multiplicative approach (e.g., 2% of 50 $\mu g/m^3$ = 1 $\mu g/m^3$). The model 1 LEZ effect estimates were similar to, but slightly more pronounced than crude LEZ effect estimates based on direct comparisons of the measurement differences at index stations and reference stations within the quadruplets while ignoring the impact

Table 2. NO_2: Quadruplets of continuous NO_2-measurements: index stations (Ind), reference stations (Ref) before (pre) and after (post) introduction of LEZ.

Statistic	Ind,pre	Ind,post	Ref,pre	Ref,post	Ind.diff.	Ref.diff
N	3038781	3038781	3038781	3038781	3038781	3038781
min	0.4	1.3	0.4	0.5	−330	−215
p5	12	11	4.0	4.0	−52	−33
p50	45.0	43.8	20.7	20.6	−1.0	0.0
mean	51.959	50.831	26.383	26.17	−1.128	−0.212
p95	115	114	68	67	49	33
max	392	436	248	434	375	317

Ind.diff and Ref.diff denote differences between index measurements and between reference measurements (negative post-pre differences indicate lower values after introduction of LEZ). Concentrations measured in $\mu g/m^3$.
N: number of quadruplets; Min: minimum, p5: 5th percentile, p50: median, mean: arithmetic average, p95: 95th percentile, max: maximum, unit:

of covariables (compare Tables 2, 5, and 8 and Table S1 in S1). All analyses point to the conclusion that on average the concentration reducing effect of LEZs was smaller than 2 $\mu g/m^3$ for each of the three components NO_2, NO, and NO_x, i.e., not higher than about 4%, when considering all investigated index stations. However, breaking down the analyses by Federal states or LEZs yielded heterogeneous estimates of effects.

The NO_2 analysis was based on 192 comparisons of index vs reference stations, among them were 31 index stations characterized as "background", one characterized as "industry" and 160 as "traffic" stations. We performed a sensitivity analysis by restricting the evaluation to the stations close to traffic. The additive linear type 2 model estimated an effect of -1.73 $\mu g/m^3$ at all index stations (see last line in Table S1 in S1). When the analysis only accounted for the traffic stations we got a slightly more pronounced LEZ effect estimate of -2.26 $\mu g/m^3$ (3,406 quadruplets, pooled data: four week averages). An analysis of the continuous data yielded almost the same result: -2.35 $\mu g/m^3$ (2,105,702 quadruplets, half-hour averages).

Discussion

In this study we analysed the effect of introducing the "Low Emission Zone (LEZ) of pollutant group 1" (which restricts from entering Diesel cars of an European emission standard below Euro 2 without particulate reduction) on NO_2, NO, and NO_x concentrations in Germany. We included as many LEZs as possible (17 out of 34 in 2009 met our inclusion criteria) into a homogeneous analysis of nitrogen oxide data measured before and after the introduction of LEZs of pollutant group 1 until the end of 2009. We used matched quadruplets of index and reference station values and analysed the changes in concentrations with fixed-effect regression models while adjusting for important covariables. We performed sensitivity analyses by applying two model structures (additive and multiplicative) with varying sets of covariables to different subsets of the data. We based our study on precisely matched quadruplets to avoid distortions and to increase validity. A potential downside of the increased validity is a loss in precision due to the reduced data set eligible for analysis. However, the loss in power was negligible in this application because P-values were small even when taking multiple testing into account [73]. The statistical approach was successfully validated in advance to the study in an analysis of simulated data from FU Berlin [67]. We checked whether the adjustment in one model that analyzed all LEZs simultaneously and assumed unknown but identical covariate coefficients was appropriate for all LEZs. To do so we evaluated each LEZ separately and performed a meta-analysis on the findings. The precision weighted mean of the effect estimates

Table 3. NO_2: Linear (additive) model 1 evaluating the quadruplets of continuous NO_2-measurements.

Ind.diff	Coef.	Std. Err.	t	p	95% Conf.	Interval
Ref.diff	0.677	0.001	623	<0.001	0.675	0.679
Ref.base	0.509	0.001	417	<0.001	0.507	0.512
Ind.base	−0.644	0.001	−845	<0.001	−0.645	−0.642
Diff 1/H	564	2.8	200	<0.001	558	569
Diff 1/V	3.12	0.025	126	<0.001	3.07	3.17
Diff 1/P	0.088	0.004	23.6	<0.001	0.081	0.095
Time.diff	−0.399	0.023	−17.7	<0.001	−0.443	−0.354
E	−1.112	0.013	−87.3	<0.001	−1.137	−1.087

Regression coefficient, robust standard errors of coefficient, t-statistic, two-sided P-value, and 95%-confidence interval of coefficient. The absolute LEZ effect estimate is given by the coefficient E in $\mu g/m^3$ (<0: concentration is lowered by LEZ).
Covariables: difference in reference stations Ref.diff in $\mu g/m^3$, centered reference baseline concentration Ref.base in $\mu g/m^3$, centered index baseline concentration Ind.base in $\mu g/m^3$, Diff 1/H = difference in 1/(height of inversion layer) in 1/m, Diff 1/V = 1/(wind velocity+0.01) in $(m/s)^{-1}$, difference in 1/P = 1/(amount of precipitation+ 0.01) $(mm/h)^{-1}$, centered difference in time Time.diff in years.

Table 4. NO_2: Log-linear (multiplicative) model 1 evaluating the quadruplets of continuous NO_2-measurements.

ln Ind.diff	Coef.	Std. Err.	t	p	95% Conf.	Interval
ln Ref.diff	0.296	0.001	582	<0.001	0.295	0.297
ln Ref.base	0.265	0.002	141	<0.001	0.262	0.269
ln Ind.base	−0.981	0.002	−412	<0.001	−0.986	−0.977
Diff ln 1/H	0.071	0.001	194	<0.001	0.070	0.071
Diff ln 1/V	0.12	0.001	301	<0.001	0.119	0.121
Diff ln 1/P	0.014	0.001	29	<0.001	0.013	0.015
ln Time.diff	−0.025	0.001	−32	<0.001	−0.027	−0.024
ln E	−0.021	0.0	−77	<0.001	−0.022	−0.021

Regression coefficient, robust standard errors of coefficient, t-statistic, two-sided P-value, and 95%-confidence interval of coefficient. The relative LEZ effect estimate is given by the coefficient E (<1: concentration is lowered by LEZ).
relative effect E = 0.979, 95% Conf. Interval = 0.979, 0.980.
Covariables, before taking logs: difference in reference stations Ref.diff in $\mu g/m^3$, centered reference baseline concentration Ref.base in $\mu g/m^3$, centered index baseline concentration Ind.base in $\mu g/m^3$, 1/H = 1/(height of inversion layer) in 1/m, 1/V = 1/(wind velocity+0.01) in $(m/s)^{-1}$, 1/P = 1/(amount of precipitation+0.01) $(mm/h)^{-1}$, centered difference in time Time.diff in years.

at all index stations (n = 192, −1.71 $\mu g/m^3$) was almost identical to the overall additive linear type 2 model (−1.73 $\mu g/m^3$, see last line in Table S1 in S1). We conclude that the fitted single model that evaluated all LEZs simultaneously was appropriate and did not suffer from an insufficient adjustment.

As an overall finding the average effect of LEZ introduction on nitrogen oxide concentrations (NO_2, NO, and $NO_x = NO_2$+NO) was not higher than 2 $\mu g/m^3$ at all index stations, i.e., not higher than about 4%. The effect was only slightly larger when we restricted the analyses to stations close to traffic. In the main analyses the coefficients describing the reductions were statistically significant on the 0.1% level, i.e., after taking multiple testing into account. We note, however, that the P-values calculated are potentially too small because autocorrelations in the data were not taken into account.

We detected a substantial heterogeneity of effects across the investigated LEZs and Federal states. However, this finding is not surprising because

- the realisation of LEZs differed between states and within states (e.g., date of introduction, covered population and area of LEZs differ (compare Table 1), some operate together with an additional restriction of van traffic)
- the degree of representativeness of monitoring stations inside the LEZs differs across LEZs (index stations: distances from centre/border of LEZ differ, used as background or hot spot stations and sometimes placed in street canyons)
- the degree of representativeness of monitoring stations outside the LEZs differs across LEZs (reference stations: distances from LEZ differ, traffic conditions differ)
- the applied measuring systems differ (continuous chemiluminescense procedure vs diffuse long-term sampling with chromatography).

The large variation of LEZ effect estimates across the LEZs should be put into perspective by considering the phenomenon of "regression-to-the-mean" [51]. Due to this phenomenon we expect that single observations with high baseline values show potentially decreasing trends – and low baseline values potentially increasing trends. This is true even under the null hypothesis of no causal LEZ effects on nitrogen oxide concentrations. "Regression-to-the-mean" has been shown to be rather pronounced in this study. Thus, the interpretation of single LEZs effect estimates is clearly limited and we will not report any details with the consent of the involved state institutions who performed the measurements (see Acknowledgements).

Table 5. NO_2: Quadruplets of pooled continuous and diffuse sampler NO_2-measurements: index stations (Ind), reference stations (Ref) before (pre) and after (post) introduction of LEZ.

Statistic	Ind,pre	Ind,post	Ref,pre	Ref,post	Ind.diff.	Ref.diff
N	4095	4095	4095	4095	4095	4095
min	13	11	6	5	−32	−23
p5	26	24	11	10	−14	−13
p50	57.31	54.23	41.04	39.78	−2.34	−1.71
mean	56.298	54.246	40.007	38.18	−2.053	−1.824
p95	84	82	66	64	12	8
max	134	136	75	76	41	25

Ind.diff and Ref.diff denote differences between index measurements and between reference measurements (negative post-pre differences indicate lower values after introduction of LEZ). Concentrations measured in $\mu g/m^3$.
N: number of quadruplets, Min: minimum, p5: 5th percentile, p50: median, mean: arithmetic average, p95: 95th percentile, max: maximum.

Table 6. NO_2: Linear (additive) model 1 evaluating the quadruplets of pooled continuous and diffuse sampler NO_2-measurements.

Ind.diff	Coef.	Std. Err.	t	p	95% Conf.	Interval
Ref.diff	0.564	0.018	31.7	<0.001	0.529	0.599
Ref.base	0.394	0.025	15.9	<0.001	0.345	0.442
Ind.base	−0.695	0.019	−36.7	<0.001	−0.732	−0.657
Diff 1/H	−65.1	20.8	−3.13	0.002	−106	−24.3
Diff 1/V	0.034	0.148	0.23	0.816	−0.256	0.325
Diff 1/P	0.009	0.028	0.30	0.764	−0.047	0.064
Time.diff	0.369	0.189	1.95	0.051	−0.001	0.739
E	−0.826	0.123	−6.71	<0.001	−1.068	−0.585

Regression coefficient, robust standard errors of coefficient, t-statistic, two-sided P-value, and 95%-confidence interval of coefficient. The absolute LEZ effect estimate is given by the coefficient E in μg/m^3 (<0: concentration is lowered by LEZ).
Covariables: difference in reference stations Ref.diff in μg/m^3, centered reference baseline concentration Ref.base in μg/m^3, centered index baseline concentration Ind.base in μg/m^3, Diff 1/H = difference in 1/(height of inversion layer) in 1/m, Diff 1/V = 1/(wind velocity+0.01) in (m/s)$^{-1}$, difference in 1/P = 1/(amount of precipitation+ 0.01) (mm/h)$^{-1}$, centered difference in time Time.diff in years.

Models of type 2 showed more instability and returned positive effect estimates in some situations (see Table 10). Regression model 2 included as additional variables time-dependent binary indicators for period of school holidays, period of environmental bonus paid and periods when trucks were not allowed to enter the area where the measurement station was located. In some LEZs these variables were highly correlated with the active LEZ periods so that unstable findings due to collinearities can be expected. Such collinearities can introduce a bias away from the null and may generate exaggerated negative or positive model coefficients even if the true effects are near to zero [74]. Log-linear models showed to be more sensitive to these distortions. This may indicate a less appropriate modelling of the data when assuming multiplicative effects of covariates.

There are evaluations available concerning potential effects of LEZs on NO_2 concentrations summarized by the German Federal Environmental Agency [7]: A total NO_2 reduction by 5% and a local traffic-related NO_2 reduction by 12% may be reached given "LEZ of pollutant group 3" so that only cars with a green sticker (Diesel vehicles of Euro 6, 5, 4 or Euro 3 with particle filter, gasoline cars with catalytic converter) are allowed to enter the LEZ [3]. This statement is based mainly on preliminary evaluations of the Berlin LEZ data by Rauterberg-Wulff and Lutz [29]. Puls and Jäger-Ambrozewicz [75] reported for the Frankfurt LEZ and an observation period until the end of 2011 effects of less than 3% which is closer to our present findings although they also cover a period of "LEZ of pollutant group 2" after Jan1, 2010. Only cars with a yellow sticker (Diesel vehicles of Euro 3 or 4 standard or Euro 2 with particle filter, gasoline cars with catalytic converter) were allowed to enter the Frankfurt LEZ after Jan 1, 2010 [3]. Bruckmann et al. [6] reported reductions of the annual average of NO_2 concentrations up to 2% associated with the introduction of "LEZs of pollutant group 1" in North-Rhine Westphalia, and an absolute LEZ effect of about -1.2 μg/m^3. In Hannover no NO_2 reduction could be shown after introducing an "LEZ of pollutant group 1" [76]. All of these statements, however, were based on crude comparisons without sufficiently adjusting for important covariates like weather conditions, and traffic restrictions etc. Only Puls and Jäger-Ambrozewicz [75] applied a more sophisticated approach. They performed a time-series analysis and fitted

Table 7. NO_2: Log-linear (multiplicative) model 1 evaluating the quadruplets of pooled continuous and diffuse sampler NO_2-measurements.

ln Ind.diff	Coef.	Std. Err.	t	p	95% Conf.	Interval
ln Ref.diff	0.359	0.015	23.9	<0.001	0.329	0.388
ln Ref.base	0.024	0.005	4.49	<0.001	0.014	0.034
ln Ind.base	−0.039	0.004	−10.1	<0.001	−0.047	−0.032
Diff ln 1/H	−0.016	0.003	−5.32	<0.001	−0.022	−0.01
Diff ln 1/V	0.013	0.003	4.21	<0.001	0.007	0.02
Diff ln 1/P	0.011	0.003	3.68	<0.001	0.005	0.017
ln Time.diff	0.005	0.006	0.95	0.34	−0.006	0.017
ln E	−0.02	0.003	−7.82	<0.001	−0.025	−0.015

Regression coefficient, robust standard errors of coefficient, t-statistic, two-sided P-value, and 95%-confidence interval of coefficient. The relative LEZ effect estimate is given by the coefficient E (<1: concentration is lowered by LEZ).
relative effect E = 0.980, 95% Conf. Interval = 0.975, 0.985.
Covariables, before taking logs: difference in reference stations Ref.diff in μg/m^3, centered reference baseline concentration Ref.base in μg/m^3, centered index baseline concentration Ind.base in μg/m^3, 1/H = 1/(height of inversion layer) in 1/m, 1/V = 1/(wind velocity+0.01) in (m/s)$^{-1}$, 1/P = 1/(amount of precipitation+0.01) (mm/h)$^{-1}$, centered difference in time Time.diff in years.

Table 8. NO: Quadruplets of pooled continuous and diffuse sampler NO-measurements: index stations (Ind), reference stations (Ref) before (pre) and after (post) introduction of LEZ.

Statistic	Ind,pre	Ind,post	Ref,pre	Ref,post	Ind.diff.	Ref.diff
N	4005	4005	4005	4005	4005	4005
min	2.7	2.5	1.0	1.0	−86	−50
p5	4.6	4.4	2.2	2.2	−22	−21
p50	48.54	43.44	31.83	26.32	−2.59	−1.26
mean	49.479	46.373	34.153	31.025	−3.105	−3.129
p95	95	95	86	81	17	13
max	230	251	139	104	77	56

Ind.diff and Ref.diff denote differences between index measurements and between reference measurements (negative post-pre differences indicate lower values after introduction of LEZ). Concentrations measured in $\mu g/m^3$.
N: number of quadruplets, Min: minimum, p5: 5th percentile, p50: median, mean: arithmetic average, p95: 95th percentile, max: maximum.

regression models for the Frankfurt LEZ. These models', however, were not correctly specified as they did not include differences of the covariables but the absolute values only, and so they could not control for potential confounding effects although this was intended by the authors. All publications cited above reported only on individual LEZs or certain Federal states in Germany and not on the LEZ effect on the national level. Generalisations from these data are problematic because of the heterogeneous configurations of LEZs. A realistic estimate should be based on a homogeneous analytical approach covering as many LEZs and Federal states simultaneously as possible, as performed in this study.

Table 11 presents an overview of other study results published in the peer-reviewed literature on forecasted or measured LEZ effects on NO_2 concentrations.

Our results are in good accordance with the prognosis study PAREST of FU Berlin [61,63]. An extensive description of the project is available [77]. The prognoses of PAREST are comparable with our estimates at all index stations because PAREST worked with an area coarseness defined by grid square of about 1 km×1 km and, thus, cannot estimate changes at single stations. Duyzer et al. [78] studied whether monitoring station data are representative for the population living in the area and concluded that the background station data are more appropriate to describe the impact on the citizens than the hot spot traffic stations. We conclude that the findings of PAREST and our results about the effect at all index stations should be preferred in an evaluation (not the effect estimates restricted to the traffic stations). PAREST predicted LEZ effects on NO_2 levels assuming that only cars with green stickers are allowed to enter (LEZ of pollutant level 3). For the Berlin LEZ the authors calculated a reduction of about 1 $\mu g/m^3$ to 1.3 $\mu g/m^3$ in the city centre (relative: 3% to 5%), for the Munich LEZ a reduction of 1 $\mu g/m^3$ in the city centre (relative: up to 5%), for the Ruhr area a reduction of 1 $\mu g/m^3$ to 1.7 $\mu g/m^3$ (relative: 3% to 4%). Setting the whole Ruhr area to a LEZ of pollutant level 3 lead to the prognosis of a reduction in NO_2 concentrations of 1 $\mu g/m^3$ to 2 $\mu g/m^3$ (relative: 3% to 6%). It needs to be taken into account that these prognoses by PAREST are based on the pollutant level 3 LEZ scenario. We do not expect, therefore, that our findings from this study may change relevantly if the LEZs are extended to cover larger areas or if stricter traffic restrictions are applied.

A very large LEZ was introduced in London as a congestion charging zone. However, only prognoses of the potential LEZ effect on nitrogen oxide concentrations are available. NO_x reductions between 3.8% in 2008 up to 7.3% in 2012 along roadways were predicted in a modelling scenario for the London LEZ with vehicles and buses required to meet Euro 4 standards

Table 9. NO: Linear (additive) model 1 evaluating the quadruplets of pooled continuous and diffuse sampler NO-measurements.

Ind.diff	Coef.	Std. Err.	t	p	95% Conf.	Interval
Ref.diff	0.666	0.022	30.5	<0.001	0.623	0.708
Ref.base	0.496	0.026	19.1	<0.001	0.445	0.547
Ind.base	−0.473	0.022	−21.5	<0.001	−0.516	−0.430
Diff 1/H	75.8	29.5	2.57	0.010	18.0	134
Diff 1/V	−0.571	0.216	−2.65	0.008	−0.994	−0.148
Diff 1/P	0.028	0.048	0.59	0.557	−0.066	0.123
Time.diff	0.245	0.313	0.78	0.433	−0.369	0.860
E	−1.128	0.218	−5.19	<0.001	−1.555	−0.702

Regression coefficient, robust standard errors of coefficient, t-statistic, two-sided P-value, and 95%-confidence interval of coefficient. The absolute LEZ effect estimate is given by the coefficient E in $\mu g/m^3$ (<0: concentration is lowered by LEZ).
Covariables: difference in reference stations Ref.diff in $\mu g/m^3$, centered reference baseline concentration Ref.base in $\mu g/m^3$, centered index baseline concentration Ind.base in $\mu g/m^3$, Diff 1/H = difference in 1/(height of inversion layer) in 1/m, Diff 1/V = 1/(wind velocity+0.01) in $(m/s)^{-1}$, difference in 1/P = 1/(amount of precipitation+0.01) $(mm/h)^{-1}$, centered difference in time Time.diff in years.

Table 10. Summarized Results on Nitrogen Oxides.

			Model 1 Estimates		Model 2 Estimates	
	Evaluation Period	Mean Index Concentration/μg/m³	Additive Effect/μg/m³	Multiplicative Effect/%	Additive Effect/μg/m³	Multiplicative Estimate/%
NO_2	0.5 h	50	−1.1	−2.1	−1.9	−3.9
	4 weeks	55	−0.8	−1.9	−1.7	−3.9
NO	4 weeks	48	−1.1	−3.2	+0.4	+1.9
NO_x	4 weeks	103	−1.7	−2.4	−0.9	+4.8

Model 1 covariables: difference in reference stations Ref.diff in μg/m³, centered reference baseline concentration Ref.base in μg/m³, centered index baseline concentration Ind.base in μg/m³, Diff 1/H = difference in 1/(height of inversion layer) in 1/m, Diff 1/V = 1/(wind velocity+0.01) in $(m/s)^{-1}$, difference in 1/P = 1/(amount of precipitation+0.01) $(mm/h)^{-1}$, centered difference in time Time.diff in years.

compared to current LEZ restrictions for Euro 3 vehicles [46]. The authors stated that despite of the large area of the London LEZ, the predicted changes in NO_2 (and PM_{10}) were generally small. Their modelled results stay partly in contrast to the prognoses published by Tonne et al. [79], who estimated for the London Congesting Charge Scheme a small decrease for NO_2 of -0.64 μg/m³ only, corresponding to -1.1%.

Table 11. Overview of studies estimating the effect of LEZs on NO_2 concentrations.

Study	Country	City/Area	Intervention	NO_2 effect/μg/m³	Comments
Builtjes et al. 2012,	Germany	Berlin	LEZ of level 3: forbidden<Euro 4	−1.3	
Stern 2013		Munich		−1.0	
		Ruhr Area		−1.7	
Tonne et al. 2008,	UK	London	Congestion Charging Zone: forbidden<Euro 4	−0.64 (−1.1%)	simulation study 2008
Kelly et al 2011				reduction up to −7.3%	simulation study 2011
Briggs 2008	Italy	Rome	2 LEZs:	−2.3/−3.0	simulation study: main effect due to the exclusion of Euro 0 vehicles
			Euro 0 forbidden 2 LEZs:		
Cesaroni et al. 2012	Italy	Rome	forbidden<Euro 4	−3.0/−4.1	
Boogaard et al. 2012	The Netherlands	Amsterdam	LEZ:	−4.5	analysis of measurements: crude comparisons, no covariates taken into account
		Den Bosch	trucks forbidden	(not statistically significant)	
		The Hague	<Euro 2 and Euro 3 trucks only allowed if retrofitted		
		Tilburg			
		Utrecht			
Johansson et al. 2009	Sweden	Stockholm	Road Pricing System: Vehicles travelling into and out of the charge cordon were charged for every passage during weekdays	NO_x:	analysis of measurements: crude comparisons, no covariates taken into account
				−0.23 (Greater Stockholm)	
				−0.81 (inner city)	
This study	Germany	17 cities	LEZ of level 1: Diesel passenger cars, trucks and buses forbidden<Euro 2 without particulate reduction system	NO_2, NO, NO_x:	analysis of measurements: construction of matched quadruplets, regression analyses with covariates, all estimated effects statistically significant at the 0.1% level
				reduction less than −2.0 (−4%)	
				[PM_{10}: reduction less than −0.2 μg/m³ (−1%)]	

The INTARESE project [80] modeled NO_2 concentration changes for both LEZs in Rome and confirmed this finding of only small additional gains by stricter traffic restrictions. The main reductions were expected to be achieved already by excluding Euro 0 cars: -2.3 µg/m^3 or -3.0 µg/m^3. If only Euro 4 cars were allowed to enter the LEZs the reductions were expected to increase only slightly to -3.0 µg/m^3 or -4.1 µg/m^3 [81].

The "Stockholm Trial" involved a road pricing system to improve the air quality and reduce traffic congestion. The test period of the trial was January 3, 2006 to July 31, 2006. Vehicles travelling into and out of the charge cordon were charged for every passage during weekdays. Annual mean contributions to total levels of nitrogen oxides from emissions from road traffic with and without charges according to the Stockholm Trial were estimated. NO_x concentrations were lowered in periods with charges, but the study showed a small decrease only: -0.23 µg/m^3 (Greater Stockholm) and -0.81 µg/m^3 (inner city) [82]. No multivariable modeling was tried.

Boogaard et al. [83] analyzed measurements of NO_2 and NO_x conducted simultaneously at eight streets, six urban background locations and four suburban background locations before (2008) and two years after implementation of an LEZ (2010) in five cities of The Netherlands (8 index stations, 4 reference stations). Index concentrations were lower in 2010 than in 2008 (NO_2: -4.5 µg/m^3, NO_x: -6.1 µg/m^3) but the differences were not statistically different. The study performed only crude comparisons and did not apply regression techniques to adjust for covariables.

The present study can be regarded as one of the most comprehensive approaches so far, analysing measurement data of nitrogen oxides concentrations in order to assess LEZ effects. The LEZ pollutant group 1 reduction effect on nitrogen oxides (NO_2, NO, and NO_x) was estimated as being no higher than 2 µg/m^3 at all index stations and index traffic stations, i.e., no higher than about 4%. This estimate based on measurement data can be rated as the most profound currently available. This result also needs to be interpreted in the light of the existing EU limit values because LEZs are often supposed to be the most effective measure that cities can take to reduce air pollution problems in their area [84]. The respective NO_2 concentration limit [24] enforced in Germany since 2010 is 40 µg/m^3 (1 year average). Values are in excess and about 69% of all German traffic stations showed annual averages higher than 40 µg/m^3 [25]. The four week averages of NO_2 concentrations at index stations after LEZ introduction were found to be 55 µg/m^3 (median and mean) or 82 µg/m^3 (95th percentile). It follows that the estimated reduction of NO_2 concentrations in the range of 2 µg/m^3 appears to be of negligible impact when the current concentration levels should be lowered to the EU limit. The same judgement seems to apply on the EU level where the NO_2 concentrations were reported to show a pronounced excess in many cities [26].

Regarding the information from the HBEFA [85] for real driving conditions in Germany, Austria and Switzerland with respect to vehicles that meet Euro 5 and 6 emission standards, no noteworthy reductions of NO_2 and NO_x immissions are to be expected until a remarkable share of vehicles with NO_x after treatment systems (Euro 5 for HD trucks and Euro 6 for passenger cars) will be on the street [85].

The Handbook of Emission Factors for Road Transport (HBEFA) was originally developed on behalf of the Environmental Protection Agencies of Germany, Switzerland and Austria. In the meantime, further countries (Sweden, Norway, France) as well as the JRC (European Research Center of the European Commission) are supporting HBEFA. HBEFA provides emission factors, i.e. the specific emission in g/km for all current vehicle categories (PC, LDV, HDV, buses and motor cycles), each divided into different categories, for a wide variety of traffic situations (http://www.hbefa.net/e/index.html).

Interestingly, remarkable differences in NO_x and NO_2 emissions from passenger cars and light duty vehicles are documented when low test cycle emissions were compared with relatively higher NO_x/NO_2 concentrations measured along roadsides [86,87].

We analysed PM_{10} concentrations additionally [32] from 19 German LEZs. From about 2005 until the end of 2009 continuous half-hour measurement values as well as gravimetrically determined daily measurements of PM_{10} were collected. Two continuous procedures were used to measure mean PM_{10} concentrations per half-hour intervals [38,88]:

- Absorption of β-radiation (BA). The particulate matter is deposited on a filter tape and the change in β-ray transmission is measured.
- Tapered Element Oscillating Microbalance (TEOM). An inertial balance directly measures the mass collected on an exchangeable filter cartridge by monitoring the corresponding frequency changes of a tapered element.

In addition, gravimetric samplers were used to measure daily averages of PM_{10} concentrations [49,88,89]. 2,110,803 quadruplets of continuous PM_{10} and 15,735 gravimetric quadruples were identified leading to 61,169 quadruplets based on daily PM_{10} averages. The analyses showed that best LEZ effect estimates were ≤0.2 µg/m^3 at all index stations, i.e., the relative PM_{10} reduction ≤1%. Best estimates at all index stations near traffic (excluding urban background and industry index stations) were below 1 µg/m^3 (less than 5%, resp). Effects were smaller than predicted prior to the introduction of LEZs. Limited data (1750 quadruplets of monthly averages) were also available to estimate the effects on soot parameters (elemental carbon, organic carbon and total carbon). The average of total carbon concentrations was estimated as 13 µg/m^3 and LEZ effect estimates were about -0.55 µg/m^3 or -4.2%. For $PM_{2.5}$ only 650 quadruplets based on half-hour data and 99 quadruplets of daily concentration averages could be analyzed. The $PM_{2.5}$ concentration mean was found at 17 µg/m^3. All LEZ effect estimates on $PM_{2.5}$ were positive, i.e., no indication of reduced concentrations after the introduction of the LEZs was found.

Due to the proven marginal reduction of nitrogen oxide concentrations (NO, NO_2, NO_x), LEZ as a regulatory action cannot be seen as an efficient measure to substantially reduce ambient nitrogen oxide exposures in the cities. Beyond that, this result is in good accordance to the effectiveness of LEZs on the reduction of PM_{10}, too [32]. As predicted [33], long-term compliance problems with ambient air NO_2 concentrations should be expected even if LEZs were introduced or enlarged for the purpose of NO_2 reductions in cities.

The approach can be extended to account for other variables that are considered relevant [45]. Such data can only be used if these data are homogeneously available at all index and reference stations and are also available before and after the introduction of LEZs. Traffic density and car fleet properties are such variables of interest that do not meet the inclusion criteria: there are almost no data available in Germany to describe differences in flow of traffic and car fleet properties between index and reference stations and across time. To put this into perspective, we like to note first that changes in traffic density and car fleet properties are potentially affected by LEZs. It follows that traffic density and fleet properties should be considered as potential outcomes of LEZ introduction and not only as confounders of LEZ effects. This means that these data must not be accounted for by covariables in regression modelling even if the data were available in such a way that the inclusion criteria were met. Anyhow, authors who described changes in traffic-flow in Berlin argued against the interpretation that LEZs caused such displacements of traffic-flow from inside the LEZ to the reference stations [31]. Second, we note that the missing information on traffic density and fleet properties can be used to argue for biases in both directions. On the ones side, traffic could be displaced from the LEZ area to the reference stations outside so that the concentrations are underestimated inside but overestimated outside the LEZ, causing a potential overestimate of the LEZ effect. On the other side, if the car fleet is renewed not only inside the LEZ but also outside at the reference stations this may lead to a potential underestimate of the LEZ effect. We cannot conclude, therefore, on the direction of the potential bias.

The data analyzed in this study are the only available longitudinal measuring data to investigate the development of nitrogen oxide concentrations before and after the introduction of LEZs in Germany. We conclude that the material used can be considered as "data best available". Interpretations are limited, however, because spatial representativeness of the measuring sites can be disputed. It is unknown whether these data can be used to reliably estimate the exposures of citizens living in the LEZs. Since this is not only a problem of German measuring networks but an issue on the European level a research project was started to investigate the representativeness of measurement sites [78]. The authors concluded that measurements at the background stations are of greater importance than the data collected at the hot spots (traffic stations). Other limitations of hot spot data result from the fact that the citizens living in the LEZ area spend most of their time indoors and that indoor pollution data differ from hot spot outdoor concentrations [90,91].

Conclusions

This is the first comprehensive approach to assess effects of LEZs on NO_2, NO and NO_x concentrations with the help of measurement data on the Federal level in Germany. Reductions due to introducing LEZs of pollutant group 1 were estimated to be limited by 2 $\mu g/m^3$ (or 4%). The 4-week averages of NO_2 concentrations at index stations after LEZ introduction were found to be 55 $\mu g/m^3$ (median and mean) or 82 $\mu g/m^3$ (95th percentile). The NO_2 concentration limit [24] enforced in Germany since 2010 is 40 $\mu g/m^3$ (1 year average). Concerning the expenditure of regulations and controls which are required to introduce and operate LEZs in cities, the proven impact of LEZs on the reduction of NO_2 ambient air concentrations with at a maximum of 4% in the first phase is very small.

Supporting Information

File S1 Contains Tables S1–S4 and Figures S1–S19. Table S1: Detailed results on NO_2 - quadruplet analyses by linear (additive) log-linear (multiplicative) regression models. **Table S2:** NO: Log-linear (multiplicative) model 1 evaluating the quadruplets of pooled continuous and diffuse sampler NO-measurements. Regression coefficient, robust standard errors of coefficient, t-statistic, two-sided P-value, and 95%-confidence interval of coefficient. The relative LEZ effect estimate is given by the coefficient E (<1: concentration is lowered by LEZ). **Table S3:** NO_x: Quadruplets of pooled continuous and diffuse sampler NO_x-measurements: index stations (Ind), reference stations (Ref) before (pre) and after (post) introduction of LEZ. Ind.diff and Ref.diff denote differences between index measurements and between reference measurements (negative post-pre differences indicate lower values after introduction of LEZ). **Table S4:** NO_x: Linear (additive) model 1 evaluating the quadruplets of pooled continuous and diffuse sampler NO_x-measurements. Regression coefficient, robust standard errors of coefficient, t-statistic, two-sided P-value, and 95%-confidence interval of coefficient. The absolute LEZ effect estimate is given by the coefficient E in $\mu g/m^3$ (<0: concentration is lowered by LEZ). **Table S5:** NO_x: Log-linear (multiplicative) model 1 evaluating the quadruplets of pooled continuous and diffuse sampler NO_x-measurements. Regression coefficient, robust standard errors of coefficient, t-statistic, two-sided P-value, and 95%-confidence interval of coefficient. The relative LEZ effect estimate is given by the coefficient E (<1: concentration is lowered by LEZ). **Figure S1:** Low emission zone **Herrenberg** (marked area), implemented in 2009-01-01 (modified from www.map24.de). One index station: 1)DEBW135 Hindenburger Straße, no NO, no NO_x. One reference station outside the low emission zone: 2)DEBW112 Gärtringen (not included in the figure since located approx. 5 km north of low emission zone). **Figure S2:** Low emission zone **Ilsfeld** (marked area), implemented in 2008-03-01 (modified from www.map24.de). One index station: 1)DEBW133 König-Wilhelm-Straße, no NO, no NO_x. One reference station outside the low emission zone: 2)DEBW034 Waiblingen (not included in the figure since located approx. 24 km south of low emission zone). **Figure S3:** Low emission zone **Karlsruhe** (marked area), implemented in 2009-01-01 (modified from www.map24.de). One index station: 1)DEBW126 Kriegsstraße, no NO_2, no NO_x. Two reference stations outside the low emission zone: 2)DEBW001 Karlsruhe-Mitte 3)DEBW004 Eggenstein (not included in the figure since located approx. 6 km north of low emission zone). **Figure S4:** Low emission zone **Ludwigsburg** (marked area), implemented in 2008-03-01 (modified from www.map24.de). Two index stations: 1)DEBW024 Weimar-/Schweizerstraße 2)DEBW017 Friedrichstraße. One reference station outside the low emission zone: 3)DEBW034 Waiblingen (not included in the figure since located approx. 7 km south east of low emission zone). **Figure S5:** Low emission zone **Mannheim** (marked area), implemented in 2008-03-01 (modified from www.map24.de). Two index stations: 1)DEBW006 Mannheim-Mitte 2)DEBW098 Friedrichsring U2. Two reference stations outside the low emission zone: 3)DEBW005 Mannheim Nord (not included in the figure since located approx. 4 km north of low emission zone) 4)DEBW007 Mannheim-Süd (not included in the figure since located approx. 5 km south of low emission zone). **Figure S6:** Low emission zone **Reutlingen** (marked area), implemented in 2008-03-01 (modified from www.map24.de). One index station: 1)DEBW027 Ebertstraße. Two reference stations outside the low emission zone: 2)DEBW042 Bernhausen (not included in the figure since located approx. 20 km north of low emission zone) 3)DEBW117 Gärtringen (not included in the figure since located approx. 28 km west of low emission zone). **Figure S7:** Low emission zone **Stuttgart** (marked area), implemented in 2008-03-01 (modified from www.map24.de). Six index stations: 1)DEBW011 Zuffenhausen 2)DEBW013 Seuberstraße 3)DEBW099 Arnulf-Klett-Platz 4)DEBW116 Hohenheimer Straße 5)DEBW118 Am Neckartor 6)DEBW134 Waiblinger Straße. Two reference stations outside the low emission zone: 7)DEBW034 Waiblingen 8)DEBW042 Bernhausen (not included in the figure since located approx. 2 km south of low emission zone). **Figure S8:** Low emission zone **Tübingen** (marked area), implemented in 2008-03-01 (modified from www.map24.de). One index station: 1)DEBW107 Derendingerstraße. One reference station outside the low emission zone: 2)DEBW112 Gärtringen (not included in the figure since located approx. 15 km north west of low emission zone). **Figure S9:** Low emission zone **Augsburg** (marked area), implemented in 2009-07-01 (modified from www.map24.de). Three index stations: 1)DEBY007

Bourges-Platz 2)DEBY110 Karlstraße 3)DEBY006 Königsplatz. One reference station outside the low emission zone: 4)DEBY099 LfU (not included in the figure since located approx. 3 km south of low emission zone). **Figure S10:** Low emission zone **Munich** (marked area), implemented in 2008-10-01 (modified from www.map24.de). Five index stations: 1)DEBY037 Stachus 2)DEBY039 Lothstraße 3)DEBY085 Luise-Kiesselbach-Platz 4)DEBY114 Prinzregentenstraße e 5)DEBY115 Landshuter Allee. Three reference stations outside the low emission zone: 6)DEBY043 Moosach, no PM_{10} 7)DEBY089Johanneskirchen 8)DEBY109 Andechs/Rothenfeld (not included in the figure since located approx. 27 km south west of low emission zone). **Figure S11a:** Low emission zone **Berlin Blume-Messnetz** (marked area), implemented in 2008-01-01 (modified from www.map24.de).Five index stations: 1)DEBE018 B Schöneberg-Belziger Straße 2)DEBE034 B Neukölln-Nansenstraße 3)DEBE064 B Neukölln-Karl-Marx-Straße 76 4)DEBE065 B Friedrichshain-Frankfurter Allee 5)DEBE067 B Hardenbergplatz. Nine reference stations outside the low emission zone: 6)DEBE061 B Steglitz-Schildhornstraße 7)DEBE062 B Frohnau, Funkturm (not included in the figure since located approx. 13 km north of low emission zone) 8)DEBE063 B Neukölln-Silbersteinstraße) 9)DEBE066 B Karlshorst-Rheingoldstraße, no PM_{10} (not included in the figure since located approx. 5 km east of low emission zone) 10)DEBE010 B Wedding-Amrumer Straße 11)DEBE027 B Marienfelde-Schichauweg (not included in the figure since located approx. 8 km south of low emission zone) 12)DEBE032 B Grunewald (not included in the figure since located approx. 4 km south west of low emission zone) 13)DEBE051 B Buch (not included in the figure since located approx. 12 km north east of low emission zone) 14)DEBE056 B Friedrichshagen (not included in the figure since located approx. 14 km south east of low emission zone). **Figure S11b:** Low emission zone **Berlin RUBIS-Messnetz** (marked area), implemented in 2008-01-01 (modified from www.map24.de). Ten index stations: 1)DEBE530 Hauptstraße 30 2)DEBE504 Beusselstraße 66 3)DEBE537 Alt Moabit 63 4)DEBE545 Sonnenallee 68 5)DEBE547 Landsberger Allee 6–8 6)DEBE517 Neukölln-Nansenstraße 7)DEBE519 Friedrichshain-Frankfurter Allee 8)DEBE555 Herrmannplatz Laterne 21 9)DEBE562 Friedrichstraße Laterne 156 10)DEBE525 Leipziger Straße 32. Twelve reference stations outside the low emission zone:11) DEBE501 Berliner Allee 118 12)DEBE577 Buch, no NO, no NO_x (not included in the figure since located approx. 12 km north of low emission zone) 13)DEBE507 Grünauer Straße 4 (not included in the figure since located approx. 9 km south east of low emission zone) 14)DEBE539 Schloßstraße 29 15)DEBE542 Tempelhofer Damm 148 16)DEBE513 Spreestraße 2 (not included in the figure since located approx. 5 km south east of low emission zone) 17)DEBE514 Alt Friedrichsfelde 8a (not included in the figure since located approx. 3 km east of low emission zone) 18)DEBE521 Steglitz-Schildhornstraße 19)DEBE559 Buschkrugallee Laterne 3 20)DEBE522 Neukölln-Silbersteinstraße1 21)DEBE573 Badstraße 22)DEBE576 Spandau, Klosterstraße 12 (not included in the figure since located approx. 6 km west of low emission zone). **Figure S12:** Low emission zone **Frankfurt a.M.** (marked area), implemented in 2008-10-01 (modified from www.map24.de). One index station: 1)DEHE041 Frankfurt-Friedb.Ldstr. Three reference stations outside the low emission zone: 2)DEHE008 Frankfurt-Ost 3)DEHE011 Hanau (not included in the figure since located approx. 13 km east of low emission zone) 4)DEHE005 Frankfurt-Höchst. **Figure S13:** Low emission zone **Hannover** (marked area), implemented in 2008-01-01 (modified from www.map24.de). One index station: 1)DENI048 Hannover Verkehr. Four reference stations outside the low emission zone: 2)DENI054 Hannover 3)DENI011 Braunschweig, Broizemer Steinberg (not included in the figure since located approx. 49 km east of low emission zone) 4)DENI041 Weserbergland/Rinteln, Brugfeldsweide (not included in the figure since located approx. 48 km south west of low emission zone) 5)DENI052 Allertal/Walsrode, Auf dem Kamp 8 (not included in the figure since located approx. 47 km north of low emission zone). **Figure S14:** Low emission zone **Dortmund** (marked area), implemented in 2008-10-01, but Brackelerstr. 2008-01-01 (modified from www.map24.de). Four index stations: 1)DENW101 Steinstraße 2)DENW136 Brackeler Straße 3)DENW184 Westfalendamm 190, no NO, no NO_x, no PM_{10} 4)DENW185 Rheinlanddamm 5–7, no NO, no NO_x, no PM_{10}. Four reference stations outside the low emission zone: 5)DENW002 Datteln-Hagem (not included in the figure since located approx. 15 km north west of low emission zone) 6)DENW008 Do-Eving 7)DENW029 Hattingen, An der Becke (not included in the figure since located approx. 19 km south west of low emission zone) 8)DENW179 Schwerte (not included in the figure since located approx. 8 km south of low emission zone). **Figure S15:** Low emission zone **Duisburg** (marked area), implemented in 2008-10-01 (modified from www.map24.de).Three index stations: 1)DENW034 Duisburg-Walsum 2)DENW040 Duisburg-Buchholz 3)DENW112 Kardinal-Galen-Straße. One reference station outside the low emission zone: 4)DENW038 45476 Mühlheim, Neustadtstraße (not included in the figure since located approx. 5 km east of low emission zone). **Figure S16:** Low emission zone **Düsseldorf** (marked area), implemented in 2009-02-15 (modified from www.map24.de). Two index stations: 1)DENW082 Corneliusstraße 2)DENW216 Düsseldorf-Bilk, no NO, no NO_x, no PM_{10}. Four reference stations outside the low emission zone: 3)DENW042 Krefeld-Linn (not included in the figure since located approx. 14 km north west of low emission zone) 4)DENW071 Düsseldorf-Lörick (not included in the figure since located approx. 3 km west of low emission zone) 5)DENW078 Ratingen-Tiefenbroich (not included in the figure since located approx. 6 km north east of low emission zone) 6)DENW116 Krefeld Hafen (not included in the figure since located approx. 12 km north west of low emission zone). **Figure S17:** Low emission zone **Essen** (marked area), implemented in 2008-10-01 (modified from www.map24.de). Eight index stations: 1)DEN-W043Ost Steeler Straße 2)DENW134 Gladbecker Straße 3)DENW135 Hombrucher Straße 4)DENW161 Alfredstraße 9/11, no NO, no NO_x, no PM_{10} 5)DENW168 Gladbecker Straße 245, no NO, no NO_x, no PM_{10} 6)DENW169 In der Baumschule 7, no NO, no NO_x, no PM_{10} 7)DENW171 Hombrucherstraße 21/23, no NO, no NO_x, no PM_{10} 8)DENW215 Hausackerstraße 11, no NO, no NO_x, no PM_{10}. Three reference stations outside the low emission zone: 9)DENW024 Essen-Vogelheim 10)DENW029 Hattingen-Blankenstein (not included in the figure since located approx. 10 km south east of low emission zone), 11) DENW162 Bruckstraße 29, no NO, no NO_x, no PM_{10} (not included in the figure since located approx. 4 km south of low emission zone). **Figure S18:** Low emission zone **Cologne** (marked area), implemented in 2008-01-01 (modified from www.map24.de). Seven index stations: 1)DENW148 Justinianstraße 13 15, no NO, no NO_x, no PM_{10} 2)DENW151 Neumarkt 25, no NO, no NO_x, no PM_{10} 3)DENW153 Tunisstraße/Elstergasse, no NO, no NO_x, no PM_{10} 4)DENW164 Hohenstaufenring 57A, no NO, no NO_x, no PM_{10} 5)DENW198 Gereonsdriesch 21, no NO, no NO_x, no PM_{10} 6)DENW211 Clevischer Ring 3 7)DENW212 Turiner Straße. Four reference stations outside the low emission zone: 8)DENW053 Cologne-Chorweiler (not included in the figure since located approx. 9 km north west of low emission zone) 9)DENW058 Hürth (not included in the figure since located approx. 7 km south west of low emission zone), 10)DENW059 Cologne-Rodenkirchen (not included in the figure since located approx. 4 km south of low emission zone), 11)DENW079 Leverkusen-Manfort (not included in the figure since located approx. 7 km north of low emission zone). **Figure S19:** Low emission zone **Wuppertal** (marked area), implemented in 2009-02-15 (modified from www.map24.de). Two index stations: 1)DENW114 Wuppertal-Langerfeld, no NO, no NO_x 2)DENW189 Wuppertal

Gathe. Two reference stations outside the low emission zone: 3)DENW029 Hattingen-Blankenstein (not included in the figure since located approx. 13 km north of low emission zone) 4)DENW080 Solingen-Wald (not included in the figure since located approx. 5 km south west of low emission zone).

Acknowledgments

We thank the project steering committee and the following state authorities for the intense and helpful discussion and the supply of the raw data. In specific: Peter Bruckmann and Reinhold Beier, Landesamt für Natur-, Umwelt- und Verbraucherschutz NRW, Abteilung 4, Luftqualität, Geräusche, Erschütterungen, Strahlenschutz; Stefan Jacobi and Wieslawa Stec-Lazaj, Hessisches Landesamt für Umwelt und Geologie (HLUG), Abteilung I Immissions- und Strahlenschutz; Michael Köster and Andreas Hainsch, Staatliches Gewerbeaufsichtsamt Hildesheim, Abteilung 4 - Zentrale Unterstützungsstelle für Luftreinhaltung und Gefahrstoffe (ZUS LG); Martin Lutz and Arnold Kettschau, Berliner Senatsverwaltung für Gesundheit, Umwelt und Verbraucherschutz, Abteilung III Umweltpolitik, Referat III D Immissionsschutz; Heinz Ott, Bayerisches Landesamt für Umwelt Abteilung 2, Referat 2, 4 Luftgütemessungen Südbayern, Luftreinhaltung beim Verkehr; Werner Scholz and Christiane Lutz-Holzhauer, Landesanstalt für Umwelt, Messungen und Naturschutz Baden-Württemberg (LUBW), Referat 33 Luftqualität; Ralf Wehrse and Jan Osmers, Bremer Senator für Umwelt, Bau, Verkehr und Europa Fachbereich Umwelt, Abteilung 2, Umweltwirtschafts-, Klima- und Ressourcenschutz, Referat 22 Immissionsschutz.

Author Contributions

Analyzed the data: PM. Contributed reagents/materials/analysis tools: PM. Contributed to the writing of the manuscript: PM DG MS. Data Organization: MS DG PM. Health Aspects: MS DG.

References

1. LEZEN (2013) Low emission zone in Europe network. Available from: http://www.lowemissionzones.eu. Accessed 2014 July 8.
2. EC (2007) Regulation (EC) No 715/2007on type approval of motor vehicles with respect to emissions from light passenger and commercial vehicles (Euro 5 and Euro 6) and on access to vehicle repair and maintenance information. 1–16. Available from: http://eur-lex.europa.eu/LexUriServ/LexUriServ.do?uri=OJ:L:2007:171:0001:0001:EN:PDF. Accessed 2014 July 8.
3. Umweltbundesamt (2014) Umweltzonen in Deutschland. Available from: http://www.umweltbundesamt.de/themen/luft/luftschadstoffe/feinstaub/umweltzonen-in-deutschland. Accessed 2014 July 8.
4. Colvile RN, Hutchinson EJ, Mindell JS, Warren RF (2001) The transport sector as a source of air pollution. Atmospheric Environment 35: 1537–1565.
5. Bruckmann P, Lutz M (2010) Verbessern Umweltzonen die Luftqualität? In: Verband der Automobilindustrie (VDA), editor. 12. Technischer Kongress, 24. und 25. März. Forum am Schlosspark, Ludwigsburg: Henrich Druck+Medien GmbH. pp. 299–311.
6. Bruckmann P, Wurzler S, Brandt A, Vogt K (2011) Erfahrungen mit Umweltzonen in Nordrhein-Westfalen. In: Bundesamt für Strahlenschutz, Bundesinstitut für Risikobewertung, Robert Koch-Institut, Umweltbundesamt, editor. UMID Umwelt und Mensch - Informationsdienst. Berlin. pp. 27–33.
7. Kacsóh L (2011) Umweltzonen in Europa und in Deutschland. In: Bundesamt für Strahlenschutz, Bundesinstitut für Risikobewertung, Robert Koch-Institut, Umweltbundesamt, editor. UMID Umwelt und Mensch - Informationsdienst. Berlin. pp. 5–10.
8. Morfeld P, Keil U, Spallek M (2013) The European "Year of the Air": fact, fake or vision? Archives of Toxicology 87: 2051–2055.
9. Turner MC, Krewski D, Pope CA, Chen Y, Gapstur SM, et al. (2011) Long-term ambient fine particulate matter air pollution and lung cancer in a large cohort of never-smokers. American Journal of Respiratory and Critical Care Medicine 184: 1374–1381.
10. Hoek G, Krishnan RM, Beelen R, Peters A, Ostro B, et al. (2013) Long-term air pollution exposure and cardio- respiratory mortality: a review. Environmental Health 12: 1–15.
11. Hystad P, Demers PA, Johnson KC, Carpiano RM, Brauer M (2013) Long-term residential exposure to air pollution and lung cancer risk. Epidemiology 24: 762–772.
12. Raaschou-Nielsen O, Andersen ZJ, Beelen R, Samoli E, Stafoggia M, et al. (2013) Air pollution and lung cancer incidence in 17 European cohorts: prospective analyses from the European Study of Cohorts for Air Pollution Effects (ESCAPE). The Lancet Oncology 14: 813–822.
13. Schuster C (2009) Umweltzonen: Nutzen weiterhin umstritten. Dtsch Arztebl 106: A228.
14. Eikmann T, Herr C (2009) Ist die Einführung von Umweltzonen tatsächlich eine sinnvolle Maßnahme zum Schutz der Gesundheit der Bevölkerung? Umweltmed Forsch Prax 14: 125–126.
15. Friedrich B (2008) Umweltzonen. Straßenverkehrstechnik 11: 673.
16. Zellner R, Kuhlbusch TAJ, Diegmann V, Herrmann H, Kasper M, et al. (2009) Feinstäube und Umweltzonen. Available from: Available from: www.processnet.org/dechema_media/Downloads/Positionspapiere/Stellungnahme__Feinstaeube.pdf. Accessed 2014 July 8.
17. Downs SH, Schindler C, Liu LJ, Keidel D, Bayer-Oglesby L, et al. (2007) Reduced exposure to PM10 and attenuated age-related decline in lung function. The New England Journal of Medicine 357: 2338–2347.
18. Gauderman WJ, Avol E, Gilliland F, Vora H, Thomas D, et al. (2004) The effect of air pollution on lung development from 10 to 18 years of age. The New England Journal of Medicine 351: 1057–1067.
19. Lim SS, Vos T, Flaxman AD, Danaei G, Shibuya K, et al. (2013) A comparative risk assessment of burden of disease and injury attributable to 67 risk factors and risk factor clusters in 21 regions, 1990–2010: A systematic analysis for the Global Burden of Disease Study 2010. The Lancet 380: 2224–2260.
20. Nawrot TS, Perez L, Kunzli N, Munters E, Nemery B (2011) Public health importance of triggers of myocardial infarction: a comparative risk assessment. The Lancet 377: 732–740.
21. Samet JM, Dominici F, Curriero FC, Coursac I, Zeger SL (2000) Fine particulate air pollution and mortality in 20 U.S. cities, 1987–1994. The New England Journal of Medicine 343: 1742–1749.
22. Latza U, Gerdes S, Baur X (2009) Effects of nitrogen dioxide on human health: systematic review of experimental and epidemiological studies conducted between 2002 and 2006. International Journal of Hygiene and Environmental Health 212: 271–287.
23. EC (2008) Directive 2008/50/EC of the European Parliament and of the councilof 21 May 2008 on ambient air quality and cleaner air for Europe. 1–44. Available from: http://eur-lex.europa.eu/LexUriServ/LexUriServ.do?uri=OJ:L:2008:152:0001:0044:EN:PDF. Accessed 2014 July 8.
24. EC (1999) Council Directive 1999/30/EC of 22 April 1999 relating to limit values for sulphur dioxide, nitrogen dioxide and oxides of nitrogen, particulate matter and lead in ambient air. 41–60. Available from: http://eur-lex.europa.eu/LexUriServ/LexUriServ.do?uri=OJ:L:1999:163:0041:0060:EN:PDF. Accessed 2014 July 8.
25. Umweltbundesamt (2012) Luftqualität 2011 - Feinstaubepisoden prägten das Bild. Dessau-Roßlau. Available from: www.umweltbundesamt.de/uba-info-medien/4211.html. Accessed 2014 July 8.
26. Giannouli M, Kalognomou E-A, Mellios G, Moussiopoulos N, Samaras Z, et al. (2011) Impact of European emission control strategies on urban and local air quality. Atmospheric Environment 45: 4753–4762.
27. European Environment Agency (2014) Indicator: Exceedance 492 of air quality limit values in urban areas. Copenhagen, Denmark: European Environment Agency. Available from: http://www.eea.europa.eu/data-and-maps/indicators/exceedance-of-air-quality-limit-1/exceedance-of-air-quality-limit-5. Accessed 2014 July 16.
28. European Environment Agency (2014) AirBase - The European air Quality database. Copenhagen, Denmark. Available from: http://acm.eionet.europa.eu/databases/airbase. Accessed 2014 July 8 http://www.eea.europa.eu/data-and-maps/data/airbase-the-european-air-quality-database-8. Accessed 2014 July 8.
29. Rauterberg-Wulff A, Lutz M (2011) Wirkungsuntersuchungen zur Umweltzone Berlin. In: Bundesamt für Strahlenschutz, Bundesinstitut für Risikobewertung, Robert Koch-Institut, Umweltbundesamt, editor. UMID Umwelt und Mensch - Informationsdienst. Berlin. pp. 11–18.
30. Lutz M, Rauterberg-Wulff A (2010) Berlin's low emission zone - top or flop? 14th ETH Conference on Combusion Generated Particles.
31. Lutz M, Rauterberg-Wulff A (2009) Ein Jahr Umweltzone Berlin: Wirkungsuntersuchungen. Berlin. 1–31. Available from: www.berlin.de/sen/umwelt/luftqualitaet/de/luftreinhalteplan/download/umweltzone_1jahr_bericht.pdf. Accessed 2014 July 8.
32. Morfeld P, Groneberg D, Spallek M (2014) Effectiveness of low emission zones: Analysis of the changes in fine dust concentrations (PM_{10}) in 19 German cities. Pneumologie 68: 173–186.
33. Vogt R, Kessler C, Schneider C (2010) Städtische NO_2 Luftqualität: Quellenanalyse und zukünftige Entwicklung. In: Verband der Automobilindustrie (VDA), editor. 12 Technischer Kongress 2010, 24 und 25 März. Forum am Schlosspark, Ludwigsburg: Henrich Druck+Medien GmbH. pp. 287–297.
34. US EPA (2008) Integrated science assessment for oxides of nitrogen - health criteria (final report). EPA/600/R-08/071. 260. Available from: http://www.epa.gov/ord/htm/whatsnew.htm. Accessed 2014 July 8.
35. EC (2001) Directive 2001/81/EC of the European Parliament and of the council of 23 October 2001 on national emission ceilings for certain atmospheric pollutants. 22–30. Available from: http://eur-lex.europa.eu/LexUriServ/LexUriServ.do?uri=OJ:L:2001:309:0022:0030:EN:PDF. Accessed 2014 July 8.

36. Atkinson RW, Barratt B, Armstrong B, Anderson HR, Beevers SD, et al. (2009) The impact of the congestion charging scheme on ambient air pollution concentrations in London. Atmospheric Environment 43: 5493–5500.
37. Bruckmann P, Brandt A, Wurzler S, Vogt K (2011) Verbessern Umweltzonen die Luftqualität? Neue Entwicklungen bei der Messung und Beurteilung der Luftqualitaet Fachtagung der Kommission Reinhaltung der Luft im VDI und DIN-Normenausschuss KRdL, Baden-Baden, 11–12 Mai. pp. 3–24.
38. Umweltbundesamt (2014) Stationsdatenbank des Umweltbundesamtes. Available from: http://www.env-it.de/stationen/public/stationList.do. Accessed 2014 July 8.
39. Pfeffer U (2010) Messtechnik für Stickstoffdioxid (NO_2). In: Kommission Reinhaltung der Luft im VDI und DIN-Normenausschuss KRdL, editor. Stickstoffdioxid und Partikel ($PM_{2,5}/PM_{10}$). pp. 113–122.
40. Umweltbundesamt (2014) Luft und Luftreinhaltung. Available from: http://www.umweltbundesamt.de/themen/luft/messenbeobachtenueberwachen/messgeraete-messverfahren. Accessed 2014 July 8.
41. Pfeffer U, Bier R, Zang T (2006) Measurements of nitrogen dioxide with diffusive samplers at traffic-related sites in North Rhine-Westphalia (Germany). Gefahrstoffe, Reinhaltung der Luft 1/2: 38–44.
42. Pfeffer U, Zang T, Rumpf E-M, Zang S (2010) Calibration of diffusive samplers for nitrogen dioxide using the reference method – Evaluation of measurement uncertainty. Gefahrstoffe, Reinhaltung der Luft 11/12: 500–506.
43. DIN EN 13528-3 Außenluftqualität (2004) Passivsammler zur Bestimmung der Konzentrationen von Gasen und Dämpfen - Teil 3: Anleitung zur Auswahl, Anwendung und Handhabung. Berlin: Beuth.
44. Rothman KJ, Greenland S, Lash TL (2008) Modern Epidemiology. 3. ed. Philadelphia: Lippincott Williams & Wilkins.
45. Morfeld P, Spallek M, Groneberg D (2011) Zur Wirksamkeit von Umweltzonen: Design einer Studie zur Ermittlung der Schadstoffkonzentrationsänderung für Staubpartikel (PM_{10}) und andere Größen durch Einfuhrung von Umweltzonen in 20 deutschen Städten. Zentralblatt für Arbeitsmedizin, Arbeitsschutz und Ergonomie 61: 148–165.
46. Kelly F, Armstrong B, Atkinson R, Anderson HR, Barratt B, et al. (2011) The London low emission zone baseline study. Health Effects Institut 163. 1–96. Available from: http://pubs.healtheffects.org/view.php?id = 366. Accessed 2014 July 8
47. Allison PD (2009) Fixed effects regression models. Los Angeles: SAGE.
48. Royston P, Sauerbrei W (2008) Multivariable model-buildung. Chichester, England: John Wiley & Sons Inc. 1–303 p.
49. Lenschow P, Abraham HJ, Kutzner K, Lutz M, Preuß JD, et al. (2001) Some ideas about the sorces of PM10. Atmospheric Environment 35: S23–S33.
50. Hoek G, Meliefste K, Cyrys J, Lewné M, Bellander T, et al. (2002) Spatial variability of fine particle concentrations in three European areas. Atmospheric Environment 36: 4077–4088.
51. Senn S (1997) Editorial: regression to the mean. Statistical Methods in Medical Research 6: 99–102.
52. Altman DG, Bland AE (1983) Measurement in medicine: the analysis of method comparison studies. The Statistician 32: 307–317.
53. Bland JM, Altman DG (1986) Statistical methods for assessing agreement between two methods of clinical measurement. The Lancet 1: 307–310.
54. Gill JS, Zezulka AV, Beevers DG, Davies P (1985) Relation between initial blood pressure and its fall with treatment. The Lancet 1: 567–569.
55. Bland JM, Altman DG (1994) Regression towards the mean. British Medical Journal 308: 1499.
56. Bland JM, Altman DG (1994) Some examples of regression towards the mean. British Medical Journal 309: 780.
57. Stigler SM (1997) Regression towards the mean, historically considered. Statistical Methods in Medical Research 6: 103–114.
58. Barnett AG, van der Pols JC, Dobson AJ (2005) Regression to the mean: what it is and how to deal with it. International Journal of Epidemiology 34: 215–220.
59. Twisk JWR (2004) Applied longitudinal data analysis for epidemiology. Cambridge: Cambridge University Press. 62–77 p.
60. Klingner M, Sähn E, Anke K, Holst T, Rost J, et al. (2006) Reduktionspotenziale verkehrsbeschränkender Maßnahmen in Bezug zu meteorologisch bedingten Schwankungen der PM_{10}- und NO_x-Immissionen. Gefahrstoffe, Reinhaltung der Luft 66: 326–334.
61. Builtjes P, Jörß W, Theloke J, Thiruchittampalam B, van der Gon HD, et al. (2012) Strategien zur Verminderung der Feinstaubbelastung. 1–160. Available from: http://www.umweltbundesamt.de/sites/default/files/medien/461/publikationen/4268.pdf. Accessed 2014 July 8
62. Reimer E, Scherer B (1992) An Operational Meteorological Diagnostic System for Regional Air Pollution Analysis and Long Term Modeling. In: Dop H, Kallos G, editors. Air Pollution Modeling and Its Application IX: Springer US. pp. 565–572.
63. Stern R (2013) Anwendung des REM-CALGRID-Modells auf die Ballungsräume Berlin, München und Ruhrgebiet. Berlin: Freie Universität Berlin, Institut für Meteorologie. Troposphärische Umweltforschung. 67/2013. 1–95. Available from: http://www.umweltbundesamt.de/sites/default/files/medien/461/publikationen/texte_67_2013_appelhans_m14_komplett_0.pdf. Accessed 2014 July 8.
64. Kerschbaumer A, Reimer E (2003) Erstellung der Meteorologischen Eingangsdaten für das REM/Calgrid-Modell: Modellregion Berlin-Brandenburg. Abschlussbericht zum UBA-Forschungsvorhaben 29943246. Freie Universität Berlin, Institut für Meteorologie. Available from: http://opus.kobv.de/zlb/volltexte/2009/7447/pdf/3729.pdf. Accessed 2014 July 17.
65. Kerschbaumer A (2010) Abhängigkeit der RCG-Simulationen von unterschiedlichen meteorologischen Treibern. Forschungs-Teilbericht an das Umweltbundesamt, im Rahmen des PAREST-Vorhabens: FKZ 206 43 200/1 "Strategien zur Verminderung der Feinstaubbelastung". Institut für Meteorologie der Freien Universität Berlin. Available from: http://www.umweltbundesamt.de/sites/default/files/medien/461/publikationen/texte_55_2013_appelhans_m02_komplett_0_0.pdf. Accessed 2014 July 17.
66. Graedel TE, Crutzen PJ (1994) Chemie der Atmosphäre. Heidelberg: Spektrum Akademischer Verlag.
67. Morfeld P, Stern R, Builtjes P, Groneberg DA, Spallek M (2013) Einrichtung einer Umweltzone und ihre Wirksamkeit auf die PM_{10}-Feinstaubkonzentration – eine Pilotanalyse am Beispiel München. Zentralblatt für Arbeitsmedizin, Arbeitsschutz und Ergonomie 63: 104–115.
68. Stern R (2003) Entwicklung und Anwendung des chemischen Transportmodells REM/CALGRID. Abschlussbericht zum Forschungs- und Entwicklungsvorhaben 298 41 252 des Umweltbundesamts "Modellierung und Prüfung von Strategien zur Verminderung der Belastung durch Ozon". Available from: http://www.umweltbundesamt.de/sites/default/files/medien/publikation/long/3604.pdf. Accessed 2014 July 17.
69. Stern R (2004) Großräumige PM_{10}-Ausbreitungsmodellierung: Abschätzung der gegenwärtigen Immissionsbelastung in Europa und Prognose bis 2010. KRdL-Experten-Forum "Staub und Staubinhaltsstoffe", 2004-11-11/10. VDI-KRdL-Schriftenreihe 33.
70. Stern R (2004) Weitere Entwicklung und Anwendung des chemischen Transportmodells REM-CALGRID für die bundeseinheitliche Umsetzung der EU-Rahmenrichtlinie Luftqualität und ihrer Tochterrichtlinien. Abschlussbericht zum FuE-Vorhaben 201 43 250 des Umweltbundesamts "Anwendung modellgestützter Beurteilungssysteme für die bundeseinheitliche Umsetzung der EU-Rahmenrichtlinie Luftqualität und ihrer Tochterrichtlinien". Available from: http://www.umweltbundesamt.de/sites/default/files/medien/publikation/long/3610.pdf. Accessed 2014 July17.
71. Benjamini Y, Hochberg Y (1995) Controlling the false discovery rate: a practical and powerful approach to multiple testing. Journal of the Royal Statistical Society, Series B 57: 289–300.
72. StataCorp (2009) Stata: Release 11. Statistical Software. College Station, TX: StataCorp LP.
73. Vickers AJ, Altman DG (2001) Statistics notes: Analysing controlled trials with baseline and follow up measurements. BMJ 323: 1123–1124.
74. Neter J, Wasserman W, Kutner MH (1985) Applied linear statistical models. Regression, Analysis of variance and experimental designs. Homewood, Illinois: Richard Dr. Irwin.
75. Puls T, Jäger-Ambrozewicz (2012) Die Auswirkungen der Umweltzone in Frankfurt auf die NO_2-Immissionen. Köln: Institut der deutschen Wirtschaft Köln. pp. 1–18.
76. ZUS LG (2010) Bewertung der Auswirkungen der Umweltzone Hannover auf Basis von Messdaten. Hildesheim: Zentrale Unterstützungsstelle Luftreinhaltung und Gefahrstoffe - Dezernat 42. 15. Available from: http://www.umwelt.niedersachsen.de/download/48880. Accessed 2014 July 8.
77. Umweltbundesamt (2014) Mediendatenbank. Available from: http://www.umweltbundesamt.de/publikationen?keys = PAREST&topic = All&series = All&sort_bef_combine = field_date_monthly_value+DESC. Accessed 2014 July 17.
78. Duyzer J, van den Hout D, Zandveld P, van Ratingen S (2013) Representativeness of air quality monitoring stations. TNO Research Report 2013 R11055. TNO Utrecht (in print).
79. Tonne C, Beevers S, Armstrong B, Kelly F, Wilkinson P (2008) Air pollution and mortality benefits of the London Congestion Charge: spatial and socioeconomic inequalities. Occupational and Environmental Medicine 65: 620–627.
80. Briggs DJ (2008) A framework for integrated environmental health impact assessment of systemic risks. Environmental Health 7: 61–77
81. Cesaroni G, Boogaard H, Jonkers S, Porta D, Badaloni C, et al. (2012) Health benefits of traffic-related air pollution reduction in different socioeconomic groups: the effect of low-emission zoning in Rome. Occupational and Environmental Medicine 69: 133–139.
82. Johansson C, Burman L, Forsberg B (2009) The effects of congestions tax on air quality and health. Atmospheric Environment 43: 4843–4854.
83. Boogaard H, Janssen NAH, Fischer PH, Kos GPA, Weijers EP, et al. (2012) Impact of low emission zones and local traffic policies on ambient air pollution concentrations. The Science of the Total Environment 435–436: 132–140.
84. EU (2013) Low emission zones in Europe. Available from: http://www.lowemissionzones.eu/. Accessed 2014 July 8.
85. HBEFA (2010) Handbook emission factors for road transport (HBEFA 3.1). Available from: http://www.hbefa.net/e/index.html. Accessed 2014 July 8.
86. Carslaw DC, Beevers SD, Tate JE, Westmoreland EJ, Williams ML (2011) Recent evidence concerning higher NO_x emissions from passenger cars and light duty vehicles. Atmospheric Environment 45: 7053–7063.
87. Beevers SD, Westmoreland E, de Jong MC, Williams ML, Carslaw DC (2012) Trends in NO_x and NO_2 emissions from road traffic in Great Britain. Atmospheric Environment 54: 107–116.
88. LANUV, Landesamt für Natur, Umwelt und Verbraucherschutz NRW (2008) Vorgehensweise des LANUV zur Korrektur kontinuierlicher PM10-Messdaten im Luftmessnetz von NRW. Available from: http://www.lanuv.nrw.de/luft/immissionen/ber_trend/erlaeuterungen.pdf. Accessed 2014 July 8.

89. LUBW, Landesanstalt für Umwelt, Messungen und Naturschutz Baden-Württemberg (2009) Untersuchung von massenrelevanten Inhaltsstoffen in Feinstaub PM_{10}. Karlsruhe. 1–72. Available from: www.lubw.baden-wuerttemberg.de/servlet/is/207409/untersuchung_massenrelevanten_inhaltsstoffen_feinstaub_pm10.pdf?command = downloadContent&filename = untersuchung_massenrelevanten_inhaltsstoffen_feinstaub_pm10.pdf. Accessed 2014 July 8.

90. Dons E, Int Panis L, Van Poppel M, Theunis J, Willems H, et al. (2011) Impact of time-activity patterns on personal exposure to black carbon. Atmospheric Environment 45: 3594–3602.

91. Fischer PH, Hoek G, van Reeuwijk H, Briggs DJ, Lebret E, et al. (2000) Traffic-related differences in outdoor and indoor concentrations of particles and volatile organic compounds in Amsterdam. Atmospheric Environment 34: 3713–3722.

Traffic Air Pollution and Oxidized LDL

Lotte Jacobs[1], Jan Emmerechts[2], Marc F. Hoylaerts[2], Chantal Mathieu[3], Peter H. Hoet[1], Benoit Nemery[1]*, Tim S. Nawrot[1,4]

1 Occupational and Environmental Medicine, Unit of Lung Toxicology, Katholieke Universiteit Leuven, Leuven, Belgium, 2 Center for Molecular and Vascular Biology, Katholieke Universiteit Leuven, Leuven, Belgium, 3 Department of Endocrinology, Katholieke Universiteit Leuven, Leuven, Belgium, 4 Centre for Environmental Sciences, Hasselt University, Diepenbeek, Belgium

Abstract

Background: Epidemiologic studies indirectly suggest that air pollution accelerates atherosclerosis. We hypothesized that individual exposure to particulate matter (PM) derived from fossil fuel would correlate with plasma concentrations of oxidized low-density lipoprotein (LDL), taken as a marker of atherosclerosis. We tested this hypothesis in patients with diabetes, who are at high risk for atherosclerosis.

Methodology/Principal Findings: In a cross-sectional study of non-smoking adult outpatients with diabetes we assessed individual chronic exposure to PM by measuring the area occupied by carbon in airway macrophages, collected by sputum induction and by determining the distance from the patient's residence to a major road, through geocoding. These exposure indices were regressed against plasma concentrations of oxidized LDL, von Willebrand factor and plasminogen activator inhibitor 1 (PAI-1). We could assess the carbon load of airway macrophages in 79 subjects (58 percent). Each doubling in the distance of residence from major roads was associated with a 0.027 μm^2 decrease (95% confidence interval (CI): -0.048 to -0.0051) in the carbon load of airway macrophages. Independently from other covariates, we found that each increase of 0.25 μm^2 [interquartile range (IQR)] in carbon load was associated with an increase of 7.3 U/L (95% CI: 1.3 to 13.3) in plasma oxidized LDL. Each doubling in distance of residence from major roads was associated with a decrease of -2.9 U/L (95% CI: -5.2 to -0.72) in oxidized LDL. Neither the carbon load of macrophages nor the distance from residence to major roads, were associated with plasma von Willebrand factor or PAI-1.

Conclusions: The observed positive association, in a susceptible group of the general population, between plasma oxidized LDL levels and either the carbon load of airway macrophages or the proximity of the subject's residence to busy roads suggests a proatherogenic effect of traffic air pollution.

Editor: Sayuri Miyamoto, Instituto de Química, Universidade de São Paulo, Brazil

Funding: The studies of the Flemish Center of Expertise on Environment and Health were commissioned, financed and steered by the Ministry of the Flemish Community (Department of Economics, Science and Innovation; Flemish Agency for Care and Health; Department of Environment, Nature and Energy). T Nawrot received a grant of the Flemish Fund for Scientific Research (FWO-Vlaanderen, krediet aan navorsers). The center for Molecular and Vascular Biology is supported by the "Excellentie financiering KULeuven" (EF/05/13). The funders had no role in study design, data collection and analysis, decision to publish, or preparation of the manuscript.

Competing Interests: The authors have declared that no competing interests exist.

* E-mail: ben.nemery@med.kuleuven.be

Introduction

Numerous epidemiological studies link various adverse health outcomes with air pollution, especially that caused by particulate matter (PM), which to a considerable extent is caused by traffic [1,2]. One of the important recent discoveries has been that exposure to PM is not only harmful to the lungs, but also to the heart and blood vessels [3–6]. This is undoubtedly true for short-term increases in PM, which are triggers for acute cardiovascular events [7], but probably also for long-lasting exposure to urban PM, which increases the risk of cardiovascular mortality and morbidity [4,6], possibly by accelerating atherosclerosis [8–10]. A cross-sectional study in Los Angeles [9] suggested a role of air pollution in intima-media thickening of the carotid artery and a follow-up study described an association between traffic proximity and the progression of intima-media thickness [10]. In a German study of more than 4000 subjects a strong relation was found between coronary artery calcification and living close to major roads [8]. These epidemiological observations strongly suggest that long-term exposure to PM exerts a proatherogenic effect. Studies in laboratory animals have begun to give experimental plausibility to these epidemiological observations [11,12]. However, so far, only few studies have provided mechanistic evidence for an effect of chronic exposure to traffic air pollution on the development of atherosclerosis in human subjects.

It is well established that persons with diabetes have a higher risk of developing cardiovascular diseases. A population-based study showed that persons with diabetes, without previous myocardial infarction, have the same risk of developing myocardial infarction as nondiabetic patients with previous myocardial infarction [13]. The metabolic abnormalities caused by diabetes induce vascular dysfunction that predispose these patients to developing atherosclerosis [14]. There is also evidence that persons with diabetes and cardiovascular disease are more sensitive to the effects of PM air pollution [15]. So it is relevant – and also probably easier – to study the effects of air pollution in this more susceptible fraction of the population. Thus, in a previous study in diabetic subjects, we showed associations between recent exposure to PM and systemic inflammation, and between recent PM and platelet activation, indicative of a prothrombotic tendency [16]. A strong point of that study is that we were also able to estimate the participants' exposure to chronic air pollution at the individual level by the carbon load of airway macrophages obtained by induced sputum. The carbon load of airway macrophage reflects a subject's exposure to soot derived from the combustion of fossil fuels, as demonstrated in children [17]. However, we also wanted to test the hypothesis that chronic air pollution would impact on indices or predictors of atherosclerosis. Therefore, we measured the concentration of oxidized LDL in plasma samples from this same population, because oxidized LDL is a well-established biomarker of (subclinical) atherosclerosis and plaque formation [18]. We also measured plasma von Willebrand factor and PAI-1, as markers for endothelial dysfunction [19,20].

We hypothesized that chronic exposure to PM, as assessed by the carbon load of airway macrophages, was associated with an increase in the concentrations of circulating oxidized LDL, in a presumably more susceptible population.

Methods

Participants

The present study population is drawn from the one previously described [16]. Briefly, non-smoking persons with either type 1 or type 2 diabetes were recruited consecutively from the diabetes outpatient clinic at the University Hospital Leuven. Of the 186 recruited subjects, 137 (74%) consented and took part in the examination. Sufficient numbers of airway macrophages, to assess the area occupied by carbon, were obtained from 80 of the 119 patients (18 of the 137 patients failed to produce sputum). Of these 80 subjects with sufficient numbers of airway macrophages, oxidized LDL was available in 79, because in one person a blood sample could not be obtained. Patients completed a questionnaire to obtain information on age, occupation, socioeconomic status, exposure to environmental tobacco smoke, alcohol use, use of medication and place of residence. Socioeconomic status was coded and condensed into a scale with scores ranging from 1 to 3, on the basis of education and occupation of the patients. Determination of underlying cardiovascular diseases was based on the tenth International Classification of Diseases (ICD-10 code: I20-I89).

Ethics

The Ethics Review Board of the Medical Faculty of the University of Leuven (K.U.Leuven) approved the study. Participants gave written informed consent at recruitment.

Exposure assessment

Carbon load of airway macrophages obtained by induced sputum. The induction of sputum in the patients and the processing of induced sputum was previously described. Briefly, nebulized saline (NaCl 3, 4 or 5%) was administered through an ultrasonic nebulizer (Ultra-NebTm2000 model 200HI, De Vilbiss Healtcare, Somerset, PA, USA) in one, two or three 7-min inhalation periods. To isolate airway macrophages, induced sputum was processed according to a standard technique [21]. Airway macrophages were visualized by light microscopy (AxioPlan 2 Imaging, Zeiss, Zaventem, Belgium). Then Scion image software (Scion Corporation, Frederick, MD, USA) was used to calculate the carbon load of airway macrophages (Figure 1A), which was defined as the median area (μm^2) occupied by carbon, in 50 randomly selected macrophages per patient [16,17,22].

Distance to major roads. Distances from the patient's residence to a major road were calculated through geocoding (the shortest distance is 10 meters). A major road was defined as an N-road (major traffic road) or an E-road (motorway/highway) [16].

Recent exposure. A portable laser-operated aerosol mass analyzer (Aerocet 531, Met One Instruments Inc, Grant Pass, OR, USA) was used to measure $PM_{2.5}$ (particles with an aerodynamic diameter of less than 2.5 µm) and PM_{10} concentrations two hours before the patient's participation in the study. The device had been previously calibrated against a local monitoring station (Flemish Environmental Agency, Borgerhout, Antwerp). The PM concentrations were measured outside, at the entrance of the hospital, as previously described [16].

Clinical measurements

Blood collection. As previously described [16], non-fasting blood samples were collected in an EDTA tube and on 0.129 M (3.8%) sodium citrate, on the same day as the sputum collection, according to standard clinical procedures. Blood cell counts and differential leukocyte counts were determined using an automated cell counter with flow differential (Cell Dyn 3500, Abbott Diagnostics, Abott Park, IL, USA). Plasma samples were kept frozen at −80°C for future analysis.

Biochemical analysis. Oxidized LDL concentration in plasma was measured by a commercially available sandwich ELISA (Mercodia, Uppsala, Sweden). Antigen levels of plasma von Willebrand factor and PAI-1 were measured with an in-house ELISA. Blood glucose levels, glycated haemoglobin, total cholesterol, high-density lipoprotein (HDL) cholesterol and triglycerides were measured according to standard clinical procedures. LDL was calculated from the Friedewald formula [23].

Statistical analysis

For database management and statistical analysis, we used SAS Software (version 9.1, SAS Institute Inc, Cary, NC). Non-normally distributed data were log transformed. We investigated associations between plasma concentrations of oxidized LDL, von Willebrand factor and PAI-1 and markers of chronic exposure (carbon load of airway macrophages and distance from residence to major roads) using stepwise linear regression in which we set $p = 0.15$ for the independent variables to enter and to stay in the model. Covariates considered for entry in the model were sex, age, body-mass index (BMI), type of diabetes, socioeconomic status, hour of blood draw, exposure to environmental tobacco smoke, physical activity, LDL, HDL, blood glucose level, glycated haemoglobin, blood leukocyte counts, use of statins, use of angiotensin-converting-enzyme (ACE) inhibitors and use of antiplatelet medication. Irrespective of selection by the stepwise regression model we forced sex, age, body-mass index, socioeconomic status, type of diabetes, glycated haemoglobin levels, statin use and blood leukocyte counts into the models. We ran three models: model 1 unadjusted analysis, model 2 adjusted for sex, age, socioeconomic status, LDL and HDL cholesterol and finally a fully adjusted model 3 for which additional covariates were selected by stepwise regression. We applied multiple logistic regression analysis to study the relation between clinical plasma levels of oxidized LDL [18] and the carbon load of airway macrophages. We defined high plasma oxidized LDL as levels above the 75th percentile (>117 U/L), which corresponds to a higher risk for moderate to large plaques [18]. Potential

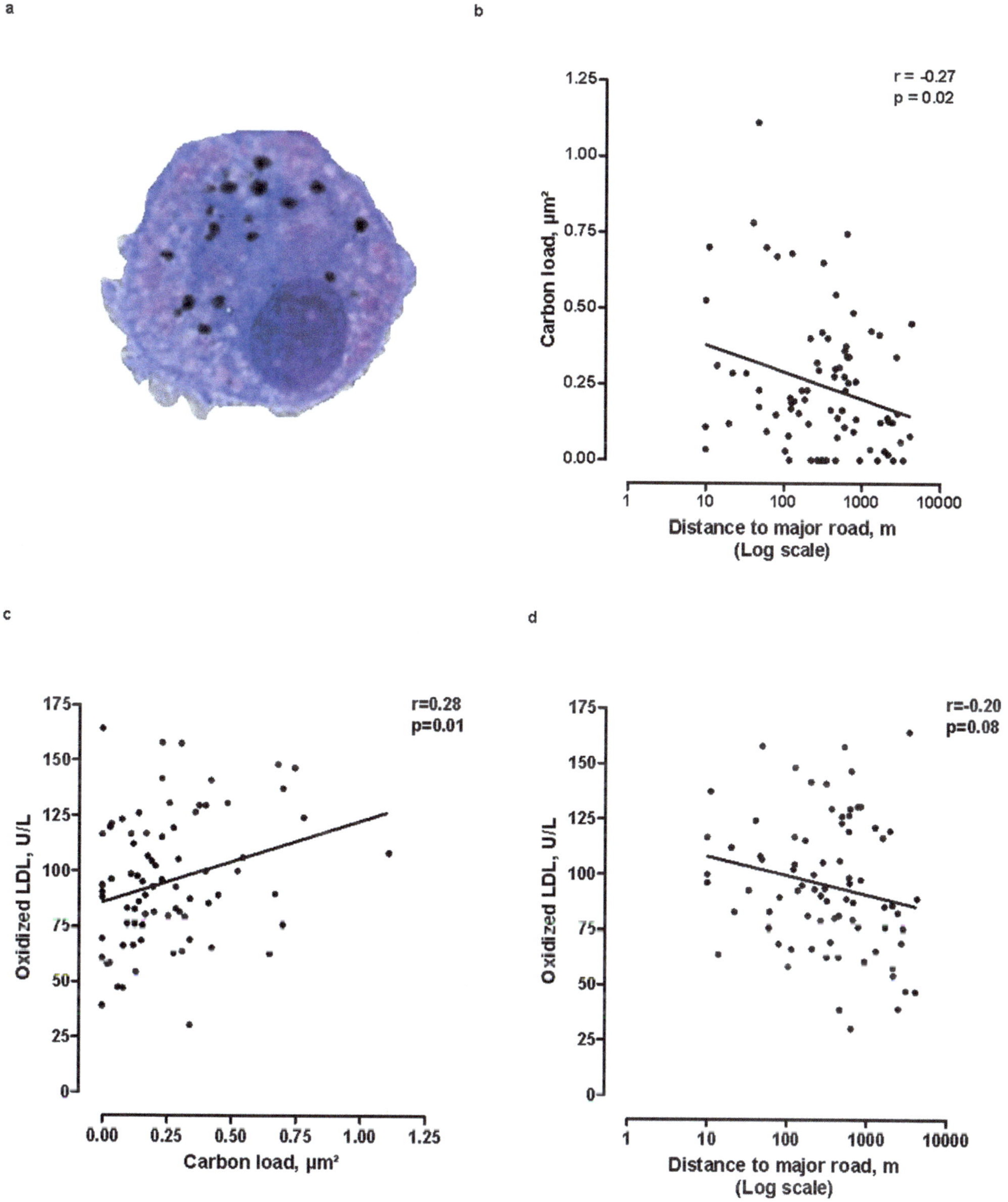

Figure 1. Traffic related exposure variables and oxidized-LDL. An airway macrophage containing carbon particles (A). We determined the surface of the macrophage occupied by carbon (in μm^2), in 50 macrophages per person. The carbon load is given as the median carbon load of 50 airway macrophages. Pearson correlation between carbon load of airway macrophages and distance from the residence to a major road (B). (Data leading to panel b have been previously published [16]). The shortest distance to a major road is 10 meters, by definition. Pearson correlation between plasma concentrations of oxidized LDL and carbon load of airway macrophages (C) and between plasma concentrations of oxidized LDL and distance from the residence to a major road (D).

interactions between carbon load of airway macrophages and type of diabetes, glycated haemoglobin and use of statins on plasma levels of oxidized LDL were investigated. Q-Q plots of the residuals were used to test the assumptions of all linear models.

Results

Characteristics of study participants

The present study population is drawn from the one previously described [16]. The characteristics of the 79 patients (age range: 22–78 years) in whom the macrophage carbon load could be determined (58%) are described in Table 1.

Carbon load of airway macrophages and distance to major road

The median carbon load of our patient's airway macrophages was 0.20 μm^2 (25th–75th percentile: 0.095 to 0.34 μm^2) and the median distance from the residence to a major road was 400 meters (25th–75th percentile: 124 to 839 meters). The relation between carbon load in airway macrophages (expressed as a surface), and distance between residence and major roads is depicted in Figure 1B. Each doubling in distance to major roads was associated with a significant decrease – by 0.027 μm^2 (95% CI, −0.048 to −0.0051; p = 0.02) – in the carbon load of airway macrophages. These data further validate the use of this novel biomarker – proposed in a study of children [17] – as an indicator of a subject's previous exposure to traffic-related air pollution.

Table 1. Patient characteristics (n = 79).

	Mean (SD) or number (%)
Women	37 (47%)
Age, years	56.5 (14.3)
BMI, kg/m^2	28.7 (5.2)
Type 1 diabetes	33 (42%)
Exposure to environmental tobacco smoke	11 (14%)
Socioeconomic status	
Low	44 (56%)
Middle	24 (30%)
High	11 (14%)
Medication use	
Antiplatelet medication	46 (58%)
Statins	48 (61%)
ACE inhibitor	31 (39%)
Insulin	75 (95%)
Oral antidiabetic medication	33 (42%)
Underlying cardiovascular disease	30 (38%)
Blood glucose, mg/dL	147 (68)
Glycated haemoglobin, %	7.4 (1.0)
Total blood leukocytes, /μL	6153 (1996)
Blood platelets, ×10^3/μL	230 (62)
von Willebrand factor, μg/mL	13.6 (5.3)
PAI-1, ng/mL	84.4 (66.5)
Cholesterol, mg/dL	158 (34)
Triglycerides, mg/dL	132 (66)
LDL, mg/dL	81.0 (26.2)
HDL, mg/dL	50.7 (18.7)
Oxidized LDL, U/L	85.6 (31.4)

Determinants of oxidized LDL

In stepwise regression analysis, plasma oxidized LDL concentration was independently and positively correlated with LDL (regression coefficient ± SE, 0.78±0.10 U/L per mg/dL; p<0.001) and inversely correlated with high-density lipoprotein (HDL) cholesterol levels (−0.38±0.14 U/L per mg/dL; p = 0.01). Although sex, age, socioeconomic status, BMI, type of diabetes, glycated haemoglobin levels, statin use and blood leukocyte counts were not significantly associated with oxidized LDL, we forced these variables, together with LDL and HDL cholesterol levels, into the regression models. Both before adjustment (Figure 1C, Table 2) and after adjustment (Table 2) for the aforementioned variables, plasma oxidized LDL concentrations were positively associated with the carbon load of airway macrophages: an interquartile (IQR) increase in carbon load (0.25 μm^2) was associated with an increase of 7.3 U/L (95% CI: 1.3 to 13.3) in oxidized LDL. Distance from residence to major roads tended to be inversely associated with oxidized LDL (Figure 1D, Table 2). After accounting for sex, age, socioeconomic status, BMI, type of diabetes, glycated haemoglobin levels, statin use, blood leukocyte counts and LDL and HDL cholesterol, each doubling in the distance from the patient's residence to a major road was associated with a decrease of 2.9 U/L (95% CI: −5.2 to −0.72) in plasma levels of oxidized LDL (Table 2).

Oxidized LDL was not associated with total blood leukocyte counts, nor with plasma von Willebrand factor or Plasminogen Activator Inhibitor-1 (PAI-1).

Determinants of von Willebrand factor and PAI-1

Neither the carbon load of macrophages nor the distance from residence to major roads, were associated with plasma von Willebrand factor or PAI-1, taken as indices of endothelial dysfunction.

Sensitivity analysis

Studies have pointed to reduced susceptibility to the effects of air pollution in those that take statins [24,25]. We therefore tested the interaction term of carbon load of macrophages by statin use. The interaction term of carbon load and oxidized LDL by statin use tended to be significant (p = 0.09) in models that did not account for LDL and HDL cholesterol but did not reach statistical significance (p = 0.64) in models that did account for plasma cholesterol levels.

There was also no effect-modification by sex (p = 0.51 for interaction), age (p = 0.55) type of diabetes (p = 0.86), percentage glycated haemoglobin (p = 0.47), and BMI (p = 0.65) on the association between carbon load and oxidized LDL. Oxidized LDL did not correlate with recent PM air pollution measured at the hospital on the day of the patient's visit.

Discussion

The key finding of our study is that plasma oxidized LDL concentration, a molecular marker of subclinical atherosclerosis, is positively associated with the carbon load of airway macrophages, a marker of chronic exposure to carbon particles derived from fossil fuel burning. This association could not be explained by sex, age, socioeconomic status, LDL and HDL cholesterol levels, BMI, type of diabetes, glycated haemoglobin levels, statin use, blood leukocyte counts or any other covariate studied.

Table 2. Estimated change in plasma oxidized LDL levels in association with carbon load or distance from residence to major roads.

Carbon load*, +0.25 μm²			
	Estimate	**95% CI**	**p-value**
Model 1	9.3	2.1 to 16.4	0.01
Model 2	6.7	1.2 to 12.2	0.02
Model 3	7.3	1.3 to 13.3	0.02
Distance to major road†, ×2			
	Estimate	**95% CI**	**p-value**
Model 1	−2.6	−5.5 to 0.31	0.08
Model 2	−2.9	−5.0 to −0.73	0.01
Model 3	−2.9	−5.2 to −0.72	0.01

Estimates reflect the change in oxidized LDL (U/L); CI = confidence interval.
Model 1: Unadjusted.
Model 2: Adjusted for sex, age, LDL and HDL cholesterol.
Model 3: Adjusted for sex, age, socioeconomic status, LDL and HDL cholesterol, BMI, type of diabetes, glycated haemoglobin, statin use and blood leukocyte counts.
*Effect size calculated for an interquartile range difference in carbon load.
†Effect size was calculated for a twofold increase in distance from residence to major road (based on a model with log distance).

Experimental work in animals has already shown associations between exposure to air pollution and oxidized LDL. Mice had increased IgM antibody titres to copper oxidized LDL after five weeks of exposure to cigarette smoke [26]. Exposure to urban air pollution for four months exacerbated the susceptibility of LDL to oxidation in hyperlipemic mice and levels of anti-oxidized LDL antibodies were significantly higher in mice on a high fat diet when exposed to urban air pollution [27].

We found no association between exposure to chronic air pollution and markers of endothelial function (von Willebrand factor and PAI-1), and also no association between these endothelial markers and oxidized LDL was found. Further, oxidized LDL was not associated with blood leukocytes, although we previously showed associations between exposure to air pollution and blood leukocyte counts [16]. This suggests that the mechanism underlying the association between chronic exposure to PM and oxidized LDL is independent of the one underlying the association between air pollution and inflammatory changes, such as increases in blood leukocyte counts. In this context, the oxidative potential of air pollutants can play a role in the observed association, since oxidized LDL has been identified as a marker of oxidative stress [28]. In susceptible apolipoprotein E-deficient mice, concentrated ultrafine particles caused systemic oxidative stress, an inhibition of the anti-inflammatory capacity of HDL, and larger early atherosclerotic lesions [29]. Studies in humans showed associations between plasma homocysteine level and exposure to $PM_{2.5}$ and black carbon [30]. Oxidative modification of LDL, together with increased blood leukocytes and platelets, contributes to the initiation and progression of atherosclerosis [31,32]. We showed that exposure to particles can have an effect on both these processes. Increased circulating levels of oxidized LDL are associated with adverse cardiovascular outcomes. In a population-based prospective study in 326 healthy men, plasma oxidized LDL levels, measured at baseline, predicted the occurrence and size of atherosclerotic plaques in the carotid arteries, three years later [18]. In our sample, a quarter of the subjects had oxidized LDL concentrations above 117 U/L, a level previously associated with large risk of having carotid plaques [18]. Here, the odds of having plasma oxidized LDL levels above that value increased by 163% for an IQR increase in the carbon load of airway macrophages. Findings from a nested case-control study suggest that a high plasma oxidized LDL/total cholesterol ratio can be a possible indicator of increased risk for acute myocardial infarction [33]. Holvoet et al. [34] also showed that patients with coronary artery disease had higher levels of oxidized LDL compared with age-matched controls without clinical evidence of cardiovascular disease. We did not find evidence of a higher sensitivity to pollution-induced effects on oxidized LDL in persons with type 2 diabetes compared with their type 1 counterparts. However, our patients had well-controlled glycated haemoglobin levels (average 7.4%) and insulin use in persons with type 2 diabetes was high (95%).

Our study has limitations. Observational studies do not prove causality, even when exposure is estimated on an individual level. Determining the carbon load in 50 randomly selected macrophages per patient is a labour-intensive technique and suitable sputum samples cannot always be obtained. Although sputum induction was successful in only 58 percent of our patients, this is unlikely to have introduced bias, since there were no differences between those in whom sputum was induced successfully and those in whom it was not [16]. We also did not observe any significant differences in exposure (carbon load or proximity to major roads) between the different socioeconomic classes. The exposure markers used here, i.e. carbon load and distances from major road, are surrogates for exposure to traffic-related air pollution. Although we focus on PM, we cannot exclude a role of gaseous pollutants associated with traffic air pollution (NO, NO2). Our study was performed on purpose in a presumably more susceptible fraction of the population, i.e.; diabetic subjects, and this means that our conclusions do not necessarily apply to healthy subjects (with or without atherosclerosis). We encourage to verify our findings in larger populations, such as the Multi-Ethnic Study of Atherosclerosis (MESA) [35].

In conclusion, we showed in a susceptible target population that individually assessed chronic exposure to air pollution is associated with plasma levels of oxidized LDL, a marker of early atherosclerosis. Our findings thus add mechanistic plausibility to the hypothesis that air pollution accelerates the development of atherosclerosis.

Author Contributions

Conceived and designed the experiments: BN TSN. Performed the experiments: LJ JE. Analyzed the data: TSN LJ. Wrote the paper: TSN LJ. Critical revision of the manuscript: CM MFH PHH.

References

1. Brunekreef B, Holgate ST (2002) Air pollution and health. Lancet 360: 1233–1242.
2. Pope CA, III, Dockery DW (2006) Health effects of fine particulate air pollution: lines that connect. J Air Waste Manag Assoc 56: 709–742.
3. Alfaro-Moreno E, Nawrot TS, Nemmar A, Nemery B (2007) Particulate matter in the environment: pulmonary and cardiovascular effects. Curr Opin Pulm Med 13: 98–106.
4. Miller KA, Siscovick DS, Sheppard L, Shepherd K, Sullivan JH, et al. (2007) Long-term exposure to air pollution and incidence of cardiovascular events in women. N Engl J Med 356: 447–458.
5. Nawrot TS, Nemmar A, Nemery B (2006) Air pollution: To the heart of the matter. Eur Heart J 27: 2269–2271.
6. Pope CA, III, Burnett RT, Thun MJ, Calle EE, Krewski D, et al. (2002) Lung cancer, cardiopulmonary mortality, and long-term exposure to fine particulate air pollution. JAMA 287: 1132–1141.
7. Peters A, von Klot S, Heier M, Trentinaglia I, Hormann A, et al. (2004) Exposure to traffic and the onset of myocardial infarction. N Engl J Med 351: 1721–1730.
8. Hoffmann B, Moebus S, Mohlenkamp S, Stang A, Lehmann N, et al. (2007) Residential exposure to traffic is associated with coronary atherosclerosis. Circulation 116: 489–496.
9. Kunzli N, Jerrett M, Mack WJ, Beckerman B, LaBree L, et al. (2005) Ambient air pollution and atherosclerosis in Los Angeles. Environ Health Perspect 113: 201–206.
10. Kunzli N, Jerrett M, Garcia-Esteban R, Basagana X, Beckermann B, et al. (2010) Ambient air pollution and the progression of atherosclerosis in adults. PLoS One 5: e9096.
11. Sun Q, Wang A, Jin X, Natanzon A, Duquaine D, et al. (2005) Long-term air pollution exposure and acceleration of atherosclerosis and vascular inflammation in an animal model. JAMA 294: 3003–3010.
12. Suwa T, Hogg JC, Quinlan KB, Ohgami A, Vincent R, et al. (2002) Particulate air pollution induces progression of atherosclerosis. J Am Coll Cardiol 39: 935–942.
13. Haffner SM, Lehto S, Ronnemaa T, Pyorala K, Laakso M (1998) Mortality from coronary heart disease in subjects with type 2 diabetes and in nondiabetic subjects with and without prior myocardial infarction. N Engl J Med 339: 229–234.
14. Beckman JA, Creager MA, Libby P (2002) Diabetes and atherosclerosis: epidemiology, pathophysiology, and management. JAMA 287: 2570–2581.
15. O'Neill MS, Veves A, Zanobetti A, Sarnat JA, Gold DR, et al. (2005) Diabetes enhances vulnerability to particulate air pollution-associated impairment in vascular reactivity and endothelial function. Circulation 111: 2913–2920.
16. Jacobs L, Emmerechts J, Mathieu C, Hoylaerts MF, Fierens F, et al. (2010) Air pollution related prothrombotic changes in persons with diabetes. Environ Health Perspect 118: 191–196.
17. Kulkarni N, Pierse N, Rushton L, Grigg J (2006) Carbon in airway macrophages and lung function in children. N Engl J Med 355: 21–30.
18. Wallenfeldt K, Fagerberg B, Wikstrand J, Hulthe J (2004) Oxidized low-density lipoprotein in plasma is a prognostic marker of subclinical atherosclerosis development in clinically healthy men. J Intern Med 256: 413–420.
19. Brodsky SV, Malinowski K, Golightly M, Jesty J, Goligorsky MS (2002) Plasminogen activator inhibitor-1 promotes formation of endothelial microparticles with procoagulant potential. Circulation 106: 2372–2378.
20. Horvath B, Hegedus D, Szapary L, Marton Z, Alexy T, et al. (2004) Measurement of von Willebrand factor as the marker of endothelial dysfunction in vascular diseases. Exp Clin Cardiol 9: 31–34.
21. Pizzichini E, Pizzichini MM, Efthimiadis A, Hargreave FE, Dolovich J (1996) Measurement of inflammatory indices in induced sputum: effects of selection of sputum to minimize salivary contamination. Eur Respir J 9: 1174–1180.
22. Kulkarni N, Prudon B, Panditi SL, Abebe Y, Grigg J (2005) Carbon loading of alveolar macrophages in adults and children exposed to biomass smoke particles. Sci Total Environ 345: 23–30.
23. Friedewald WT, Levy RI, Fredrickson DS (1972) Estimation of the concentration of low-density lipoprotein cholesterol in plasma, without use of the preparative ultracentrifuge. Clin Chem 18: 499–502.
24. Schwartz J, Park SK, O'Neill MS, Vokonas PS, Sparrow D, et al. (2005) Glutathione-S-transferase M1, obesity, statins, and autonomic effects of particles: gene-by-drug-by-environment interaction. Am J Respir Crit Care Med 172: 1529–1533.
25. O'Neill MS, Veves A, Sarnat JA, Zanobetti A, Gold DR, et al. (2007) Air pollution and inflammation in type 2 diabetes: a mechanism for susceptibility. Occup Environ Med 64: 373–379.
26. Tani S, Dimayuga PC, Anazawa T, Chyu KY, Li H, et al. (2004) Aberrant antibody responses to oxidized LDL and increased intimal thickening in apoE−/− mice exposed to cigarette smoke. Atherosclerosis 175: 7–14.
27. Soares SR, Carvalho-Oliveira R, Ramos-Sanchez E, Catanozi S, da Silva LF, et al. (2009) Air pollution and antibodies against modified lipoproteins are associated with atherosclerosis and vascular remodeling in hyperlipemic mice. Atherosclerosis 207: 368–373.
28. Egert S, Bosy-Westphal A, Seiberl J, Kurbitz C, Settler U, et al. (2009) Quercetin reduces systolic blood pressure and plasma oxidised low-density lipoprotein concentrations in overweight subjects with a high-cardiovascular disease risk phenotype: a double-blinded, placebo-controlled cross-over study. Br J Nutr 102: 1065–1074.
29. Araujo JA, Barajas B, Kleinman M, Wang X, Bennett BJ, et al. (2008) Ambient particulate pollutants in the ultrafine range promote early atherosclerosis and systemic oxidative stress. Circ Res 102: 589–596.
30. Ren C, Park SK, Vokonas PS, Sparrow D, Wilker E, et al. (2010) Air pollution and homocysteine: more evidence that oxidative stress-related genes modify effects of particulate air pollution. Epidemiology 21: 198–206.
31. Stocker R, Keaney JF Jr (2004) Role of oxidative modifications in atherosclerosis. Physiol Rev 84: 1381–1478.
32. Ross R (1999) Atherosclerosis-an inflammatory disease. N Engl J Med 340: 115–126.
33. Nordin FG, Hedblad B, Berglund G, Nilsson J (2003) Plasma oxidized LDL: a predictor for acute myocardial infarction? J Intern Med 253: 425–429.
34. Holvoet P, Mertens A, Verhamme P, Bogaerts K, Beyens G, et al. (2001) Circulating oxidized LDL is a useful marker for identifying patients with coronary artery disease. Arterioscler Thromb Vasc Biol 21: 844–848.
35. Holvoet P, Jenny NS, Schreiner PJ, Tracy RP, Jacobs DR (2007) The relationship between oxidized LDL and other cardiovascular risk factors and subclinical CVD in different ethnic groups: the Multi-Ethnic Study of Atherosclerosis (MESA). Atherosclerosis 194: 245–252.

Gender Differences and Effect of Air Pollution on Asthma in Children with and without Allergic Predisposition: Northeast Chinese Children Health Study

Guang-Hui Dong[1,2]*, Tao Chen[3], Miao-Miao Liu[1], Da Wang[1], Ya-Nan Ma[1], Wan-Hui Ren[4], Yungling Leo Lee[5]*, Ya-Dong Zhao[6], Qin-Cheng He[2]*

1 Department of Biostatistics and Epidemiology, School of Public Health, China Medical University, Shenyang, Liaoning Province, People's Republic of China, **2** Department of Occupational and Environmental Health, School of Public Health, China Medical University, Shenyang, Liaoning Province, People's Republic of China, **3** Department of Science and Technology, Educational Department of Liaoning Province, Shenyang, Liaoning Province, People's Republic of China, **4** Department of Ambient Air Pollution Monitor, Shenyang Environmental Monitoring Center, Shenyang, Liaoning Province, People's Republic of China, **5** Institute of Epidemiology and Preventive Medicine, College of Public Health, National Taiwan University, Taipei, Taiwan, **6** Institute of Respiratory Diseases, The First Affiliated Hospital of China Medical University, Shenyang, Liaoning Province, People's Republic of China

Abstract

Background: Males and females exhibit different health responses to air pollution, but little is known about how exposure to air pollution affects juvenile respiratory health after analysis stratified by allergic predisposition. The aim of the present study was to assess the relationship between air pollutants and asthmatic symptoms in Chinese children selected from multiple sites in a heavily industrialized province of China, and investigate whether allergic predisposition modifies this relationship.

Methodology/Principal Findings: 30139 Chinese children aged 3-to-12 years were selected from 25 districts of seven cities in northeast China in 2009. Information on respiratory health was obtained using a standard questionnaire from the American Thoracic Society. Routine air-pollution monitoring data was used for particles with an aerodynamic diameter $\leq$10 µm (PM_{10}), sulfur dioxide (SO_2), nitrogen dioxides (NO_2), ozone (O_3) and carbon monoxide (CO). A two-stage regression approach was applied in data analyses. The effect estimates were presented as odds ratios (ORs) per interquartile changes for PM_{10}, SO_2, NO_2, O_3, and CO. The results showed that children with allergic predisposition were more susceptible to air pollutants than children without allergic predisposition. Amongst children without an allergic predisposition, air pollution effects on asthma were stronger in males compared to females; Current asthma prevalence was related to PM_{10} (ORs = 1.36 per 31 µg/m^3; 95% CI, 1.08–1.72), SO_2 (ORs = 1.38 per 21 µg/m^3; 95%CI, 1.12–1.69) only among males. However, among children with allergic predisposition, more positively associations between air pollutants and respiratory symptoms and diseases were detected in females; An increased prevalence of doctor-diagnosed asthma was significantly associated with SO_2 (ORs = 1.48 per 21 µg/m^3; 95%CI, 1.21–1.80), NO_2 (ORs = 1.26 per 10 µg/m^3; 95%CI, 1.01–1.56), and current asthma with O_3 (ORs = 1.55 per 23 µg/m^3; 95%CI, 1.18–2.04) only among females.

Conclusion/Significance: Ambient air pollutions were more evident in males without an allergic predisposition and more associations were detected in females with allergic predisposition.

Editor: Claire Thorne, UCL Institute of Child Health, University College London, United Kingdom

Funding: The funding for this study was supported by Grants from China Environmental Protection Foundation (CEPF2008-123-1-5). The funders had no role in study design, data collection and analysis, decision to publish, or preparation of the manuscript.

Competing Interests: The authors have declared that no competing interests exist.

* E-mail: ghdong@mail.cmu.edu.cn (G-HD); qche@mail.cmu.edu.cn (Q-CH); leolee@ntu.edu.tw (YLL)

Introduction

Recent epidemiologic studies have revealed a wide variation in the increasing prevalence of asthma and allergies in many Asian countries, including mainland China [1,2]. Comparative studies of the population of the same ethnic background living in different environments revealed important environmental risk factors for asthma, especially for ambient air pollution, which may play an important role in the development of asthma symptoms [3–6]. The contribution of ambient air pollution exposure to human respiratory disease seems to vary in different parts of the world, perhaps because there is a difference in spatial and temporal variability of air pollutants sources and composition between different regions [7,8].

China is undergoing urbanization and westernization at an unprecedented rate, and the levels and patterns of outdoor and indoor air pollutants have altered dramatically. Ambient air quality has improved since the beginning of the 21st century, and levels of indoor coal smoke pollution have decreased rapidly as people move to new houses with gas or electric power. But in many Chinese cities, the levels of thoracic particles (less than 10 µm in diameter [PM_{10}]) and sulfur dioxide (SO_2) are still higher than the air quality guidance value of the World Health

Organization (WHO). The rapid increase in motor vehicle ownership and use has been accompanied by a concomitant increase in traffic-related air pollution, which poses an increasingly serious problem in urban areas of China. Consequently, urban air pollution in China has changed from the coal combustion type to a compound air pollution type due to the coexistence of coal smoke and motor vehicle emissions type. However, few studies have evaluated the health effects of this type of compound air pollution in China in the past decade, and to the best of our knowledge, no studies have considered the modified effects of not only gender difference but also the status of allergic predisposition when in the evaluating of the relationships between ambient air pollution and respiratory symptoms. Therefore, the effects of compound air pollution on human health require further study.

In the past, the outdoor levels of PM_{10} and SO_2 in the province of Liaoning were among the highest in China. Current levels still exceed clean air guidelines of WHO, and, simultaneously, nitrogen oxides (NO_2), carbon monoxide (CO), and ozone (O_3) concentrations have significantly increased, albeit with substantial geographical variation. All the five pollutants were consistently low in one area near the coast of Yingkou and Dalian cities. This heavy but non-uniform air pollution in Liaoning province offers a valuable epidemiologic opportunity for the relationship between exposure and response to be assessed. In the present study, we investigated the associations between exposures to urban compound air pollution and respiratory health in school children, and also investigated whether allergic predisposition and gender modifies these relationships.

Methods

Ethics statement

The procedures followed were in accordance with ethical standards of the responsible committee on human experimentation of China Medical University. All human experiments performed herein were approved by the China Medical University Institutional Human Ethics Committee and were consistent with the principles outlined in NIH guidelines on the ethical conduct of human research. A written informed consent was obtained from the parent/guardian of each participant before data collection.

Site Selection and Subject Recruitment

Liaoning province is located in northeast of China, and includes 14 cities. In order to maximize the inter- and intra-city gradients of the pollutants of interest and to minimize the correlation between district-specific ambient pollutants, in April 2009 we selected seven cities (Shenyang, Dalian, Anshan, Fushun, Benxi, Liaoyang, Yingkou) from Liaoning province based on the results of air pollution measurements between 2006 and 2008. There are five districts in Shenyang, four districts in Dalian and Fushun, and three districts in Anshan, Benxi, Liaoyang and Yingkou, respectively. An elementary school and two kindergartens within 1 km of air monitoring sites were randomly selected from each district. The resulting 25 elementary schools and 50 kindergartens were included with the 25 districts of the selected cities. Our entry criterion was "the subject should have lived at that place for at least 3 years". The procedures followed were in accordance with ethical standards of the responsible committee on human experimentation of the China Medical University, and it complied with the principles outlined in the Helsinki Declaration. Endorsed consent forms were obtained from the parents or guardians of the children before their participation in the present study.

Assessment of Ambient Air Pollution Exposure

Concentrations of PM_{10}, SO_2, NO_2, O_3, and CO in 2006–2008 were obtained from municipal air pollution monitoring stations. PM_{10}, SO_2, NO_2, O_3 and CO measurements at the municipal monitoring stations strictly followed the standard methods set by the State Environmental Protection Administration of China. Concentrations of each pollutant are measured continuously and reported hourly-PM_{10} by beta-attenuation, SO_2 by ultraviolet fluorescence, NO_2 by chemiluminescence, O_3 by ultraviolet photometry, and CO by non-dispersive infrared spectrometry. A daily (24-hourly) averaged concentration was calculated when at least 13 valid hourly values were available with not more than 6 successive hourly values missing and an 8-hourly time-weighted average concentration was calculated when at least 6 valid hourly values were available. Exposure parameters in the present study were 3-year averages (2006–2008) and the yearly deviations from the 3-year average concentrations in each districts, calculated from the 24-hour PM_{10}, SO_2, NO_2, CO, and 10:00 AM to 6:00 PM 8-hour O_3.

Questionnaire Survey

We included respiratory health related questions from the Epidemiologic Standardization Project Questionnaire of the American Thoracic Society (ATS-DLD-78-A) [9], which has been well validated in previous studies [3,5,6,10,11]. This questionnaire (ATS) has also been translated into Chinese for previous studies in other Chinese cities [3,6,11]. The questionnaire included detailed questions on date of birth, sex, birth weight, breast-feeding, number of years lived in each residence, types and characteristics of dwelling, method of cooking and heating, location of kitchen, types of home ventilation devices (if any), history and current status of children's respiratory illnesses and symptoms, parental education levels, smoking status of parents and other household members, and parental respiratory health histories. Local study staff informed students about the survey at the participating schools. After obtaining the written parental consent, parents were invited to a parents' night with the attendance of study staff where teachers explained the study and the conditions of consent. Parents who wished to complete the questionnaire at home returned it (via their child) to the teacher in an envelope. All questionnaire responses were recorded electronically in a database according to a standardized code and file structure.

Definitions of Respiratory Symptoms and Illnesses

The following specific respiratory symptoms and illnesses were determined from questionnaire responses: ***a) Doctor diagnosed asthma:*** a yes answer to the question "Has a doctor ever diagnosed with asthma in this child?" ***b) Current asthma:*** under the condition that the child has been diagnosed asthma, a yes answer to the question "Has this child been in a paroxysm of asthma in last two years?" or a yes answer to the question "Does this child take medicine or treatment for asthma or asthmatic bronchitis?". ***c) Current Wheeze:*** a yes answer to the question "Does this child's chest ever sound wheezy or whistling, including times when he or she had a cold?" and a yes answer to the question "Has this child had 2 or more such episodes in last 12 months?" ***d) Persistent cough:*** the answers to several cough-related questions which indicate that the study child has cough on most days (4 or more days per week) for at least 3 months a year either with or without cold during the last 12 months. ***e) Persistent phlegm:*** the answers to several phlegm-related questions which indicate that the study child has been seem congested or brought up phlegm, sputum, or mucus from the chest on most days (4 or more days per week) for at least 3 months a year either with or without cold during the last 12 months. ***f) Allergic rhinitis:*** a "yes" answer to the question "Has a doctor ever diagnosed allergic rhinitis in this child?"

Family history of allergies was defined as any biological parent or grandparent in whom hay fever or allergies (including allergic dermatitis, allergic conjunctivitis, and eczema) had been diagnosed. ***Family history of asthma*** was defined as any biological parent or grandparents in whom asthma or bronchial asthma had been diagnosed. If subjects answer 'yes' to family history of allergy or family history of asthma, then they were defined as having an ***allergic predisposition***. ***Personal allergic history*** was defined by reports of the father or mother of the index child's ever having received a diagnosis of allergic constitution, allergic conjunctivitis or atopic eczema, and included any history of hay fever, allergies to food or medicine, inhaled dusts, pollen, molds, animal fur or dander, or skin allergies not including poison ivy and oak.

Statistical Methods

The data was analyzed using SAS software (version 9.1; SAS Institute, Cary, N.C., USA). We studied the relationship between district-specific ambient levels of pollutants and prevalence rates of the following questionnaire-based morbidity end points: persistent cough, persistent phlegm, symptoms of asthma, current asthma, wheeze, and symptoms of wheeze. We estimated odds ratios in a two-stage hierarchical model using logistic and ecologic model analyses. The models assume two sources of variation: the variation among subjects in the first stage, part of which could be explained by the individual confounders, and the variation of air pollution between districts in the second stage, part of which could be explained by variables measured at the municipal level. In the analyses we assumed that 1) the outcome variable follows

Table 1. Characteristics of the children with (n = 4135) or without (n = 26004) allergic predisposition at 25 districts in northeast of China.

Characteristic	Children without Allergic Predisposition		Children with Allergic Predisposition	
	Male	Female	Male	Female
Asthma and asthma-related symptoms				
Persistent cough	1186(9.0)	1088(8.5)	311(15.0)	318(15.5)
Persistent phlegm	574(4.4)	478(3.7)*	164(7.9)	164(8.0)
Doctor-diagnosed asthma	822(6.3)	580(4.5)*	332(16.0)	254(12.4)*
Current asthma	263(2.0)	176(1.4)*	139(6.7)	102(5.0)*
Current wheeze	757(5.8)	579(4.5)*	293(14.1)	272(13.2)
Allergy rhinitis	651(5.0)	443(3.4)*	215(10.4)	172(8.4)*
Age (years)				
3~6	3817(29.1)	3692(28.7)	599(28.8)	599(29.1)
7~9	4879(37.2)	4901(38.1)	767(36.9)	771(37.5)
10~12	4429(33.7)	4286(33.3)	712(34.3)	687(33.4)
Parental education<higher school	3637(28.0)	3472(27.0)	562(27.1)	546(26.5)
Breast feeding	11306(86.1)	11234(87.2)*	1840(88.6)	1876(91.2)*
Low birth weight	388(3.0)	429(3.3)	80(3.9)	63(3.1)
Obesity	816(6.2)	594(4.6)*	116(5.6)	68(3.3)*
Respiratory disease before 2 years old	2774(10.7)	2386(18.5)*	733(35.3)	611(29.7)*
Personal allergic history	1921(14.9)	2140(16.3)*	679(32.7)	641(31.2)
Numbers or room <3	7154(54.5)	6877(53.4)	1093(52.6)	1105(53.7)
House close to main road	2442(18.6)	2440(19.0)	440(21.2)	472(23.0)
House close to factory or chimney	2550(19.4)	2513(19.5)	508(24.5)	525(25.5)
Home decoration in recent 2 years	4416(33.7)	4346(33.7)	766(36.9)	709(34.5)
Home coal use	855(6.5)	837(6.5)	170(8.2)	165(8.0)
Ventilation device in kitchen	10878(82.9)	10706(83.1)	1688(81.2)	1703(82.8)
Air exchange in winter	7837(59.7)	7682(59.7)	1196(57.6)	1126(54.7)
Humidator use	1915(14.6)	1865(14.5)	305(14.7)	293(14.2)
Home carpet use	212(1.6)	157(1.2)*	36(1.7)	35(1.7)
House pets	1901(14.5)	2136(16.6)*	376(18.1)	390(19.0)
Passive smoking exposure				
Father	4845(36.9)	4705(36.5)	967(46.5)	956(46.5)
Mother	117(0.9)	91(0.7)	21(1.0)	32(1.6)
Anyone	6325(48.2)	6138(47.7)	1237(59.5)	1195(58.1)
Parents as responders	11690(89.1)	11652(90.5)*	1914(92.1)	1899(92.3)

Values are n (%).
*The difference between male and female is significant at the 0.05 level.

Table 2. Distribution of 3-year average air pollutant concentrations (μg/m^3) between 25 districts in seven cities of Liaoning Province 2006–2008.

	PM_{10}	SO_2	NO_2	O_3	CO
Mean ± SD	124.2±24.1	50.3±16.8	36.7±7.6	54.8±16.1	2045±578
Min	79	20	21	34	929
Max	171	80	51	89	2911
Interquartile Range*	31	21	10	23	1001
National standard†	100	60	40	160	4000
% of >NS	76	32	36	0	0
WHO guideline‡	20	20	40	100	-
% of >WHO	100	96	36	0	-

*Range from 25th to 75th percentile of district-specific concentrations.
†China national ambient air quality standard.
‡WHO air quality guidelines, 2005.
Abbreviations: PM_{10}, particles with aerodynamic diameter 10 μm or less; SO_2, sulfur dioxide; NO_2, nitrogen dioxide; CO, carbon monoxide; O_3, ozone.

the Bernoulli distribution; 2) intercept terms are random at the municipal level; 3) all the explanatory variables are fixed effects. A logistic regression model was fitted in the first stage for the risk of asthma and asthma related symptoms as a function of site-specific intercepts, j, where $\alpha j = 1, \ldots, 25$, and personal covariates. The adjusted site-specific intercepts and prevalence rates are related by $Pj = e^{\alpha j}/(1+e^{\alpha j})$. In the second stage, these intercept terms representing the logit of the site-specific prevalence rates (Pj; $j = 1, \ldots, 25$), adjusted for personal covariates, were regressed on each site-specific ambient pollutant level by using a linear "ecologic" regression, i.e., logit $\alpha j = \alpha + Uj + \beta Zj$, where Uj denotes the random departure from the predicated prevalence αj on the logit scale for site j; Zj denotes the ambient pollution level for site j. Thus, β can be interpreted as the log odds ratio (per interquartile changes) for each pollutant, adjusted for personal characteristics. The results from the models are presented as odds ratios (ORs), along with their 95% confidence intervals (CIs). The goodness of fit was assessed with likelihood ratio tests (LR) to determine whether a variable contributed significantly to the model. First, we fitted a full model with a complete set of covariates. To elaborate sources of confounding, we fitted models with different combinations of covariates and compared the effect from models with and without the covariate of interest. If the covariate was with a probability value of less than 0.15 level for a given symptom, that covariate was retained and included in the final model. A two-tailed probability value of ≤0.05 was considered to be statistically significant.

Table 3. Correlation coefficients of air pollutants across 25 districts.

	PM_{10}	SO_2	NO_2	CO	O_3
PM_{10}	1	0.78*	0.70*	0.47*	0.74*
SO_2		1	0.52*	0.32	0.67*
NO_2			1	0.23	0.66*
CO				1	0.26
O_3					1

Abbreviations: PM_{10}, particles with aerodynamic diameter 10 μm or less; SO_2, sulfur dioxide; NO_2, nitrogen dioxide; CO, carbon monoxide; O_3, ozone.
*$p<0.05$.

Results

Characteristics of the Participants

There were a total of 35527 children in 25 elementary schools and 50 kindergartens, of which 31649 returned the questionnaire to give an overall response rate of 89.1%. The participation rates varied from 81.3% in Yingkou to 94.7% in Dalian, which did not correlated with either pollution levels or disease prevalence. In total, 1510 children (275 aged <3, 189 aged >13, 1046 residing in the current district for <3 years) were excluded from further analyses (exclusion rate of 4.8%). Among the 30139 children analyzed, the average age was 8.5 years (SD = 2.4 years) and 15203 (50.4%) were males. The gender ratios (male/female) varied from 48.1% in Yingkou to 50.5% in Benxi, and this variance was not significant ($\chi^2 = 5.51$, $p = 0.48$). For all children, the prevalence rates of doctor-diagnosed asthma, current asthma, current wheeze, allergic rhinitis, persistent cough, and persistent phlegm were 6.6%, 2.3%, 6.3%, 4.9%, 9.6%, and 4.6%, respectively. A total of 13.7% (4135/30139) of children were fitted for allergic predisposition, and the prevalence of allergic predisposition varied from 11.7% in Liaoyang to 17.4% in Benxi ($\chi^2 = 70.8$, $p<0.01$), whereas, this variance did not correlated with the pollutions levels both in males and in females. The characteristics of the participants with and without allergic predisposition were shown in Table 1. Compared with the children with allergic predisposition, children without allergic predisposition spent more time outdoors (17.6 hours/week vs. 8.4 hours/week; $p<0.01$).

Air Pollution Levels

Table 2 summarizes the distribution of the annual mean air pollutants concentrations in the 25 monitoring stations in the year 2006–2008. 72% to 100% of PM_{10} measurements, 32% to 96% of SO_2, and 36% of NO_2 exceeded the National Standard of China (State Environmental Protection Administration 1996) and WHO Air Quality Guidelines (WHO 2005) [12,13]. There are wide variations between districts gradients for PM_{10} (79–171 μg/m^3), SO_2 (20–80 μg/m^3), O_3 (34–89 μg/m^3) and CO (929–2911 μg/m^3). The correlations between different pollutants between districts are shown in Table 3. The correlations among CO and NO_2 (0.23), or O_3 (0.26) tended to be relatively low across the 25

Table 4. Adjusted odds ratios (OR) for personal and household covariates associated with respiratory morbidity.

Variable	Persistent cough	Persistent phlegm	Doctor-diagnosed asthma	Current asthma	Current wheeze	Allergic rhinitis
Age (ref: 3–6 years)						
7–9 years	0.67**	0.84**	1.04	0.69**	0.52**	2.34**
10–12 years	0.58**	0.84**	0.93	0.52**	0.39**	2.83**
Female (ref: male)	0.99	0.92	0.77**	0.76**	0.87**	0.73**
Parental education (ref: ≥junior high school)	1.18**	1.26**	1.34**	1.17*	1.01	0.92
Breast feeding (ref: not breast feeding)	0.83**	0.72**	0.78**	0.82*	0.95	0.68**
Low birth weight (ref: normal birth weight)	1.44**	1.49**	1.22*	1.03	1.16	1.30**
Obesity (ref: not obesity)	1.26**	1.37**	0.99	1.15	1.23**	1.28**
Respiratory disease history before 2 years old (ref: no)	2.72**	2.90**	8.68**	5.90**	4.97**	1.57**
Family history atopy (ref: no history)	1.46**	1.60**	1.49**	1.50**	1.78**	1.93**
Personal allergic history (ref: no history)	1.87**	2.05**	3.66**	5.43**	3.33**	7.39**
Numbers or room <3 (ref: ≥3 rooms)	1.13**	1.27**	1.12**	0.93	1.07	0.79**
House close to main road (ref: distance ≥20 m)	1.20**	1.14*	1.08	1.24**	1.11	0.98
House close to factory or chimney (ref: distance ≥100 m)	1.14**	1.06	1.11*	1.05	1.17**	0.98
Home decoration in recent 2 years (ref: not decoration in recent 2 yr)	1.28**	1.49**	1.19**	1.15*	1.16**	1.29**
Home coal use (ref: not coal use)	0.97	0.84*	1.07	0.88	1.12	0.85
Ventilation device in kitchen (ref: no device in kitchen)	0.78**	0.70**	0.91	0.91	0.99	1.02
Air exchange in winter (ref: no exchange in winter)	0.93*	0.91*	1.03	1.02	0.87**	1.10*
Humidator use (ref: no use)	1.26**	1.36**	1.30**	1.19*	1.15**	1.49**
Bedroom carpet use (ref: no carpet)	1.32*	1.02	1.34*	1.97**	1.30*	1.48**
House pets (ref: no pets in home)	1.48**	1.81**	1.27**	0.92	1.08	1.25**
Passive tobacco exposure at home (ref: no exposure)	1.33**	1.42**	1.23**	1.19**	1.40**	1.09*
Parents as responders (ref: others as responders)	0.86**	0.86*	1.01	1.07	1.10	0.72
Time spent outdoor (hour)	0.94	1.02	0.89*	0.75**	0.91	0.98

*$p<0.15$;
**$p<0.05$.
Items with asterisks are included in the final adjustment model for this measurement. These items are adjusted for each other; remaining variables are adjusted only for the footnoted items, as well as for districts.

districts, with a higher correlation between PM_{10} and SO_2 (0.78), and between PM_{10} and O_3 (0.74). The district-specific ambient pollutant levels were stable over several years before and during the study.

Health Effects

Table 4 presents multivariate relationships between the risk of respiratory morbidity and potentially relevant personal or residential characteristics in the first-stage logistic regression. Respiratory diseases history at an early age, family history of atopic diseases, home decoration in recent the last 2 years, humidifier use at home, and passive tobacco smoke exposure at home significantly increased the OR for all six symptoms/illnesses. Significantly negative associations were observed between elder age, the gender, breast feeding, use of a ventilation device in the kitchen, air exchange in winter and some symptoms/illnesses.

For all participating children, we observed significant associations between ambient air pollutants and some symptoms/illness; when stratified by gender, there was little difference in these associations between males and females (data not shown). However, amongst children without an allergic predisposition, ambient air pollutants were associated with more respiratory symptoms and diseases in males than in females (Table 5). For instance, in the single pollutant model analysis, positive associations were identified for current asthma with interquartile range of PM_{10} (OR = 1.36, 95%CI: 1.08–1.72), SO_2 (OR = 1.38, 95%CI: 1.12–1.69), and doctor diagnosed asthma with NO_2 (OR – 1.19, 95%CI: 1.06–1.34) only in males, but not in females. Among children with an allergic predisposition, females were more susceptible to ambient air pollutants than males (Table 6). In females, an increased prevalence of doctor diagnosed asthma was significantly associated with an interquartile range of SO_2 (OR = 1.48; 95%CI: 1.21–1.80), NO_2 (OR = 1.26; 95%CI: 1.01–1.56), and current asthma with O_3 (OR = 1.55; 95%CI: 1.18–2.04), but in males, these relationships were not identified. Even for the PM_{10}, the adjusted OR in females were higher than that in males.

Discussion

In present study, the average levels of PM_{10} and SO_2 were around 6 and 2.5 times higher than WHO recommended limit respectively, substantially extending the upper end of the pollution ranges of previous epidemiological studies conducted in Europe [14] and North America [5,10], and Taiwan [3]. These wide gradients and 'natural' experiment offered us the opportunity to

Table 5. Adjusted OR and 95% CIs of respiratory diseases with respect to ambient air pollutants (2006–2008) among children without allergic predisposition (n = 26004)†.

Pollutant		PM_{10}	SO_2	NO_2	O_3	CO
Single pollutant model‡						
Males	Persistent cough	**1.12(1.02–1.23)**	1.03(0.95–1.12)	**1.28(1.16–1.41)**	**1.14(1.04–1.26)**	**1.20(1.06–1.35)**
	Persistent phlegm	1.11(0.96–1.27)	1.02(0.91–1.14)	**1.16(1.02–1.33)**	1.08(0.95–1.24)	**1.24(1.04–1.48)**
	Doctor-diagnosed asthma	**1.27(1.12–1.45)**	**1.19(1.06–1.33)**	**1.19(1.06–1.34)**	**1.21(1.08–1.36)**	1.11(0.94–1.31)
	Current asthma	**1.36(1.08–1.72)**	**1.38(1.12–1.69)**	1.08(0.89–1.30)	1.09(0.90–1.33)	1.21(0.90–1.63)
	Current wheeze	1.11(0.98–1.26)	**1.10(1.00–1.24)**	1.00(0.89–1.12)	1.06(0.93–1.19)	**1.15(1.00–1.30)**
	Allergy rhinitis	**1.49(1.29–1.73)**	**1.42(1.24–1.62)**	**1.56(1.35–1.81)**	**1.49(1.30–1.72)**	1.11(0.71–1.72)
Females	Persistent cough	**1.12(1.00–1.25)**	1.03(0.94–1.12)	**1.21(1.09–1.33)**	1.06(0.96–1.17)	**1.15(1.02–1.29)**
	Persistent phlegm	1.01(0.89–1.15)	1.01(0.89–1.15)	1.13(0.98–1.30)	1.06(0.92–1.23)	1.15(0.98–1.34)
	Doctor-diagnosed asthma	**1.18(1.05–1.33)**	**1.15(1.04–1.28)**	1.14(0.99–1.30)	**1.17(1.02–1.33)**	0.96(0.84–1.10)
	Current asthma	1.16(0.95–1.41)	1.06(0.89–1.25)	1.11(0.88–1.40)	1.15(0.91–1.45)	0.94(0.74–1.18)
	Current wheeze	1.05(0.94–1.18)	1.02(0.92–1.13)	0.96(0.85–1.10)	1.01(0.88–1.16)	1.13(0.95–1.33)
	Allergy rhinitis	**1.39(1.22–1.57)**	**1.23(1.10–1.37)**	**1.51(1.33–1.71)**	**1.30(1.15–1.46)**	1.02(0.79–1.30)
Multi-pollutant (five pollutants) model‡						
Males	Persistent cough	0.96(0.80–1.15)	0.86(0.75–0.98)	**1.36(1.19–1.56)**	1.06(0.90–1.24)	**1.16(1.02–1.32)**
	Persistent phlegm	0.74(0.58–0.96)	0.99(0.82–1.19)	**1.26(1.05–1.51)**	1.15(0.91–1.44)	**1.23(1.03–1.47)**
	Doctor-diagnosed asthma	**1.36(1.06–1.76)**	1.05(0.87–1.27)	0.97(0.80–1.16)	0.91(0.72–1.15)	0.95(0.79–1.15)
	Current asthma	**1.54(1.09–2.19)**	1.46(0.94–2.27)	0.88(0.63–1.23)	0.64(0.40–1.00)	0.95(0.68–1.33)
	Current wheeze	1.03(0.82–1.28)	0.94(0.79–1.11)	0.92(0.78–1.09)	1.13(0.91–1.40)	1.14(0.97–1.34)
	Allergy rhinitis	**1.46(1.13–1.89)**	1.01(0.85–1.20)	**1.36(1.15–1.61)**	0.86(0.71–1.04)	0.97(0.80–1.19)
Females	Persistent cough	1.11(0.92–1.33)	0.88(0.76–1.01)	**1.30(1.13–1.50)**	0.88(0.74–1.05)	**1.18(1.03–1.35)**
	Persistent phlegm	1.08(0.83–1.42)	0.86(0.70–1.06)	1.15(0.94–1.42)	0.98(0.76–1.27)	**1.24(1.02–1.52)**
	Doctor-diagnosed asthma	**1.38(1.01–1.94)**	0.93(0.71–1.22)	0.97(0.74–1.26)	0.96(0.69–1.34)	0.83(0.63–1.09)
	Current asthma	1.08(0.87–1.36)	1.05(0.89–1.22)	1.08(0.93–1.27)	1.07(0.88–1.30)	0.87(0.75–1.02)
	Current wheeze	1.19(0.94–1.52)	1.17(0.96–1.43)	0.85(0.71–1.03)	0.84(0.65–1.09)	1.04(0.86–1.26)
	Allergy rhinitis	1.31(0.96–1.79)	1.16(0.94–1.43)	**1.31(1.07–1.59)**	0.99(0.78–1.24)	0.98(0.77–1.23)

†Models were adjusted for the variables with asterisks in table 4.
‡OR were scaled to the interquartile range for each pollutant (31 $\mu g/m^3$ for PM_{10}, 21 $\mu g/m^3$ for SO_2, 10 $\mu g/m^3$ for NO_2, 1001 $\mu g/m^3$ for CO, and 23 $\mu g/m^3$ for O_3).

study the effect of exposure to compound air pollution on the prevalence of respiratory symptoms among children within the high pollution range. Also, the ecologic exposure assessment had many advantages in our study. Compared to European and American children, the residential mobility of Chinese children is low and the density of elementary schools in China is very high. Monitoring stations located near the schools were also likely to be near the students' homes and thus provided good indicators for both school and home exposure.

A number of studies have evaluated the effects of ambient air pollution on children's respiratory health in the past two decades, however, the inconsistent results from these studies do not provide a clear overall picture of health damage [3,5,6,10,11,14,15]. In the early Harvard trial performed in six cities, respiratory morbidity outcomes in children were positively associated with ambient levels of PM_{10}, SO_2 and O_3 [10]. A recent population-based nested case-control study also found that the prevalence of respiratory symptoms was associated with increased concentration of NO, NO_2 and CO [15]. Even at relatively low ambient concentrations, ambient air pollution also independently contributed to the burden of emergency department visits for pediatric asthma [16]. On the other hand, a study of 12 communities of Southern California did not show any association between ambient air pollutants (PM_{10}, O_3, and NO_2) and respiratory illness in children except for acid with wheezing (OR = 1.26 per 1.7 ppb increase in acid) [5]. A study of four Chinese cities, which was conducted between 1993 and 1996, also showed a significantly increased prevalence of persistent phlegm (OR = 3.21, 95% CI:1.55–6.67) for an interquartile increment of PM_{10} (87 $\mu g/m^3$) among children, but did not reveal a relationship between ambient NO_2 or SO_2 levels and the prevalence of respiratory symptoms [6]. Also, a more recent one-city study did not show any significant relationship between the outdoor levels of NO_2, SO_2, and O_3 and increased of wheeze and asthma in children from Taiyuan city of China [11]. By selecting 25 districts from 7 cities, we have demonstrated the characteristics of the overall harmful effects which could not have been evaluated in one-city studies.

There is growing epidemiological evidence of the differing associations between air pollution and respiratory health for males and females, but the literature is far from consistent [17,18]. The gender analyses are more common in occupational epidemiology than in environmental health, because persistent job stratification by gender has produced marked differences in occupational exposures to chemical agents, ergonomic demands, injury, and psychosocial stressors [17]. Compared with the studies among adults, disentangling gender effects in air pollution–health

Table 6. Adjusted OR and 95% CIs of respiratory diseases with respect to ambient air pollutants (2006–2008) among children with allergic predisposition (n = 4135)[†].

Pollutant		PM_{10}	SO_2	NO_2	O_3	CO
Single pollutant model[‡]						
Males	Persistent cough	**1.31(1.08–1.59)**	**1.28(1.07–1.52)**	**1.26(1.04–1.53)**	**1.32(1.09–1.59)**	1.10(0.86–1.40)
	Persistent phlegm	**1.40(1.08–1.82)**	**1.30(1.04–1.63)**	1.24(0.96–1.61)	**1.37(1.07–1.75)**	1.30(0.95–1.79)
	Doctor-diagnosed asthma	**1.31(1.08–1.60)**	1.18(0.99–1.41)	1.23(0.99–1.52)	**1.30(1.05–1.61)**	0.98(0.77–1.25)
	Current asthma	**1.52(1.14–2.05)**	**1.37(1.05–1.77)**	**1.38(1.04–1.82)**	1.10(0.79–1.51)	0.95(0.67–1.35)
	Current wheeze	1.13(0.92–1.38)	1.03(0.86–1.24)	0.99(0.81–1.20)	0.99(0.81–1.23)	1.01(0.78–1.30)
	Allergy rhinitis	**1.47(1.16–1.85)**	**1.47(1.19–1.82)**	**1.47(1.17–1.83)**	**1.52(1.22–1.89)**	0.87(0.66–1.14)
Females	Persistent cough	**1.37(1.13–1.66)**	**1.28(1.08–1.53)**	**1.40(1.15–1.70)**	**1.43(1.19–1.75)**	1.16(0.91–1.47)
	Persistent phlegm	**1.44(1.12–1.87)**	**1.34(1.06–1.69)**	**1.35(1.05–1.74)**	**1.64(1.27–2.11)**	**1.35(1.01–1.82)**
	Doctor-diagnosed asthma	**1.46(1.17–1.83)**	**1.48(1.21–1.80)**	**1.26(1.01–1.56)**	**1.41(1.16–1.72)**	0.98(0.75–1.27)
	Current asthma	**1.59(1.15–2.22)**	**1.61(1.20–2.15)**	**1.44(1.04–1.98)**	**1.55(1.18–2.04)**	1.16(0.78–1.73)
	Current wheeze	1.15(0.93–1.42)	1.07(0.89–1.29)	1.08(0.87–1.33)	**1.25(1.02–1.53)**	1.07(0.83–1.39)
	Allergy rhinitis	**1.74(1.34–2.26)**	**1.55(1.23–1.96)**	**1.73(1.35–2.21)**	**1.68(1.33–2.11)**	1.23(0.87–1.74)
Multi-pollutant (five pollutants) model[‡]						
Males	Persistent cough	0.92(0.64–1.34)	1.07(0.81–1.40)	1.18(0.90–1.56)	1.31(0.93–1.83)	0.99(0.75–1.31)
	Persistent phlegm	1.08(0.64–1.82)	1.05(0.73–1.52)	1.14(0.79–1.65)	1.11(0.71–1.72)	1.14(0.78–1.65)
	Doctor-diagnosed asthma	1.27(0.87–1.87)	0.88(0.66–1.16)	0.91(0.69–1.21)	**1.42(1.01–1.99)**	0.85(0.64–1.12)
	Current asthma	1.40(0.80–2.44)	1.05(0.71–1.54)	0.94(0.63–1.40)	1.28(0.79–2.07)	0.75(0.50–1.11)
	Current wheeze	1.39(0.93–2.08)	1.02(0.75–1.39)	1.07(0.79–1.46)	0.73(0.50–1.06)	0.95(0.70–1.29)
	Allergy rhinitis	1.17(0.73–1.86)	1.28(0.93–1.76)	1.16(0.84–1.58)	1.11(0.77–1.58)	070(0.51–0.96)
Females	Persistent cough	1.24(0.84–1.81)	1.08(0.82–1.42)	1.03(0.78–1.36)	1.04(0.74–1.46)	0.97(0.73–1.28)
	Persistent phlegm	0.96(0.58–1.61)	0.97(0.67–1.38)	0.79(0.54–1.14)	**1.98(1.25–3.13)**	1.22(0.84–1.77)
	Doctor-diagnosed asthma	**1.54(1.01–2.39)**	**1.45(1.06–1.98)**	0.97(0.72–1.33)	0.78(0.53–1.13)	0.69(0.51–0.95)
	Current asthma	**2.04(1.24–3.35)**	**1.91(1.00–3.67)**	1.34(0.84–2.15)	0.33(0.19–0.60)	0.72(0.45–1.16)
	Current wheeze	1.13(0.77–1.67)	0.82(0.62–1.09)	0.73(0.54–0.97)	**1.66(1.16–2.38)**	0.97(0.73–1.30)
	Allergy rhinitis	**1.84(1.06–3.19)**	1.15(0.81–1.63)	1.19(0.84–1.69)	0.99(0.67–1.45)	0.50(0.35–0.71)

[†]Models were adjusted for the variables with asterisks in table 4.
[‡]OR were scaled to the interquartile range for each pollutant (31 $\mu g/m^3$ for PM_{10}, 21 $\mu g/m^3$ for SO_2, 10 $\mu g/m^3$ for NO_2, 1001 $\mu g/m^3$ for CO, and 23 $\mu g/m^3$ for O_3).

associations among children may be more complicated, because lung function growth rates (critical periods for pollution effects) differ by sex [19]. In our study, the vulnerability towards exposure to ambient air pollutants appeared to differ between males and females after stratified analyses by the status of allergic predisposition (Table 5 and Table 6). Consistent with our findings in children without an allergic predisposition, several studies indicated that the airways of males and females respond differently to exposure to air pollutants [5,20–22]. This is plausible as there are differences between male and female airways from early in fetal lung development and throughout life [23,24], for example female lungs mature earlier with regard to surfactant production. Throughout life women have smaller lungs than men, but their lung architecture is more advantageous with a greater airway diameter in relation to the volume of the lung parenchyma. Thus, in childhood, airway hyper-responsiveness and asthma is more common among boys than girls. As shown in the present study, among the children without an allergic predisposition, the stronger association between ambient air pollutants exposure and respiratory symptoms and diseases in males (Table 5), could be related to males having less mature lungs and relatively narrower airways during childhood. These factors are believed to contribute to the generally higher rates of pulmonary morbidity in boys than in girls, and could possibly also explain a higher susceptibility for damage by exposure to air pollutants during this age window.

Because there is little information in the literature regarding the gender-specific effects of ambient air pollution on respiratory symptoms/illness among children with an allergic predisposition, it is difficult to compare the results of our current studies with those of other investigations. First, children with an allergic predisposition spend more time at home compared with children without an allergic predisposition, so their exposure to air pollutants may be different. In the six Cities study, Neas et al. [25] reported stronger associations between home indoor NO_2 and respiratory symptoms among girls (OR = 1.7, 95%CI: 1.3–2.2) than among boys (OR = 1.2, 95%CI: 0.9–1.5) 7–11 years of age. Results from an Italian study showed there was no association between time spent in the kitchen and lung function level in boys [26], but a reduction in lung function was detected in girls which was statistically significant for FEF_{75} (sometimes –10.3%, often –11.1%), and after stratifying boys and girls into four groups on the basis of the IgE serum level (below and above the median value of IgE), the reduction in lung function was significant in girls with a high IgE value whereas no significant deleterious effects were evident in girls with a low IgE value or in boys with either a low or high IgE. So, our results (Table 5 and Table 6), in combination with those from Corbo and colleagues, suggest that when exposed to indoor air pollutants, females with an allergic predisposition may be more susceptible than females without an allergic predisposition or males either with or without an allergic predisposition. Secondly, compared with allergic asthmatics, non-allergic asthmatics have a higher sensibility of nasal and bronchial epithelia to non-allergenic stimuli like air pollutants, strong smells, cold air, wind or respiratory viruses [27]. As previously reported [28], clinical history of atopy was significantly more frequent in allergic asthmatics, and hay fever was the factor that most lowered the risk of displaying non-allergic asthma. Among the asthmatics with allergic predisposition of this study, we found 40.1% of personal allergic history and 17.3% of history hay fever in females, which was significantly lower than that in males (59.9% and 35.1%, respectively). Also, Romanet-Manent et al. [27] reported that the female sex is associated with an increased risk of non-allergic asthma as compared to allergic asthma. Therefore, we may hypothesize that, amongst children with an allergic predisposition, females may be more likely to display a non-allergic type of asthma comparing with males, and therefore may be more sensitive to air pollutants than males. The difference of health response to air pollution between males and females remains unclear and needs to be explored further. Careful consideration of gender effects and exploration of nascent methods for quantitative gender analysis may help to elucidate sources of difference.

The strong correlations between PM_{10} and SO_2 ($r=0.78$), PM_{10} and NO_2 ($r=0.70$), PM_{10} and O_3 ($r=0.74$), SO_2 and O_3 ($r=0.67$), NO_2 and O_3 ($r=0.66$) make it difficult to distinguish the effects of individual air pollutant. The increased ORs for PM_{10} and decreased ORs for SO_2, NO_2 and O_3 in multi-pollutant models may reflect the dominated influence of particulate material pollution in the 'coal smoke' style air pollution in the seven cities of northeast China [29]. Rather than being itself responsible for the increased risks of respiratory symptoms, PM_{10} maybe operating as the best surrogates of the mixed air pollutants in this context. The decreased ORs for SO_2, NO_2 and O_3 may be caused by the limited effectiveness of multi-pollutant regression models in controlling for confounding by co-pollutants [30].

Our study has some limitations. This was a study of respiratory prevalence in children between 3 and 12 years of age, not of respiratory symptom incidence; for example, some of the factors that we studied might have affected respiratory symptoms prevalence through effects on diseases duration rather than disease incidence. Also, this is a cross-sectional study and a temporal relationship between exposure and outcome could not be established.

In conclusion, our data confirm that ambient compound air pollution was associated with respiratory symptoms and diseases in young children. Among children without an allergic predisposition, males might be more susceptible to ambient air pollution than females; whereas among children with an allergic predisposition, more associations were detected in females in this group. These observations further support the results of published studies that ambient air pollution may cause respiratory symptoms and allergic diseases in children. Additionally, researchers should track these responses across multiple generations to identify potential long-term risks of exposures and the influence of gender on that risk.

Acknowledgments

We would like to thank our subjects for their time and enthusiastic participation.

Author Contributions

Conceived and designed the experiments: G-HD Q-CH YLL. Performed the experiments: G-HD. Analyzed the data: G-HD YLL. Contributed reagents/materials/analysis tools: G-HD TC Y-NM M-ML DW W-HR Y-DZ Q-CH. Wrote the paper: G-HD YLL. Revised the paper: G-HD YLL. Contributed the investigation: Y-NM M-ML DW W-HR Y-DZ.

References

1. Gerez IF, Lee BW, van Bever HP, Shek LP (2010) Allergies in Asia: differences in prevalence and management compared with western populations. Expert Rev Clin Immunol 6: 279–289.
2. Leung TF, Wong GW (2008) The Asian side of asthma and allergy. Curr Opin Allergy Clin Immunol 8: 384–390.
3. Hwang BF, Lee YL (2010) Air Pollution and Prevalence of Bronchitic Symptoms among Children in Taiwan. Chest 138: 956–964.
4. Millman A, Tang D, Perera FP (2008) Air pollution threatens the health of children in China. Pediatrics 122: 620–628.
5. Peters JM, Avol E, Navidi W, London SJ, Gauderman WJ, et al. (1999) A study of twelve Southern California communities with differing levels and types of air pollution. I. Prevalence of respiratory morbidity. Am J Respir Crit Care Med 159: 760–767.
6. Zhang JJ, Hu W, Wei F, Wu G, Korn LR, et al. (2002) Children's Respiratory Morbidity Prevalence in Relation to Air Pollution in Four Chinese Cities. Environ Health Perspect 110: 961–967.
7. Kim SB, Temiyasathit C, Chen VC, Park SK, Sattler M, et al. (2008) Characterization of spatially homogeneous regions based on temporal patterns of fine particulate matter in the continental United States. J Air Waste Manag Assoc 58: 965–975.
8. Kim E, Hopke PK, Pinto JP, Wilson WE (2005) Spatial variability of fine particle mass, components, and source contributions during the regional air pollution study in St. Louis. Environ Sci Technol 39: 4172–4179.
9. Ferris BG, Jr. (1978) Epidemiology Standardization Project (American Thoracic Society). Am Rev Respir Dis118: 1–120.
10. Schwartz J, Dockery DW, Neas LM, Wypij D, Ware JH, et al. (1994) Acute effects of summer air pollution on respiratory symptom reporting in children. Am J Respir Crit Care Med 150: 1234–1242.
11. Zhao Z, Zhang Z, Wang Z, Ferm M, Liang Y, et al. (2008) Asthmatic symptoms among pupils in relation to winter indoor and outdoor air pollution in schools in Taiyuan, China. Environ Health Perspect 116: 90–97.
12. State Environmental Protection Administration of China (1996) Ambient air quality standards of China. GB3095-1996. 18 January.
13. World Health Organization (2005) WHO air quality guidelines global update 2005. Report on a Working group meeting. Bonn, Germany, 18–20 October.
14. Braun-Fahrländer C, Vuille JC, Sennhauser FH, Neu U, Künzle T, et al. (1997) Respiratory health and long-term exposure to air pollutants in Swiss schoolchildren. SCARPOL Team. Swiss Study on Childhood Allergy and Respiratory Symptoms with Respect to Air Pollution, Climate and Pollen. Am J Respir Crit Care Med 55: 1042–1049.
15. Clark NA, Demers PA, Karr CJ, Koehoorn M, Lencar C, et al. (2010) Effect of early life exposure to air pollution on development of childhood asthma. Environ Health Perspect 118: 284–290.
16. Strickland MJ, Darrow LA, Klein M, Flanders WD, Sarnatt JA, et al. (2010) Short-term associations between ambient air pollutants and pediatric asthma emergency department visits. Am J Respir Crit Care Med 182: 307–316.
17. Clougherty JE (2010) A Growing Role for Gender Analysis in Air Pollution Epidemiology. Environ Health Perspect 118: 167–176.
18. Keitt SK, Fagan TF, Marts SA (2004) Understanding sex differences in environmental health: a thought leaders' roundtable. Environ Health Perspect 112: 604–609.
19. Berhane K, McConnell R, Gilliland F, Islam T, Gauderman WJ, et al. (2000) Sex-specific effects of asthma on pulmonary function in children. Am J Respir Crit Care Med 162: 1723–1730.
20. Gehring U, Cyrys J, Sedlmeir G, Brunekreef B, Bellander T, Fischer P, et al. (2002) Traffic-related air pollution and respiratory health during the first 2 yrs of life. Eur Respir J19: 690–698.
21. Gauderman WJ, Vora H, McConnell R, Berhane K, Gilliland F, et al. (2007) Effect of exposure to traffic on lung development from 10 to 18 years of age: a cohort study. Lancet 369: 571–577.
22. Rojas-Martinez R, Perez-Padilla R, Olaiz-Fernandez G, Mendoza-Alvarado L, Moreno-Macias H, et al. (2007) Lung function growth in children with long-term exposure to air pollutants in Mexico City. Am J Respir Crit Care Med 176: 377–384.
23. Becklake MR, Kauffmann F (1999) Gender differences in airway behaviour over the human life span. Thorax 54: 1119–1138.

24. Polgar G, Weng TR (1979) The functional development of the respiratory system from the period of gestation to adulthood. Am Rev Respir Dis 120: 625–695.
25. Neas LM, Dockery DW, Ware JH, Spengler JD, Speizer FE, et al. (1991) Association of indoor nitrogen dioxide with respiratory symptoms and pulmonary function in children. Am J Epidemiol 134: 204–219.
26. Corbo GM, Forastiere F, Agabiti N, Dell'Orco V, Pistelli R, et al. (2001) Effect of gas cooking on lung function in adolescents: modifying role of sex and immunoglobulin E. Thorax 56: 536–540.
27. Romanet-Manent S, Charpin D, Magnan A, Lanteaume A, Vervloet D, et al. (2002) Allergic vs nonallergic asthma: what makes the difference? Allergy 57: 607–613.
28. Inouye T, Tarlo S, Broder I, Corey P, Davies G, et al. (1985) Severity of asthma in skin test-negative and skin test-positive patients. J Allergy Clin Immunol 75: 313–319.
29. Du G, Zhang JX (2006) Study on air pollution status and distribution characteristics of ambient particulate in typical cities of Liaoning province. Modem. Scientific. Instruments 2: 67–38.
30. Kim JY, Burnett RT, Neas L, Thurston GD, Schwartz J, et al. (2007) Panel discussion review: session two–interpretation of observed associations between multiple ambient air pollutants and health effects in epidemiologic analyses. J Expo Sci Environ Epidemiol 17(Suppl 2): S83–89.

24

Activity Change in Response to Bad Air Quality, National Health and Nutrition Examination Survey, 2007–2010

Ellen M. Wells[1,2]*, Dorr G. Dearborn[1], Leila W. Jackson[2]

1 Department of Environmental Health Sciences, Case Western Reserve University School of Medicine, Cleveland, Ohio, United States of America, **2** Department of Epidemiology and Biostatistics, Case Western Reserve University School of Medicine, Cleveland, Ohio, United States of America

Abstract

Air pollution contributes to poor respiratory and cardiovascular health. Susceptible individuals may be advised to mitigate effects of air pollution through actions such as reducing outdoor physical activity on days with high pollution. Our analysis identifies the extent to which susceptible individuals changed activities due to bad air quality. This cross-sectional study included 10,898 adults from the National Health and Nutrition Examination Survey (NHANES) 2007–2010. Participants reported if they did something differently when air quality was bad. Susceptible categories included respiratory conditions, cardiovascular conditions and older age (≥65 years). Analyses accounted for complex survey design; logistic regression models controlled for gender, race, education, smoking, and body mass index. 1305 individuals reported doing something differently (12.0%, 95% confidence interval (CI): 10.9, 13.1). This percentage was 14.2% (95% CI: 11.6, 16.8), 25.1% (95% CI: 21.7, 28.6), and 15.5% (95% CI: 12.2, 18.9) among older adults, those with a respiratory condition, and those with a cardiovascular condition, respectively. In adjusted regression models the following were significantly more likely to have changed activity compared to those who did not belong to any susceptible group: respiratory conditions (adjusted odds ratio (aOR): 2.61, 95% CI: 2.03, 3.35); respiratory and cardiovascular conditions (aOR: 4.36, 95% CI: 2.47, 7.69); respiratory conditions and older age (aOR: 3.83; 95% CI: 2.47, 5.96); or all three groups (aOR: 3.52; 95% CI: (2.33, 5.32). Having cardiovascular conditions alone was not statistically significant. Some individuals, especially those with a respiratory condition, reported changing activities due to poor air quality. However, efforts should continue to educate the public about air quality and health.

Editor: Mauricio Rojas, University of Pittsburgh, United States of America

Funding: EMW and DGD were supported by the Swetland Center for Environmental Health. The funders had no role in study design, data collection and analysis, decision to publish, or preparation of the manuscript.

Competing Interests: The authors have read the journal's policy and have the following competing interests: EMW is a former member of the Greater Cleveland Clean Air Campaign and DGD is a former member of the American Thoracic Society's Environmental Health Policy Committee and is currently a co-chair of the Greater Cleveland Asthma Coalition.

* E-mail: ellen.wells@case.edu

Introduction

Ample evidence links ambient air pollution with decreased health outcomes. Fine particulate matter and ozone have been linked with increased respiratory and cardiovascular morbidity and mortality [1–3]. Air pollution exposure is a major public health concern: using air quality data from 2005, Fann et al. estimated 130,000 deaths in the United States were related to particulate matter exposure and 4700 deaths were related to ozone exposure [4].

Moreover, reductions in these pollutants have been shown to reduce rates of related conditions [5,6]. Reduced exposure to air pollutants may occur regionally via reduced emissions; however, individuals may also reduce their personal exposures through actions such as shortening the duration and intensity of outdoor activities on days with elevated air pollution. For example, Langrish et al. showed in an open randomized case-crossover trial that wearing a mask while walking outdoors was associated with improved cardiovascular function among patients with coronary heart disease [7].

The United States Environmental Protection Agency (US EPA) recommends that individuals susceptible to the health effects of air pollution take precautions at lower concentrations than the general population, as susceptible individuals may be affected at these concentrations [8]. Susceptible populations include infants and children, the elderly, and those with respiratory or cardiovascular health conditions [9–13]. Physicians are also urged to encourage their patients, especially those in susceptible groups, to limit their exposure to ambient air on days with elevated air pollution [14–17].

The objective of this study was to determine the extent to which individuals take action to limit their exposure to ambient air pollution based on knowledge of poor air quality, and whether those in a susceptible population were more likely to do so. Susceptible groups considered in the present analysis include those with a self-reported respiratory condition (asthma, emphysema, and chronic bronchitis), those with a self-reported cardiovascular condition (congestive heart failure, coronary heart disease, angina, heart attack or stroke) and the elderly (≥65 years of age).

Materials and Methods

Study Population

This is a cross-sectional study using data from the 2007–2010 National Health and Nutrition Examination Survey (NHANES) conducted by the National Center for Health Statistics (NCHS) within the United States Centers for Disease Control and Prevention (CDC) [18]. Additional details about NHANES are available at http://www.cdc.gov/nchs/nhanes.htm. Participants provided written informed consent, and the study was operated under approval from the NCHS Research Ethics Review Board. This analysis had approval from the Case Western Reserve University Institutional Review Board.

Inclusion criteria included participation in both the household interview and the health examination for NHANES 2007–2010 (n = 20,015). Those less than 20 years of age (n = 8249) were excluded because key self-reported medical conditions were only asked among those 20 years or older. Other individuals were excluded due to missing data on key variables including activity change due to air quality (n = 2); educational status (n = 20); body mass index (BMI) (n = 159); or blood cotinine (n = 687). A total of 10,898 individuals were included in analyses.

Activity Change

Within the air quality questionnaire, activity change was determined from the question "During the past 12 months, when you thought or were informed air quality was bad, did you do anything differently?" [18]. Respondents could answer "yes", "no" or "never thought/not informed about air quality"; hereafter referred to as "not informed" for brevity. For analyses, those in the not informed category were included with those who said no, as these individuals did not change their activities based upon knowledge about air quality. The supporting information includes a sensitivity analysis, where models were also constructed when excluding individuals in the not informed category.

Individuals who responded yes to the air quality question were then asked if they made any of the following changes: wore a mask; spent less time outdoors; avoided roads that have heavy traffic; did less strenuous activities; took medication; closed windows of your house; drove your car less; canceled outdoor activities; exercised indoors instead of outside; used buses, trains or subways; or other. During data editing, a new category, "used or changed air filter or air cleaner", was created for responses that mentioned doing so. In a second sensitivity analysis, we created additional models where activity change was limited to those who identified making changes which could have resulted in reducing their exposure to ambient air pollution or reducing the severity of health effects resulting from their exposure. These included activities which may reduce exposure (spent less time outdoors, closed windows of your house, canceled outdoor activities, exercised indoors instead of outside, wore a mask, did less strenuous activities) and activities which may reduce the impact of exposure (took medication).

Susceptible Groups

Three groups susceptible to the health effects of air pollution were considered in analyses: 1) the elderly; 2) those with a respiratory condition; and 3) those with a cardiovascular condition. The elderly were defined as those 65 years of age or older. Respiratory condition was based on self-report of at least one of the following from the medical conditions questionnaire: current asthma, emphysema, or current bronchitis. Similarly, cardiovascular condition was based on self-report of at least one of the following from the medical conditions questionnaire: congestive heart failure, coronary heart disease, angina, heart attack or stroke. As there is substantial overlap among these three variables, we created a susceptible category variable which classified participants according to their status for each of the three susceptible groups.

Additional Covariates

The demographics questionnaire provided data on age, education level and race/ethnicity. Age was categorized into those 20–34, 35–49, 50–64 and 65 and older for descriptive statistics. Education was categorized as less than high school, high school, some college or a 2-year degree, and a 4-year degree or higher. Race/ethnicity was categorized as Hispanic, non-Hispanic white, non-Hispanic black, and other or mixed race.

Serum cotinine levels were measured as an indicator of smoking status. Whole blood specimens were collected by trained medical staff and analyzed for cotinine using dilution-high performance liquid chromatography/atmospheric pressure chemical ionization tandem mass spectrometry. Limit of detection was 0.015 ng/mL. Smoking status was based on serum cotinine levels, where <1 ng/mL indicates a nonsmoker, 1–10 ng/mL indicates a passive smoker and >10 ng/mL indicates an active smoker [19].

Height and weight were was obtained by trained health technicians during the health examination and used to calculate BMI (kg/m^2). BMI was categorized using standard definitions (<18.5 kg/m^2 = underweight; 18.5–24.9 kg/m^2 = normal weight; 25–29.9 kg/m^2 = overweight and ≥ 30 kg/m^2 = obese) [20]; the underweight category was then combined with the normal weight category due to the small number of underweight individuals (population weighted proportion: 1.5%).

Data Analysis

All data were analyzed using Stata 11.2 (College Station, TX). Appropriate survey weights were used to account for the complex design and non-response. Sampling error was estimated using the Taylor series linearized method. Reported sample sizes are those of the study; however, proportions, percentages and odds ratios are all population-based estimates.

Descriptive statistics were used to assess the frequencies and distribution of individual variables, particularly in relationship to activity change and susceptible groups. Pearson's chi-square test was used to evaluate differences across groups. The association between activity change (dependent variable) and susceptibility category (independent variable) was created using logistic regression models. Covariates were considered for inclusion in the model based on *a priori* hypotheses and strength of the relationship with activity change in bivariate analyses. The final logistic regression model was adjusted for gender, educational level, race/ethnicity, smoking and body mass index.

Results

Population demographics are presented in Table 1. Mean age of the study population was 46.9 years (95% confidence interval (CI): 46.3, 47.6); 17.0% (95% CI: 15.9, 18.1) were at least 65 years old. Slightly more than half of the participants were women (51.8%, 95% CI: 50.9, 52.6), and the majority (69.3%, 95% CI: 64.3, 74.4) were non-Hispanic white. Mean body mass index was 28.6 kg/m^2 (95% CI: 28.4, 28.8). Serum cotinine levels were highly right skewed, with a geometric mean of 0.36 ng/mL (95% CI: 0.30, 0.44). The range of cotinine was from below the limit of detection to 1438 ng/mL. Active smokers comprised 25.8% (95% CI: 23.9, 27.6) of the population.

Current asthma was more prevalent than emphysema or current bronchitis (7.4% versus 1.8% and 2.5%, respectively)

Table 1. Population characteristics by activity change status, NHANES 2007–2010, N = 10,898.

Characteristic		Changed activity		Did not change activity		Not informed about air quality	
		N[a]	Percent (95% CI)[b]	N[a]	Percent (95% CI)[b]	N[a]	Percent (95% CI)[b]
Entire population		1305	12.0 (10.9, 13.1)	8895	81.5 (79.5, 83.5)	698	6.5 (4.5, 8.6)
Age[c]	20–34 years	230	8.3 (6.6, 9.9)	2244	85.6 (85.6, 88.6)	2635	6.2 (4.0, 8.3)
	35–49 years	336	12.3 (10.5, 14.0)	2329	80.9 (78.2, 83.6)	2845	6.8 (4.4, 9.3)
	50–64 years	398	14.2 (12.0, 16.4)	2186	79.7 (76.7, 82.6)	2752	6.1 (3.7, 8.5)
	≥65 years	341	14.2 (11.6, 16.8)	2136	79.7 (76.5, 82.6)	2666	7.0 (4.3, 9.8)
Gender[c]	Male	513	9.3 (8.2, 10.5)	4453	84.4 (82.2, 86.7)	335	6.3 (4.1, 8.5)
	Female	792	14.5(13.0, 15.9)	4442	78.8 (76.6, 80.9)	363	6.7 (4.7, 8.8)
Education[c]	Less than high school	243	7.9 (6.5, 9.3)	2737	84.4 (81.8, 87.0)	246	7.8 (4.8, 10.7)
	High school	281	10.6 (8.8, 12.4)	2181	84.5 (82.7, 86.4)	124	4.9 (3.4, 6.4)
	Some college or 2-year degree	436	13.3 (11.7, 15.0)	2315	79.5 (82.7, 86.4)	188	7.2 (4.6, 9.9)
	4-year college degree or higher	345	14.8 (12.1, 17.5)	1662	78.9 (75.9, 81.9)	140	6.3 (4.0, 8.6)
Race/ethnicity[c]	Non-Hipanic white	659	12.3 (10.8, 13.7)	4222	80.9 (78.6, 83.2)	367	6.9 (4.4, 9.3)
	Non-Hispanic black	293	14.2 (10.4, 17.9)	1637	83.8 (79.7, 87.9)	43	2.0 (1.2, 2.8)
	Hispanic	276	8.0 (6.0, 10.0)	2616	83.8 (79.5, 88.0)	249	8.2 (4.2, 12.2)
	Other/mixed	77	14.0 (10.3, 17.7)	420	79.5 (74.0, 85.0)	39	6.5 (3.1, 9.9)
Serum cotinine[c]	<1 ng/mL	958	12.8 (11.5, 14.0)	6115	80.2 (78.1, 82.3)	538	7.1 (4.9, 9.2)
	1 to 10 ng/mL	50	10.2 (6.2, 14.2)	393	81.8 (76.5, 87.2)	30	8.0 (4.2, 11.7)
	>10 ng/mL	297	10.2 (8.7, 11.8)	2387	85.0 (82.7, 87.2)	130	4.8 (2.9, 6.8)
Body mass index	<25 kg/m^2	354	11.5 (10.1, 12.8)	2552	82.1 (79.1, 85.0)	193	6.5 (4.2, 8.7)
	25–29 kg/m^2	414	11.6 (9.9, 13.3)	3083	81.6 (79.6, 83.7)	247	6.8 (4.7, 8.9)
	≥30 kg/m^2	537	12.9 (11.2, 14.6)	3260	80.8 (78.5, 83.2)	258	6.3 (4.0, 8.6)
Respiratory condition[c]	No	1018	10.6 (9.6, 11.6)	8120	82.8 (80.7, 85.0)	637	6.6 (4.5, 8.7)
	Yes	287	25.1 (21.7, 28.6)	775	69.1 (65.1, 73.0)	61	5.8 (3.1, 8.5)
Cardiovascular condition[c]	No	1121	11.7 (10.6, 12.8)	7971	81.8 (79.6, 84.0)	618	6.5 (4.5, 8.6)
	Yes	184	15.5 (12.2, 18.9)	924	78.0 (74.5, 81.4)	80	6.5 (3.7, 9.3)

NHANES = National Health and Nutrition Examination Survey; CI = confidence interval.
[a]Unweighted sample N.
[b]Percents are corrected for survey design, are row percents, and may not sum to 100 due to rounding.
[c]Significant ($p<0.05$) Pearson's chi-squared test corrected for survey design, comparing the characteristic to activity change.

and 9.7% (95% CI: 8.4%, 10.9%) had at least one respiratory condition. The proportion with cardiovascular conditions ranged from 2.0% (angina) to 3.3% (heart attack), with 8.1% (95% CI: 7.2, 9.1) having at least one cardiovascular condition. Additional data regarding health outcomes is included in Table S1.

A total of 1305 (12.0%, 95% CI: 10.9, 13.1) individuals responded that they did something differently due to bad air quality (Table 1). Among those who reported changing an activity, the most commonly reported activities changed were spending less time outdoors (69.4% of those who had changed an activity) and closing the windows of your house (25.5%) (Table S2).

In bivariate comparisons, those more likely to change activities were older, women, those with more education, those with a respiratory condition, and those with a cardiovascular condition. Hispanics and active smokers were less likely to have changed activities.

Twenty-seven percent of the population belongs to at least one of the three susceptible groups (Table 2). The most common susceptible categories were a) those only >65 years old (10.9%), b) those only with a respiratory condition (6.4%), and c) those ≥65 with a cardiovascular condition (3.8%); all other susceptible group categories contained less than two percent of the study population.

Table 2. Prevalence of individuals susceptible to health effects from air pollution, NHANES 2007–2010, N = 10,898.

Susceptible category	N[a]	Percent (95% CI)[a]
None	7135	73.0 (71.4, 74.6)
Respiratory only	642	6.4 (5.5, 7.2)
Cardiovascular only	319	2.6 (2.3, 2.9)
≥65 years only	1713	10.9 (10.1, 11.8)
Respiratory and cardiovascular	136	1.0 (0.7, 1.3)
Respiratory and ≥65 years	220	1.6 (1.3, 1.8)
Cardiovascular and ≥65 years	608	3.8 (3.3, 4.3)
All three groups	125	0.7 (0.5, 0.9)

NHANES = National Health and Nutrition Examination Survey; CI = confidence interval.
[a]N is the unweighted sample N; percents are corrected for survey design and may not sum to 100 due to rounding.

Logistic regression model results are presented in Table 3, Table S3, and Figure S1. In adjusted regression models, there were significantly increased odds of having changed activity for those with a respiratory condition, either alone or in combination with another susceptible condition. The strongest relationships were among those with a cardiovascular and respiratory condition (adjusted odds ratio (aOR): 4.36, 95% CI: 2.47, 7.69) and those ≥65 years old with a respiratory condition (aOR: 3.83, 95% CI: 2.47, 5.96). Results were similar in the sensitivity analyses exploring the relationship when excluding the population not informed about air quality or limiting those who changed activities to those who changed an activity which may reduce exposure or the impact of exposure (Tables S3, S4).

Discussion

This research demonstrates that some individuals, particularly those with respiratory disease, changed activities in response to bad air quality. Our results are consistent with the limited body of literature exploring activity change in response to poor air quality. Wen et al. looked at change in outdoor activity based on individual perception of air quality as well as awareness of medical alerts using data from the 2005 Behavioral Risk Factor Surveillance System [21]. They found that 12.0% of those without asthma and 25.6% of those with lifetime asthma changed activities based on personal perceptions of bad air quality. Semenza et al. report 10–15% of respondents changed activities due to poor air quality based on a telephone survey during July-September 2005 in Houston, Texas and Portland, Oregon [22]. In our analysis, 12.0% of the study population changed activities due to bad air quality, and 25% of those with a respiratory condition changed activities (Table 1).

There was no association between having a cardiovascular condition or being at least 65 years old, without also having a respiratory condition, and changing activities due to bad air quality. Although the cause for this cannot be definitively determined from this study, it is possible that individuals with a respiratory condition are more likely to understand the connection between ambient air pollution and their personal health, as decreased respiratory function may be easier to detect in comparison with decreased cardiovascular function. In a survey of five United Kingdom neighborhoods, Howel et al. found between 82–89% of respondents thought asthma was related to air pollution and 69–78% of respondents thought bronchitis was related to air pollution; additionally, having the condition in question meant a person was more likely to perceive air pollution as affecting it [23]. In contrast, in a survey among patients of a cardiology outpatient clinic in Michigan, only 43% were aware that air pollution negatively affects the heart, and only 8.2% of patients had ever discussed health risks from outdoor air pollution with their doctors [24]. Wen and colleagues noted that advice from a professional, such as a physician, had an impact on the percentage of individuals changing activity [21]; this suggests that current recommendations that physicians discuss outdoor air pollution with cardiology patients [14,17] may be effective in increasing activity change.

It is important to recognize that the results presented here are specific to those who changed activities on days with bad air quality; not merely those who were aware of or concerned about air quality. In order to change an activity due to bad air quality one must have planned an activity that could be changed on a day which had high ambient air pollution. Therefore the proportion of individuals who changed activities may be an underestimate of the proportion of individuals who would potentially have made changes if they had planned activities that were suitable to change. These results should be interpreted accordingly.

This study has a few limitations. The analysis relies on self-reported data for medical conditions. As a result, the prevalence of those with a respiratory or cardiovascular condition are likely to be underestimated, as not all specific respiratory or cardiovascular illnesses were included in the respiratory or cardiovascular condition variables, and it is possible that some individuals with these illnesses are undiagnosed. If this were a substantial factor in analysis, it would have served to weaken our ability to detect a difference in changing actions due to air quality between these groups. Recall bias related to reporting of activity change within the past 12 months may also be a concern. If recall bias were present in this analysis, it could have increased the strength of the association between being in a susceptible group and changing activity. Furthermore, given that this is a cross-sectional survey, we are unable to definitively establish that individuals were part of a susceptible group prior to changing activity due to bad air quality.

Local media may share information on local air quality alerts to their audience; however, the extent to which this was carried out during the study period may have varied considerably. Individuals may have also obtained information about local air quality directly from federal websites [8]. Wen and colleagues demonstrated that media alerts significantly contributed to an individual's changing outdoor activity [21]. A limitation of the current analysis is that we were unable to assess the impact of the frequency and awareness of media alerts on individuals' knowledge of air quality or change in activities, as these data were not collected within NHANES.

This study also has several strengths. It has a large sample size, which allows for comparisons among several demographic and susceptible groups. Additionally, NHANES is based on a representative sample of the United States population; therefore the proportions and odds ratios from this study are representative of the United States population.

Table 3. Odds ratio (95% confidence interval)[a] for changing activity due to poor air quality by susceptible category, NHANES 2007–2010, N = 10,898.

Susceptible category	Unadjusted	Adjusted[b]
None (referent)		
Respiratory only	2.64 (2.06, 3.37)	2.61 (2.03, 3.35)
Cardiovascular only	1.16 (0.76, 1.77)	1.33 (0.86, 2.04)
≥65 years only	1.20 (0.93, 1.54)	1.22 (0.95, 1.57)
Respiratory and cardiovascular	4.06 (2.31, 7.15)	4.36 (2.47, 7.69)
Respiratory and ≥65 years	3.64 (2.35, 5.64)	3.83 (2.47, 5.96)
Cardiovascular and ≥65 years	1.23 (0.78, 1.91)	1.38 (0.89, 2.13)
All three groups	2.80 (1.94, 4.04)	3.52 (2.33, 5.32)

NHANES = National Health and Nutrition Examination Survey.
[a]The model incorporates complex survey design and survey weights.
[b]Model adjusted for gender, education, race/ethnicity, smoking status (based on serum cotinine), and body mass index category.

Conclusions

This analysis demonstrates that those with a respiratory condition are more likely to change activities based on poor air quality; however, more can be done by health and public health professionals to encourage persons susceptible to the effects of air pollution to make changes that will minimize their exposure to air pollution.

Supporting Information

Figure S1 Odds ratio (95% confidence interval) for changing activities, by susceptible group category. NHANES = National Health and Nutrition Examination Survey; aOR = adjusted odds ratio; 95% CI = 95% confidence interval. Probability of changing activity due to poor air quality by susceptible category, based on the adjusted regression model from Table 3 (main article); NHANES 2007–2010, N = 10,898. Model adjusted for gender, education, race/ethnicity, smoking status, and body mass index.

Author Contributions

Conceived and designed the experiments: EMW LWJ. Analyzed the data: EMW. Wrote the paper: EMW. Interpretation of data and results: EMW DGD LWJ. Edited and critically revised manuscript: EMW DGD LWJ. Approval of final version: EMW DGD LWJ.

References

1. Bell ML, McDermott A, Zeger SL, Samet JM, Dominici F (2004) Ozone and short-term mortality in 95 US urban communities, 1987–2000. JAMA 292: 2372–2378.
2. Dominici F, Peng RD, Bell ML, Pham L, McDermott A, et al. (2006) Fine particulate air pollution and hospital admission for cardiovascular and respiratory diseases. JAMA 295: 1127–1134.
3. Peel JL, Tolbert PE, Klein M, Metzger KB, Flanders WD, et al. (2005) Ambient air pollution and respiratory emergency department visits. Epidemiology 16: 164–174.
4. Fann N, Lamson AD, Anenberg SC, Wesson K, Risley D, et al. (2012) Estimating the national public health burden associated with exposure to ambient PM2.5 and ozone. Risk Anal 32: 81–95.
5. Friedman MS, Powell KE, Hutwagner L, Graham LM, Teague WG (2001) Impact of changes in transportation and commuting behaviors during the 1996 Summer Olympic Games in Atlanta on air quality and childhood asthma. JAMA 285: 897–905.
6. Laden F, Schwartz J, Speizer FE, Dockery DW (2006) Reduction in fine particulate air pollution and mortality: Extended follow-up of the Harvard Six Cities study. Am J Respir Crit Care Med 173: 667–672.
7. Langrish JP, Li X, Wang S, Lee MM, Barnes GD, et al. (2012) Reducing personal exposure to particulate air pollution improves cardiovascular health in patients with coronary heart disease. Environ Health Perspect 120: 367–372.
8. US Environmental Protection Agency, National Oceanic and Atmospheric Adminsistration, National Aeronautics and Space Administration, National Park Service, National Association of Clean Air Agencies, et al. (2011) AIRNow. US Environmental Protection Agency,.
9. Bateson TF, Schwartz J (2004) Who is sensitive to the effects of particulate air pollution on mortality? A case-crossover analysis of effect modifiers. Epidemiology 15: 143–149.
10. Medina-Ramon M, Schwartz J (2008) Who is more vulnerable to die from ozone air pollution? Epidemiology 19: 672–679.
11. Pope CA 3rd (2000) Epidemiology of fine particulate air pollution and human health: biologic mechanisms and who's at risk? Environ Health Perspect 108 Suppl 4: 713–723.
12. Sacks JD, Stanek LW, Luben TJ, Johns DO, Buckley BJ, et al. (2011) Particulate matter-induced health effects: who is susceptible? Environ Health Perspect 119: 446–454.
13. Stafoggia M, Forastiere F, Faustini A, Biggeri A, Bisanti L, et al. (2010) Susceptibility factors to ozone-related mortality: a population-based case-crossover analysis. Am J Respir Crit Care Med 182: 376–384.
14. Abelsohn A, Stieb DM (2011) Health effects of outdoor air pollution: approach to counseling patients using the Air Quality Health Index. Can Fam Physician 57: 881–887, e280–887.
15. Kelly FJ, Fuller GW, Walton HA, Fussell JC (2012) Monitoring air pollution: use of early warning systems for public health. Respirology 17: 7–19.
16. Laumbach RJ (2010) Outdoor air pollutants and patient health. Am Fam Physician 81: 175–180.
17. Shofer S, Chen TM, Gokhale J, Kuschner WG (2007) Outdoor air pollution: counseling and exposure risk reduction. Am J Med Sci 333: 257–260.
18. Centers for Disease Control and Prevention C (2007–2008) National Health and Nutrition Examination Survey. Hyattsville, MD: U.S. Department of Health and Human Services, Centers for Disease Control and Prevention, National Center for Health Statistics.
19. Hukkanen J, Jacob P 3rd, Benowitz NL (2005) Metabolism and disposition kinetics of nicotine. Pharmacol Rev 57: 79–115.
20. National Heart Lung and Blood Institute (1998) Clinical guidelines on the identification, evaluation and treatment of overweight and obesity in adults: the evidence report. Washington, D.C.: National Institutes of Health.
21. Wen XJ, Balluz L, Mokdad A (2009) Association between media alerts of air quality index and change of outdoor activity among adult asthma in six states, BRFSS, 2005. J Community Health 34: 40–46.
22. Semenza JC, Wilson DJ, Parra J, Bontempo BD, Hart M, et al. (2008) Public perception and behavior change in relationship to hot weather and air pollution. Environ Res 107: 401–411.
23. Howel D, Moffatt S, Bush J, Dunn CE, Prince H (2003) Public views on the links between air pollution and health in Northeast England. Environ Res 91: 163–171.
24. Nowka MR, Bard RL, Rubenfire M, Jackson EA, Brook RD (2011) Patient awareness of the risks for heart disease posed by air pollution. Prog Cardiovasc Dis 53: 379–384.

Permissions

All chapters in this book were first published in PLOS ONE, by The Public Library of Science; hereby published with permission under the Creative Commons Attribution License or equivalent. Every chapter published in this book has been scrutinized by our experts. Their significance has been extensively debated. The topics covered herein carry significant findings which will fuel the growth of the discipline. They may even be implemented as practical applications or may be referred to as a beginning point for another development.

The contributors of this book come from diverse backgrounds, making this book a truly international effort. This book will bring forth new frontiers with its revolutionizing research information and detailed analysis of the nascent developments around the world.

We would like to thank all the contributing authors for lending their expertise to make the book truly unique. They have played a crucial role in the development of this book. Without their invaluable contributions this book wouldn't have been possible. They have made vital efforts to compile up to date information on the varied aspects of this subject to make this book a valuable addition to the collection of many professionals and students.

This book was conceptualized with the vision of imparting up-to-date information and advanced data in this field. To ensure the same, a matchless editorial board was set up. Every individual on the board went through rigorous rounds of assessment to prove their worth. After which they invested a large part of their time researching and compiling the most relevant data for our readers.

The editorial board has been involved in producing this book since its inception. They have spent rigorous hours researching and exploring the diverse topics which have resulted in the successful publishing of this book. They have passed on their knowledge of decades through this book. To expedite this challenging task, the publisher supported the team at every step. A small team of assistant editors was also appointed to further simplify the editing procedure and attain best results for the readers.

Apart from the editorial board, the designing team has also invested a significant amount of their time in understanding the subject and creating the most relevant covers. They scrutinized every image to scout for the most suitable representation of the subject and create an appropriate cover for the book.

The publishing team has been an ardent support to the editorial, designing and production team. Their endless efforts to recruit the best for this project, has resulted in the accomplishment of this book. They are a veteran in the field of academics and their pool of knowledge is as vast as their experience in printing. Their expertise and guidance has proved useful at every step. Their uncompromising quality standards have made this book an exceptional effort. Their encouragement from time to time has been an inspiration for everyone.

The publisher and the editorial board hope that this book will prove to be a valuable piece of knowledge for researchers, students, practitioners and scholars across the globe.

List of Contributors

Chang-Sheng Qu
State Key Laboratory of Pollution Control and Resource Reuse, School of the Environment, Nanjing University, Nanjing, China
Jiangsu Provincial Academy of Environmental Science, Nanjing, China

Zong-Wei Ma, Jun B and Lei Huang
State Key Laboratory of Pollution Control and Resource Reuse, School of the Environment, Nanjing University, Nanjing, China

Jin Yang
Department of Environmental Science & Engineering, Fudan University, Shanghai, China

Yang Liu
Department of Environmental Health, Rollins School of Public Health, Emory University, Atlanta, Georgia, United States of America

Christine T. Cowie, Wafaa Ezz, Wei Xuan, Adriana Cortes-Waterman and Guy B. Marks
Woolcock Institute of Medical Research, Sydney, Australia
Cooperative Research Centre for Asthma and Airways, Sydney, Australia

Nectarios Rose
Woolcock Institute of Medical Research, Sydney, Australia
NSW Health Department, Sydney, Australia

Elena Belousova and Brett G. Toelle
Woolcock Institute of Medical Research, Sydney, Australia

Vicky Sheppeard
Western Sydney and Nepean Blue Mountains Public Health Unit, Sydney, Australia

An Zhang, Qingwen Qi, Lili Jiang, Fang Zhou and Jinfeng Wang
State Key Laboratory of Resources and Environmental Information System, Institute of Geographical Sciences and Natural Resources Research, Chinese Academy of Sciences, Beijing, China

Gilaad G. Kaplan
Division of Gastroenterology, University of Calgary, Calgary, Alberta, Canada
Departments of Medicine, University of Calgary, Calgary, Alberta, Canada
Community Health Sciences, University of Calgary, Calgary, Alberta, Canada

Mieczyslaw Szyszkowicz, Eugeniusz Porada and Renaud Vincent
Environmental Health Sciences and Research Bureau, Health Canada, Ottawa, Ontario, Canada

Jakub Fichna
Department of Biomolecular Chemistry, Medical University of Lodz, Lodz, Poland

Brian H. Rowe
Department of Emergency Medicine and School of Public Health, University of Alberta, Edmonton, Alberta, Canada

Karen Madsen
Department of Medicine, Division of Gastroenterology, University of Alberta, Edmonton, Alberta, Canada

Subrata Ghosh
Division of Gastroenterology, University of Calgary, Calgary, Alberta, Canada
Departments of Medicine, University of Calgary, Calgary, Alberta, Canada

Martin Storr
Division of Gastroenterology, University of Calgary, Calgary, Alberta, Canada
Departments of Medicine, University of Calgary, Calgary, Alberta, Canada
Department of Medicine, Ludwigs-Maximilians University, Munich, Germany

Tapan Dey, Kabita Gogoi, Bala Gopalan Unni, Munmi Kalita, Moonmee Bharadwaz, Pranab Kumar Boruah, Thaneswar Bora, Dibyajyoti Ozah and Manoj Kalita
Biotechnology Division, Council of Scientific & Industrial Research (CSIR) - North-East Institute of Science and Technology, Jorhat, Assam, India

Minakshi Bhattacharjee
Biotechnology Department, University of Science and Technology Meghalaya (USTM), 9th Mile, Meghalaya, India

Hui-Hsien Pan, Hai-Lun Sun, Min-Sho Ku, Ko-Hsiu Lu, Ji-Nan Sheu and Ko-Huang Lue
Department of Pediatrics, Chung Shan Medical University Hospital, Taichung City, Taiwan R.O.C
School of Medicine, Chung Shan Medical University, Taichung City, Taiwan R.O.C

Chun-Tzu Chen
Department of Pediatrics, Chung Shan Medical University Hospital, Taichung City, Taiwan R.O.C
Department of Health Policy and Management, Chung Shan Medical University, Taichung City, Taiwan R.O.C

Pei-Fen Liao
Department of Pediatrics, Chung Shan Medical University Hospital, Taichung City, Taiwan R.O.C

Jing-Yang Huang
Institute of Public Health, Department of Public Health, Chung Shan Medical University, Taichung City, Taiwan R.O.C

Jar-Yuan Pai
Department of Health Policy and Management, Chung Shan Medical University, Taichung City, Taiwan R.O.C

Marie-Abele Bind
Department of Environmental Health, Harvard School of Public Health, Boston, Massachusetts, United States of America
Department of Biostatistics, Harvard School of Public Health, Boston, Massachusetts, United States of America

Brent Coull
Department of Biostatistics, Harvard School of Public Health, Boston, Massachusetts, United States of America

Helen Suh and Robert Wright
Department of Environmental Health, Harvard School of Public Health, Boston, Massachusetts, United States of America

Andrea Baccarelli and Joel Schwartz
Department of Environmental Health, Harvard School of Public Health, Boston, Massachusetts, United States of America
Department of Epidemiology, Harvard School of Public Health, Boston, Massachusetts, United States of America

Pantel Vokonas
VA Normative Aging Study, Boston University School of Medicine, Boston, Massachusetts, United States of America

Chien-Cheng Jung, Hsiu-Hao Liang, Nai-Yun Hsu and Huey-Jen Su
Department of Environmental and Occupational Health, College of Medicine, National Cheng Kung University, Tainan city, Taiwan

Hui-Ling Lee
Department of Chemistry, College of Science and Engineering, Fu Jen Catholic University, New Taipei city, Taiwan

B. Rey deCastro
Centers for Disease Control and Prevention, National Center for Environmental Health, Division of Laboratory Sciences, Atlanta, Georgia, United States

Elisabetta Ceretti, Donatella Feretti, Gaia C. V. Viola, Ilaria Zerbini, Rosa M. Limina, Claudia Zani, Francesco Donato and Umberto Gelatti
Unit of Hygiene, Epidemiology and Public Health, Department of Medical and Surgical Specialities, Radiological Sciences and Public Health, University of Brescia, Brescia, Italy

Michela Capelli and Rossella Lamera
Post-Graduate School of Public Health, University of Brescia, Brescia, Italy

Woo Hyun Lee, Jee Hye Wee, Dong-Kyu Kim, Chae-Seo Rhee, Chul Hee Lee and Jeong-Whun Kim
Department of Otorhinolaryngology, Seoul National University Bundang Hospital, Seongnam; Seoul National University College of Medicine, Seoul, South Korea

Soyeon Ahn and Ju Hyun Lee
Medical Research Collaborating Center, Seoul National University Bundang Hospital, Seongnam, South Korea

Yang-Sun Cho
Department of Otorhinolaryngology-Head and Neck Surgery, Sungkyunkwan University School of Medicine, Seoul, South Korea

Kun Hee Lee
Department of Otorhinolaryngology-Head and Neck Surgery, Kyung Hee University School of Medicine, Seoul, South Korea

Kyung Soo Kim
Department of Otorhinolaryngology-Head and Neck Surgery, Yonsei University College of Medicine, Seoul, South Korea

Si Whan Kim
Department of Otorhinolaryngology-Head and Neck Surgery, Hallym University College of Medicine, Anyang, South Korea

Ari Lee
Division of Chronic Disease Surveillance, Korea Centers for Disease Control & Prevention, Seoul, South Korea

Ludmilla da Silva Viana Jacobson
Department of Statistics, Fluminense Federal University, Niteró i, Brazil

Sandra de Souza Hacon and Hermano Albuquerque de Castro
National School of Public Health, Oswaldo Cruz Foundation, Rio de Janeiro, Brazil

Eliane Ignotti
Institute of Natural and Technological Science, Mato Grosso State University, Ca´ceres, Brazil

Paulo Artaxo
Physics Institute, University of São Paulo, São Paulo, Brazil

Paulo Hilário Nascimento Saldiva
Department of Pathology, University of São Paulo, São Paulo, Brazil

Antonio Carlos Monteiro Ponce de Leon
Department of Epidemiology, Rio de Janeiro State University, Rio de Janeiro, Brazil

Tsung-Ming Tsao
The Experimental Forest, National Taiwan University, Nantou, Taiwan

Ming-Jer Tsai and Ya-Nan Wang
The Experimental Forest, National Taiwan University, Nantou, Taiwan
The School of Forestry and Resource Conservation, National Taiwan University, Taipei, Taiwan

Heng-Lun Lin and Ta-Chen Su
Department of Internal Medicine and Cardiovascular Center, National Taiwan University Hospital, Taipei, Taiwan

Chang-Fu Wu
Department of Public Health, National Taiwan University, Taipei, Taiwan

Jing-Shiang Hwang
Institute of Statistical Science, Academia Sinica, Taipei, Taiwan

Sandy-H.J. Hsu
Laboratory Medicine, National Taiwan University Hospital, Taipei, Taiwan

Hsing Chao and Kai-Jen Chuang
School of Public Health, Taipei Medical University, Taipei, Taiwan

Charles- C.K. Chou
Research Center for Environmental Changes, Academia Sinica, Taipei, Taiwan

Kuang-Hsi Chang
Department of Public Health, China Medical University, Taichung, Taiwan
Department of Medical Research, Taichung Veterans General Hospital, Taichung, Taiwan

Mei-Yin Chang
Department of Medical Laboratory Science and Biotechnology, School of Medical and Health Sciences, Fooyin University, Kaohsiung, Taiwan

Chih-Hsin Muo
Management Office for Health Data, China Medical University Hospital, Taichung, Taiwan

Trong-Neng Wu and Chiu-Ying Chen
Department of Public Health, China Medical University, Taichung, Taiwan

Chia-Hung Kao
Graduate Institute of Clinical Medical Science, College of Medicine, China Medical University, Taiwan
Department of Nuclear Medicine and PET Center, China Medical University Hospital, Taichung, Taiwan

Yunpeng Luo, Qiu'an Zhu, Gang Yang, Yanzheng Yang and Yao Zhang
State Key Laboratory of Soil Erosion and Dryland Farming on the Loess Plateau, College of Forestry, Northwest A&F University, Yangling, Shaanxi, China

Huai Chen
State Key Laboratory of Soil Erosion and Dryland Farming on the Loess Plateau, College of Forestry, Northwest A&F University, Yangling, Shaanxi, China
Chengdu Institute of Biology, Chinese Academy of Sciences, Chengdu, China

Changhui Peng
State Key Laboratory of Soil Erosion and Dryland Farming on the Loess Plateau, College of Forestry, Northwest A&F University, Yangling, Shaanxi, China
Center of CEF/ESCER, Department of Biology Science, University of Quebec at Montreal, Montreal, Canada

Owen F. Price
School of Biological Sciences, University of Wollongong, Wollongong, New South Wales, Australia,

Grant J. Williamson and David M. J. S. Bowman
School of Plant Sciences, University of Tasmania, Hobart, Tasmania, Australia

Sarah B. Henderson
Environmental Health Services, British Columbia Centre for Disease Control, Vancouver, British Columbia, Canada

Fay Johnston
School of Plant Sciences, University of Tasmania, Hobart, Tasmania, Australia
Menzies Research Institute, University of Tasmania, Hobart, Tasmania, Australia

Maciej Strak, Flemming R. Cassee and Erik Lebret
Centre for Environmental Health (MGO), National Institute for Public Health and the Environment (RIVM), Bilthoven, The Netherlands
Division of Environmental Epidemiology and Division of Toxicology, Institute for Risk Assessment Sciences (IRAS), Utrecht University, Utrecht, The Netherlands

Gerard Hoek and Maaike Steenhof
Division of Environmental Epidemiology and Division of Toxicology, Institute for Risk Assessment Sciences (IRAS), Utrecht University, Utrecht, The Netherlands

Krystal J. Godri
Environmental Research Group, MRC-HPA Centre for Environmental Health, King's College London, London, United Kingdom
Division of Environmental Health and Risk Management, School of Geography, Earth & Environmental Sciences, University of Birmingham, Birmingham, United Kingdom

Ilse Gosens and Nicole A. H. Janssen
Centre for Environmental Health (MGO), National Institute for Public Health and the Environment (RIVM), Bilthoven, The Netherlands

Ian S. Mudway and Frank J. Kelly
Environmental Research Group, MRC-HPA Centre for Environmental Health, King's College London, London, United Kingdom

Renévan Oerle and Henri M. H. Spronk
Laboratory for Clinical Thrombosis and Haemostasis, Department of Internal Medicine, Cardiovascular Research Institute Maastricht, Maastricht University Medical Center, Maastricht, The Netherlands

Roy M. Harrison
Division of Environmental Health and Risk Management, School of Geography, Earth & Environmental Sciences, University of Birmingham, Birmingham, United Kingdom
Department of Environmental Sciences / Center of Excellence in Environmental Studies, King Abdulaziz University, Jeddah, Saudi Arabia

Bert Brunekreef
Division of Environmental Epidemiology and Division of Toxicology, Institute for Risk Assessment Sciences (IRAS), Utrecht University, Utrecht, The Netherlands
Julius Centre for Health Sciences and Primary Care, University Medical Centre Utrecht, Utrecht, The Netherlands

Janine Wichmann, Lars Barregard and Gerd Sallsten
Department of Occupational and Environmental Medicine, Sahlgrenska University Hospital and University of Gothenburg, Gothenborg, Sweden

Annika Rosengren
Department of Molecular and Clinical Medicine, Sahlgrenska University Hospital and University of Gothenburg, Gothenborg, Sweden

Karin Sjöberg
IVL Swedish Environmental Research Institute, Gothenborg, Sweden

Longxiang Li and Jieping Zhou
Institute of Remote Sensing and Digital Earth, Chinese Academy of Sciences, Olympic Science & Technology Park of CAS, Beijing, China

Jianhua Gong
Institute of Remote Sensing and Digital Earth, Chinese Academy of Sciences, Olympic Science & Technology Park of CAS, Beijing, China and Zhejiang-CAS Application Center for Geoinformatics, Jiashan, Zhejiang, China

Hao Xiang
Department of Epidemiology and Biostatistics, School of Public Health, Wuhan University, Wuhan, Hubei, China
Wuhan University Global Health Institute, Wuhan University, Wuhan, Hubei, China

Kristen J. Mertz and Luann L. Brink
Department of Epidemiology, Graduate School of Public Health, University of Pittsburgh, Pittsburgh, Pennsylvania, United States of America

Vincent C. Arena
Department of Biostatistics, Graduate School of Public Health, University of Pittsburgh, Pittsburgh, Pennsylvania, United States of America

Xiaohui Xu
Department of Epidemiology, College of Public Health and Health Professions, College of Medicine, University of Florida, Gainesville, Florida, United States of America

Yongyi Bi
Wuhan University Global Health Institute, Wuhan University, Wuhan, Hubei, China
Department of Labor and Environmental Hygiene, School of Public Health, Wuhan University, Wuhan, Hubei, China

Evelyn O. Talbott
Department of Epidemiology, Graduate School of Public Health, University of Pittsburgh, Pittsburgh, Pennsylvania, United States of America

Peter Morfeld
Institute for Occupational Epidemiology and Risk Assessment (IERA) of Evonik Industries, Essen, Germany
Institute and Policlinic for Occupational Medicine, Environmental Medicine and Preventive Research, University of Cologne, Cologne, Germany

David A. Groneberg
Institute of Occupational Medicine, Social Medicine and Environmental Medicine, Goethe-University, Frankfurt am Main, Germany

Michael F. Spallek
Institute of Occupational Medicine, Social Medicine and Environmental Medicine, Goethe-University, Frankfurt am Main, Germany
European Research Group on Environment and Health in the Transport Sector (EUGT), Berlin, Germany

Lotte Jacobs, Peter H. Hoet and Benoit Nemery
Occupational and Environmental Medicine, Unit of Lung Toxicology, Katholieke Universiteit Leuven, Leuven, Belgium

Jan Emmerechts and Marc F. Hoylaerts
Center for Molecular and Vascular Biology, Katholieke Universiteit Leuven, Leuven, Belgium

Chantal Mathieu
Department of Endocrinology, Katholieke Universiteit Leuven, Leuven, Belgium

Tim S. Nawrot
Occupational and Environmental Medicine, Unit of Lung Toxicology, Katholieke Universiteit Leuven, Leuven, Belgium
Centre for Environmental Sciences, Hasselt University, Diepenbeek, Belgium

Guang-Hui Dong
Department of Biostatistics and Epidemiology, School of Public Health, China Medical University, Shenyang, Liaoning Province, People's Republic of China
Department of Occupational and Environmental Health, School of Public Health, China Medical University, Shenyang, Liaoning Province, People's Republic of China

Tao Chen
Department of Science and Technology, Educational Department of Liaoning Province, Shenyang, Liaoning Province, People's Republic of China

Miao-Miao Liu, Da Wang and Ya-Nan Ma
Department of Biostatistics and Epidemiology, School of Public Health, China Medical University, Shenyang, Liaoning Province, People's Republic of China

Wan-Hui Ren
Department of Ambient Air Pollution Monitor, Shenyang Environmental Monitoring Center, Shenyang, Liaoning Province, People's Republic of China

Yungling Leo Lee
Institute of Epidemiology and Preventive Medicine, College of Public Health, National Taiwan University, Taipei, Taiwan

Ya-Dong Zhao
Institute of Respiratory Diseases, The First Affiliated Hospital of China Medical University, Shenyang, Liaoning Province, People's Republic of China

Qin-Cheng He
Department of Occupational and Environmental Health, School of Public Health, China Medical University, Shenyang, Liaoning Province, People's Republic of China

Ellen M. Wells
Department of Environmental Health Sciences, Case Western Reserve University School of Medicine, Cleveland, Ohio, United States of America
Department of Epidemiology and Biostatistics, Case Western Reserve University School of Medicine, Cleveland, Ohio, United States of America

Dorr G. Dearborn
Department of Environmental Health Sciences, Case Western Reserve University School of Medicine, Cleveland, Ohio, United States of America

Leila W. Jackson
Department of Epidemiology and Biostatistics, Case Western Reserve University School of Medicine, Cleveland, Ohio, United States of America

Index